T0174151

CATALOGUE OF RADIAL VELOCITIES OF GALAXIES

CATALOGUE OF RADIAL VELOCITIES
OF GALAXIES ·

G. G. C. Palumbo
*Istituto Tecnologie Studio
delle Radiazioni Extraterrestri
Consiglio Nazionale delle Ricerche
Bologna, Italy*

G. Tanzella-Nitti
*Istituto di Radioastronomia
Consiglio Nazionale delle Ricerche
Bologna, Italy*

G. Vettolani
*Istituto di Radioastronomia
Consiglio Nazionale delle Ricerche
Bologna, Italy*

CRC Press
Taylor & Francis Group
Boca Raton London New York

CRC Press is an imprint of the
Taylor & Francis Group, an **informa** business

First published 1983 by Gordon and Breach, Science Publishers, Inc.

Published 2021 by CRC Press
Taylor & Francis Group
6000 Broken Sound Parkway NW, Suite 300
Boca Raton, FL 33487-2742

© 1983 by Taylor & Francis Group, LLC
CRC Press is an imprint of Taylor & Francis Group, an Informa business

No claim to original U.S. Government works

ISBN 13: 978-0-677-06090-3 (hbk)

This book contains information obtained from authentic and highly regarded sources. Reasonable efforts have been made to publish reliable data and information, but the author and publisher cannot assume responsibility for the validity of all materials or the consequences of their use. The authors and publishers have attempted to trace the copyright holders of all material reproduced in this publication and apologize to copyright holders if permission to publish in this form has not been obtained. If any copyright material has not been acknowledged please write and let us know so we may rectify in any future reprint.

Except as permitted under U.S. Copyright Law, no part of this book may be reprinted, reproduced, transmitted, or utilized in any form by any electronic, mechanical, or other means, now known or hereafter invented, including photocopying, microfilming, and recording, or in any information storage or retrieval system, without written permission from the publishers.

For permission to photocopy or use material electronically from this work, please access www.copyright.com (http://www.copyright.com/) or contact the Copyright Clearance Center, Inc. (CCC), 222 Rosewood Drive, Danvers, MA 01923, 978-750-8400. CCC is a not-for-profit organization that provides licenses and registration for users. For organizations that have been granted a photocopy license by the CCC, a separate system of payment has been arranged.

Trademark Notice: Product or corporate names may be trademarks or registered trademarks, and are used only for identification and explanation without intent to infringe.

Visit the Taylor & Francis Web site at
http://www.taylorandfrancis.com

and the CRC Press Web site at
http://www.crcpress.com

Library of Congress Cataloging in Publication Data

Palumbo, G. G. C.
Catalogue of radial velocities of galaxies.

Bibliography: p.
1. Galaxies—Motion in line of sight—Catalogs.
I. Tanzella-Nitti, G. II. Vettolani, G. III. Title.
QB857.P34   1983   523.1'12   82-20926

Can we actually "Know" the Universe?
My God, it's hard enough finding your way around Chinatown

Woody Allen

TABLE OF CONTENTS

FOREWORD

The Catalogue of Radial Velocities of 8,250 galaxies compiled by Drs. G. G. C. Palumbo, G. Tanzella-Nitti and G. Vettolani will be invaluable to all extragalactic astronomers. It presents a major effort to produce a substantially complete list of galaxy velocities published before the end of 1980. The authors have not only compiled available data, but have, as often as seemed required, checked identifications, remeasured precise coordinates, and flagged redshift values which are discordant.

We have followed with interest the progress of their work during the past several years and can testify that, although no catalogue can claim perfection, theirs represents a thorough search and careful evaluation which has gone through several cycles of corrections, revisions and additions.

While the user will still need to evaluate the raw data and apply appropriate corrections and weights as, say, in 2RCBG, or select the "best" as in Rood's catalogue, the availability of a single source listing all original measurements will save researchers incalculable time and effort.

However, as the authors point out, the user should be cautioned that erroneous measurements or misidentifications cleared up long ago, e.g., in the Notes to 1RCBG and 2RCBG, are listed here again as in the original papers; some, but not all, are recorded in the notes appended to the Catalogue. This is particularly serious when the radial velocity of a star (VH $\simeq$ 0), close to or superimposed on a galactic nucleus has been measured by the original observer. Evidently a simple or weighted average of published values, particularly discordant values, should not be used to estimate the most probable radial velocity of a galaxy.

The present collection of raw data clearly demonstrates the need for repeated, independent redshift determinations. The old adage "testis unus, testis nullus" is as valid as ever, particularly in a field long dominated by just a few observers who, no matter how competent or careful, were not immune to misidentifications, coordinate errors, marginal spectra, etc. In general, the redshifts of bright galaxies with just one old optical determination should be reobserved. A recent example is the case of the large spiral galaxy NGC 4212 which was long assigned a velocity of +2125 km s^{-1} (as in this catalogue) on the authority of a single optical spectrum, but which turned out to have a velocity of -83 km s^{-1} (G. Helou, private communication) when no HI emission could be detected near the optical velocity and the search was extended to the lower velocity range suggested by its luminosity. Optical observers are not the only ones subject to such errors. In their push toward higher redshifts and lower fluxes radio astronomers too have occasionally measured noisy (or nonexistent) signals producing fictitious redshifts to which a no less fictitiously small error was attached. In this respect the general catalogue of 21 cm redshifts and line widths recently published by Bottinelli, Gouguenheim and Paturel (Astron. and Ap. Suppl. 47, 182, 1982) will be a useful addition to the present catalogue.

We hope that the authors will continue their survey of the redshift literature and keep it up to date. Publication of a yearly supplement would be extremely useful. It is interesting to note that the numbers of known galaxy radial velocities from Slipher's early measurements of 13 objects in 1914, through Strömberg's 1925 list of 45, Humason's famous 1935 list of 100, and the HMS catalogue of ~800 in 1955 to the present catalogue of ~8000 in 1980, delineate a steady exponential growth at the nearly constant rate of 10% per annum, or a doubling time of 6.9 years. This growth rate is far in excess of the growth rate (~6% p.a.) of most other types of astronomical data. To sustain it until the year 2000 we would need to measure some 52,000 new redshifts between 1980 and 2000!

Gérard de Vaucouleurs and Antoinette de Vaucouleurs
Department of Astronomy and McDonald Observatory
University of Texas

ACKNOWLEDGMENTS

In compiling a list of this size a large number of people in various institutes around the world have been asked, at some stage, for references, reprints, information and support of one kind or another. We cannot possibly list them all here, but we express to them collectively our warmest thanks. The staff of the Istituto di Radioastronomia of the Italian National Research Council (CNR) has been of help on innumerable occasions. A special acknowledgment goes to them all.

The staff of the Osservatorio Astronomico of the University of Bologna gave us free access to their extensive Astronomical Library. Dr. B. Marano helped us in the early days of compilation and successively found, through careful checking, a number of errors and omissions.

Professor G. de Vaucouleurs and Mrs. A. de Vaucouleurs were of invaluable assistance in checking the manuscript tables, in filling gaps and removing inconsistencies and in suggesting improvements. Without their collaboration this catalogue would not have been completed. Of course, any errors and omissions that may still be present are solely the authors' responsibility. Drs. G. Gisler, J. Huchra and H. J. Rood kindly provided copies of their catalogues prior to publication. Last, but not least, we are grateful to Dr. Charlotte Gordon for making possible the publication of this book.

SYMBOLS AND ABBREVIATIONS

A number of symbols have been used to simplify the text of this Catalogue. These abbreviations are described below.

Catalogues

CGCG = Catalogue of Galaxies and of Clusters of Galaxies (Zwicky *et al.* 1961-1968)

CGRV = Catalog of Galaxy Radial Velocities (Huchra *et al.*, 1981)

CSCG = Catalogue of Selected Compact Galaxies and of Posteruptive Galaxies (Zwicky and Zwicky, 1971)

GRC = Catalog of Galaxy Redshifts (Rood, 1980)

IC = Index Catalogue of Nebulae (Dreyer, 1895, 1910)

IGS = Index of Galaxy Spectra (Gisler and Friel, 1979)

MCG = Morphological Catalogues of Galaxies (Vorontsov-Velyaminov *et al.*, 1962-1968)

NGC = A New General Catalogue of Nebulae and Clusters of Stars (Dreyer, 1888)

1RCBG = Reference Catalogue of Bright Galaxies (de Vaucouleurs and de Vaucouleurs, 1964)

2RCBG = The Second Reference Catalogue of Bright Galaxies (de Vaucouleurs *et al.*, 1976)

RNGC = The Revised New General Catalogue of Nonstellar Astronomical Objects (Sulentic and Tifft, 1973)

RS-AC = A Revised Shapley-Ames Catalogue (Sandage and Tammann, 1981)

UGC = Uppsala General Catalogue of Galaxies (Nilson, 1973)

Journal Abbreviations Cited in the References

A.J.	=	The Astronomical Journal
Ap. J.	=	The Astrophysical Journal
Ap. J. L....	=	The Astrophysical Journal Letters (The L for Letters appears only before the page number)
Ap. J. Suppl.	=	The Astrophysical Journal Supplement Series
Ann. Astrophys.	=	Annales d'Astrophysique
Ann. Rev. Astron. and Ap.	=	Annual Review of Astronomy and Astrophysics
Astr. Circ. (USSR)	=	Astronomicheskii Circular Akademia Nauk (USSR)
Astron. and Ap.	=	Astronomy and Astrophysics
Astron. and Ap. Suppl.	=	Astronomy and Astrophysics Supplements
Astrofizika	=	Astrofizika (English edition: Astrophysics)
Astron. Nachr.	=	Astronomische Nachrichten
Astrophys. Lett.	=	Astrophysical Letters
Astrophys. and Space Sci.	=	Astrophysics and Space Science
Austral. J. Phys.	=	Australian Journal of Physics
BAN	=	Bulletin of the Astronomical Institutes of the Netherlands
Bol. Assoc. Arg. Astron.	=	Boletin de la Asociacion Argentina de Astronomia
Bull. Am. Astr. Soc.	=	Bulletin of the American Astronomical Society
Comptes Rendus Ac. Sci.	=	Comptes Rendus de l'Academie des Sciences de Paris
Contr. Oss. Astrof. Asiago	=	Contributi dell'Osservatorio Astrofisico di Asiago
IAU Circ.	=	International Astronomical Union Circulars
IAU Symp.	=	International Astronomical Union Symposium
J. Astroph. Astron.	=	Journal of Astrophysics and Astronomy
Kodaikanal Obs. Bull.	=	Kodaikanal Observatory Bulletin
Lick Obs. Bull.	=	Lick Observatory Bulletin
Lund Ann.	=	Lund Annaler
Mem. Soc. Astr. Italians	=	Memorie della Societa' Astronomica Italiana
Mitt. Astron. Ges.	=	Mitteilungen Astronomische Gesellschaft

MNASSA	=	Monthly Notices of the Astronomical Society of South Africa
MNRAS	=	Monthly Notices of the Royal Astronomical Society
Nature	=	Nature
Nature Phys. Sci.	=	Nature Physical Science
Observatory	=	The Observatory
PASP	=	Publications of the Astronomical Society of the Pacific
Proc. Austral. Astr. Soc.	=	Proceedings of the Australian Astronomical Society
Pub. Astr. Soc. Japan	=	Publications of the Astronomical Society of Japan
Pub. Obs. Haute Provence	=	Publications de l'Observatoire de Haute-Provence
Soob. Byurakan Obs.	=	Soobshcheniya Byurakanskoj Observatorii, Akademya Nauk Armyanskoj SSR, Erevan
Soob. Spez. Astrof. Obs.	=	Soobshcheniya Spetsialnoj Astrofizicheskoj Observatorii. Izdanie Spetsialnoj Astrofizicheskoj Observatorii AN, SSR
Soviet Astron.	=	Soviet Astronomy
Soviet Astron. Lett.	=	Soviet Astronomy Letters

References Cited in the Text

Arakelian, M. A., 1975, Soob. Byurakan Obs., 47, 3

Arp, H., 1966, Ap. J. Suppl., 14, 123

Berg van den, S., 1959, Pub. David Dunlap Obs., 2, 147

Berg van den, S., 1966, A. J., 71, 922

de Vaucouleurs, G., de Vaucouleurs, A., 1964, Reference Catalogue of Bright Galaxies, Texas University Press, Austin, Texas

de Vaucouleurs, G., de Vaucouleurs, A., Corwin, G. Jr., 1976, Second Reference Catalogue of Bright Galaxies, Texas University Press, Austin, Texas

Gisler, G. R., Friel, E. D., 1979, Index of Galaxy Spectra, Pachart Publishing House, Tucson, Arizona

Holmberg, E., 1937, A Study of Double and Multiple Galaxies, Lund Ann. N.6

Huchra, J., *et al.*, 1981, Catalog of Galaxy Radial Velocity, private communication

Humason, M. L., 1931, Ap. J., 71, 35

Karachentsev, I. D., 1972, Soob. Spez. Astrof. Obs. N.7

Khachikian, E. E., Weedman, D. W., 1974, Ap. J., 192, 581

Kojoian, G., Elliott, R., Tovmassian, H. R., 1978, A. J., 83, 1545

Kojoian, G., Elliott, R., Tovmassian, H. R., 1981, preprint

Markarian, B. E., 1977, Astron. and Ap., 58, 139

Nilson, P., 1973, Uppsala General Catalogue of Galaxies, Nova Acta Regiae Societatis Scientiarum Upsaliensis SER.V:A. Vol. 1

Peterson, S. D., 1973, A. J., 78, 811

Rood, J. H., Baum, W. A., 1967, A. J., 72, 398

Rood, J. H., Sastry, G. N., 1972, A. J., 77, 451

Rood, J. H., 1980, A Catalog of Galaxy Redshifts, private communication

Sandage, A., Tammann, G. A., 1981, A Revised Shapley-Ames Catalogue, Carnegie Institution of Washington, Pub. 653

Stromberg, G., 1925, Ap. J., 61, 352

Sulentic, J. W., Tifft, W. G., 1973, The Revised New General Catalogue of Nonstellar Astronomical Objects, The University of Arizona Press, Tucson, Arizona

Vorontsov-Velyaminov, B. A., 1956, Atlas and Catalogue of Interacting Galaxies, Part 1, Moscow State University, Moscow

Vorontsov-Velyaminov, B. A., 1977, Astron. and Ap. Suppl., 28, 1

Vorontsov-Velyaminov, B. A., Krasnogorskaja, A. A., Arhipova, V. P., 1962-1968 Morphological Catalogue of Galaxies, 4 Volumes, Moscow State University, Moscow

Weedman, D. W., 1977, Ann. Rev. Astron. and Ap., 15, 69

Weedman, D. W., 1978, MNRAS, 184, 11P

Zwicky, F., Zwicky, M. A., 1971, Catalogue of Selected Compact Galaxies and of Posteruptive Galaxies, Zurich

Zwicky, F., Herzog, E., Wild, P., Karpowicz, M., Kowal, C., 1961-1968, Catalogue of Galaxies and of Clusters of Galaxies, 6 Volumes, California Institute of Technology, Pasadena

Zwicky, F., Sargent, W. L. W., Kowal, C. T., 1975, A. J., 80, 545

CHAPTER 1

Introduction

The present Catalogue of Radial Velocities of Galaxies (CRVG) was started as a private data collection by G. Vettolani. It was soon realized, however, that there was need for a standard of reference in the copious but scattered literature of the astronomical community. This need is apparent from the changing trend in measuring radial velocities in recent years, an increasing tendency to systematically and uniformly survey galaxies. An indication of this trend is seen in the recently published "Revised Shapley–Ames Catalogue" (RS-AC) by Sandage and Tammann (1981) and in the new "Galaxy Redshift Survey" (GRS) almost completed at the Smithsonian Center for Astrophysics, as well as in the 21-cm radio survey at NRAO. This new systematic trend has become possible thanks to the large technological advances in data acquisition and reduction which have become available to astronomers. Hence it seemed appropriate to compile a comprehensive summary, as complete as possible, of what has been done so far. Completion of this rather formidable task resulted in the present work.

This catalogue substantially differs from existing published lists, for example the "Second Reference Catalogue of Bright Galaxies" (de Vaucouleurs et al., 1976, 2RCBG) or the "Index of Galaxy Spectra" (Gisler and Friel, 1979, IGS) because it presents all individual data separately, not weighted mean values as in 2RCBG, and is not literature-, magnitude-, velocity- and declination-limited as the IGS.

It is also substantially different from the as yet unpublished "Bright Galaxy Redshift Catalogue" by Huchra et al. (1981, BGRC) and the "Catalog of Galaxy Redshifts" by Rood (1980, CGR) which present "best" or reliable values and weighted mean values respectively for each

1

object and do not claim to be complete either in literature or in sky coverage. The major objection to taking different measurements by different observers using different instruments under different circumstances and averaging them in order to obtain the "definitive" radial velocity of a galaxy is that the user of such a value has no way of knowing whether and/or how measurements of rather questionable pedigree have been taken into account.

The present CRVG is limited, however, when compared to the BGRC or the CGR, as it does not contain any unpublished data. Every measurement contained in the CRVG has a reference that can be reviewed by the user. On the other hand, a compilation like the present one may be complementary in many ways to the above cited lists. The massive set of references generated in collecting the present data set makes the CRVG a bibliographical source of spectroscopic information as well as of other data on galaxies. As far as we are aware it is unique in containing all published data pertaining to northern as well as southern galaxies. Some attempts to provide comprehensive lists of references containing redshifts of galaxies have recently appeared in the literature. Apart from typographical errors, many of the papers quoted as "original" redshift sources are instead compilations of previously published data. Often the original papers are quoted as well, creating confusion. From the experience gained in compiling the present CRVG we warn readers against blindly taking values from papers cited. The original source should be checked whenever possible.

In this catalogue galaxies have been ordered by increasing right ascension. A personal contribution to the tedious work of compiling other people's measurements has been added: galaxy coordinates are given as accurately as possible from sources other than the original papers and not simply reiterated. The task of attributing different velocity measurements to the same object required scrupulous identification work. This was especially true when the same galaxy was labelled by different observers using different criteria, such as lists for the author's own purpose or numbers from catalogues with inaccurate coordinates. Survey lists for particular sky regions, such as Rood and Baum (1967) for the Coma Cluster region and Rood and Sastry (1972) for Abell 2199, were of great help for this purpose.

An accurate reading of additional literature was also necessary to find out when and where original spectra were taken and published for the first time. Moreover, for galaxies in groups and clusters for which a radial velocity is published but for which the authors provide only finding charts, we have measured, whenever possible, accurate positions from the Palomar Observatory Sky Survey (POSS) prints with the coordinate measuring machine of the Istituto di Radioastronomia in Bologna.

We take here the opportunity to recommend to all observers to list accurate coordinates in their papers rather than identifying the objects under study by name only, or worse, publishing only finding charts.

In principle, the present CRVG should contain ALL published radial velocities up to the end of December, 1980. However, for work published prior to 1930 we have relied entirely on the lists by Stromberg (1925, Ref. 1), Humason (1931, Ref. 2) and Moore (1932, Ref. 3). The present catalogue was compiled with the greatest possible care keeping in mind that "quidquid recte factum quamvis humile praeclarum." However, given the amount of data and the number of references surveyed, omissions and errors are not only possible but inevitable. We would be grateful to all those who would point them out to us.

CHAPTER 2

Description of the Catalogue

The catalogue lists coordinates, names and radial velocities for 8250 galaxies scattered over the whole sky. A summary of the data contained in this book is listed below:

Total number of galaxies with measured radial velocity	8,250
Total number of radial velocities	13,672
Total number of optical measurements	11,132
Total number of radio measurements	2,540
Galaxies with more than one measurement	2,500
Galaxies with both radio and optical measurements	1,217
Total number of references cited	967

We have listed both optical and radio measurements of radial velocities. We list 552 radiogalaxies and 144 Seyfert galaxies. Our catalogue includes and updates both the Burbidge and Crowne (1979, Ref. 783) list of radiogalaxies and Weedman's (1977 and 1978) lists of Seyfert galaxies.

A detailed explanation of the catalogue columns follows this section. The catalogue's main list is followed by a number of comments on individual measurements and/or identifications. These comments are explained in the NOTES section at the end of the main catalogue. Our aim in writing these notes is to provide a guide to a correct reading of individual measurements and to give additional information about particular galaxies where necessary. Although some care was taken in listing the available names of each galaxy we do not claim to be complete in name designation.

A number of notes about the references here surveyed have also been included separately in order to point out misprints, misidentifications, errors and other confusing points which have emerged during the compilation. Some of these points were already known and reference is given to the work which cites them. The appendices present some tables compiling peculiar or historical names of objects listed in the CRVG, as well as finding lists of Messier, Arp, Vorontsov-Velyaminov and Markarian galaxies. The 967 references are listed in chronological and in alphabetical order.

COLUMNS 1 and 2   Right Ascension (RA) (HH MM SS) and Declination (DEC.) (DD MM.M)

Each object is identified by its coordinates, printed on the same line of the galaxy's first name (Column 3) if present. Independent measurements are listed without repeating coordinates. In the case of tight double or interacting galaxies identical coordinates may occur for lack of more accurate positions, but appropriate names and/or identifying letters (e.g., A,B, . . .N,S, . . .) as well as a note in the comments column (Column 9) should remove ambiguity. Galaxies are listed in increasing RA, and when RA is the same, in increasing DEC. The positive sign (+) before positive DEC has been omitted for clarity of the text. RA and DEC refer to the 1950.0 equinox. Whenever coordinates provided in the reference were not accurate enough we resorted mainly to the 2RCBG but also to the following catalogues: UGC (Nilson, 1973), CSCG (Zwicky and Zwicky 1971), CGCG (Zwicky et al. 1961, 1963, 1965, 1966, 1968a, 1968b), the Revised NGC (Sulentic and Tifft, 1973), MCG (Vorontsov-Velyaminov et al., 1962, 1963, 1964, 1968) and Arakelian (1975). Optical positions of Markarian galaxies were generally taken from Peterson (1973) and Kojoian et al. (1978, 1981). In particular we must point out that in Refs. 268, 306, 308 and 437 we have assumed that the Soviet authors have taken into consideration Peterson's (1973) corrections concerning the numbering of Markarian galaxies from MRK229 to MRK237. Optical positions of radiogalaxies come mostly from Burbidge and Crowne (1979, Ref. 783). When two decimal places are printed after the DEC arcminutes these coordinates were measured by us using the standard Bologna measuring machine procedure.

COLUMN 3   Name(s) of the Galaxy

The NGC number is always given first if present. The Index Catalogue (IC) number follows whenever applicable. Other names listed: Zwicky lists of compact galaxies (ZW), (Zwicky and Zwicky, 1971, Zwicky et al. 1975) (ZW-S indicates Zwicky's southern hemisphere survey), Markarian galaxies (MRK), David Dunlap Observatory galaxies (DDO) (van den Berg, 1959, 1966), Arp peculiar galaxies (ARP) (Arp, 1966) Vorontsov-Velyaminov interacting objects (VV) (Vorontsov-Velyaminov, 1959, 1977) Karachentsev double galaxies (KDG) (Karachentsev, 1972) and Holmberg double and multiple galaxies (HOL) (Holmberg, 1937). Messier's numbers (M) are listed in Column 9. If KDG doubles are identified by separate coordinates, then the KDG number is reported for each component; if coordinates are the same, capital letters A,B, . . ., distinguish the components. A slash (/) after the name indicates that the radial velocity on that line refers to a multiple system as a whole. Formal coordinates assigned to the system are generally those of the western component. Measurements made on single components are identified instead by capital letters (A,B, ...,N,S,E,W, . .). Radiogalaxies are identified by one of their radio catalogue names (see Burbidge and Crowne, 1979, Ref. 783). Specifically, 3CR radiosources come from Smith and Spinrad (1980, Ref. 946). In a very few cases anonymous classification (A) has been used as in the 1RCBG (de Vaucouleurs and de Vaucouleurs, 1964). When a galaxy has no name this column is blank.

COLUMN 4   Type of Observation

When available, radio measurements (R) of recession velocities of the HI 21-cm line are listed first; velocities obtained from optical spectra follow (O).

COLUMN 5   Heliocentric Radial Velocity of the Galaxy

The heliocentric radial velocity VH is given in km/s. If the published value was redshift z, the formal recession velocity $VH = c*z$ has been computed. Corrections of any kind for solar motion or zero scale adjustment between different observatories have been removed. Values in brackets indicate author's low confidence in quoted velocity, generally measurements on single lines in the case of optical spectra. A "W" following the velocity value indicates that the measurement was successively recognized as wrong by the author(s). These values, although unreliable, have been included in order to prevent the reader from using them. Radio 21-cm line redshifts come from published tables of values. With very few exceptions, no attempt was made to read radial velocities from hydrogen profiles. The reason is that estimates from very small and distorted graphs taken from journals would have resulted in quite unreliable values. Because of this decision on our part papers sometimes rich in 21-cm redshifts (e.g., Fisher, J. R., and Tully, R. B., 1976, Astron. Astrophys., 54, 661) do not appear in our reference list. For completeness, values of radial velocities appearing from time to time in papers cited as "private communication" have also been included. They are referred to as "quoted in Ref. . . ." There is, of course, no way for us to establish the accuracy and/or reliability of these measurements. The catalogue user should therefore take these values "cum grano salis."

COLUMN 6   Mean Error Associated to VH

These entries are not comparable with each other because of different meaning attached to errors by different authors. The only purpose in quoting these errors is to specify the quality of the measurements claimed by the author(s). A value in this column is only present when it was explicitly given in the quoted reference.

COLUMN 7   Galactic Correction

Galactic correction for the motion of the Sun around the Galaxy was computed from $\Delta VH = 300 \sin \mathit{lII} \cos \mathit{bII}$ km/s, where lII and bII are the new galactic coordinates (see for details 2RCBG and IAU Transactions, 1976, 16B, pp. 201 and 28).

COLUMN 8   Reference Number

References are identified by numbers which indicate the chronological order of publication. References are listed in both chronological and alphabetical order at the end of the main catalogue. If more than one velocity appears with the same reference number, referring to the same object, then more than one independent spectrum was taken or independent methods of estimate of VH were used. A note may be present (column 9).

There are some authors and/or editors who occasionally drop initials from names. Some people therefore appear with one initial in one journal and two or three in another. In order to avoid confusion we have standardized initials, especially for first authors. Russian names present a serious problem in transliteration. This is evident even in the Russian journals where in the English abstracts names are transliterated differently in different issues. For the above mentioned reason we have arbitrarily adopted the most common transliteration and standardized Russian names.

COLUMN 9   Comments

The following compact comments may occur in this column.

SEYF   Seyfert galaxy. Sources for this classification are Khachikian and Weedman (1974), Weedman (1977), Markarian (1977) as well as more recent spectral information given in some of the references quoted.

*   When an asterisk appears a note is present at the end of the catalogue.

DIS   Discrepant velocity. It indicates that the quoted velocity differs by more than 750 km/s (i.e., five times a formal optical error of 150 km/s) from the first reported velocity, usually a 21-cm high accuracy measurement. If the velocity of the galaxy is greater than 10,000 km/s the flag indicates a 3000 km/s discrepancy. Some of the discrepancies in *VH* can be ascribed to serious errors such as inappropriate line identification in the spectra or misidentification of the object.

HII REG or
KNOT   This is not a morphological comment about the structure of the galaxy. It simply indicates that the reported spectrum was taken on a bright knotty region of the galaxy rather than on its nucleus. These comments may be quite incomplete due to lack of details in published papers. If this comment appears on the first line it implies that all radial velocities listed under those coordinates refer to the same HII

region or Knot. Radial velocities referring to the nucleus of the galaxy containing that HII region or Knot appear under a separate heading.

F G   This label identifies galaxies in the optical field of a radiosource that are not associated with the radiosource itself, although their coordinates in this catalogue could be the same as those of the radiosource.

XYZ   Peculiar names. In order not to overcrowd Column 3 some peculiar names given to galaxies either historically or by their discoverer are given in this column. All Messier's numbers for instance, are given here. All names (shortened in this column) are listed in full in Appendix 9.1. In the case of groups, only the western component is identified; the other components follow. This also holds for groups given in the appendices.

CHAPTER 3

The Catalogue

R.A. (1950)	DEC. (1950)	NAME	OBS	HEL VEL (C*Z)	ERR	GAL CORR	REF	COMMENTS
0 0 36	21 40.9	MRK 334	O	6900		219	307	
0 0 42	15 51.0	NGC 7814	R	1063	20	202	885	
			R	1050	4		914	
			O	1047	50		13	
0 1 9	14 56.4	DDO 222	R	885	10	199	492	
0 1 15	7 11.7	NGC 7816	O	5141	48	173	741	
0 1 18	-57 38.7		O	19340	200	-110	794	
0 1 24	-57 36.2		O	(20180)	200	-109	794	
0 1 24	20 28.0	NGC 7817	R	2312	9	215	914	
			O	2383	49		813	
			O	(1800)	75		555	
0 1 48	31 11.0	NGC 7819	R	4958	10	241	582	
			O	4858	75		582	
0 2 15	- 1 46.6	MRK 544	O	7211		139	569	
0 2 36	6 34.0	NGC 7824A	O	3120	86	170	594	
0 2 36	6 34.0	NGC 7824B	O	3065	56	170	594	
0 2 50	15 56.4		O	34900		201	726	
0 3 22	15 51.4		O	33730		201	726	
0 3 24	-13 41.0	NGC 7828 ARP 144 VV 272	O	5660		88	440	
0 3 24	-13 41.0	NGC 7829 ARP 144 VV 272	O	5792		88	440	
0 3 45	19 55.5	MRK 335	O	7500		212	307	SEYF
0 3 48	17 9.0	KDG 1	O	5286	10	204	908	
0 3 48	17 10.0	KDG 1	O	5510	23	204	908	
0 4 10	- 6 55.0	ARP 146A/B	O	22615	58	117	789	
0 4 30	-41 44.0		O	1435		-40	190	
0 4 36	27 26.0	NGC 1 KDG 2	R	4548	10	231	829	
			R	4534	6		914	
			O	4483	73		812	
			O	4628	77		500	
			O	(4475)			473	

R.A. (1950)		DEC. (1950)		NAME		OBS	HEL VEL (C×Z)	ERR	GAL CORR	REF	COMMENTS
0	4 42	8	2.0	NGC	3	0	3900		174	475	
0	4 42	27	24.0	NGC	2	0	7366	54	231	812	
				KDG	2						
0	5 26	32	47.5	MRK	336	0	5100		242	307	
0	5 45	-34	51.0			R	207	7	-9	620	
0	6 12	4	20.0	NGC	12	R	3940	20	160	582	
						0	3863	50		582	HII REG
						0	3946	50		582	HII REG
0	6 12	15	32.0	NGC	14	R	867	9	198	914	
				ARP	235						
				VV	80						
0	6 30	27	27.0	NGC	16	0	3110	50	230	13	
0	7 18	25	39.0	NGC	23	R	4566	15	225	829	
				MRK	545	0	4568	100		13	
0	7 24	-25	15.0	NGC	24	R	550		34	363	
						0	595	52		148	
0	7 48	28	43.0	KDG	3	0	6899	30	232	908	
0	7 51	25	33.3	NGC	26	R	4583	15	225	829	
						R	4589	7		914	
0	7 54	28	44.0	KDG	3	0	7075	20	232	908	
0	7 56	10	40.9			0	25500	600	181	702	
						0	25960			262	
0	7 57	10	41.8	3 ZW	2	0	26780		182	168	SEYF
						0	26700			186	*
						0	26783			262	
						0	28000	300		702	
0	8 6	13	25.9			R	1736	10	190	860	
0	8 9	10	44.0			0	27000	600	182	702	
0	8 33	2	24.0			0	12752	59	151	741	
0	8 48	6	6.4	NGC	36	0	6106	64	165	741	
0	8 45	32	0.5	B2		0	32170		239	595	
0	8 34	-12	23.1	MRK	938	0	5753		91	874	
0	11 24	47	57.0	NGC	48	R	1776		260	744	

13

R.A. (1950)	DEC. (1950)	NAME	OBS	HEL VEL (C*Z)	ERR	GAL CORR	REF	COMMENTS
0 11 31	-23 27.6	NGC 45	R	495		40	330	
		DDO 223	R	471	10		492	
			R	470			156	
			R	467	5		489	
			O	450	30		13	KNOT
0 11 54	7 59.0	MRK 1138	O	5020		170	921	
0 12 2	- 1 49.7	MRK 546	O	3917		133	569	
0 12 12	-60 35.0		O	37339		-125	878	
0 12 12	-60 36.0	NGC 53	O	4565		-125	878	
0 12 24	-39 28.0	NGC 55	R	130		-33	133	
			R	131	5		93	
			R	133			156	
			R	133			121	
			R	116			671	
			O	113	10		49	
			O	110	10		48	
			O	233			145	
			O	162	10		555	
			O	210	50		13	KNOT
0 12 47	-26 19.5		O	36300		26	717	
0 12 48	24 12.0	3 ZW 3	O	10201	60	219	262	
			O	12000			28	
0 13 6	46 27.0	VV 769E/W	O	5090	80	258	776	
0 13 12	15 48.0		R	4203	9	195	914	
0 13 30	21 8.0	MRK 1139	O	5760		210	920	
0 13 35	79 0.2	3CR 6.1	O	251900	120	245	849	
0 13 36	24 32.0	MRK 1140	O	19811		219	921	
0 14 14	12 4.0		R	1127	10	182	860	
0 15 0	22 13.0		R	3695	8	212	859	LGS 1
0 15 6	11 10.0	NGC 63	R	1172	26	179	914	
0 15 30	22 14.0	MRK 1141	O	5188		212	921	
0 15 38	29 47.1	NGC 67	O	6832		230	448	
		ARP 113						
		VV 166						
0 15 42	29 47.6	NGC 68	O	5705	100	230	299	
		ARP 113	O	5787	65		13	
		VV 166	O	5877			448	

R.A. (1950)	DEC. (1950)	NAME	OBS	HEL VEL (C*Z)	EPR	GAL CORR	REF	COMMENTS
0 15 44	29 45.7	NGC 69	O	6937		230	448	
		ARP 113	O	6637	150		13	
		VV 166						
0 15 48	29 47.0	NGC 71	O	6591	150	230	13	
		ARP 113	O	6913			448	
		VV 166						
0 15 48	29 49.0	NGC 70	O	7356		230	448	
		ARP 113	O	7055	100		299	
		IC 1539						
		VV 166						
0 15 54	29 46.0	NGC 72	O	7161		230	448	
		ARP 113	O	6976	150		13	
		VV 166						
0 16 0	29 45.0	NGC 72A	O	6807	130	230	13	
0 16 10	-19 7.1	DDO 1	R	2060	75	57	424	
0 16 24	46 3.0	5 ZW 11NW	O	25649		256	321	
0 16 24	46 3.0	5 ZW 11SE	O	25453		256	321	
0 16 42	3 51.0	3 ZW 4NW	O	13076		152	321	
0 16 42	3 51.0	3 ZW 4SE	O	12984		152	321	
0 17 18	19 42.0	KDG 4	O	7767	10	204	908	
0 17 24	19 43.0	KDG 4	O	7755	12	204	908	
0 17 30	6 12.0	KDG 5A	O	10382	47	160	908	
0 17 30	6 12.0	KDG 5B	O	10161	15	160	908	
0 17 30	10 36.1	DDO 2	R	1144	10	175	492	
0 17 40	59 0.8	IC 10	R	-347	7	262	93	
			R	-344	3		944	
			R	-348			744	
			R	-341	3		216	
			R	-340			171	
			R	-350			156	
			R	-346	10		462	
			R	-346	9		489	
			R	-346	2		787	
			R	-346	3		809	
			R	-345	7		73	
			O	-343	2		11	
			O	-343	12		13	

15

R.A. (1950)	DEC. (1950)	NAME	OBS	HEL VEL (C*Z)	ERR	GAL CORR	REF	COMMENTS
0 17 48	0 33.0	NGC 78A KDG 6	O	5098	60	139	812	
0 17 54	0 33.3	NGC 78B KDG 6 MRK 547	O O	(5108) 5212	50	138	447 812	
0 17 54	47 9.4		R	5136	15	256	829	
0 18 12	21 41.0	MRK 1142	O	5971		209	920	
0 18 28	6 9.1	MRK 548	O	(12131)		159	447	
0 18 35	22 4.8	NGC 80	O	5586	100	210	13	
0 18 36	30 11.0	KDG 7	O	4679	15	230	908	
0 18 42	30 14.0	KDG 7	O	4713	15	230	908	
0 18 46	-48 55.0	NGC 87	O O	3435 3603	47	-78	836 613	
0 18 46	22 9.4	NGC 83	O	6541	150	210	13	
0 18 52	-48 58.0	NGC 89	O O	3283 3824	43	-79	836 613	
0 18 52	-48 56.0	NGC 88	O O	3383 3533	16 50	-78	836 613	
0 19 0	22 12.9		O	81800		210	868	
0 19 4	-48 55.0	NGC 92	O O	3362 3498	11 45	-78	836 613	
0 19 12	22 7.0	NGC 91 ARP 65	O	5168	190	210	582	
0 19 20	- 1 53.4	MRK 549	O	11712		128	569	
0 19 39	10 12.9	NGC 95	O	4886	48	173	148	
0 19 48	23 0.9	4C 23.01	O	39930		212	400	
0 21 13	14 24.6	MRK 338	O	5400		186	307	
0 21 25	15 29.4	NGC 99	O	5184	78	189	741	
0 21 27	16 12.5	NGC 100 HOL 9A	R	836		191	744	
0 22 0	29 0.0		O	4750		225	449	

R.A. (1950)	DEC. (1950)	NAME	OBS	HEL VEL (C*Z)	ERR	GAL CORR	REF	COMMENTS
0 22 7	14 32.7	MRK 339	0	5400		185	307	
0 22 36	29 46.0		0	4770		227	449	
0 22 41	12 36.4	NGC 105	0	5290	26	179	741	
0 23 2	-33 19.3	PKS	0	14910	60	-11	355	
0 23 4	25 27.0	VV 622	0	(10500)		216	776	
0 23 14	40 40.7		0	12418		246	13	
0 23 22	- 3 41.8	MRK 945	0	4398		118	874	
0 23 24	21 31.0	KDG 8	0	5549	53	205	908	
0 23 30	13 22.0	KDG 9	0	5161	21	181	812	
0 23 33	-62 28.8		0	5640		-136	896	
0 23 36	13 22.0	KDG 9	0	5141	31	181	812	
0 23 36	21 32.0	KDG 8	0	5512	56	205	908	
0 24 0	16 53.4		0	117300	300	192	529	★
0 24 12	48 51.0	5 ZW 20	0	5720	200	255	820	
0 24 18	11 18.0	KDG 10	0	2074	20	173	908	
0 24 27	1 10.9	MRK 550	0	12903		137	569	
0 24 30	11 18.0	KDG 10	0	2178	10	173	908	
0 24 42	22 23.0		0	47796	75	207	13	
0 24 42	39 31.0	4 ZW 20	0	10736	60	244	262	
0 24 43	-57 15.3	NGC 119	0	7340		-116	574	
0 24 54	22 23.0		0	47479	40	207	13	
0 25 0	8 36.0	IC 1551	0	13170	66	164	813	
0 25 12	22 25.0		0	37052	60	207	13	
0 25 42	30 32.0		0	6420		226	449	
0 25 56	30 52.7	MRK 340	0	6400		227	307	
			0	6118	45		382	
0 26 0	0 38.6	4C 00.03	0	31500		134	471	

17

R.A. (1950)	DEC. (1950)	NAME	OBS	HEL VEL (C*Z)	ERR	GAL CORR	REF	COMMENTS
0 26 6	2 40.0		O	4460	50	141	13	
0 26 12	3 9.0		O	4500		143	475	
0 26 16	2 33.7	NGC 125	O	5289	50	141	13	
0 26 19	7 33.4	OB045	O	83600		159	717	
0 26 28	39 12.1	4 ZW 22	O	11080	200	242	820	
			O	11029			13	
0 26 32	15 37.4		R	758	10	186	860	
0 26 34	2 32.0	NGC 126	O	4252	50	140	130	
0 26 35	-31 6.9		O	7244	100	-3	714	
0 26 38	2 35.8	NGC 127	O	4094	40	141	13	
			O	4045	30		609	
0 26 41	2 35.3	NGC 128	O	4242	50	141	609	
			O	4250	50		13	
			O	4255			106	
0 26 45	2 35.7	NGC 130	O	4516	70	141	130	
			O	4425	50		609	
0 26 48	30 17.0	4 ZW 23A	O	15119	47	225	500	
		MRK 551A	O	14967			321	
0 26 48	30 17.0	4 ZW 23B	O	15008		225	321	
		MRK 551B						
0 27 6	-33 33.0	NGC 131	O	1415	63	-14	789	
0 27 21	10 41.7	MC 2	O	16500		169	652	
0 27 29	3 14.0		R	1349	10	142	860	
0 27 52	-13 12.5		O	42000		75	717	
0 27 54	-33 32.0	NGC 134	R	1566		-14	671	
			O	1600			574	
			O	1681			13	
0 27 54	-13 11.5		O	47700		75	717	
0 27 59	13 5.4		O	9934	77	177	741	
0 28 12	-28 59.3		O	7304	100	5	714	
0 28 17	-29 37.2		O	7216	100	2	714	

R.A. (1950)	DEC. (1950)	NAME	OBS	HEL VEL (C*Z)	ERR	GAL. CORR	REF	COMMENTS
0 28 30	39 25.3	WK 10	O	21610		242	318	
0 28 42	-10 45.0		O	3480	30	85	445	
0 28 42	8 11.0	HOL 11A	O	4753	184	160	594	
0 28 44	8 11.0	HOL 11B	O	4341	34	160	594	
		MRK 552	O	4037			447	
0 28 45	-28 41.3		O	6879	100	6	714	
0 29 12	- 5 26.0	NGC 145	O	4155	19	107	148	
		ARP 19	O	4154			375	
0 30 24	48 14.0	NGC 147	O	-263	80	252	555	
		DDO 3						
0 30 36	7 38.0	KDG 12	O	13527	30	157	908	
0 30 48	7 35.0	KDG 12	O	13497	30	156	908	
0 31 6	-67 41.3		O	12080		-157	896	
0 31 30	- 9 59.0	NGC 151	R	3742		87	744	
		NGC 153	O	3654	42		148	
0 31 33	39 7.6	3CR 13	O	314800		240	947	★
0 31 43	-31 2.9	DDO 224	R	1585	10	-5	492	
			O	1619	100		714	
0 31 47	-28 4.8	NGC 150	O	1571	53	7	789	
			O	1523	100		714	
			O	1555	20		741	
			O	1470			574	
0 31 48	-32 3.7	NGC 148	O	1897	9	-10	741	
			O	1481	52		148	
			O	1396	85		555	
0 32 5	-30 17.5		O	1575	100	-2	714	
0 32 5	30 9.5		O	52280		222	899	★
0 32 14	- 8 40.4	NGC 157	R	1651	20	92	544	
			R	1672	25		752	
			O	1826	100		13	
			O	1682	45		554	
			O	1710			44	
0 32 21	10 11.4	4C 10.01	O	17100		164	600	
0 32 59	10 11.4		O	24000		164	652	

19

R.A. (1950)	DEC. (1950)	NAME	OBS	HEL VEL (C∗Z)	ERR	GAL CORR	REF	COMMENTS
0 33 18	-25 39.1	IC 1558 DDO 225	R	1555	10	17	492	
0 33 19	-28 45.6		O	7056	100	3	714	
0 33 26	23 40.9	NGC 160	O O	5255 2600	50	205	13 3	DIS
0 33 34	12 21.8		O	10064	85	171	741	
0 33 36	-10 10.0		O	4870		85	449	
0 33 42	45 24.0		O	14470	200	247	820	
0 33 51	2 54.6	MRK 553	O	4513		137	569	
0 33 54	-10 23.0	NGC 165	R O	5874 5944	10 25	83	582 582	
0 34 0	-28 3.7		O	10426	100	6	714	
0 34 10	-28 3.9		O	6905	100	6	714	
0 34 14	23 42.6	IC 1559 MRK 341 ARP 282 KDG 13	O O O O O	4894 4812 4500 4291 4610	100 30 30	205	789 500 307 908 910	
0 34 14	23 42.9	NGC 169 ARP 282 KDG 13	R O O O O	4477 4508 4469 4558 5487	20 45 30 40	205	779 789 908 910 500	DIS
0 34 24	-28 38.6		O	7065	100	3	714	
0 34 26	25 25.4	B2	O	9410		209	485	
0 34 31	-29 45.1	NGC 174	O	3470	100	-1	714	
0 34 31	- 1 25.7	3CR 15	O	21900		120	143	
0 34 38	1 40.2	NGC 173	R O O	4366 4478 4231	10 50 34	132	582 582 741	
0 34 52	-20 12.6	NGC 175	O O O	3713 3890 1694	100 60	40	741 728 555	DIS
0 34 53	39 33.8		O	16841		239	13	

R.A. (1950)	DEC. (1950)	NAME	OBS	HFL VEL (C*Z)	ERR	GAL CORR	REF	COMMENTS
0 35 2	0 0.3	MRK 955	O	10424		125	874	
0 35 9	39 38.3		O	27128		239	13	
0 35 14	-33 59.4		R	8934	30	-20	649	CRW GAL
			O	9179			634	*
			O	9212			634	
0 35 16	-33 58.7		O	8639		-20	634	
0 35 18	-33 58.7		O	9104		-20	634	
0 35 23	8 21.8	NGC 180	O	5251	29	156	741	
0 35 28	-29 11.9		O	3604	100	0	714	
0 35 38	2 27.2	NGC 182	O	5234	50	134	13	
			O	5222	121		741	
0 35 43	13 15.5	MRK 342	O	11100		172	307	
0 35 47	- 2 24.1	3CR 17	O	65860		115	120	
0 35 47	14 45.9	MRK 343	O	5400		177	307	
			O	5479	45		382	
0 35 49	41 12.4		O	21730	60	241	826	
			O	21590			539	
0 36 0	41 43.0		O	5590	26	241	813	
0 36 11	48 3.7	NGC 185	O	-266	75	249	13	
			O	-263	39		72	
			O	-241			13	
			O	-305	36		11	
			O	(-60)	240		555	
0 36 16	6 46.5		O	12573		150	321	
0 36 18	6 45.3		O	12401		150	321	
0 36 18	6 45.7		O	12149		150	321	
0 36 18	6 46.0	NGC 190 3 ZW 10	O	12107		150	321	
0 36 30	- 9 17.4	IC 1563 ARP 127	O	(13652)	141	86	789	
0 36 30	- 9 16.6	NGC 191 ARP 127	O	(5065)	141	86	789	
0 36 37	-14 26.8	NGC 178 IC 39	O	1479	18	64	741	
			O	1517	34		789	

R.A. (1950)	DEC. (1950)	NAME	OBS	HEL VEL (C*Z)	ERR	GAL CORR	REF	COMMENTS
0 36 44	2 45.8	NGC 194	O	5105	50	134	13	
0 36 44	3 41.0	MRK 554	O	(12602)		138	569	
0 36 48	3 3.0	NGC 193 4C 03.01	O	4200		136	166	
0 36 50	6 27.8	IC 1565	O	11150		148	255	
0 37 0	8 41.0	KDG 14	O	4545	20	156	908	
0 37 1	2 36.7	NGC 200	O	5140	39	134	741	
0 37 6	8 43.0	KDG 14	O	4656	32	156	908	
0 37 36	-20 20.0		O	3954	100	38	714	
0 37 37	-30 52.2		O	14554	100	-8	714	
0 37 38	41 24.9	NGC 205	R	-236		240	27	M 110
			O	-300	50		2	
			O	-239	3		11	
			O	-239			223	
			O	-268	63		554	
			O	-300			3	
			O	-239	12		13	
			O	-233			13	
			O	-330	100		555	
0 37 51	24 45.1	MRK 345	O	4500		205	307	
0 38 0	-14 9.0	NGC 210	O	1768		65	13	
0 38 14	32 53.7	3CR 19	O	144500		224	717	
0 38 42	29 22.0	IC 43	O	4842		216	777	
0 38 48	- 1 59.3		O	5100		115	465	
0 38 49	25 13.5	NGC 214	R	4466	50	206	544	
			O	4535	50		13	
			O	4485			13	
0 38 51	9 47.0	3CR 18	O	56360	300	159	585	
0 38 58	-21 19.6	NGC 216	R	1558		33	366	
			O	1456	100		714	
			O	1606	77		246	
0 39 10	40 4.9	5C 3.10	O	21330		237	240	
0 39 16	-79 30.9		O	9000		-197	867	

R.A. (1950)	DEC. (1950)	NAME	OBS	HEL VEL (C*Z)	ERR	GAL CORR	REF	COMMENTS
0 39 30	40 3.0	4 ZW 29	O	30820		237	266	ZW SEY
0 39 42	29 22.0	IC 43	R	4860	10	216	582	
			O	4963	50		582	
0 39 48	2 59.0	MRK 1143	O	11047		133	920	
0 39 54	2 58.0	MRK 1144	O	11167		133	920	
0 39 58	40 35.5	NGC 221	R	-200		238	27	M 32
		ARP 168	O	-259	41		554	
			O	-300	50		1	
			O	-195	2		658	
			O	-214	10		13	
			O	-267			223	
			O	-214	2		11	
			O	-193			13	
			O	-148	35		555	
0 40 0	41 0.1	NGC 224	R	-311	6	238	247	M 31
			R	-299	7		247	AND NEB
			R	-319	4		247	
			R	-303	5		247	
			R	-289	5		129	
			R	-296	2		489	
			R	-300	3		93	
			R	-296			100	
			R	-310	5		132	
			R	-300	10		127	
			R	-313	6		354	
			R	-296			27	
			R	-298			81	
			R	-296			16	
			O	-300			4	
			O	-254	20		555	
			O	-300	50		1	
			O	-320	50		1	
			O	-290			13	
			O	-266	15		13	
			O	-311	21		223	
			O	-294	6		658	
			O	-314			207	
			O	-300			259	
			O	-266	4		11	
			O	-300	8		842	
			O	-310	11		490	
			O	-300	11		490	
			O	-220			3	
			O	-304			3	
			O	-329			3	
			O	-350			3	
0 40 4	- 1 48.3	NGC 227	O	5315	65	114	13	

R.A. (1950)	DEC. (1950)	NAME	OBS	HEL VEL (C*Z)	ERR	GAL CORR	REF	COMMENTS
0 40 13	-22 2.8		O	18254	100	29	714	
0 40 20	51 47.1	3CR 20	O	104900		250	913	
0 40 35	-22 31.5	IC 1574	R	368	10	26	492	
0 40 54	- 0 23.8	NGC 237	O	4139	27	119	741	
0 40 54	14 4.0	NGC 234	R	4448	16	172	914	
0 41 24	1 35.0	IC 49	R	4562	10	127	582	
			O	4571	20		582	
0 42 17	-76 48.3		O	7850		-189	896	
0 42 17	27 10.6	MRK 346	O	5400		209	307	
0 42 54	-61 2.0		O	10500	130	-137	670	
0 43 17	-15 52.2	NGC 244	R	941		54	366	
			O	967			163	
			O	971	40		72	
			O	910	46		246	
0 43 26	20 18.63		O	30131	100	189	633	
0 43 28	-20 53.0		O	4113	100	32	714	
0 43 32	-11 47.0	DDO 5	R	1617	10	71	492	
0 43 32	- 1 59.7	NGC 245	O	4236		111	447	
		MRK 555	O	4114	20		741	
0 43 32	20 12.81		O	30565	100	189	633	
0 43 34	20 18.60		O	17523	100	189	633	
0 43 36	19 13.0	KDG 15	O	2662	20	186	908	
0 43 43	20 12.78		O	30387	100	189	633	
0 43 44	20 20.85		O	30752	100	189	633	
0 43 48	20 22.39		O	32609	100	189	633	
0 43 48	36 3.0	KDG 16	O	11054	20	228	908	
0 43 49	20 11.45		O	31199	100	189	633	
0 43 49	20 20.36		O	29512	100	189	633	
0 43 50	20 16.95		O	31330	100	189	633	

R.A. (1950)	DEC. (1950)	NAME	OBS	HEL VEL (C*Z)	ERR	GAL CORR	REF	COMMENTS
0 43 52	20 11.93	3C 21	O	30805	100	189	633	
			O	31100	150		529	
0 43 54	36 4.0	KDG 16	O	11148	47	228	908	
0 43 55	-42 24.2	PKS	O	15830	389	-62	355	
0 43 58	20 12.25		O	32166	100	189	633	
0 44 2	20 19.58		O	31373	100	189	633	
0 44 3	19 59.56		O	36354	100	188	633	
0 44 12	32 24.0	KDG 17	O	5090	30	220	908	
0 44 24	30 4.0		R	2931	20	214	752	
0 44 24	32 25.0	KDG 17	O	5023	45	220	908	
0 44 28	-22 7.2		O	6564	100	26	714	
0 44 30	-21 2.0	NGC 247	R	156		30	671	
			R	182			219	
			R	156	15		93	
			R	157	9		489	
			O	-28	35		13	KNOT
0 44 36	50 37.0		R	5176	10	248	582	
			O	5206	25		582	
0 44 42	14 26.0	MRK 1146	O	11672		170	921	SEYF
0 44 47	-52 19.4		O	8231	25	-104	613	
0 44 47	39 48.0	5C 3.175	O	37385		234	509	
0 44 49	-52 19.4		O	8091	30	-104	613	
0 44 51	-52 19.4		O	8192	28	-104	613	
0 45 3	-31 41.7	NGC 254	R	1624	65	-16	885	
			O	1441	86		741	
0 45 6	-20 42.1	A 45A	O	6071	47	31	789	
0 45 6	-20 45.0	A 45C	O	6315	82	31	789	
			O	5957	100		714	
0 45 8	-25 33.7	NGC 253	R	250	15	10	217	
			R	257			219	
			R	239			671	
			R	250	10		326	
			R	250	10		625	
			R	-88			27	
			O	255	10		755	*
			O	250			66	

0ʰ45ᵐ

R.A. (1950)	DEC. (1950)	NAME	OBS	HEL VEL (C*Z)	ERR	GAL CORR	REF	COMMENTS
0 45 8	-25 33.7	NGC 253	O	210	22	10	245	*
			O	245			778	
			O	236	5		725	
			O	(-68)			12	
			O	240			66	KNOT
			O	-81	35		13	* KNOT
0 45 9	-20 47.7		O	6351	100	31	714	
0 45 12	-21 46.0		O	6402	100	27	714	
0 45 12	19 18.0	KDG 15	O	4621	70	185	908	
0 45 18	-11 44.0	NGC 255	R	(1594)		70	693	
			R	1608			744	
			O	1921			13	
0 45 18	22 6.0	IC 1586	R	5730		193	389	
		3 ZW 12	R	5812	62		622	
		MRK 347	O	5819	60		262	
			O	5820			389	
			O	6000			307	
0 45 26	8 1.4	NGC 257	O	5302	40	148	741	
0 45 54	27 25.4	NGC 260 HOL 23C	R	5206	25	207	829	
0 46 0	-13 5.0		O	6536	34	64	72	
0 46 0	-12 59.0		O	6528		64	163	
			O	6300	270		445	
0 46 0	10 4.0	MRK 1147	O	10665		155	921	
0 46 4	31 41.0	NGC 262 MRK 348	R	4534		217	707	SEYF
			R	4453			610	
			O	4700			570	
			O	4200			307	
0 46 12	14 5.0	NGC 263	O	6425	25	168	246	
0 46 17	4 3.6	MRK 556	O	12457		133	569	
0 46 24	- 2 58.0	MRK 557	O	4020		105	438	
			O	3793			569	
0 47 6	42 19.0		O	60980	250	236	13	BAD A
			O	16804	25		283	DIS
0 47 6	42 20.0		O	23908	30	237	13	BAD B
			O	27196	25		283	DIS

R.A. (1950)	DEC. (1950)	NAME	OBS	HEL VEL (C*Z)	ERR	GAL CORR	REF	COMMENTS
0 47 20	-21 17.5	DDO 6	R	299	10	27	492	
			R	307	20		424	
0 47 37	- 5 28.1	NGC 268	O	5515	33	94	741	
0 47 42	24 15.0		O	10200		198	475	
0 48 30	- 7 19.7	NGC 274/5 ARP 140	R	1750	15	86	455	
0 48 30	- 7 19.7	NGC 274 VV 81 HOL 26B	O	1890	45	86	72	
0 48 30	- 7 20.0	NGC 275 VV 81 HOL 26A	R R O O	1750 1749 1824 1773	 24	86	744 693 375 72	*
0 49 0	-13 7.0	IC 56	O	6090		62	445	
0 49 0	- 1 30.0		O	1500		109	475	
0 49 12	47 18.0	NGC 278	R R R O O O	640 641 655 (650) 622 656	 9 30	242	693 956 363 1 13 13	
0 49 24	-13 1.0		O	12570	150	62	445	
0 49 30	- 0 44.0		O	1800		112	475	
0 49 36	- 2 29.0	NGC 279 MRK 558	O	3942		105	447	
0 50 12	28 45.0		O	5100		208	475	
0 50 18	-23 14.9		O	15962		17	714	
0 50 24	-31 29.2	NGC 289	O O	1928 1740		-18	13 574	
0 50 50	-20 37.5		O	14087	100	28	714	
0 50 54	21 14.4	MRK 349	O O	6900 6934	 45	187	307 382	
0 51 0	-73 6.0		R O O	160 168 168		-180	10 11 1	SMC

R.A. (1950)	DEC. (1950)	NAME	OBS	HEL VEL (C*Z)	ERR	GAL CORR	REF	COMMENTS
0 51 0	12 25.0	1 ZW 1	O	17600		159	579	SEYF
			O	18200			825	*
			O	17500			825	
			O	18230			579	
			O	18300			186	
			O	18150			187	
0 51 35	- 3 49.7	3C 26	O	63090		98	120	
0 51 40	-21 59.2		O	16959	100	21	714	
0 51 48	42 0.0	5 ZW 40	O	5759	200	234	820	
0 51 53	-23 49.5		O	20156	100	13	714	
0 52 0	28 30.0		O	6900		206	475	
0 52 24	31 16.0	NGC 295	R	5430	34	212	914	
0 52 31	-37 57.4	NGC 300	R	146		-48	671	
			R	145	2		835	
			R	145	2		162	
			R	137			121	
			R	140			156	
			R	145	10		93	
			O	248	40		13	KNOT
			O	88	30		789	
0 53 9	26 8.4	3CR 28	O	58560		199	120	
0 53 42	-14 33.0	MRK 1149	O	6338		52	921	*
0 53 52	- 1 31.5		O	11539	51	106	741	
0 54 11	-10 11.0	NGC 309	R	5665		70	744	
		HOL 27A	R	5665	10		752	
			O	5679	16		741	
0 54 23	- 1 28.7		O	15048	41	106	741	
0 54 41	-22 11.6		O	15887	100	19	714	
0 54 39	23 37.1	MRK 350	O	5100		191	307	
0 54 50	43 31.0	NGC 317B	O	5324	110	234	567	
		KDG 19	O	5106	30		812	
			O	5666			748	
0 54 51	43 32.0	NGC 317A	O	5866		234	748	
		KDG 19	O	5305	78		812	
0 54 54	43 26.0		R	5422	10	234	582	
			O	5397	25		582	

R.A. (1950)	DEC. (1950)	NAME	OBS	HEL VEL (C*Z)	ERR	GAL CORR	REF	COMMENTS
0 55 0	- 5 16.0	NGC 321	O	5770		90	449	
0 55 0	33 5.0		O	5400		215	475	
0 55 1	- 1 39.7	3CR 29	O	13400	180	104	159	
			O	13414	127		741	
			O	13340	200		350	
0 55 6	30 4.9	NGC 315	O	4800		208	485	
		B2	O	4949			842	
		HOL 29A						
0 55 11	-63 42.0		O	11540		-150	896	*
			O	10845	90		680	
0 55 15	-63 41.5		O	9230		-150	896	*
			O	10390			680	DIS
0 55 30	48 23.2		R	6812	15	240	582	
			O	6903	75		582	
0 55 36	36 28.0		R	6140	10	222	582	
			O	6127	20		582	
0 55 40	26 35.5	NGC 326	O	14490		199	400	*
		4C 26.03	O	13940			485	
		4 ZW 35						
0 56 21	-67 4.5		O	(9800)		-162	896	
0 56 42	6 40.0	MRK 559	O	13200		135	438	
0 57 9	31 33.5	MRK 352	R	4448		210	707	SEYF
			R	4477	20		779	
			O	4500			307	
0 57 18	- 7 50.7	NGC 337	R	1646		78	744	
			O	1690	33		148	
0 57 36	-33 58.0		O	196		-33	704	SCP SYS
0 58 34	-68 20.3		O	7015		-167	896	
0 58 35	-40 25.0 N		O	6783		-61	878	
0 58 35	-40 25.4 S		O	16397		-61	878	
0 58 48	7 21.3	DDO 7	R	2211	10	136	492	
0 59 3	- 7 51.4	NGC 337A	O	388	60	76	72	
0 59 27	14 27.3	3C 30	O	56400		160	868	

29

R.A. (1950)	DEC. (1950)	NAME	OBS	HEL VEL (C*Z)	ERR	GAL CORR	REF	COMMENTS
0 59 42	30 27.0	1 ZW 2	O	10370		206	303	
0 59 48	34 51.0	MRK 1150	O	24064		216	921	
1 0 6	25 35.7	4C 25.03	O	23520		193	400	
1 0 14	-67 30.8		O	19840		-165	896	
1 0 35	22 4.4	NGC 354	O	4800		183	307	
		MRK 353	O	4876	45		382	
1 0 36	13 47.0	KDG 21	O	12314	50	157	812	
1 0 42	- 7 16.0	NGC 356	O	5822		78	774	
		VV 486						
1 0 42	13 46.0	KDG 21	O	12681	38	157	812	
1 0 42	31 58.0		O	5418	66	209	653	
1 0 48	- 6 37.0	NGC 357	O	2541	50	80	13	
1 1 12	21 37.0		R	-280	8	181	859	LGS 3
1 1 12	29 52.0	1 ZW 3	O	24360	108	203	98	
1 1 26	20 9.9	MRK 354	O	14100		177	307	
1 1 59	-80 12.2	PKS	O	17100		-202	711	
1 2 13	1 51.0	IC 1613	R	-236	3	113	93	
		DDO 8	R	-234			58	
			R	-228			934	
			R	-242			27	
			O	-238	2		11	
			O	-238	10		13	KNOT
1 2 32	-64 23.4		O	5942	27	-155	763	
1 2 36	- 6 29.0	A 103	R	1096		79	744	
			O	983	48		72	
1 3 12	32 9.0	IC 1618	O	4706	6	208	653	
1 3 19	44 41.3		R	707	10	231	860	
1 3 24	31 8.0		O	9322		205	777	
			O	(4500)			54	DIS
1 3 50	- 0 24.7	MRK 560	O	10400		103	438	
1 4 18	32 5.0	NGC 375	O	6011	40	207	13	
		ARP 331						

R.A. (1950)	DEC. (1950)	NAME	OBS	HEL VEL (C*Z)	ERR	GAL CORR	REF	COMMENTS
1 4 18	32 32.0	NGC 374	O	5067	161	208	653	
1 4 24	13 42.0	KDG 22	O	11674	43	154	908	
1 4 30	13 42.0	IC 1620 KDG 22	O	11461	30	154	908	
1 4 30	32 13.0	NGC 380 ARP 331	O O	4341 4400	150 75	207	13 2	
1 4 30	32 15.0	NGC 379 ARP 331	O	5374	65	207	13	
1 4 35	32 9.1	NGC 383 ARP 331 VV 193 3CR 31 KDG 23B	O O O O	4888 4777 5147 4500	50 128 100	207	13 500 748 2	
1 4 36	32 2.0	NGC 384 ARP 331	O O	4401 4500	100	206	13 475	
1 4 36	32 48.0		O	9600		208	475	
1 4 38	32 8.2	NGC 382 ARP 331 4 ZW 38 VV 193 KDG 23A	O O	5156 4879	50	207	13 748	
1 4 42	32 3.0	NGC 385 ARP 331	O O	4845 4900	150 100	206	13 2	
1 4 42	32 6.0	NGC 386 ARP 331	O	5555	150	207	13	
1 4 44	39 7.9		R	5869	20	221	829	
1 4 58	1 55.0		O	12950		111	726	
1 5 0	-13 30.0		O	16020		49	449	
1 5 0	32 3.0	NGC 388 ARP 331	O	5114	100	206	13	
1 5 0	41 43.0	5 ZW 51	O	11200	200	226	820	
1 5 18	33 11.0		O	4719	68	209	653	
1 5 34	1 55.0		O	13100		111	255	
1 5 36	32 52.0	NGC 392	O	4672		208	653	

1ʰ5ᵐ

R.A. (1950)	DEC. (1950)	NAME	OBS	HEL VEL (C*Z)	ERR	GAL CORR	REF	COMMENTS
1 5 38	1 55.6		O	12590		111	726	
1 5 40	-70 9.0	NGC 406	O	1468	26	-174	741	
			O	1570	100		844	
1 5 40	1 54.6		O	13610		110	726	
1 6 0	33 12.0		O	12624	73	208	653	
1 6 6	-13 5.0	MRK 1151	O	16260		50	921	
1 6 6	1 23.0		O	2010		108	320	
1 6 6	72 56.0		O	54260		243	657	*
1 6 12	32 22.0	NGC 399	O	5167	121	206	653	
1 6 15	13 4.3	3CR 33	O	17750	60	151	846	
			O	17850			120	
1 6 18	-15 36.0	A	O	15440	60	39	13	
1 6 18	-15 36.0	B	O	16057	60	39	13	
1 6 24	-16 13.0	IC 79	O	12567		36	13	
1 6 30	32 29.0	NGC 403	O	4977	212	206	653	
1 6 36	35 27.0	NGC 404	R	-35	18	213	956	
			R	-63			537	
			O	-26			106	
			O	-25	50		1	
			O	14	32		72	
			O	-55	30		13	
1 7 12	32 5.0		O	5246		205	777	
			O	5414			653	
1 7 19	42 50.6	KDG 24	R	5050	10	226	829	
			O	4986	24		908	
1 7 36	43 1.4	KDG 24	R	4910	20	227	829	
			O	4922	10		908	
1 7 45	49 20.1	DDO 9	R	646	10	235	492	
			R	639	10		956	
1 7 48	32 51.0	NGC 407	O	5610		206	653	
1 7 59	17 15.24		O	19362	100	163	633	
1 8 4	17 25.05		O	19231	100	163	633	

R.A. (1950)	DEC. (1950)	NAME	OBS	HEL VEL (C*Z)	ERR	GAL CORR	REF	COMMENTS
1 8 7	17 23.99		0	18714	100	163	633	
1 8 9	17 23.11		0	20659	100	163	633	
1 8 12	-30 29.0	NGC 418	0	5684	25	-25	582	
1 8 12	32 53.0	NGC 410	0	5238		206	653	
1 8 19	17 21.77		0	18334	100	163	633	
1 8 22	17 23.64		0	13512	100	163	633	
1 8 23	17 23.88	IC 1634	0	20405	100	163	633	
1 8 24	17 23.22	IC 1635	0	18284	100	163	633	
1 8 29	17 35.07		0	20947	100	164	633	
1 8 30	17 25.65		0	19700	100	163	633	
1 8 30	31 37.0		0	5229		203	653	
1 8 30	32 50.0	NGC 414A KDG 25A	0	5748	51	206	812	
1 8 30	32 50.0	NGC 414B KDG 25B	0	5298	32	206	812	
1 8 36	-30 41.0	IC 1637	0	6002	50	-26	582	
1 8 37	25 49.6	4C 25.04	0	19920		188	400	
1 8 39	-14 13.5	PKS	0	15480	180	43	355	
1 8 46	17 24.68		0	19692	100	163	633	
1 8 48	1 1.0	KDG 26	0	6780	10	105	908	
1 8 51	17 20.70		0	20116	100	163	633	
1 9 0	1 4.0	KDG 26	0 0	6734 6900	17	105	908 475	
1 9 5	49 12.8	3CR 35	0	20090		235	318	
1 9 13	- 0 56.0	IC 1639 MRK 562	0	(2453)		97	447	
1 9 18	- 1 55.0	MRK 563	0	4975		93	438	
1 9 18	31 52.0	NGC 420	0	5199		203	653	
1 9 40	-68 48.8		0	21670		-171	896	

33

R.A. (1950)	DEC. (1950)	NAME	OBS	HEL VEL (C*Z)	ERR	GAL CORR	REF	COMMENTS
1 10 12	-58 31.0	NGC 434	O O	4855 4960	97	-138	203 574	
1 10 18	15 14.0	PKS	O	13340			155	595
1 10 24	0 43.0	NGC 428	R R R O	1130 1175 1153 1078	40 25	102	216 183 744 13	
1 10 30	15 15.5		O	13655			155	595
1 10 54	-58 33.0	NGC 440	O O	5144 4935	19	-138	203 574	
1 11 6	0 36.7		R	1164	10	101	860	
1 11 9	2 6.4	PKS	O	14100		107	600	
1 11 12	7 31.2	MRK 564	O	5575		127	438	
1 11 18	33 27.0	NGC 431	O	5786		206	653	
1 11 24	-15 7.0	MRK 1152	O	15643		37	921	SEYF
1 11 25	-32 0.9	NGC 439	O	5644	62	-34	741	
1 11 36	42 17.0		R O	5914 5914	10 25	223	582 582	
1 11 57	0 3.61		O	12890	100	99	633	
1 12 12	31 42.0	IC 1652	O	5317		201	653	
1 12 12	50 5.0	5 ZW 58	O	7280	200	234	820	
1 12 15	- 0 45.6		O	10143	42	95	741	
1 12 15	- 0 2.05		O	13975	100	98	633	
1 12 18	-32 32.0		O	5262	30	-37	582	
1 12 20	0 2.35		O	13298	100	98	633	
1 12 24	0 10.00		O	13439	100	99	633	
1 12 26	- 0 0.30		O	14621	100	98	633	
1 12 32	0 3.48		O	12625	100	98	633	
1 12 34	0 12.10		O	12560	100	99	633	
1 12 39	0 4.69		O	13453	100	98	633	

R.A. (1950)	DEC. (1950)	NAME	OBS	HEL VEL (C*Z)	ERR	GAL CORR	REF	COMMENTS
1 12 42	− 0 3.00		0	13567	100	98	633	
1 12 43	− 0 4.70		0	13420	100	97	633	
1 12 47	− 0 0.28		0	13039	100	98	633	
1 12 51	0 1.07		0	14083	100	98	633	
1 12 57	− 1 7.5	NGC 450	R	1761	9	93	744	
		KDG 27A	0	1897	58		789	
			0	1914	53		741	
1 12 57	0 2.11		0	14040	100	98	633	
1 13 6	30 49.0	KDG 28	0	4889	27	198	908	
		IC 1658	0	(5771)	100		567	DIS
1 13 19	32 49.5	MRK 1	R	4780		203	707	* SEYF
			R	4847			779	
			0	4800			264	
			0	4800			280	
			0	4800			192	
			0	4793			723	
1 13 28	4 2.0	IC 89	0	(6333)		113	447	
		MRK 565						
1 13 30	30 46.0	NGC 452	0	4955	20	198	908	
		KDG 28						
1 13 36	−32 44.0		0	6041	25	−38	582	
1 15 31	11 7.1		0	5062	50	137	741	
1 16 0	−44 57.0		0	6771	70	−90	603	
1 16 24	− 1 16.0		0	4855		90	124	
			0	13471			623	DIS
1 16 24	8 14.1	PKS	0	177960		126	588	
1 16 30	4 4.0	MRK 566	0	9680		110	438	
1 16 36	3 3.0	NGC 467	0	5568	45	107	148	
1 16 42	1 57.3		0	13339	79	102	741	
1 16 43	4 19.0	MRK 567	0	9932		111	447	
1 16 45	12 11.1	KDG 29A	0	14528	38	140	908	
		MRK 984	0	14243			874	
		ARP 119	0	14211	187		812	
		VV 347	0	14345	111		812	

35

1ʰ16ᵐ

R.A. (1950)	DEC. (1950)	NAME	OBS	HEL VEL (C*Z)	ERR	GAL CORR	REF	COMMENTS
1 16 45	12 12.0	KDG 29B	O	14429	46	140	908	
		VV 347	O	14806	83		812	
		ARP 119						
1 16 47	31 55.1	NE	O	17570	210	198	263	
		4C 31.04	O	17750			400	
			O	17510			318	
1 16 47	31 55.1	SW	O	16430	150	198	263	
		4C 31.04						
1 17 0	7 54.7		O	9469	41	124	741	
1 17 10	3 8.9	NGC 470	R	2555)	9	107	581	
		ARP 227	R	2370	10		829	
			O	2557	40		148	
			O	2362	38		741	
1 17 15	16 17.0	NGC 473	O	2252	36	153	741	
1 17 32	3 9.0	NGC 474	R	2385)	9	106	581	
		ARP 227	O	2306	40		13	
1 17 38	-58 47.9	NGC 484	O	5200		-142	574	
1 17 42	-41 28.0		O	5052	35	-77	719	★
			O	5572	70		603	
1 17 48	5 33.8		R	2166	10	115	860	
1 17 48	29 22.0	KDG 30	O	4404	15	191	908	
1 17 54	29 26.0	IC 1672	O	7154	35	191	908	
		KDG 30						
1 18 24	22 31.0	3 ZW 23	O	13220	60	172	262	
1 18 24	40 13.0	NGC 477	R	5859	15	215	752	
1 18 36	16 46.0	3 ZW 25	O	21715	60	154	262	
1 18 40	12 9.0	DDO 10	R	643	10	138	492	
			R	646	10		956	
1 18 48	- 0 48.0		O	5189		90	623	
1 18 48	- 0 10.0		O	3882		92	623	
1 19 3	-34 19.4	NGC 491	O	3899	30	-48	741	
1 19 11	4 59.6	NGC 488	R	2268		112	693	
			O	2260	20		924	
			O	2180	150		13	

R.A. (1950)	DEC. (1950)	NAME	OBS	HEL VEL (C*Z)	ERR	GAL CORR	REF	COMMENTS
1 19 16	2 44.3		0	40500		103	717	
1 19 24	33 0.0	IC 1682	0	18900		199	475	
1 19 24	34 25.0		0	5400		203	475	
1 19 30	- 1 18.0	2 ZW 1	0	16164	60	87	262	SEYF
1 19 36	0 41.0	NGC 493	0	2338		95	777	
1 19 42	- 2 49.0		0	16204		81	623	
1 19 49	8 47.6	MRK 568	0	(5640)		125	557	
			0	5425			569	
1 19 52	- 1 8.2	NGC 497	0	8140	73	88	741	
		ARP 8	0	8030	100		99	
1 19 55	26 36.4	MRK 355	0	9175	45	182	382	
1 19 57	22 54.5	MRK 357	0	16200		172	307	
1 19 57	26 36.3	MRK 356	0	9000		182	307	
			0	9055	45		382	
1 20 0	- 2 40.0		0	5152		82	623	
1 20 2	1 38.0	MRK 569	0	9945		98	447	
1 20 3	33 15.4		0	4132	100	199	519	
1 20 6	- 0 50.0		0	7615		89	623	
1 20 7	32 54.8	NGC 494	0	5314	100	199	519	
1 20 7	33 12.6	NGC 495	0	4114	50	199	13	
1 20 12	- 1 39.0		0	5651	150	86	99	
1 20 18	- 0 13.0		0	13159		91	623	
1 20 22	33 13.7	NGC 498	0	6151	100	199	519	
1 20 23	33 12.0	NGC 499	0	4375	50	199	13	
1 20 30	33 1.0	IC 1687	0	4881	100	199	519	
1 20 33	33 10.4	NGC 501	0	4887	100	199	519	
1 20 36	34 19.0	1 ZW 4	0	7200		202	282	
		2 ZW 2	0	6851	60		262	
1 20 36	34 19.0	1 ZW 4N	0	7160		202	214	

R.A. (1950)	DEC. (1950)	NAME	OBS	HEL VEL (C*Z)	ERR	GAL CORR	REF	COMMENTS
1 20 36	34 19.0	1 ZW 4S	O	6868		202	98	
1 20 39	32 56.6	NGC 504	O	4090	100	198	519	
1 20 39	33 4.3	NGC 503	O	5978	100	199	519	
1 20 42	- 1 9.0		O	7707	200	87	99	
1 20 42	- 0 57.0	MRK 1153	O	1772		88	921	
1 20 42	- 0 54.0		O	8014		88	623	
1 20 42	- 0 39.0		O	7610		89	623	
1 20 48	- 2 14.0		O	4730	30	83	99	
1 20 51	32 59.7	NGC 507 ARP 229 VV 207 B2	O O	4737 4929	100 50	198	519 13	
1 20 52	33 1.2	NGC 508 ARP 229 VV 207	O	5476	100	198	519	
1 20 53	32 56.5		O	5158	100	198	519	
1 21 1	32 53.8	IC 1690	O	4537	100	198	519	
1 21 6	- 2 5.0		O	5648	150	83	99	
1 21 9	33 3.2		O	4995	100	198	519	
1 21 24	- 1 53.0		O	5932	100	84	99	
1 21 24	13 46.0	3 ZW 27	O	16355	60	142	262	
1 21 25	12 39.5	NGC 514	R R R O O	2473 2473 2470 2602 2487	10 6 10 60	138	544 914 752 13 13	
1 21 30	- 1 54.0		O	5919	100	84	99	
1 21 42	33 33.0	NGC 513	O	6000		199	475	
1 21 48	- 2 0.0		O	5275	30	83	99	
1 21 49	3 37.4		R	2150	10	105	860	
1 21 51	-59 3.9		O O O	13974 13735 13500	90 120	-144	761 800 695	* SEYF FAI 9

38

R.A. (1950)	DEC. (1950)	NAME	OBS	HEL VEL (C*Z)	ERR	GAL CORR	REF	COMMENTS
1 21 54	- 1 53.0		O	5332	150	83	99	
1 21 54	- 1 39.0		O	4855	150	84	99	
1 21 54	31 55.0		O	10800		195	475	
1 22 0	1 28.2	NGC 521	R	5018	4	96	914	
			O	5100			777	
			O	4999	27		741	
1 22 0	3 28.0		O	34800		104	423	
1 22 0	3 32.0	NGC 520	R	2162		104	693	*
		ARP 157	R	2260	30		455	
		VV 231	R	2280	40		390	
			R	2193			744	
			O	2110			951	
			O	1972			387	
			O	2326	33		72	
			O	2084			13	
1 22 6	- 1 50.0	NGC 530 IC 106	O	5016	40	83	99	
1 22 12	- 1 49.0	MRK 1154	O	5647		83	921	
1 22 18	- 1 56.0		O	5229	200	83	99	
1 22 18	- 1 52.0	IC 1696	O	5768	30	83	99	
1 22 18	- 1 45.0		O	4788	100	84	99	
1 22 18	9 17.0	NGC 524	O	2470	65	125	13	
1 22 30	0 10.0	IC 1697	O	8699		91	623	
1 22 30	33 46.0	NGC 523 ARP 158	O	4840		199	369	
1 22 30	33 46.0	NGC 523E	O	4762		199	321	
1 22 30	33 46.0	NGC 523W	O	4701		199	321	
1 22 31	-68 52.7		O	(14240)		-175	791	
1 22 36	14 35.0	IC 1698	O	5365	110	143	594	
1 22 41	9 0.3	NGC 532	R	2375	7	124	914	
1 22 42	14 36.0	IC 1700	O	5747	192	143	594	
1 22 45	-68 52.8		O	10610		-175	791	

R.A. (1950)	DEC. (1950)	NAME		OBS	HEL VEL (C*Z)	ERR	GAL CORR	REF	COMMENTS
1 22 45	-68 52.5			0	(10970)		-175	791	
1 22 45	-68 52.0			0	10640		-175	791	
1 22 48	- 1 46.0			0	5118	50	83	99	
1 22 50	-68 50.8			0	(10580)		-175	791	
1 22 54	-18 26.0	NGC	539	0	9657	25	15	582	
1 22 54	- 1 48.0	NGC	538	0	5398	30	83	99	
				0	5100			623	
1 22 55	-68 52.3			0	11330		-175	791	
1 22 55	-68 51.7			0	12260		-175	791	
1 22 57	1 30.0	NGC	533	0	5506	21	96	741	
				0	5003	42		148	
1 23 0	- 2 3.0			0	6500		82	623	
1 23 0	- 1 45.0			0	6591	25	83	99	
				0	5570			623	DIS
1 23 0	- 1 39.0	NGC	535	0	4939	50	83	99	
1 23 1	-68 53.2			0	(10490)		-175	791	
1 23 6	0 55.0			0	16800		93	475	
1 23 12	- 1 44.0			0	5078	150	83	99	
1 23 12	- 1 42.0			0	5321	100	83	99	
1 23 12	- 1 37.0	NGC ARP	541 133	0	5392		83	99	
1 23 12	- 1 35.0			0	5469	30	83	99	
1 23 18	- 1 34.0			0	5374	30	83	99	
1 23 18	- 1 32.0	NGC	543	0	5238	30	84	99	
1 23 24	11 11.6	IC	112	0	5700		131	534	
1 23 26	- 1 36.0	NGC ARP 3CR	545 308 40	0 0 0	5316 5354 5499	30 45	83	99 748 148	
1 23 28	- 1 36.3	NGC ARP 3CR	547 308 40	0 0 0	5472 5492 5361	30 48	83	99 748 148	

R.A. (1950)	DEC. (1950)	NAME	OBS	HEL VEL (C★Z)	ERR	GAL CORR	REF	COMMENTS
1 23 30	- 1 29.0	NGC 548	O	5332	40	84	99	
1 23 30	33 9.0	MRK 1155	O	4873		197	921	
1 23 31	34 26.6	NGC 536	O	5200		200	426	
1 23 36	48 8.0	5 ZW 68	O	10420	200	226	820	
1 23 42	- 2 2.0		O	5110	200	81	99	
1 23 42	31 22.0		O	13525		192	777	
1 23 42	39 37.0	5 ZW 69	O	2610	200	211	820	
1 23 45	31 21.2	MRK 358	O	12600		192	307	SEYF
			O	13585	45		382	
1 23 54	- 1 53.0	IC 1703	O	5688	50	82	99	
1 24 0	-23 38.0		O	5574		-7	728	
1 24 0	- 1 21.0		O	5956	150	84	99	
1 24 6	- 1 33.0		O	5013		83	623	
1 24 6	1 45.0	NGC 550	O	5962		96	777	
1 24 12	- 1 14.0		O	4800		84	475	
1 24 12	- 1 13.0		O	5005	250	84	99	
1 24 12	18 57.5	IC 115 4C 18.06	O	13040	120	157	350	
1 24 30	14 31.3	IC 1706	O	6349	86	142	741	
1 24 42	- 2 16.0	NGC 558	O	5018	200	80	99	
1 24 42	- 1 31.0		O	4794	100	83	99	
1 24 50	18 55.1	MRK 359	O	5100		156	307	
1 24 53	- 2 10.4	NGC 560	O	5503	150	80	13	
			O	5405	30		99	
1 25 0	- 1 20.0		O	4661	150	83	99	
1 25 6	- 1 22.0		O	5174	150	83	99	
1 25 8	62 51.0		O	5490	90	239	852	
1 25 18	- 2 7.0	NGC 564	O	5851	150	80	13	
			O	5556	50		99	

1ʰ25ᵐ

R.A. (1950)	DEC. (1950)	NAME	OBS	HEL VEL (C*Z)	ERR	GAL CORR	REF	COMMENTS
1 25 24	− 2 17.0	IC 119	O	6203	300	79	99	
			O	5705			623	
1 25 30	− 1 0.0		O	5211		84	623	
1 25 30	48 8.0	NGC 562	R	10254	20	225	582	
			O	10268	20		582	
1 25 36	− 1 32.0	NGC 565	O	4464	40	82	99	
1 25 42	− 2 9.0	IC 120	O	4900	100	79	99	
1 25 42	16 11.0	KDG 33	O	11344	21	147	908	
1 25 42	16 12.0	KDG 33	O	11571	10	147	908	
1 25 43	28 47.5	3CR 42	O	118420	300	184	585	
1 25 54	− 2 51.0		O	5071		76	623	
1 26 6	− 1 58.0		O	6271	150	80	99	
1 26 24	− 1 11.0	NGC 570	O	5502	30	83	99	
1 26 24	− 0 49.0		O	5325		84	623	
1 26 30	10 53.0	NGC 569 KDG 34A	O	5768	20	128	812	
1 26 36	− 2 7.0		O	5173	40	79	99	
1 26 36	10 53.0	KDG 34B	O	5775	30	128	812	
1 26 48	45 20.4	KDG 35	R	5239	20	220	829	
			O	5180	30		908	
1 27 4	40 42.9		R	2806	15	211	829	
1 27 12	− 2 14.0	IC 126	O	5704		78	623	
1 27 12	− 1 30.0		O	5129		81	623	
			O	900			475	DIS
1 27 12	45 22.0	KDG 35	O	12940	30	220	908	
1 27 41	−42 39.9		O	7607	50	−87	680	
1 27 53	41 0.0	NGC 573	R	2796	15	211	829	
1 28 5	−22 55.5	NGC 578	R	1688	20	−7	544	
			R	1619	10		752	
			O	2017			13	
			O	1603			728	
			O	1590			574	

R.A. (1950)	DEC. (1950)	NAME	OBS	HEL VEL (C*Z)	ERR	GAL CORR	REF	COMMENTS
1 28 6	- 2 15.0	NGC 577	0	6050		77	623	
1 28 6	21 11.0	IC 1710	0	3161	43	161	813	
1 28 36	- 1 45.0		0	5281		79	623	
1 28 52	- 7 7.6	NGC 584	0	1875	12	57	658	
		IC 1712	0	1827	75		13	
			0	(1800)			1	
			0	1917	53		554	
1 28 55	33 21.6	NGC 579	R	4981	15	194	829	
1 29 6	- 1 11.0	NGC 585	0	5204		80	623	
1 29 6	18 20.0		0	25200		151	475	
1 29 7	33 13.3	NGC 582	R	4354	10	193	829	
1 29 12	32 55.0	MRK 1156	0	10207		193	921	
1 29 18	31 51.0	ARP 98E 4 ZW 51	0	12542		190	321	
1 29 18	31 51.0	ARP 98W 4 ZW 51	0	12577		190	321	
1 29 25	-13 52.2		0	63900		29	717	
1 29 50	30 23.9	NGC 588	0	-174		186	104	HII REG
1 30 0	4 19.0	KDG 36	0	1911	20	101	908	
1 30 0	4 20.0	KDG 36	0	1786	20	101	908	
1 30 20	30 23.0	NGC 592	0	-150		185	104	HII REG
1 30 24	- 7 17.3	NGC 596	0	1903	6	55	658	
			0	1888	13		938	
			0	1854			842	
			0	2049	65		13	
1 30 24	- 0 57.0	IC 138	0	4581		80	623	
1 30 24	44 41.0	KDG 37	0	5112	60	217	908	
1 30 31	- 7 34.1	NGC 600	0	1867	71	54	554	
1 30 36	44 40.0	KDG 37	0	5058	30	217	908	
1 30 42	35 25.0	NGC 591 MRK 1157	0	4332		198	921	SEYF

R.A. (1950)	DEC. (1950)	NAME	OBS	HEL VEL (C*Z)	ERR	GAL CORR	REF	COMMENTS
1 31 0	- 1 21.0		0	4862		78	623	
1 31 3	30 23.9	NGC 598	R	-180	3	185	284	M 33
			R	-180	1		414	
			R	-180	1		580	
			R	-180	2		384	
			R	-181	1		277	
			R	-170			131	
			R	-184	2		489	
			R	-175			15	
			R	-176	2		26	
			R	-179			81	
			R	-184			81	
			R	-186	5		67	
			R	-186	5		93	
			0	-260	50		1	
			0	-167	5		6	
			0	-189	5		11	
			0	-189	15		13	
			0	-330			3	
			0	-195	7		104	
			0	43)	150		555	
			0	-70	50		1	KNOT
			0	-278			3	KNOT
1 31 7	-72 9.2		0	18900		-186	675	
1 31 30	- 1 20.0		0	4690		78	623	
1 31 36	- 1 17.0		0	4926		78	623	
1 31 42	30 32.0	NGC 604	0	-241	1	185	650	HII REG
			0	-235			104	
			0	-244			13	
			0	-226	12		13	
1 31 44	-36 41.1	NGC 612	0	9008	30	-66	138	
		PKS	0	8900	50		897	
1 32 0	-29 40.0	NGC 613	R	1528	50	-38	544	
			0	1558			13	
			0	1525			92	
1 32 6	34 48.0	MRK 1158	0	4305		195	921	*
1 32 27	4 7.4	DDO 12	R	1966	10	98	492	
1 32 30	17 27.0	3 ZW 31	0	19592	60	146	262	
1 32 34	37 38.8	3CR 46	0	131100		202	946	
1 32 36	- 7 35.8	NGC 615	0	1946	43	52	72	
			R	1857			693	

R.A. (1950)	DEC. (1950)	NAME	OBS	HEL VEL (C★Z)	ERR	GAL CORR	REF	COMMENTS
1 32 36	33 25.0	NGC 608 KDG 38A	O	5157	50	192	812	
1 32 55	-41 41.4	NGC 625	O O	410 405	40 5	-86	844 741	
1 33 0	33 26.0	NGC 614 KDG 38B	O	5059	45	191	812	
1 33 26	0 25.0	NGC 622 MRK 571	O O	5313 5166	69	83	447 813	
1 33 54	35 16.0	NGC 621	O	3305		195	774	
1 34 0	15 32.0	NGC 628	R R R R R R R R R R O	654 655 659 662 659 654 654 659 660 655 561	4 10 5 5 4 50	138	489 886 956 934 744 93 74 216 171 156 13	M 74
1 34 12	-37 33.6	NGC 633	O	5160		-71	574	
1 34 12	5 35.0	NGC 631	O	4010	52	102	594	
1 34 42	5 37.0	NGC 632	O	3310	36	102	594	
1 35 6	-13 15.0		O	64800	300	27	529	
1 35 28	-13 4.3		O	62400	300	27	583	
1 35 34	-13 0.6		O	61500	300	28	529	
1 35 37	-12 59.9		O	14700	300	28	583	
1 35 42	34 56.0	MRK 1160	O	4787		193	921	
1 35 43	7 16.9		O O	4229 4322	54 20	107	741 813	
1 35 48	-65 10.0	NGC 646	O	8230	18	-169	434	
1 35 54	-65 10.0		O	8123	27	-169	434	
1 35 54	- 9 25.0	MRK 1161	O	12038		42	920	
1 36 0	29 23.0	MRK 1162	O	13011		179	920	

1ʰ36ᵐ

R.A. (1950)	DEC. (1950)	NAME	OBS	HEL VEL (C∗Z)	ERR	GAL CORR	REF	COMMENTS
1 36 33	39 41.8	4C 39.04	O	63210		203	400	
1 36 36	- 7 46.0	NGC 636	O	1941	50	48	13	
			O	1851	12		938	
1 37 18	31 0.0	MRK 1163	O	8148		182	920	
1 37 28	15 39.1	DDO 13	R	634	10	136	492	
			R	633	10		956	
			R	632	10		860	
1 37 36	5 28.0	NGC 645	R	3312	10	99	752	
			R	3304	7		914	
1 37 36	32 0.0	5 ZW 85	O	19620	200	185	820	
1 37 54	18 40.0		O	51773	75	145	13	
1 38 12	32 38.0	MRK 1164	O	11894		186	920	
1 39 30	12 20.0	NGC 658	R	2977	20	123	752	
			R	2988	6		914	
			O	2985	42		741	
1 39 36	25 53.0	NGC 656	O	4000	133	166	813	
1 39 46	13 43.5		R	766	15	127	956	
1 39 48	- 2 23.0	KDG 39A	O	10112	38	67	908	
1 39 48	- 2 23.0	KDG 39B	O	10070	17	67	908	
1 40 12	7 50.0		O	13800		106	475	
1 40 12	27 58.0	MRK 1165	O	10738		172	921	
1 40 21	13 23.3	NGC 660	R	854	10	126	956	
			R	852			934	
			R	855	20		885	
			R	891			886	
			O	805	20		542	
			O	802	130		555	
1 40 36	19 44.0		R	500	12	147	956	
1 41 5	11 55.0	MRK 572	O	5126		120	447	
1 41 12	3 59.0	NGC 664	O	5412	49	91	741	
1 41 14	16 48.8	MRK 360	R	8028	11	137	745	
		3 ZW 33	O	7922	60		262	
			O	8100			307	

R.A. (1950)	DEC. (1950)	NAME	OBS	HEL VEL (C*Z)	ERR	GAL CORR	REF	COMMENTS
1 41 23	2 6.0	MRK 573	O	5013		83	447	SEYF
			O	5081			874	
1 41 36	37 26.0	NGC 662	O	5660	36	195	813	
1 42 4	16 51.4	MRK 361 3 ZW 35	O	8100		136	307	
1 42 5	26 39.8	4C 26.04	O	111200		167	899	
1 42 6	16 51.4		O	8130		136	28	
1 42 42	4 22.0	IC 1726	O	5500		91	289	
1 43 13	23 13.2		O	12809		156	321	
1 43 14	23 12.8		O	12838		156	321	
1 43 16	23 12.5		O	13287		156	321	
1 43 18	23 12.0		O	12712		156	321	
1 43 24	36 12.0	NGC 668	R	4515	15	191	752	
			O	5151	32		653	
1 44 18	35 18.0	NGC 669	O	4756	22	188	653	
1 44 37	27 38.1	NGC 670	O	(3200)	240	168	555	
			O	3788	18		741	
1 44 42	27 5.1	IC 1727 VV 338 KDG 40	R	343		166	817	
			R	340	20		956	
			R	391			886	
			O	333			890	
			R	360			934	
			O	415	17		908	
			O	362	50		13	KNOT
1 45 0	-33 50.0	IC 1728	O	8748		-63	728	
1 45 6	27 10.7	NGC 672 VV 338 KDG 40	R	420		166	156	
			R	396			647	
			R	420			886	
			R	425	20		956	
			R	410			934	★
			R	428			890	
			R	428			817	
			R	408			693	
			R	409			744	
			O	425	9		908	
			O	340			13	
1 45 24	-16 58.0		R	(5160)	20	4	582	
			O	5231	20		582	

47

R.A. (1950)	DEC. (1950)	NAME	OBS	HEL VEL (C⋆Z)	ERR	GAL CORR	REF	COMMENTS
1 45 42	11 17.0	NGC 673	R	5173	15	114	752	
			O	5241	45		741	
1 45 53	12 22.0	MRK 575	R	5315	100	118	885	
			O	(6028)			447	
			O	5295	63		741	
1 45 54	-53 2.0	NGC 685	O	1420	100	-135	844	
			O	1407	50		741	
1 45 59	-12 37.9	DDO 14	R	1617	10	21	492	
		ARP 4	R	1620	20		455	
1 46 0	10 14.0	VV 54	O	4811	40	110	594	
1 46 0	19 59.0	KDG 41	O	9105	32	144	908	
1 46 0	20 0.0	KDG 41	O	8848	23	144	908	
1 46 6	10 15.0	IC 162	O	5019	83	110	594	
1 46 18	34 50.0		O	4864	172	186	653	
			O	17400			475	DIS
1 46 24	12 50.0	MRK 1166	O	25841		119	921	
1 46 24	34 43.6		O	3900		186	534	
1 46 30	20 27.8	IC 163	R	2735	15	145	829	
			R	2746	4		914	
1 46 30	35 12.0		O	4238	14	187	653	
1 46 34	5 23.0	MRK 576	O	5304		92	447	
1 46 39	21 45.0	NGC 678	R	2836	28	149	914	
1 46 39	32 20.5		R	164	10	179	956	
1 46 42	-10 40.0	NGC 681	R	1757		28	693	
			O	1705			108	
			O	1888	64		72	
			O	1750	60		13	
1 46 48	35 32.0	NGC 679	O	4853	80	187	653	
			O	5026	100		633	
1 46 50	12 16.0	MRK 577	O	12600		117	438	SEYF
			O	11725)			447	
			O	5162			874	DIS
1 47 0	-32 58.0	IC 1734	O	4986		-61	728	

R.A. (1950)	DEC. (1950)	NAME	OBS	HEL VEL (C*Z)	ERR	GAL CORR	REF	COMMENTS
1 47 24	27 23.9	NGC 684 / IC 165	R	3533	6	165	914	
1 47 36	36 7.0	NGC 687	O / O	5094 / 5147	191 / 100	188	653 / 633	
1 47 48	33 30.0	5 ZW120	O	5940	200	182	820	
1 47 48	35 2.0	NGC 688	O	4096	149	185	653	
1 47 54	35 40.0		O	4889	100	187	633	
1 47 54	36 1.0		O / O	5034 / 5312	155 / 100	188	653 / 633	
1 47 56	21 30.8	NGC 691	R	2664	9	147	914	
1 48 12	21 45.0	NGC 694 / 5 ZW122 / MRK 363	R / O / O	2963 / 3000 / 2824	17 / / 45	147	914 / 307 / 382	
1 48 17	28 56.1		O	9069	124	169	246	
1 48 18	28 56.0		O	12964	40	169	246	
1 48 22	21 40.0	IC 167 / ARP 31	R	2928	10	147	829	
1 48 30	35 49.0	KDG 43	O	5375	50	187	812	
1 48 36	- 9 57.0	NGC 701	R / O / O / O	1829 / 1795 / 1898 / 1808	/ 10 / 53 / 47	30	744 / 936 / 789 / 741	
1 48 36	- 1 40.0		O	5700		63	475	
1 48 47	- 4 18.2	NGC 702 / ARP 75	O	10607	71	52	789	
1 48 54	8 0.0	KDG 44A	O	5146	105	100	594	
1 49 0	8 3.0	KDG 44B	O	4992	82	100	594	
1 49 12	35 10.0		O	4877		185	653	
1 49 12	35 50.0	NGC 700	O	4364	42	187	653	
1 49 12	39 8.0	5 ZW132	O	6910	200	194	820	
1 49 14	6 3.0	NGC 706	O	4911	54	92	741	
1 49 18	37 59.0	6 ZW 81	O	5530	200	192	820	

1ʰ49ᵐ

R.A. (1950)	DEC. (1950)	NAME	OBS	HEL VEL (C*Z)	ERR	GAL CORR	REF	COMMENTS
1 49 21	7 2.0	MRK 579	O	19540		96	447	
1 49 30	35 36.0		O	4099	100	186	633	
1 49 36	35 52.0		O	4085	167	186	651	
1 49 42	35 52.0	NGC 704	O	4618	42	186	653	
			O	4941	30		651	
1 49 42	36 15.0		R	(4155)		187	967	
			O	4428	78		653	
			O	4398	100		633	
1 49 48	35 54.0	NGC 705	O	4645	217	186	653	
			O	4526	100		633	
			O	4570	200		651	
1 49 48	36 22.0		R	5524		187	967	
			O	4996	122		653	
			O	5514			777	
1 49 50	35 56.0	NGC 703 B2	O	4663		186	651	
			O	5175	59		653	
			O	5592	100		633	
1 49 54	35 55.0	NGC 708	O	4640		186	255	
			O	4827	100		633	
			O	4939	120		651	
			O	5047	47		653	
			O	4610			485	
1 49 54	35 59.0	NGC 709	O	3338	78	186	653	
			O	3419	38		651	
1 50 0	35 51.0		O	2381	84	186	651	
1 50 0	36 15.0		R	4975		187	967	
			O	5244	100		633	
1 50 12	36 34.0	NGC 712	O	5303	147	188	653	
			O	5286	100		633	
1 50 18	36 5.0		O	4984	173	186	651	
1 50 19	12 27.9	IC 1743	R	4597	30	115	885	
			O	4561	53		741	
1 50 21	6 42.9	MRK 580	O	7996		94	569	
1 50 24	35 46.0		O	5113	28	186	653	
1 50 24	36 43.0		O	5055	173	188	653	

R.A. (1950)	DEC. (1950)	NAME	ORS	HEL VEL (C*Z)	ERR	GAL CORR	REF	COMMENTS
1 50 36	-13 59.1	NGC 720	0	1808	100	12	13	
			0	1719	24		938	
			0	1749			842	
			0	1661	14		658	
1 50 36	35 58.0	NGC 714	0	4534	63	186	653	
			0	4470	100		633	
			0	4280	44		651	
1 50 42	3 57.0	NGC 718	0	1802		83	13	
1 50 54	36 6.0		0	4114	172	186	653	
			0	4124	100		633	
1 50 57	6 25.7	MRK 581	0	9798.		92	569	
1 51 0	35 59.0	NGC 717	0	4968	142	186	653	
1 51 54	33 36.7		0	24985		179	342	
1 51 56	36 40.2	MRK 2	0	5400		187	192	
			0	5520			280	
			0	(5400)	170		555	
			0	5458	69		653	
1 52 0	33 41.7		0	26873		179	342	
1 52 6	35 11.0		0	16200		183	534	
1 52 15	33 40.8		0	25539		179	342	
1 52 16	33 40.6		0	26626		179	342	
1 52 17	33 39.6		0	26236		179	342	
1 52 18	35 2.0	IC 171	0	5362	228	182	653	
1 52 19	33 38.0		0	26304		179	342	
1 52 19	33 38.4		0	26170		179	342	
1 52 20	33 37.6		0	26452		179	342	
1 52 23	33 37.5		0	26654		179	342	
1 52 24	33 41.3		0	25331		179	342	
1 52 25	33 37.8		0	(25925)		179	342	
1 52 32	33 35.0		0	24780		179	342	
1 52 32	33 35.8		0	27632		179	342	

1ʰ52ᵐ

R.A. (1950)	DEC. (1950)	NAME		OBS	HEL VEL (C*Z)	ERR	GAL CORR	REF	COMMENTS
1 52 35	33 40.4			O	26662		179	342	
1 52 36	21 2.0	KDG	45	O	12320	56	142	908	
1 52 45	6 22.0			O	5190	12	90	741	
1 52 48	21 5.0	KDG	45	O	4903	55	142	908	
1 53 24	1 2.0	IC	173	O	13907	77	69	741	
				O	13916	30		582	
1 53 30	36 34.0	NGC	732	O	5813	173	185	653	
				O	6000			534	
1 53 36	5 21.0			R	5460		86	389	
				O	5559			389	
				O	900			534	DTS
1 53 42	33 56.0	NGC	735	O	4748		179	777	
1 53 44	5 23.1	NGC	741	O	5397	60	86	262	
		VV	175	O	5622	27		658	
		4C 05.1		O	5559	50		13	
				O	5535	25		658	
1 53 48	5 23.0	NGC	741/2	O	5554		86	227	
		VV	175						
1 53 48	32 48.0	NGC	736	O	4366	40	176	13	
1 53 48	36 8.0			O	4796	58	184	653	
1 53 54	31 27.0	MRK 1167		O	5138		172	920	
1 54 19	32 1.1			O	26890		173	255	
1 54 20	28 37.1	3CR 55		O	72000		164	946	
1 54 22	32 0.3			O	26500	150	173	529	
1 54 24	28 20.0	IC 1753		O	10200		163	475	
1 54 24	28 23.0	6 ZW118		O	10300	200	163	820	
1 54 24	32 0.1	4C31.06B		O	26440		173	595	*
1 54 30	28 21.0	6 ZW122		O	10100	200	163	820	
1 54 38	32 57.6	NGC	751	O	5126	60	175	13	
		ARP	166						
		VV	189						
1 54 38	32 58.0	NGC	750	O	5130	40	176	13	
		ARP	166						
		VV	189						

R.A. (1950)	DEC. (1950)	NAME	OBS	HEL VEL (C*Z)	ERR	GAL CORR	REF	COMMENTS
1 54 45	35 40.3	NGC 753	R	4902		182	744	
			R	4868	50		544	
			O	4895	25		936	
			O	4766			13	
			O	4902	75		653	
1 54 45	44 40.5	NGC 746	R	710	12	203	956	
1 54 54	36 5.0	NGC 759	O	4714	232	183	653	
1 54 58	27 37.3	MRK 364	O	8100		160	307	
		5 ZW155	O	9630	200		820	DIS
1 55 0	3 13.0	MRK 1168	O	12633			77	920	
1 55 6	2 11.0	MRK 1169	O	4608		72	920	
1 55 6	37 20.0	MRK 1170	O	4614		186	920	
			O	4669	84		653	
1 55 36	2 50.0	KDG 47B	O	5587	27	75	812	
		MRK 582	O	5921			447	
		ARP 126						
		VV 122						
1 55 36	2 50.0	KDG 47A	O	5511	41	75	812	
			O	5408	60		473	
1 56 0	- 1 42.0	KDG 48A	O	8314	90	57	908	
1 56 6	-58 1.0	NGC 782	O	5975	33	-155	741	
1 56 6	- 1 42.0	KDG 48B	O	4809	48	57	908	*
			O	8100			475	DIS
1 56 17	18 42.8		O	20174		131	234	
1 56 23	18 42.8		O	19680		131	234	
1 56 24	- 8 24.0		R	4754	20	30	582	
			O	4814	25		582	
1 56 24	36 35.0		O	4817	24	184	653	
1 56 29	18 42.8	NGC 770	R	2419		131	893	
			O	2477	124		594	
			O	2454			234	
1 56 36	18 46.0	NGC 772	R	2415		131	886	
		ARP 78	R	2430	25		183	
			R	2475			934	
			R	2449			893	
			R	2435	14		251	
			R	2459			693	
			O	2437			234	
			O	2458	87		594	

53

1ʰ56ᵐ

R.A. (1950)	DEC. (1950)	NAME	OBS	HEL VEL (C∗Z)	ERR	GAL CORR	REF	COMMENTS
1 56 36	18 46.0	NGC 772	O	2431	150	131	13	
1 56 42	-56 30.0	A/B	O	5976	57	-151	724	
1 56 42	30 40.0	NGC 769	O	4500		168	475	
1 56 54	9 41.4	MRK 583	O	6051		99	569	
1 57 6	23 24.0	NGC 776	R	4920	10	146	752	
1 57 12	- 6 12.3	NGC 779	R O	1386 1423	 54	38	693 148	
1 57 12	37 47.0	IC 179	O	4062	50	186	653	
1 57 16	0 9.2	MRK 1014	O	48902		63	874	
1 57 21	31 11.2	NGC 777	O O O	(5000) 4997 5019	280 43 28	169	555 938 741	
1 57 51	2 25.8	MRK 584	O O O	23558 23680 23309		71	874 557 569	SEYF
1 57 52	39 20.7	4C 39.05	O	21510		189	400	
1 57 54	37 58.0		O	4249	43	186	653	
1 58 3	31 32.0	IC 1766 4C 31.07	O	52420		169	400	
1 58 12	31 38.0	NGC 783 MRK 1171	R O	5195 4901	5	169	752 921	
1 58 20	8 4.2		O	4756	36	92	741	
1 58 24	44 46.0		O	7163		201	708	
1 58 25	28 35.8	NGC 784	R R	200 201	 10	161	934 956	
1 58 27	33 5.3		R	5087		173	901	
1 58 36	- 7 3.0	NGC 788	O	4137	65	33	13	
1 58 36	15 25.0	NGC 786A/B KDG 50	O	4482	49	118	812	
1 58 43	29 19.3	4C 29.05	O O	44460 44460		163	595 400	
1 58 56	26 18.2		R O	5009 5100	15	153	829 475	

R.A. (1950)	DEC. (1950)	NAME	OBS	HEL VEL (C*Z)	ERR	GAL CORR	REF	COMMENTS
1 59 30	31 50.0	NGC 789	O	5100		169	475	
1 59 42	- 0 22.0	NGC 800 KDG 52	O	5934	18	59	908	
1 59 42	- 0 20.0	NGC 799 KDG 52	O	5844	69	59	908	
2 0 28	37 52.7	NGC 797	R	5647	20	184	829	
2 0 33	21 48.0	DDO 17	R O	2635 2670	15 20	138	492 813	
2 0 36	18 23.0		R O	2374 2377	10 25	127	582 582	
2 0 42	38 1.0	NGC 801	R O O	5764 5716 5763	10 174 10	184	829 653 740	
2 0 54	2 19.0	MRK 585	O	6325		68	438	
2 1 0	14 28.0	IC 195	O	8600	34	113	594	
2 1 2	15 47.5	NGC 803	R	2110		118	693	
2 1 6	14 30.0	IC 196	O	8563	69	113	594	
2 1 8	-69 41.3		O	7723	60	-188	680	
2 1 11	-69 41.3		O	978	70	-188	680	
2 1 12	23 58.0	5 ZW173	R	594		145	956	
2 1 26	28 25.1	MRK 365	O O	4800 4472	 90	158	307 562	
2 2 30	30 56.0		R O	5274 5245	15 56	164	752 813	
2 2 54	44 58.0		O	5688		198	708	
2 3 0	31 57.0	6 ZW165	O	4390	100	167	820	
2 3 6	- 8 28.0	MRK 1172	O	12206		24	920	
2 3 7	-55 27.3		O	5981	38	-151	763	
2 3 22	5 12.4	PKS	O	38900		77	318	
2 3 24	9 3.0	IC 198	O	9414	74	92	741	
2 3 42	44 20.0	NGC 812	O	5351		197	708	

R.A. (1950)	DEC. (1950)	NAME	OBS	HEL VEL (C*Z)	ERR	GAL CORR	REF	COMMENTS
2 3 43	- 0 31.5	MRK 1018	O	12693		55	874	SEYF
2 4 10	29 16.7	4C 29.06	O	32700		158	400	SEYF
			O	32860			595	
2 4 12	32 44.0	KDG 54	O	11982	30	168	908	
2 4 24	32 44.0	KDG 54	O	11126	30	168	908	
2 4 50	-55 27.0		O	16485	43	-151	763	
2 4 54	1 52.0		O	7144		63	713	
2 5 0	1 56.0	N	O	6284		63	713	
2 5 0	1 56.0	S	O	6280		63	713	
2 5 6	1 54.0		O	7277		63	713	
2 5 14	2 28.5	MRK 586	O	46885		65	557	
2 5 18	1 55.0		O	7069		63	713	
2 5 18	20 8.0	MRK 1173	O	8231		129	921	
2 5 30	1 39.0	MRK 1174	O	4688		62	920	
2 5 30	1 52.0		O	7084		63	713	
2 5 36	14 6.0	NGC 820	R	4426	20	108	752	
2 5 42	10 45.0	NGC 821	O	1778	100	96	13	
			O	1721	17		938	
2 5 43	38 32.3	NGC 818	R	4245	10	182	829	
2 5 54	6 5.0	NGC 825	O	3121	107	78	594	
2 6 12	- 8 2.0	NGC 829	R	4056		23	744	
2 6 39	35 33.7	4C 35.03	O	11190		174	400	
			O	11060			485	
			O	11060	200 ·		350	
2 6 53	-10 22.2	NGC 833 ARP 318	O	3901	53	13	554	
2 6 57	-10 22.3	NGC 835 ARP 318	O	4066	41	13	554	
			O	4010			728	
2 7 7	38 57.4	NGC 828 B2	R	5374	15	182	829	
			O	5240			485	

R.A. (1950)	DEC. (1950)	NAME		OBS	HEL VEL (C*Z)	ERR	GAL CORR	REF	COMMENTS
2 7 11	-10 23.0	NGC	838	O	3870		13	728	
		ARP	318	O	3787	45		554	
2 7 16	-10 25.2	NGC	839	O	3811	47	13	554	
		ARP	318						
2 7 36	7 36.0	NGC	840	O	7143	86	83	741	
2 7 50	-10 33.4	NGC	848	O	3970	47	12	554	
2 7 57	-33 10.5	IC	1783	O	3299	31	-77	741	
2 8 0	5 38.0	MRK	587	O	4572		75	447	
2 8 1	37 25.9	NGC	834	R	4553	25	178	829	
				O	4800			475	
2 8 7	6 32.1	DDO	18	R	1611	10	78	492	
2 8 30	3 38.0	IC	211	R	3242	15	67	582	
				O	3236	20		582	
				O	3075	115		594	
				O	3318	41		741	
2 8 36	3 32.8	NGC	851	O	3199	84	66	594	
		MRK	588	O	3382			447	
				O	3045	35		500	
2 8 51	13 40.9	MRK	366	O	7800		104	307	
2 9 52	37 34.9			O	5130		177	485	
2 10 0	86 6.1			O	55800	999	226	391	*
2 10 12	-22 42.0	NGC	858	O	12356	35	-38	582	
2 10 25	-32 11.0	NGC	857	O	3230		-75	574	
2 10 52	16 51.0	MRK	367	O	11100		113	307	
2 10 54	31 38.0	MRK	1175	O	5630		160	921	
		5 ZW202							
2 11 1	86 5.3			O	54000	900	226	391	*
2 11 2	86 6.7			O	56400	600	226	391	*
2 11 9	3 52.1	3 ZW 43		R	3546		66	389	
		MRK	589	R	(3429)			622	
				O	3234			569	
				O	3364	44		197	
				O	4319	60		262	DIS
				O	3347			311	
				O	3530	100		521	
				O	3500			289	

2ʰ11ᵐ

R.A. (1950)	DFC. (1950)	NAME	OBS	HEL VEL (C*Z)	ERR	GAL CORR	REF	COMMENTS
2 11 28	4 56.5	MRK 1027A IC 214	O	8941	82	69	929	*
2 11 28	4 56.5	MRK 1027B IC 214	O	9001	88	69	929	*
2 11 30	-47 58.0	PKS	O	66000		-131	691	
2 11 30	4 56.0	IC 214	O	8531		69	774	
2 12 0	- 1 0.0	NGC 863 MRK 590	R O	7910 (8198)		46	707 447	SEYF
2 12 8	86 6.4		O	33900	999	226	391	*
2 12 48	5 46.0	NGC 864	R R O	1568 1559 1583	20	71	544 744 13	
2 13 12	32 26.0	IC 1784 KDG 61	O	4772	14	161	908	
2 13 24	32 26.0	IC 1785 KDG 61	O	4652	47	161	908	
2 13 39	-31 26.0	IC 1788	O	3388	20	-74	741	
2 14 6	27 22.0	5 ZW208	O	14018	60	145	262	
2 14 18	38 11.0		O	6000		176	475	SEYF
2 14 27	14 19.1	NGC 871	R R R O O	3728 3740 3726 3757 3705	10 20 44	102	693 829 544 13 72	
2 14 30	1 0.0	NGC 875 KDG 62	O	6491	50	52	908	
2 14 30	1 2.0	IC 218 KDG 62	O	6448	55	52	908	
2 14 36	29 18.0	5 ZW212	O	5700	200	151	820	
2 14 37	1 28.0	MRK 591	O	(12388)		53	447	
2 14 53	-48 3.0	PKS	O	18574		-134	627	
2 15 3	37 50.8		R	5159	25	174	829	
2 15 12	14 17.0	NGC 876	R	3905		101	744	

R.A. (1950)	DEC. (1950)	NAME	OBS	HEL VEL (C★Z)	ERR	GAL CORR	REF	COMMENTS
2 15 15	14 19.0	NGC 877	R	3909		101	693	
			R	3914	25		829	
			O	4016			13	
2 17 7	- 0 29.0	MRK 592	O	7451		44	447	
2 17 23	1 41.7	PKS	O	12300		52	868	
2 17 36	28 56.0	5 ZW220	O	13575	60	148	262	
2 18 0	42 33.0		R	630		184	956	
2 18 22	- 2 10.5	3CR 63	O	52500		36	946	
2 18 24	39 9.0	5 ZW223A	O	7623	150	175	567	
		KDG 64	O	7497	21		908	
		ARP 273						
		VV 323						
2 18 30	39 8.0	5 ZW223B	O	7778	150	175	567	
		KDG 64	O	7356	10		908	
		ARP 273						
		VV 323						
2 18 36	16 20.0	KDG 65A/B	R	4098	9	105	914	
2 18 36	16 20.0	KDG 65B	O	4133	40	105	812	
2 18 54	-33 56.0	NGC 897	O	4860		-87	574	
2 19 2	42 50.9		O	6000	63	184	169	
2 19 6	- 5 45.0	NGC 895	R	2286		21	693	
		NGC 894	R	2294			744	
			O	2344	27		148	
2 19 18	33 2.0	NGC 890	O	4043	65	158	13	
2 19 25	42 7.2	NGC 891	R	525		182	693	
			R	530	10		509	
			R	530	5		839	
			R	527	4		944	
			R	529			956	
			R	531			934	
			O	72	100		13	
2 19 36	-21 2.0	NGC 899	O	1800		-39	360	
2 19 42	-20 58.0	IC 223	O	1600		-38	360	
2 19 48	28 2.0	IC 221	R	5079	15	143	752	
2 19 48	42 46.0		O	6578	80	183	169	
			O	6366			708	

2ʰ20ᵐ

R.A. (1950)	DEC. (1950)	NAME	OBS	HEL VEL (C*Z)	ERR	GAL CORR	REF	COMMENTS
2 20 0	41 8.8	5 ZW229	R	5380		179	389	
		ARP 145	R (	5425)			622	
			O	5178	43		789	
			O	5239	60		262	
			O	5240			385	
			O	5306	50		141	
2 20 2	42 44.3		O	6706	6	183	169	
2 20 2	42 45.9	3CR 66	O	6196		183	708	
			O	6397	80		169	
2 20 12	41 44.0	NGC 898	O	5518		181	708	
2 20 18	31 57.0	5 ZW233A	O	9985	72	154	500	
		KDG 67A	O	10097	10		812	
			O	10500			475	
2 20 24	31 58.0	5 ZW233B	O	10055	24	154	500	SEYF
		KDG 67B	O	10142	10		812	
			O	10500			475	
2 20 45	-20 56.4	NGC 907	O	1600		-39	360	
2 20 46	-21 27.7	NGC 908	R	1498	50	-41	544	
			O	1734			13	
2 21 18	27 36.6	3CR 67	O	93070		141	657	
2 21 31	- 8 49.2		O	81800	300	7	529	
2 21 36	42 24.0		O	6087		181	708	
2 21 42	43 6.0		O	5065		183	708	
2 21 57	35 48.8	DDO 19	R	575	10	164	492	
			R	577			956	
			R	592	20		424	
2 22 6	41 52.0	NGC 906	O	4586	50	180	582	
2 22 12	41 49.0	NGC 909	O	4899		180	708	
2 22 18	41 36.0	NGC 910	O	5126		179	708	
2 22 24	36 57.0	B2	O	9640		167	485	
2 22 36	41 45.0	NGC 911	O	5514		179	708	
2 22 42	20 10.0	IC 1797	O	3900		115	475	
2 22 49	-25 1.1	NGC 922	O	3028	39	-57	789	
			O	3120	34		732	
			O	3067	15		741	

R.A. (1950)	DEC. (1950)	NAME	OBS	HEL VFL (C*Z)	ERR	GAL CORR	REF	COMMENTS
2 22 54	39 15.0		O	4816	58	173	813	
2 22 54	41 55.0	NGC 914	R	5548	15	179	752	
			O	(5415)			708	
2 23 6	-40 41.0		O	5891	70	-114	603	
2 23 6	18 16.0	NGC 918	R	1512	10	108	582	
			O	1502	20		582	
2 23 12	42 38.0		O	26540		181	449	
2 23 36	- 0 33.0	NGC 926	O	6515	29	38	741	
2 23 36	41 37.0		O	5723		178	708	
2 23 41	-21 38.9	DDO 21	R	1555	10	-44	492	
2 23 54	11 56.0	NGC 927	R	8270	30	85	582	
		MRK 593	O	8252	25		582	
2 23 54	34 18.0	6 ZW211	O	8280	200	159	820	
2 24 12	40 50.0		O	5372		176	708	
2 24 17	33 21.3	NGC 925	R	553		156	744	
			R	572			886	
			R	550			817	
			R	537			647	
			R	560			934	
			R	550			693	
			R	574	30		114	
			R	545	1		898	
			R	553			956	
			R	570			171	
			R	565	10		216	
			R	555			156	
			R	557	20		489	
			O	587			13	
			O	420	200		13	
			O	576	15		741	
			O	580	50		94	
2 24 24	41 45.0	NGC 923	O	5625		178	708	
2 24 24	41 47.0	MRK 1176	O	5338		178	921	
2 24 30	22 52.0	5 ZW242	O	9765		123	26?	
2 24 36	-13 21.0	MRK 1178	O	6972		-12	¥21	
2 24 36	-13 20.0	MRK 1177	O	6942		-12	921	

61

2ʰ25ᵐ

R.A. (1950)	DEC. (1950)	NAME	OBS	HEL VEL (C∗Z)	ERR	GAL CORR	REF	COMMENTS
2 25 6	− 1 22.7	NGC 936	R	1430		33	538	
			O	1300	50		1	
			O	1343	50		13	
2 25 17	31 5.4	NGC 931	O	4910	60	148	760	
2 25 17	31 25.0		O	5070	60	149	760	
2 25 23	19 22.6	NGC 935	O	4173	20	110	611	
		KDG 68	O	4184	25		908	
2 25 24	19 21.0	IC 1801	O	4013	7	110	611	
		KDG 68	O	4055	18		908	
			O	4323	50		500	
2 25 55	− 1 22.5	NGC 941	R	1430		33	538	
			O	1610	49		741	
2 26 19	−10 44.1	NGC 948	O	4511	71	−4	789	
2 26 24	31 15.0		O	5098		148	777	
2 26 45	28 45.3	B2	O	14000		140	595	
2 26 54	−41 38.0	NGC 954	O	5353	20	−120	582	
2 27 30	42 1.0	NGC 946	O	5655		177	708	
2 27 39	36 54.7	NGC 949	R	610		163	956	
			R	602			900	
			O	638	10		813	
			O	593	31		741	
			O	622			375	
2 28 0	− 1 19.8	NGC 955	O	1534	17	31	741	
2 28 0	22 15.0	5 ZW249	O	9520	200	118	820	
2 28 6	43 15.0		O	5859		179	708	
2 28 10	− 3 9.8	NGC 958	R	5738	15	24	752	
			R	5747	50		544	
			O	(5694)	45		148	
2 28 30	39 10.0		O	8005		169	777	
2 28 52	1 7.4		O	7390	21	40	741	
2 29 0	2 36.0	IC 233	O	8221	22	45	908	
		KDG 69						
2 29 0	32 29.0	5 ZW251	O	10050	200	150	820	

R.A. (1950)	DEC. (1950)	NAME		OBS	HFL VEL (C*Z)	ERR	GAL CORR	REF	COMMENTS
2 29 0	43 14.0			O	6162		179	708	
2 29 6	2 35.0	KDG	69	O	8258	10	45	908	
2 29 12	1 2.0			O	13800		39	475	
2 29 12	41 59.0			O	6989		175	708	
2 29 18	35 17.0	NGC	959	R	601	15	158	752	
				R	608			956	
2 29 34	- 1 35.0			O	11269	50	29	741	
2 29 36	-58 8.5			O	9590		-171	896	
2 29 39	-58 9.0			O	9410		-171	896	
2 29 48	38 27.5	DDO	22	R	567	10	166	492	
				R	570			956	
2 30 0	28 36.0			R	1017		137	956	
2 30 1	20 25.5	IC	235	O	9000		110	307	
		MRK	368	O	8674	4		562	
2 30 17	33 16.1	DDO	25	R	615	10	151	492	
				R	611			956	
2 30 34	40 18.4	DDO	24	R	581	10	170	492	
				R	581			956	
2 30 36	27 45.0	MRK	1179	O	10956		134	921	SEYF
2 30 42	-52 43.5			O	6443	30	-156	613	
2 31 11	20 45.4	NGC	976	O	4362	23	110	741	
2 31 12	11 0.0	IC 1817A KDG 70A		O	7282	17	75	812	
2 31 12	11 0.0	IC 1817B KDG 70B		O	7321	30	75	812	
2 31 17	29 5.5	NGC	972	R	1550		138	744	
				O	1593	53		72	
				O	1538	60		13	
				O	1544			111	
				O	1545			109	
2 31 31	29 31.8	DDO	26	R	1035	10	139	492	
				R	1034	20		424	
				R	1030			956	

R.A. (1950)	DEC. (1950)	NAME	OBS	HEL VEL (C*Z)	ERR	GAL CORR	REF	COMMENTS
2 31 36	-39 15.0	NGC 986	0	2073		-115	84	
2 31 48	32 38.0	NGC 978 KDG 71A	0	4686	27	148	908	
2 31 48	32 38.0	KDG 71B	0	4412	30	148	908	
2 31 54	23 12.0	NGC 984	0	11182		118	774	
2 32 12	- 9 0.3	NGC 985A VV 285	0 0	12934 12950	22 50	-2	789 487	
2 32 12	- 9 0.3	NGC 985B VV 285	0	12646	47	-2	789	
2 32 30	37 25.0	VV 96	0	3660		161	449	
2 32 36	59 26.0		0	-10	50	209	298	MAF I
2 33 0	- 9 34.5	NGC 988	0	1455	100	-5	789	
2 33 3	- 7 22.0	NGC 991	0	1515	33	3	741	
2 33 18	23 41.0		R 0	5648 5616	10 25	119	582 582	
2 33 20	38 45.1	IC 239	R R R	903 907 897	9 4	164	744 956 771	
2 33 23	25 12.4		R 0	708 716	 20	124	956 813	
2 33 24	31 30.0		0	5100		144	475	
2 33 48	33 6.0	NGC 987 MRK 1180	0	4442		148	921	
2 33 48	35 54.0		R	5126	15	156	752	
2 34 36	20 53.0	NGC 992	R 0	4135 4500	30	108	622 475	
2 34 37	34 12.9		0 0	5100 4800	 150	151	475 13	
2 34 38	20 55.4	MRK 369 3 ZW 50 2 ZW 4	0 0	3900 3750	 29	108	307 28	
2 34 38	34 13.0		R	4915	50	151	829	
2 35 30	29 32.0		R 0	5078 5068	15 20	136	582 582	

R.A. (1950)	DEC. (1950)	NAME	OBS	HEL VEL (C*Z)	ERR	GAL CORR	REF	COMMENTS
2 35 48	1 42.0	NGC 1019	O	7251	13	36	741	
2 36 0	3 50.0	MRK 1181	O	7986		44	920	
2 36 6	- 6 53.0	NGC 1022	O	1498	23	2	148	
2 36 7	40 39.4	NGC 1003	R	626		168	744	
			R	625			956	
			O	585	60		13	
2 36 17	29 56.2	NGC 1012	R	983		137	956	
2 36 24	18 9.0	ARP 258 VV 143	O	4037		97	13	
2 36 30	10 38.0	NGC 1024 ARP 333	O	3572	81	70	594	
2 36 42	12 58.0	3 ZW 51	O	13240	60	78	262	
2 36 46	-62 12.0		O	15620		-183	896	
2 36 51	-27 39.3	IC 1826	R	1447		-77	366	
			O	1393	40		246	
2 36 54	10 35.0	NGC 1029	O	3513	103	69	594	
2 37 0	- 8 20.0	NGC 1035	R	1234		-3	744	
			R	1210	20		884	
			O	1230	10		936	
			O	1317	27		148	
2 37 12	38 51.0	NGC 1023 ARP 135	R	600	160	162	216	
			R	670			682	
			R	685	40		771	
			R	607			900	
			R	590	50		455	
			R	680	20		775	
			O	300	50		1	
			O	557	60		13	
			O	734	41		72	
2 37 20	- 1 47.8		O	111800		21	717	
2 37 31	1 17.7	NGC 1038	R	4372	25	33	829	
2 37 36	-27 33.0	IC 1830	O	1474	32	-78	72	
2 37 40	-34 29.0	NGC 1049	O	20		-103	574	
			O	39	7		11	
			O	40	30		13	
2 37 40	19 5.0	NGC 1036 MRK 370	R	787		99	480	
			O	777	45		382	
			O	748			480	
			O	900			307	

R.A. (1950)	DEC. (1950)	NAME	OBS	HEL VEL (C*Z)	ERR	GAL CORR	REF	COMMENTS
2 37 49	1 0.8	DDO 27	R	(1186)		32	492	
2 37 54	16 36.0	MRK 1182	O	8664		90	921	
2 38 0	- 8 40.0	NGC 1042	R	1374	11	-5	700	
			R	1369	20		884	
			R	1368			744	
			O	1407	67		789	
			O	355			13	DIS
2 38 5	- 8 21.6	NGC 1047	R	1340	30	-4	884	
2 38 6	-34 41.0		O	35	60	-104	13	FOR SYS
2 38 8	59 23.4		R	-10	15	206	331	MAF II
			R	-28	10		297	
			R	-15	10		544	
			R	-15	7		390	
			R	20	5		417	
			R	25	10		271	
2 38 24	-15 21.0		R	7756	30	-32	582	
			O	7653			582	
2 38 25	8 31.4	NGC 1044 4C 08.11	O	6420	120	60	350	
2 38 37	- 8 28.1	NGC 1052	R	1540	40	-5	700	
			R	1590	30		884	
			P	1520	10		733	
			R	1300			671	
			R	1470	70		716	
			O	1458			106	
			O	1507	19		938	
			O	1468			842	
			O	1407	10		700	
			O	1439	40		13	
			O	1471	14		658	
			O	1488	5		24	
			O	1523			13	
			O	1430			281	
2 38 56	6 58.5	MRK 595	O	8250		54	557	SEYF
			O	7746			569	
2 39 6	0 13.0	NGC 1055	R	992		28	693	
			R	994			934	
			R	1050	110		216	
			R	1070			171	
			O	790	180		555	
			O	1041			916	
2 39 30	34 34.0	NGC 1050	O	3910	15	148	813	

R.A. (1950)	DEC. (1950)	NAME	OBS	HEL VEL (C*Z)	ERR	GAL CORR	REF	COMMENTS
2 39 54	28 21.0	NGC 1056 MRK 1183	O	1521		129	921	SEYF
2 40 6	- 0 14.0	NGC 1068 ARP 37 3CR 71	R	1140		25	707	SEYF
			R	1144			693	M 77
			R	1145	25		267	
			R	1130			671	
			R	1133	22		390	
			O	765	50		3	
			O	1076	45		112	
			O	1120	50		1	
			O	1128	26		112	
			O	1079	17		50	
			O	1073	27		50	
			O	1203			20	
			O	1100	50		112	
			O	1123			102	
			O	1107	12		191	
			O	864			3	
			O	1050			574	
			O	1065			227	
			O	1020	40		13	
			O	1121			13	
2 40 22	30 7.8	NGC 1058	R	520	2	134	769	
			R	541	10		956	
			R	517	1		390	
			R	518			693	
			O	521	37		148	
			O	439	25		97	
			O	80			13	*
2 40 36	16 26.0	KDG 73	O	8047	45	88	500	
			O	7708	57		908	
2 40 36	16 27.0	KDG 73	O	7734	97	88	500	
			O	7604	27		908	
2 40 48	32 18.0	NGC 1067	O	4535	25	141	582	
2 41 6	1 9.9	NGC 1073	R	1218		29	283	
			O	1874			13	
			O	1209	25		283	
2 41 36	-29 13.0	NGC 1079	O	2252	250	-87	13	
2 42 42	- 4 55.0	NGC 1080	R	7848	10	4	582	
			O	7843	20		582	
2 42 54	-72 29.5		O	30500		-205	675	SEYF
2 42 59	36 41.8		O	14450		152	255	
			O	14500	300		529	

2ʰ43ᵐ

R.A. (1950)	DEC. (1950)	NAME	OBS	HEL VEL (C*Z)	ERR	GAL CORR	REF	COMMENTS
2 43 18	15 38.5	MRK 597	O	7563		82	447	
2 43 24	- 5 51.0	MRK 1184	O	8930		0	921	
2 43 28	-74 51.0		O	9860		-209	896	
2 43 32	- 7 47.1	NGC 1084	R	1410		-7	693	
			R	1402			744	
			R	1407	20		544	
			O	1558	100		13	
			O	1442			78	
2 43 48	3 24.0	NGC 1085	O	6980	72	36	741	
2 43 52	- 0 42.5	NGC 1087	R	(1523)		20	693	
			R	1395			363	
			O	1449	55		881	
			O	1503	10		936	
			O	1824	200		13	
			O	1536	43		554	
			O	1522	47		554	HIT REG
2 43 52	7 11.6	MRK 598	O	5385		50	557	
2 44 1	- 0 27.4	NGC 1090	O	2708		21	777	
			O	2699	54		881	
			O	2729	37		741	
2 44 11	-30 29.1	NGC 1097	R	1245	285	-93	216	
		ARP 77	R	1255			171	
			O	1326	100		13	
			O	1307			29	
			O	1424			13	
			O	1229	55		506	
2 44 12	15 35.0	MRK 1185	O	7569		81	921	
2 44 14	-69 31.7		O	7261	35	-201	680	
2 44 42	41 3.0	NGC 1086	R	4037	10	163	752	
2 44 48	37 20.0	KDG 77	R	588	15	153	956	
			O	659	29		908	
2 44 48	37 22.0	KDG 77	O	9324	44	153	908	
2 44 48	50 36.0	KDG 78	O	4903	30	186	908	
2 44 54	- 0 30.0	NGC 1094	R	6464	23	20	885	
			O	6294	58		741	
			O	6560	53		881	
2 44 54	- 0 29.0		O	6418	111	20	881	

R.A. (1950)	DEC. (1950)	NAME	OBS	HEL VEL (C*Z)	ERR	GAL CORR	REF	COMMENTS
2 45 0	15 43.0	MRK 1186	O	7599		81	921	
2 45 0	50 33.0	KDG 78	O	4937	16	186	908	
2 45 6	2 57.0	MRK 599	O	8845		33	438	
2 45 20	3 40.5	DDO 28	R	1030	10	36	492	
2 45 36	13 44.0	MRK 1187	O	13357		73	921	SEYF
2 46 18	- 0 44.0	1 ZW 9	O	7340		18	303	
2 46 24	18 7.0		O	10010	100	89	521	
2 46 31	19 5.9	IC 1854 MRK 372	O	9300		92	307	SEYF
2 46 33	1 55.1	DDO 29	R	1108	10	28	492	
2 46 47	34 46.9	MRK 1058	O	5190		144	874	
2 47 41	-71 39.3		O	17781	68	-205	613	
2 47 46	-71 39.3		O	17783	97	-205	613	
2 48 7	-52 35.7		O	13040	30	-165	680	
2 48 11	-52 35.2		O	13695		-165	680	
2 48 30	4 15.0	MRK 600	O	990		35	438	
2 48 54	-32 25.0		O	13500		-103	439	
2 49 12	36 32.0	5 ZW284	O	3370	100	147	820	
2 49 21	- 1 22.8	DDO 30	R	1508	10	12	492	
2 49 24	41 9.0		O	4305	65	160	902	
2 49 24	41 13.0		O	3924	65	160	902	
2 49 24	41 25.0		O	4524	65	161	902	
2 50 12	25 17.0	IC 1861	R	6688		111	901	
2 50 18	41 42.0		O	7156	65	161	902	
2 50 24	41 32.0		O	6149	65	161	902	
2 50 26	-66 23.8		O	8090		-197	896	
2 50 36	41 16.0		O	4581	65	160	902	

R.A. (1950)	DEC. (1950)	NAME	OBS	HEL VEL (C∗Z)	ERR	GAL CORR	REF	COMMENTS
2 50 43	41 33.0		O	4510		160	950	
2 51 7	41 11.0		O	6030		159	950	
2 51 12	14 46.0	KDG 82A	O	9354	37	73	812	
2 51 12	14 47.0		O	(9218)	100	73	812	
2 51 12	41 25.0	NGC 1130	O	5900		160	950	
			O	5849	65		902	
			O	6162	22		940	
2 51 18	41 23.0	NGC 1129	O	5150		160	950	
			O	5273	20		940	
			O	5240	65		902	
2 51 18	41 23.0	NGC 1129A	O	5056	24	160	940	
			O	4863	65		902	
			O	4950			950	
2 51 18	41 28.0		O	5650		160	950	
2 51 24	41 22.0	NGC 1131	O	5185		159	950	
			O	5198	65		902	
			O	5349	23		940	
2 51 30	41 20.0		O	4310		159	950	
			O	4365	65		902	
			O	4456	26		940	
2 51 30	41 28.0	IC 265	O	5225		160	950	
			O	5408	65		902	
			O	5296	28		940	
2 51 30	41 40.0		O	5804	65	160	902	
2 51 32	41 12.0		O	5460		159	950	
2 51 36	41 7.0		O	5825	65	159	902	
2 51 44	41 25.0		O	4140		159	950	
			O	4026	65		902	
			O	4192	29		940	
2 51 48	- 8 40.7		O	5361	100	-17	714	
2 52 12	-10 14.0	NGC 1140	R	1506		-24	744	
			R	1507			693	
			R	1528	25		183	
			O	1486	100		714	
			O	1544	40		13	
2 52 36	- 0 22.8	NGC 1143	O	8487		13	748	
		KDG 83A	O	8368	88		812	
		VV 331						
		ARP 118						

R.A. (1950)	DEC. (1950)	NAME	OBS	HEL VEL (C*Z)	ERR	GAL CORR	REF	COMMENTS
2 52 39	- 0 23.1	NGC 1144	O	8887		13	748	
		KDG 83B	O	8710	54		812	
		ARP 118	O	8919	47		789	
		VV 331	O	8529	80		812	
2 52 42	41 23.0		O	4710	65	159	902	
2 52 42	41 32.0		O	6607	65	159	902	
2 52 54	41 26.0		O	5996	65	159	902	
			O	5981	65		902	
2 53 10	-10 20.4		O	9734	100	-25	714	
2 53 15	-10 41.4		O	9887	100	-26	714	
2 53 24	-22 5.9	PKS	O	33820	150	-70	709	
2 53 24	41 8.0		O	4605	65	158	902	
2 53 30	-27 38.0		O	5272	40	-90	582	
2 54 1	- 2 58.0	MRK 601	O	7043		2	447	
2 54 51	9 56.1		R	757	10	51	860	
2 54 53	-11 11.3		O	7686	100	-30	714	
2 55 3	5 49.6	3CR75A/B KDG 84A/B	O	6890		35	255	
2 55 3	5 49.6	KDG 84A	O	7180	35	35	444	
2 55 3	5 49.6	KDG 84B	O	6821		35	444	
2 55 10	12 50.3		O	21298	240	62	909	
2 55 10	12 53.7		O	21548	130	62	909	
2 55 13	-10 40.5		O	10073	100	-28	714	
2 55 18	12 58.0		O	21578		62	909	
2 55 20	12 57.3		O	22298	210	62	909	
2 55 23	12 47.4		O	23358	160	62	909	
2 55 24	-10 22.0		O	4502	100	-27	714	
2 55 25	13 6.2		O	20627		63	909	
2 55 26	12 53.3		O	20338	95	62	909	

71

R.A. (1950)	DEC. (1950)	NAME	OBS	HEL VEL (C*Z)	ERR	GAL CORR	REF	COMMENTS
2 55 26	13 6.0		O	20477	105	63	909	
2 55 26	13 28.7		O	22340	150	64	639	
2 55 42	-10 33.7	NGC 1154	O	4634	100	-28	714	
2 55 42	13 15.2		O	22900	150	63	639	
2 55 42	41 6.0	5 ZW297	O	4920	60	156	262	
2 55 47	-10 32.9	NGC 1155	O	4660	100	-28	714	
2 55 47	13 22.4	PKS	O	19500	150	63	639	
2 55 48	-35 54.0		R	6208	70	-120	752	
2 55 51	13 5.0		O	20738		62	909	
2 55 57	13 20.0		O	21840	150	63	639	
2 56 0	-36 54.0		O	6170		-124	728	*
2 56 5	3 14.1		R	7089	25	24	829	
2 56 6	39 28.0	5 ZW300	O	5290	200	151	820	
2 56 9	13 33.5		O	24900	150	64	639	
2 56 12	13 23.8		O	21450	150	63	639	
2 56 13	13 23.0		O	22340	150	63	639	
			O	(22393)			444	
			O	22706			444	
2 56 14	13 24.1		O	22450	150	63	639	
2 56 16	13 24.7		O	24090	150	63	639	
2 56 18	6 4.0		O	6900		35	595	
2 56 30	13 15.3		O	23270	150	62	639	
2 56 31	13 20.3		O	22100	150	63	639	
2 56 31	13 24.4		O	23220	150	63	639	
2 56 37	3 17.9		O	32100		24	717	
2 56 40	36 37.3	MRK 1066	O	3574		142	874	
2 56 45	13 30.0		O	20240	153	63	639	
2 56 47	25 2.3	NGC 1156	R	372		105	693	
			R	335			610	
			R	345			780	
			R	374			744	

R.A. (1950)	DEC. (1950)	NAME	OBS	HEL VEL (C*7)	ERR	GAL CORR	REF	COMMENTS
2 56 47	25 2.3	NGC 1156	R	380	25	105	183	
			O	405	40		13	
2 56 56	-36 48.6		O	6124	105	-124	680	*
2 57 12	2 34.0	IC 277	R	2852	15	21	752	
		MRK 602	O	2679			447	
2 57 42	13 25.4		O	22660	150	62	639	
2 57 48	11 39.0	NGC 1166	O	7693	34	55	908	
		KDG 85						
2 57 53	44 45.5	NGC 1160	O	2494	30	164	908	
		KDG 86						
2 57 54	44 42.0	NGC 1161	O	1836	40	164	908	
		KDG 86	O	1998			910	
2 58 0	11 35.0	NGC 1168	O	7627	23	55	908	
		KDG 85						
2 58 6	-11 36.8		O	9046	100	-34	714	
2 58 19	-11 1.2		O	10126	100	-32	714	
2 58 35	35 0.2	NGC 1167	O	5100		136	600	
		4C 34.09	O	4800			400	
			O	5950		166		DIS
2 58 43	35 38.5	4C 35.06	O	13980		138	400	
			O	14060			255	*
2 58 47	-11 5.9		O	9277	100	-33	714	
2 59 15	-15 2.0	NGC 1172	O	1531	7	-48	658	
			O	1712	11		741	
2 59 26	-10 49.1		O	10052	100	-32	714	
2 59 28	-73 22.0		O	34770		-211	896	
2 59 10	-10 6.6		O	9711	100	-29	714	
3 0 10	46 11.1	NGC 1169	O	2372	20	167	741	
3 0 23	-23 3.6	NGC 1187	R	1426	40	-79	238	
			O	1579			13	
3 0 27	16 14.7	3CR 76.1	O	9720		70	143	
			O	9778	150		741	
			O	9770	150		159	

3ʰ0ᵐ

R.A. (1950)	DEC. (1950)	NAME	OBS	HEL VEL (C*Z)	ERR	GAL CORR	REF	COMMENTS
3 0 34	-18 36.1	PKS	O	26710	150	-63	709	
3 0 34	- 9 19.7	NGC 1185	O	4812	100	-27	714	
3 0 42	43 12.0	NGC 1171	R	2742	15	159	752	
3 1 0	31 11.0	5 ZW317	O	17468	44	122	197	
3 1 3	- 9 52.0	NGC 1182	O	3248	100	-30	714	
3 1 16	42 8.7	NGC 1175	O	5458	24	155	741	
3 1 18	-11 48.0	NGC 1199	O	2581	50	-37	13	
			O	2682	11		938	
3 1 24	- 1 5.0	MRK 1188	O	19877		3	921	
3 2 0	-26 15.0	NGC 1201	O	1722	50	-92	13	
3 2 10	79 56.4		R	2377	30	221	829	
3 2 42	- 2 32.0	MRK 1189	O	12203		-3	921	
3 2 50	-27 42.0		R	6536		-98	366	
			O	6551	21		246	
3 3 43	-15 48.1	NGC 1209	O	2720	23	-55	741	
			O	2578			106	
			O	2568	150		13	
3 3 44	- 9 1.6		O	10251	100	-29	714	
3 3 47	- 9 44.1	NGC 1214	O	4607	100	-32	714	
3 4 3	- 9 5.6		O	10235	100	-29	714	
3 4 31	- 9 44.3	NGC 1215	O	5136	100	-32	714	
3 4 36	- 2 18.0	MRK 1190	O	12894		-4	921	
3 4 42	29 35.0	5 ZW321	O	11056		114	197	
3 4 44	- 9 47.2	NGC 1216	O	5016	100	-33	714	
3 5 0	-31 38.0		O	4907	200	-113	582	
3 5 0	38 11.0	NGC 1207B	O	4769	150	141	567	
3 5 49	3 55.2	NGC 1218 3CR 78	O	8640		19	120	
3 5 59	-23 8.9		O	10676	15	-84	741	

R.A. (1950)		DEC. (1950)		NAME	OBS	HEL VEL (C*Z)	ERR	GAL CORR	REF	COMMENTS
3 5 59		-23 6.4			O	10461	84	-84	741	
3 6 14		-23 14.3		IC 1892	O	3016	24	-84	741	
3 6 16		-23 48.0			O	12268	40	-86	741	
3 6 25		-67 5.0		NGC 1246	O	5310		-204	896	
3 6 26		- 3 8.5		MRK 603	O	2708		-8	447	
					O	2455			557	
3 6 44		-10 29.2			O	3246	100	-37	714	
3 7 0		40 35.0		IC 292	O	2993		147	276	
3 7 9		-10 0.6			O	4316	100	-36	714	
3 7 11		16 54.5		3CR 79	O	76780		67	120	
3 7 28		-10 14.6			O	4018	100	-37	714	
3 7 30		-20 46.0		NGC 1232	O	1820		-76	13	
				ARP 41	O	1723	42		148	
3 7 46		-20 47.4		NGC 1232A	O	6496	58	-77	789	★
				ARP 41	O	1780			528	DIS
3 7 52		-10 56.2			O	5147	100	-40	714	
3 8 0		41 11.0		NGC 1224	O	5026		148	276	
3 8 8		-72 33.8			O	9590		-212	896	
3 8 23		-21 9.1			O	14400		-79	672	
3 8 36		-53 32.0		NGC 1249	O	1090	100	-178	844	
					O	1007	26		741	
3 8 48		1 8.0		IC 298A/B	O	9655		5	591	
				1 ZW 11	O	9671	39		789	
				ARP 147	O	9420			262	
3 8 49		- 9 6.7		NGC 1241	O	4336	71	-34	789	
				ARP 304	O	2168	65		555	DIS
				VV 334	O	3816	42		741	
3 8 53		- 9 5.5		NGC 1242	O	3919	71	-34	789	
				ARP 304						
				VV 334						
3 9 12		39 5.3		4C 39.11	O	48270	120	141	602	
3 9 18		39 8.0		NGC 1233	O	4740		141	166	

75

3ʰ9ᵐ

R.A. (1950)	DEC. (1950)	NAME	OBS	HEL VEL (C*Z)	ERR	GAL CORR	REF	COMMENTS
3 9 19	-74 9.0		O	-1290		-214	896	
3 9 42	36 5.0	5 ZW326	O	21440	60	131	262	
3 10 12	4 31.0	IC 302	P	5905	10	17	582	
3 10 18	- 5 27.3	MRK 604	O	2265		-21	557	
			O	2121			569	
3 11 23	-25 54.7	NGC 1255	R	1690	25	-98	752	
			O	1619			728	
			O	1831	39		148	
3 11 36	- 3 0.0	NGC 1253	R	1710	10	-12	455	
		VV 587W	O	1823	50		873	
		ARP 279	O	1666	71		789	
			O	1742	64		543	
			O	1822	34		741	
3 11 43	41 51.0	MRK 1073	O	6976		148	874	
3 11 48	-57 32.6		O	1150	225	-189	598	
3 11 52	- 2 59.3	NGC 1253A	O	1820	47	-13	789	
		DDO 31	O	1837	37		543	
		VV 587F	O	1818	50		873	
		ARP 279						
3 12 4	41 10.3	NGC 1250	O	6172		145	276	
3 12 9	- 4 57.8	DDO 32	R	2218	10	-20	492	
3 12 13	41 4.0		O	5435		145	753	
3 12 40	41 1.9		O	5711		145	753	
3 12 48	40 38.0	IC 309	O	4214		143	276	
3 12 58	40 59.8		O	6267		144	276	
3 13 7	- 3 39.0	MRK 605	O	8410		-16	447	
			O	(8490)			557	
3 13 8	41 20.8		O	4722		145	276	
3 13 9	41 26.6		O	4957		145	753	
3 13 20	-58 16.7	A	O	19500	150	-191	670	
3 13 20	-58 16.7	B	O	19550	160	-191	670	
3 13 24	41 8.0	IC 310	O	5184		144	276	
			O	(5756)			169	

76

R.A. (1950)	DEC. (1950)	NAME	OBS	HEL VEL (C*Z)	ERR	GAL CORR	REF	COMMENTS
3 13 42	41 10.4		O	6333		144	276	
3 13 45	41 27.0		O	7206		145	276	
3 13 54	31 23.0	KDG 88	R (	4238)	140	113	582	
			O	4209	36		908	
			O	4201	100		582	
3 13 55	41 28.1		O	5930		145	276	
3 13 56	41 15.2		O	4784		144	753	
3 13 59	41 12.1		O	5816		144	276	
3 14 0	31 24.0	KDG 88	O	4126	30	113	908	
3 14 10	41 18.3		O	4639		144	276	
3 14 10	41 25.8		O	6135		145	753	
3 14 11	41 11.6		O	6417		144	276	
3 14 12	41 13.0	NGC 1260	O	5496		144	276	
3 14 14	41 13.7		O	5933		144	276	
3 14 16	41 6.9		O	5836		144	753	
3 14 31	41 47.1		O	4033		146	753	
3 14 33	41 16.1		O	4279		144	276	
3 14 39	41 18.9		O	4717		144	276	
3 14 39	41 43.1		O	7812		145	753	
3 14 41	41 20.3		O	3221		144	276	
3 14 42	3 26.0		O	900		9	475	
3 14 43	41 24.6		O	4249		144	753	
3 14 48	- 0 14.0		O	6600		-4	475	
3 14 48	40 11.0		O	5388		141	777	
3 14 50	41 34.3	IC 312	O	4778		145	276	
3 14 57	41 40.5	NGC 1265 3CR 83.1	O	7511		145	276	
3 14 58	41 33.2		O	6685		145	753	

77

R.A. (1950)	DEC. (1950)	NAME	OBS	HEL VEL (C★Z)	ERR	GAL CORR	REF	COMMENTS
3 15 1	41 17.2		O	3385		144	276	
3 15 1	41 40.4		O	8493		145	753	
3 15 3	41 16.9		O	4909		144	276	
3 15 4	41 13.7		O	6327		144	276	
3 15 5	41 5.2		O	3974		143	753	
3 15 12	-32 45.6	NGC 1288	R	4533	15	-124	752	
			O	4515			728	
			O	4497			528	
			O	4370	31		741	
3 15 14	41 5.5		O	5319		143	753	
3 15 18	41 17.7		O	3313		144	276	
3 15 20	41 48.9		O	8088		145	753	
3 15 21	41 29.2		O	4319		144	753	
3 15 26	40 53.5		O	5487		142	753	
3 15 26	41 17.1	NGC 1267	O	5086		144	276	
3 15 27	41 18.4	NGC 1268	O	3099		144	276	
3 15 29	41 14.8		O	5930		143	753	
3 15 30	-41 18.5	NGC 1291	R	833	10	-150	250	
			R	842	5		823	
			R	835			538	
			R	837	15		752	
			R	836			671	
			O	810	60		486	
			O	756	30		486	
			O	906			84	
			O	765	15		49	
			O	774	180		555	
3 15 37	41 37.3		O	4463		144	753	
3 15 39	41 17.3	NGC 1270	O	4905	65	143	13	
			O	4800	100		2	
3 15 40	41 31.3		O	4099		144	753	
3 15 44	41 22.1		O	4160		144	753	
3 15 44	41 47.6		O	4615		145	753	

R.A. (1950)	DEC. (1950)	NAME	OBS	HEL VEL (C*Z)	ERR	GAL CORR	REF	COMMENTS
3 15 46	41 3.1		0	5481		143	753	
3 15 46	41 17.2		0	3633		143	753	
3 15 48	40 54.1		0	8166		142	753	
3 15 53	41 10.3	NGC 1271	0	5712		143	276	
3 15 54	40 54.5		0	4948		142	753	
3 15 59	41 27.8		0	6179		144	276	
3 16 0	40 58.0		0	7865		142	276	
3 16 0	41 19.0	NGC 1272	0	4172	74	143	72	
3 16 0	41 23.0	NGC 1274	0	6447	80	143	72	
3 16 4	41 14.9		0	5233		143	753	
3 16 8	-27 47.6	NGC 1292	0	1452		-108	528	
			0	1430	15		741	
3 16 8	41 21.6	NGC 1273	0	5354	50	143	13	
			0	5800	75		2	
3 16 9	41 27.2		0	7836		144	276	
3 16 16	41 24.0		0	4425		143	276	
3 16 18	40 58.2		0	5851		142	753	
3 16 19	41 18.3		0	6661		143	753	
3 16 19	41 27.1		0	8422		143	276	
3 16 22	41 8.9		0	6170		142	753	
3 16 23	41 19.1		0	4991		143	753	
3 16 24	41 25.2		0	3331		143	753	
3 16 26	41 5.8		0	5376		142	276	
3 16 26	41 16.0		0	4165		143	753	
3 16 29	41 25.0		0	7330		143	276	
3 16 30	-57 38.0		0	8570	200	-191	598	
3 16 30	41 19.8	NGC 1275 3CR 84	0	5160	40	143	13	SEYF
			0	5265			105	
			0	5178			739	
			0	5200	25		2	PER A
			0	5300			604	

3ʰ16ᵐ

R.A. (1950)	DEC. (1950)	NAME	OBS	HEL VEL (C*Z)	ERR	GAL CORR	REF	COMMENTS
3 16 31	-74 54.1		O	21600		-217	675	
3 16 33	41 21.3		O	5821		143	753	
3 16 33	41 23.5	NGC 1277	O	4974	50	143	13	
			O	5200	75		2	
3 16 34	41 22.2		O	6211		143	753	
3 16 35	41 7.3		O	3568		142	276	
3 16 36	41 23.0	NGC 1278	O	6115	50	143	13	
3 16 37	41 20.5		O	5328		143	753	
3 16 41	41 18.0		O	7247		143	276	
3 16 42	41 22.4		O	5173		143	753	
3 16 43	41 4.3		O	4195		142	276	
3 16 44	41 19.7		O	4823		143	753	
3 16 48	41 27.0	NGC 1281	O	4228		143	276	
3 16 52	41 10.3		O	2069		142	753	
3 16 54	41 11.2	NGC 1282	O	2179		142	276	
3 16 55	41 6.4		O	7079		142	753	
3 16 57	41 13.0	NGC 1283	O	6703		142	276	
3 16 59	-19 16.8	NGC 1297	O	1599	32	-79	741	
3 16 59	41 10.1		O	5013		142	753	
3 17 3	41 11.6		O	7020		142	753	
3 17 10	41 18.5		O	5261		142	753	
3 17 11	41 20.1		O	6621		143	753	
3 17 12	3 58.0	MRK 606	O	9000		9	438	
3 17 12	41 19.8		O	4484		142	753	
3 17 14	41 23.7		O	3072		143	753	
3 17 24	41 13.5		O	3669		142	753	
3 17 25	-19 35.6	NGC 1300	R	1537	10	-80	238	
			R	1502	20		544	
			O	1625			13	

R.A. (1950)	DEC. (1950)	NAME	OBS	HEL VEL (C*Z)	ERR	GAL CORR	REF	COMMENTS
3 17 27	41 33.3		O	5359		143	753	
3 17 30	-32 39.0	IC 1913	O	1289		-125	528	
3 17 31	41 11.5		O	4725		142	753	
3 17 32	41 25.3		O	5001		143	753	
3 17 34	17 6.9		R	351	10	59	860	
3 17 39	-66 40.7	NGC 1313	O	470	30	-207	501	*
			O	453	10		474	
			O	200W	15		431	
3 17 39	-66 40.7	NGC 1313I	O	513		-207	474	HII REG
			O	514			501	
			O	516	85		488	
			O	252W	10		49	
3 17 39	41 19.6		O	4965		142	753	
3 17 39	41 42.4	IC 313	O	4405		143	276	
3 17 42	-26 14.4	NGC 1302	R	1696		-104	645	
			R	1698	20		775	
			O	1730	75		13	
3 17 42	- 2 17.0	NGC 1298	O	300		-15	475	
3 17 42	41 23.0		O	4821		142	753	
3 17 44	41 13.8		O	4891		142	753	
3 17 45	41 22.6		O	4540		142	753	
3 17 51	41 4.4		O	4754		141	753	
3 18 0	15 46.0	3 ZW 53	O	11886	60	54	262	
3 18 2	41 16.9		O	4404		142	753	
3 18 6	40 38.0		R	277	10	140	956	
3 18 10	41 21.5		O	5944		142	753	
3 18 14	-51 13.7		O	10368		-178	680	
3 18 15	41 19.6		O	7029		142	753	
3 18 18	41 12.8	NGC 1293	O	4106		141	276	
3 18 21	41 10.9	NGC 1294	O	6526		141	276	

3ʰ18ᵐ

R.A. (1950)	DEC. (1950)	NAME	OBS	HEL VEL (C*Z)	ERR	GAL CORR	REF	COMMENTS
3 18 39	41 22.4		O	5813		142	753	
3 19 0	41 16.0		O	6057		141	753	
3 19 6	-37 19.0	NGC 1310	O	1718		-141	528	
3 19 6	41 18.1		O	5346		141	753	
3 19 47	-15 34.7	NGC 1309	R	2132		-67	744	
			R	2137			693	
			O	2257	30		148	
3 20 47	-37 23.1	NGC 1316 ARP 154 PKS	O	1835		-142	528	FOR A
			O	1908			84	
			O	1796	70		49	
			O	1758			106	
			O	1782			941	
			O	1857			227	
			O	1878	75		13	
3 20 51	-37 16.9	NGC 1317	O	1900		-142	574	
			O	2060	100		13	
			O	1952			941	
3 20 54	-21 33.0	NGC 1315	O	1730		-90	528	
3 21 30	17 53.0		O	1826	12	59	246	
3 21 41	-35 57.3		O	1823	60	-139	907	
3 21 42	-21 41.0	NGC 1319	O	4112		-91	528	
3 21 43	-37 4.8		O	12234	90	-142	907	
3 21 51	-64 32.9		O	(6700)		-206	896	
3 22 1	-36 38.4	NGC 1326	R	1360		-141	538	
			R	1349	50		544	
			R	1364	10		823	
			R	1381	45		238	
			O	1560			528	
			O	1332	23		741	
			O	1461	89		789	
3 22 12	-21 43.3	NGC 1325	O	1670	30	-92	741	
			O	1868			528	
3 22 18	- 3 13.0	NGC 1320 MRK 607	O	(3023)		-23	447	
3 22 55	-35 36.6		O	1433	120	-139	907	
3 22 58	- 6 19.0	MRK 609	O	(9029)		-35	447	SEYF
			O	10390			557	DIS

R.A. (1950)	DEC. (1950)	NAME	OBS	HEL VEL (C*Z)	ERR	GAL CORR	REF	COMMENTS
3 23 0	-37 12.0	NGC 1316C	O	2069		-143	528	
3 23 3	- 6 18.0	MRK 610	O	10078		-35	447	
			O	10395			557	
3 23 6	-36 31.0	NGC 1326A	R	1818		-141	957	
			O	1823	150		907	
			O	1881	47		789	KNOT
3 23 18	-36 31.0	NGC 1326B	R	1003		-142	957	
			O	885	58		789	
			O	1104	150		907	
			O	785			528	
3 23 41	- 0 23.0	MRK 611	O	8857		-13	447	
			O	8900			868	
3 24 1	-21 31.0	NGC 1332	R	1381	35	-93	238	
			R	1524	35		775	
			O	1609	50		13	
			O	1573			13	
3 24 15	-21 32.0	NGC 1331	O	1408		-93	13	
3 24 18	-52 57.0	IC 1933	O	1063	39	-185	741	
3 24 46	-35 12.9		O	15623	120	-139	907	
3 24 54	21 10.0	5 ZW346	O	12992	60	68	262	
3 25 15	-76 40.5		O	113320		-220	851	
3 25 18	-33 39.5		O	1642	60	-134	907	
3 25 19	2 23.4	3CR 88	O	9060		-3	120	
3 25 40	- 8 33.8	NGC 1337	R	1235		-46	693	
			R	1238			744	
			O	1209	41		741	
3 25 41	-33 43.8	NGC 1351A	O	9353	30	-135	907	
3 25 58	-17 35.3		R	1881		-80	366	
3 26 6	-37 19.3	NGC 1341	O	1894	26	-146	741	
			O	1833			528	
3 26 6	-32 27.3	NGC 1339	O	1268	37	-131	741	
			O	1328			528	
3 26 18	-31 14.5	NGC 1344	O	1257		-127	528	
			O	1271	15		741	
			O	1155			203	

R.A. (1950)	DEC. (1950)	NAME	OBS	HEL VEL (C*Z)	ERR	GAL CORR	REF	COMMENTS
3 26 30	39 51.0		O	6818		132	777	
3 26 52	-31 21.0		O	1330	120	-128	907	
3 27 13	-17 57.1	NGC 1345	R	1523		-82	366	
			O	1540	33		246	
3 27 33	24 37.5	B2	O	32080		79	595	
3 27 45	-33 43.7		O	1702	150	-136	907	
3 27 48	- 4 25.0		R	8395	20	-32	582	
			O	8376	25		582	
3 28 10	- 3 18.6	MRK 612	O	6324		-28	447	
			O	6195			557	
			O	6002			874	
3 28 36	-35 2.0	NGC 1351	O	1589	60	-141	72	
3 28 48	4 12.6	NGC 1349	R	6593		0	901	
3 29 10	-33 47.8	NGC 1350	R	1868	50	-137	885	
			O	1940	41		72	
			O	1883	90		907	
			O	1780			39	
3 29 13	-36 27.5		O	1916	120	-146	907	
3 29 25	-35 39.7		O	16583	60	-143	907	
3 29 29	-35 30.0		O	1403	120	-143	907	
3 29 42	-52 41.3		O	17247		-187	526	
3 29 45	0 5.0	IC 331	O	6717	48	-16	594	
3 29 49	-20 59.3	NGC 1353	O	1700		-95	528	
			O	1448	46		741	
3 30 12	-52 5.0	IC 1954	O	980	100	-186	844	
			O	1116	36		741	
3 30 50	-34 24.4		O	2123	120	-141	907	
3 30 56	-13 59.8	NGC 1357	O	2095	25	-71	741	
			O	1993	75		555	
3 30 58	-35 18.5		O	15233	120	-143	907	
3 31 1	39 11.4	4C 39.12	O	6130		127	166	
			O	5930			485	

R.A. (1950)	DEC. (1950)	NAME	OBS	HEL VEL (C*Z)	ERR	GAL CORR	REF	COMMENTS
3 31 2	-35 41.8		O	5125	60	-145	907	
3 31 10	-34 58.5		O	1552	120	-142	907	
3 31 11	- 5 15.4	NGC 1358	O	4071	36	-38	741	
3 31 30	-21 38.7	IC 1953	R	1883	20	-99	752	
			O	1858			528	
			O	1962	76		741	
3 31 30	-19 39.5	NGC 1359	O	1992		-92	13	
3 31 36	-33 44.4		O	1553	90	-139	907	
3 31 39	-37 18.2		O	19196	120	-150	907	
3 31 42	-36 18.3	NGC 1365	R	1595		-147	171	
			R	1610	90		216	
			R	1623			957	
			O	1790			17	
			O	1763	90		907	
			O	1660	20		64	
			O	1658			29	
			O	1617	60		49	
			O	1750	78		72	
3 31 43	- 1 21.4	3CR 89	O	41550	300	-23	585	
3 31 52	-31 21.6	NGC 1366	O	1257		-132	528	
			O	891	47		741	
3 32 20	-37 27.7		O	14037	120	-151	907	
3 32 24	72 24.0	NGC 1343 7 ZW 8	O	300		207	282	
3 32 31	-36 0.4		O	14395	60	-147	907	
3 32 34	-35 42.8		O	1194	150	-146	907	
3 32 34	-34 27.9		O	1254	120	-142	907	
3 32 53	-25 6.0	NGC 1371	O	1510		-112	528	
3 32 53	-25 6.0	NGC 1367	O	1378	15	-112	741	
3 32 58	-32 48.3		O	1853	30	-137	907	
3 33 13	-33 32.3		O	13974	60	-140	907	
3 33 18	-35 26.0	NGC 1375	O	(83)	90	-145	72	
			O	671			528	

85

R.A. (1950)	DEC. (1950)	NAME	OBS	HEL VEL (C*Z)	ERR	GAL CORR	REF	COMMENTS
3 33 24	-35 24.0	NGC 1374	O	1289	50	-145	72	
3 33 34	-34 36.7	IC 1963	O	1733	90	-143	907	
3 33 58	- 4 51.9	MRK 613	O O	6315 6639		-39	557 569	
3 34 12	-35 37.0	NGC 1379	O	1457	58	-147	72	
3 34 36	-35 9.0	NGC 1380	O O	1872 1856	38 75	-145	72 13	
3 34 37	- 5 12.5	NGC 1376	R O	4155 4471	50	-41	744 741	
3 34 42	-35 28.0	NGC 1381	O	1871	59	-146	72	
3 34 46	71 14.7		R	1172	10	204	860	
3 34 48	-44 7.0	IC 1970	O	1078		-170	528	
3 34 48	-34 53.0	NGC 1380A	O	(1619)		-145	528	
3 34 48	-34 5.7		O	15922	60	-142	907	
3 34 51	-36 25.2	NGC 1392	O	1405	90	-149	907	
3 34 59	-35 32.3		O	1823	90	-147	907	
3 35 0	-36 10.0	NGC 1386	O O	924 797	14	-149	741 528	
3 35 6	-35 41.0	NGC 1387	O	1274	53	-147	72	
3 35 6	-35 21.0	NGC 1380B	O	1927		-146	528	
3 35 6	4 48.0	KDG 92	O	9723	11	-2	908	
3 35 13	-32 9.8		O	11873	60	-137	907	
3 35 18	-35 55.0	NGC 1389	O	1070	77	-148	72	
3 35 18	4 47.0	KDG 92	O	9624	46	-3	908	
3 35 20	-24 39.9	NGC 1385	R O	1488 2012	20	-113	544 13	
3 36 11	-35 36.2		O	894	120	-148	907	
3 36 13	-34 40.9		O	1041	150	-145	907	
3 36 15	-33 17.4		O	1371	60	-141	907	

R.A. (1950)	DEC. (1950)	NAME	OBS	HEL VEL (C*Z)	ERR	GAL CORR	REF	COMMENTS
3 36 18	-23 11.0	NGC 1395	O	1697	26	-108	658	
			O	1820			13	
			O	1690	40		13	
3 36 36	-35 37.0	NGC 1399	O	1458	200	-148	13	
		PKS	O	1500	60		355	
			O	1494	60		907	
			O	1468	58		72	
			O	1473			528	
3 36 39	19 37.5		R	1282	10	53	860	
3 36 48	-26 30.0	NGC 1398	O	1524		-120	13	
3 37 0	-35 45.0	NGC 1404	O	2044	200	-149	13	
			O	1977	46		72	
3 37 6	-44 15.0	NGC 1411	O	991	50	-172	741	
			O	1290	96		203	
			O	977			528	
3 37 12	-22 53.0	NGC 1401	O	1569		-108	528	
3 37 13	-33 41.7		O	15991	90	-143	907	
3 37 16	71 9.0		R	1378	10	203	860	
3 37 18	-35 32.0		O	952	150	-148	907	
3 37 18	-18 51.0	NGC 1400	O	483	40	-94	13	
			O	585	24		938	
3 37 23	-31 29.0	NGC 1406	O	882		-136	528	
			O	1112	12		741	
3 37 48	15 36.0	SW	O	9587		36	321	
3 37 48	15 36.0	NE	O	9905		36	321	
3 37 53	-33 32.4		O	18527	60	-143	907	
3 37 54	-18 44.0	NGC 1407	O	1811	50	-94	13	
			O	1769	19		938	
3 38 1	-71 13.3		O	14554	25	-219	763	
3 38 18	-35 47.3	NGC 1427A	R	2008		-150	957	
			O	2123	120		907	
3 38 42	- 1 27.0	NGC 1409 KDG 93A 3 ZW 55	O O	7403 7448	60 63	-30	262 812	SEYF

R.A. (1950)	DEC. (1950)	NAME	OBS	HEL VEL (C*Z)	ERR	GAL CORR	REF	COMMENTS
3 38 42	- 1 27.0	NGC 1410	O	7606	100	-30	567	
		KDG 93B	O	7412	93		812	
		3 ZW 55						
3 38 48	-22 42.0	NGC 1415	R	1617	75	-109	775	
			O	1508	50		13	
3 39 30	- 4 52.0	NGC 1417	O	4101	50	-44	13	
3 39 47	- 4 53.5	NGC 1418	O	3723	100	-44	789	
3 40 10	-30 3.2	NGC 1425	O	1640		-134	574	
			O	1415	74		741	
			O	1630			528	
3 40 11	-36 58.2		O	15177	120	-155	907	
3 40 12	-13 40.0	NGC 1421	R	2079		-78	744	
			O	2140	32		148	
			O	2063	15		936	
3 40 12	- 6 32.0	MRK 1191	O	6527		-51	921	
3 40 21	- 6 30.0		O	67800		-51	717	
3 40 24	-35 34.0	NGC 1427	O	1681	77	-151	72	
3 40 27	-47 22.8	NGC 1433	R	1069		-182	671	
			O	1072			680	
			O	974	25		49	
3 40 30	-35 19.0	NGC 1428	O	(9)	78	-150	72	
			O	230			528	
			O	1642	90		907	DIS
3 40 34	39 8.9		O	4950		120	255	
3 40 39	-22 16.0	NGC 1426	O	1358	50	-109	13	
			O	1444	6		658	
3 40 45	- 4 53.4	NGC 1424	O	5640	58	-45	789	
3 41 9	-36 25.9	NGC 1437A	R	888		-154	957	
3 41 19	-53 48.4		O	16204	80	-196	706	
3 41 20	-53 47.6		O	17070	107	-196	706	
3 41 26	-34 5.8		O	1189	60	-147	907	
3 41 31	-53 35.5		O	17706	120	-195	706	
3 41 35	-53 35.3		O	17941	161	-195	706	

R.A. (1950)	DEC. (1950)	NAME	OBS	HEL VEL (C*Z)	ERR	GAL CORR	REF	COMMENTS
3 41 35	-53 47.6		O	17227	56	-196	706	
3 41 42	-53 47.9		O	17239		-196	706	
3 41 42	-53 45.7		O	17920	33	-196	706	
3 41 42	-36 1.0	NGC 1437	O	1134	36	-153	741	
			O	1234			528	
3 41 45	-53 47.9		O	16371	70	-196	706	
3 41 45	-53 47.6		O	19374	105	-196	706	
3 41 49	-53 49.0		O	16889	98	-196	706	
3 41 53	-53 47.2		O	17013	49	-196	706	
3 41 55	-53 47.3		O	16438	37	-196	706	
3 41 57	67 56.5	IC 342	R	25	5	196	397	
			R	35			934	
			R	35	5		837	
			R	56	5		93	
			R	25	3		923	
			R	56			68	
			O	-10	20		13	
			O	-10	26		11	
			O	8	25		283	
			O	34			13	
3 42 2	-53 49.0		O	19562	121	-196	706	
3 42 6	-53 50.6		O	18337	138	-196	706	
3 42 9	-53 51.3		O	18433	160	-196	706	
3 42 10	-53 45.1		O	15854	144	-196	706	
3 42 10	-53 44.1		O	18316	137	-196	706	
3 42 31	-53 47.0		O	15940	80	-196	706	
3 42 35	-53 50.5		O	25063	109	-196	706	
3 42 39	-22 4.6	NGC 1439	O	1997	100	-110	13	
			O	1667	10		658	
3 42 40	-53 50.9		O	17702	60	-197	706	
3 42 48	-18 25.1	NGC 1440	O	1534	27	-97	741	
3 42 54	-44 48.1	NGC 1448	R	1161		-177	671	
		NGC 1457	O	1185			528	
			O	1222	45		741	

89

R.A. (1950)	DEC. (1950)	NAME	OBS	HEL VEL (C★Z)	ERR	GAL CORR	REF	COMMENTS
3 43 8	-18 47.3	NGC 1452	O	1904	17	-99	741	
3 43 12	-73 3.0		O	14690		-222	896	
3 43 13	- 4 14.9	NGC 1441	O	4207	34	-45	741	
			O	4262	150		13	
3 43 18	70 0.0	4C69.051	R	1201	10	200	752	
			O	1200			595	
3 43 33	- 4 17.7	NGC 1449	O	4049	29	-45	741	
			O	4176	100		13	
3 43 42	- 4 13.0	NGC 1451	O	3927	75	-45	13	
3 43 57	- 4 7.6	NGC 1453	O	3919	40	-45	13	
			O	4035			13	
			O	3882			106	
3 44 22	-36 51.0	NGC 1460	O	1525	90	-157	907	
3 44 24	-35 5.8		R	1908		-152	957	
			O	1940	90		907	
3 44 48	13 5.0	KDG 94	O	6644	32	21	812	
3 45 0	-33 50.0	IC 1993	O	1050		-149	728	
3 45 15	-37 31.9		O	5758	150	-160	907	
3 45 36	33 44.1	3CR 93.1	O	72850		98	657	
3 46 0	-36 37.5		O	1856	150	-158	907	
3 46 0	12 58.0	KDG 94	O	6805	100	19	812	
3 46 11	-16 32.4	NGC 1461	O	1450	36	-93	741	
3 46 15	-36 51.2		O	5245	60	-159	907	
3 47 10	-36 1.6		O	5393	60	-157	907	
3 47 56	-36 5.8		O	1326	120	-158	907	
3 48 12	6 10.0		O	25662	100	-8	13	
3 49 32	-27 53.5	PKS	O	19800		-134	189	
3 49 45	21 17.5	4C 21.13	O	39680		49	318	
3 49 48	35 27.0		R	5815	15	101	752	
			O	5785	25		936	

R.A. (1950)	DEC. (1950)	NAME	OBS	HEL VEL (C*Z)	ERR	GAL CORR	REF	COMMENTS
3 50 5	-33 37.0		O	1636	120	-152	907	
3 51 12	-47 38.0	NGC 1483	O	1005		-189	528	
3 52 12	-36 8.0	IC 2006	O	1374	42	-161	741	
			O	1394			528	
3 51 18	15 50.0	KDG 95	O	6583	31	26	908	
3 51 24	15 47.0	KDG 95	O	6675	22	26	908	
3 52 30	-20 39.0	NGC 1482	O	1655		-113	528	
3 53 18	- 9 42.0	MRK 1192	O	8944		-74	921	
3 53 23	2 47.7	4C 02.11	O	180650		-26	584	
3 54 6	-42 31.0	NGC 1487	O	728	40	-179	719	*
		VV 78	O	881	16		741	
			O	720			528	
3 55 0	67 0.0		R	85	5	190	837	
			R	85			934	
3 55 54	-46 21.1	NGC 1493	R	1051		-188	671	
			O	1027			528	
			O	1022	20		741	
3 56 10	10 17.5	3CR 98	O	9180		0	120	
			O	9080	100		846	
3 56 12	-49 3.0	NGC 1494	O	1106	83	-194	741	
			O	1105			528	
3 56 18	78 9.0	VV 793	O	2151	50	210	873	
3 58 5	79 41.9		R	2184	10	212	860	
3 58 33	0 28.2	3CR 99	O	127560		-40	584	
			O	127700			868	
3 58 36	-10 25.0		R	11060	25	-81	582	
			O	10985	50		582	
3 59 18	-67 46.0	NGC 1511	O	1702		-222	154	
			O	1349	22		741	
4 0 2	-64 21.3	PKS	O	142700		-220	851	
4 0 12	1 49.0		O	3812	24	-36	813	
4 1 38	-46 16.3		O	21000		-192	675	

4ʰ1ᵐ

4 1 54	-43 33.0	NGC 1510	O	958		-186	528	
			O	976	27		789	
			O	1000			377	
4 1 56	- 2 19.5	NGC 1507A/B	R	840		-53	216	
			O	898	27		148	
4 1 56	- 2 19.5	NGC 1507A KDG 97A	O O	957 893	15 26	-53	812 812	
4 1 56	- 2 19.5	NGC 1507B KDG 97B	O	868	22	-53	812	
4 2 18	-43 29.0	NGC 1512	R	900	5	-186	802	
			O	734	40		49	
			O	765	71		789	
4 2 34	69 40.7	IC 356 ARP 213	R R	743 870	15 15	193	390 455	
4 2 51	-54 14.0	NGC 1515	O	1131	20	-207	741	
4 2 54	4 17.0	KDG 98	O	5357	17	-28	908	
4 2 54	4 19.0	KDG 98	O	5214	60	-28	908	
4 4 38	-21 18.7	NGC 1518	R	933	20	-125	544	
			O	972	10		238	
			O	1027			13	
4 4 39	3 34.5	3CR105	O	26650		-33	946	
4 4 42	-10 18.0	MRK 1193	O	9946		-86	921	
4 6 6	-21 11.0	NGC 1521	O	4222	50	-125	13	
4 6 30	8 31.0	NGC 1517	R	3483	10	-15	752	
4 6 54	-48 1.0	NGC 1527	O	1008		-198	154	
			O	1097	35		741	
			O	960			574	
4 7 29	-45 38.8	IC 2035	O	1458	40	-194	741	
4 8 18	26 45.0		R	1447	15	55	752	
4 8 29	- 7 24.6	MRK 614	O	3255		-78	557	
			O	3378			569	
4 8 50	-56 15.0	NGC 1533	O	600		-213	574	
			O	812	29		741	
4 10 4	-32 58.7	NGC 1531	O	1308		-164	145	
			O	1141	32		741	

R.A. (1950)	DEC. (1950)	NAME	OBS	HEL VEL (C*Z)	ERR	GAL CORR	REF	COMMENTS
4 10 8	-33 0.1	NGC 1532	R	1220	60	-165	216	
			O	1321	26		741	
			O	958	82		789	
			O	900			528	
			O	1760			145	
4 10 29	-68 9.3		O	(4050)		-226	896	
4 10 36	25 21.0		O	8052	34	48	813	
4 10 41	10 20.4		O	27000		-11	606	
4 10 48	29 2.0	5 ZW372	O	5290		62	389	
			O	5295			262	
4 10 54	10 20.6		O	27000		-11	606	
4 10 54	11 4.7	3CR109	O	91690		-8	143	
			O	91650			160	
4 10 57	-56 36.9	NGC 1536	O	1565		-215	154	
4 11 44	-31 46.2	NGC 1537	O	1378	45	-162	741	
			O	1300			528	
4 11 47	-57 52.0	NGC 1543	O	1400		-217	136	
4 13 33	-56 11.0	NGC 1546	O	1190	37	-215	741	
4 13 58	-51 4.2		O	3818	15	-208	680	
4 14 39	-55 42.9	NGC 1549	O	1142		-215	136	
			O	1266			84	
			O	1176			136	
4 15 1	37 54.3	3CR111	O	14550		93	657	
			O	14500			868	
			O	15000			727	
4 15 1	37 54.3		O	15400	2	93	717	
4 15 5	-55 54.2	NGC 1553	O	1293		-216	136	
			O	1231	30		49	
			O	1035			84	
4 15 39	-60 20.0	IC 2056	O	1129	41	-221	741	
4 17 1	-62 54.3	NGC 1559	R	1291		-225	671	
			O	1384	67		203	
			O	1450	100		844	
			O	1284			145	
4 17 3	75 10.7	NGC 1530	R	2470		201	934	
		7 ZW12	O	2514	9		813	

93

4ʰ18ᵐ

R.A. (1950)	DEC. (1950)	NAME	OBS	HEL VEL (C*Z)	ERR	GAL CORR	REF	COMMENTS
4 18 53	-55 3.4	NGC 1566	R	1497		-216	671	SEYF
			O	1450			136	
			O	1430			574	
			O	1616	120		113	
			O	1330			136	
			O	1421	100		49	
4 19 24	1 43.0		O	13200		-52	534	
4 20 59	-57 5.4	NGC 1574	O	890		-220	136	
4 21 24	-56 27.0	A/B	O	13240	178	-219	724	
4 22 0	- 0 52.0	MRK 615	O	4645		-65	438	
4 23 18	70 14.0		R	3058	10	189	582	
			O	2947	50		582	
4 23 24	69 29.0	7 ZW 14	O	10280		187	282	
4 24 48	-53 55.5		O	12000		-218	627	
4 25 12	21 33.0		R	2407	15	22	752	
4 26 6	64 44.4	NGC 1569	R	-90		174	156	
		ARP 210	R	-95	25		183	
			R	-77	1		933	
			O	-34	30		13	
			O	-58			13	
4 26 34	-48 1.4		O	4950		-209	801	
4 26 35	-55 8.1	NGC 1596	O	1526	36	-220	741	
			O	1420			574	
4 26 48	-55 10.0	NGC 1602	O	1740	100	-220	844	
4 26 56	-47 55.5	NGC 1595	O	(4740)		-209	801	
4 27 6	71 46.2	NGC 1560	R	-43	6	191	390	
		IC 2062	R	-41			934	
			O	-242	90		555	
4 27 8	-47 53.5	NGC 1598	O	5130		-209	801	
4 27 22	-47 4.4		O	4433	90	-208	680	
4 27 30	6 50.0	VV 555	O	4359	50	-39	873	
			O	4280	150		776	
			O	4300			776	KNOT
4 27 58	-53 56.1	IC 2082	O	11690		-219	627	
		PKS	O	11750			317	
			O	11719	200		680	

94

R.A. (1950)	DEC. (1950)	NAME	OBS	HEL VEL (C*Z)	ERR	GAL CORR	REF	COMMENTS
4 28 6	0 33.0	NGC 1587/8	R	4019		-64	389	
4 28 6	0 33.3	NGC 1587	O	3890	75	-64	13	
		2 ZW 12N	O	3890			389	
		KDG 99	O	3864	50		908	
			O	3876			748	
4 28 7	20 31.2	PKS	O	55710		15	850	
4 28 10	-53 55.5		O	12000		-219	627	
4 28 12	0 33.0	NGC 1588	O	3328		-64	262	
		2 ZW 12S	O	3339			748	
		MRK 616	O	3030			389	
		KDG 99	O	3559	90		908	
4 28 24	- 5 54.0	IC 2075	R	4328	15	-89	582	
			O	4315	25		582	
4 28 30	7 31.0	NGC 1590	R	3927	45	-37	622	
		2 ZW 13	O	3769			262	
4 29 0	73 9.0	NGC 1573	O	4246	40	194	938	
4 29 12	- 5 11.5	NGC 1600	O	4830	100	-87	13	
			O	4687			842	
4 29 18	1 5.0		R	3538	10	-63	582	
			O	3482	20		582	
4 29 22	- 5 12.0	NGC 1601	O	4997	100	-87	13	
4 29 43	-17 10.5	PKS	O	43620		-130	851	
4 30 31	5 15.0	2 ZW 14	O	9966		-47	262	SEYF
		3C 120	O	9890	90		345	
			O	9950			168	
			O	9690			161	
			O	9900			186	
			O	9900			880	*
			O	9790			166	
			O	10020			143	
4 30 52	-61 33.9		O	18249		-229	526	
4 30 57	-54 42.5	NGC 1617	O	1040	20	-222	741	
4 31 36	- 8 40.7	NGC 1614	R	(4778)		-102	622	
		2 ZW 15	O	4740	30		353	
		MRK 617	O	4720			557	
		ARP 186	O	6791			262	DIS
			O	4744	27		148	

4ʰ31ᵐ

R.A. (1950)	DEC. (1950)	NAME	OBS	HEL VEL (C∗Z)	ERR	GAL CORR	REF	COMMENTS
4 31 43	−13 24.0		O	9079		−119	651	
4 31 49	−13 29.0	PKS	O	10910	270	−119	355	
			O	9900			651	
4 31 54	−13 24.0		O	9826		−119	651	
4 31 54	8 3.0	KDG 100	O	7459	75	−37	908	
4 32 0	8 4.0	KDG 100	O	8121	58	−37	908	
4 32 42	− 1 50.0	2 ZW 17	O	9718		−77	262	
4 33 36	− 3 15.1	NGC 1618	O	(5350)	85	−83	555	
4 33 55	29 34.2	3CR123	O	65400		48	717	
			O	65270	90		517	
			O	191000W			517	
4 34 0	−10 28.0	MRK 618	O	10840		−110	438	SEYF
4 34 0	− 0 15.0	NGC 1620	R	3509	10	−72	752	
			R	3508			744	
			O	3496	10		740	
4 34 18	− 3 8.0		O	4713		−83	777	
4 34 36	− 3 24.0	NGC 1625	O	3033	48	−85	148	
4 34 54	− 0 23.0		O	20100		−73	475	
4 35 42	67 38.0	7 ZW 19	O	4830	200	178	820	
4 35 54	11 9.0	2 ZW 18	O	4419		−28	262	
4 37 24	7 14.0	NGC 1634	O	7500		−45	534	
		KDG 101A	O	4411	25		908	DIS
4 37 24	7 15.0	NGC 1633	O	4799	55	−45	908	
		KDG 101B						
4 38 6	4 9.0		O	4600	50	−58	13	
4 38 58	− 2 57.1	NGC 1637	R	731	10	−86	238	
			R	711	20		544	
			R	712	25		183	
			R	717			744	
			R	716			283	
			O	716	25		283	
			O	528			13	
			O	695	50		13	
4 39 5	− 1 54.1	NGC 1638	O	3306	34	−83	741	

R.A. (1950)	DEC. (1950)	NAME	OBS	HEL VEL (C∗Z)	ERR	GAL CORR	REF	COMMENTS
4 39 18	- 7 11.0	IC 387	R O	4530 4663	20 50	-103	582 582	
4 40 4	-20 31.8	NGC 1640	O	1676		-149	13	
4 41 29	74 50.1	DDO 33	R	1635	10	195	492	
4 42 17	-62 43.0		O	(1320)		-234	896	
4 42 44	-18 28.2		O	84440		-145	851	
4 43 18	- 2 30.0	NGC 1653	O	4351	20	-88	938	
4 43 18	- 2 10.4	NGC 1654	O	4577		-87	426	
4 43 36	3 25.0	IC 392 VV 665	O	4265		-65	774	
4 44 0	- 4 53.0	NGC 1659	O	4537	37	-98	148	
4 44 55	-59 20.2	NGC 1672	R O O O O O	1338 1373 1267 1343 1300 1269	10	-233	671 453 155 741 574 145	
4 45 21	44 56.8	3CR129	O	6240	100	100	516	
4 45 28	-76 39.8		O	(24000)		-235	675	
4 46 0	0 9.1	DDO 34	R	669	10	-80	492	
4 46 10	- 6 24.4	NGC 1667	O O O	4578 4622 (4520)	24 63 70	-105	741 789 555	
4 46 30	44 58.2	3CR129.1	O	6675	200	100	516	
4 46 49	21 35.9		R	3918	10	6	860	
4 47 0	3 14.0	KDG 103A	O O	8539 8383	190 10	-69	567 908	
4 47 12	3 15.0	KDG 103B 2 ZW 23	O O O O	8184 8377 8306 8429	160 18 64	-69	567 262 908 197	
4 47 13	-29 17.9	DDO 228	R	1471	10	-180	492	
4 47 41	-59 53.3	NGC 1688	O	1266	15	-234	741	

97

4ʰ47ᵐ

R.A. (1950)	DEC. (1950)	NAME	OBS	HEL VEL (C*Z)	ERR	GAL CORR	REF	COMMENTS
4 47 58	-57 44.8		0	7040		-233	896	
4 48 57	51 59.9	3CR130	0	32680	300	124	585	
4 49 5	-17 35.2	PKS	0	9530	60	-147	355	
4 49 6	- 2 39.0		0	2100		-93	475	
4 49 42	78 7.0	IC 391	R	1565		201	744	
			0	1607			13	
4 50 12	1 10.0		0	17700		-79	475	
4 50 37	-59 19.0		0	16200	70	-235	680	
4 50 49	-25 19.8	DDO 229	R	1374	10	-172	492	
4 51 12	80 6.2		0	16040		205	255	
4 51 27	2 48.7		0	61000		-74	529	
4 51 42	1 35.0		0	2100		-79	475	
4 51 58	-63 2.5	NGC 1706	0	5130		-238	896	
4 52 48	68 14.0	IC 396A	0	762	140	175	567	
4 53 7	-53 26.5	NGC 1705	0	640	16	-231	741	
			0	831			154	
4 53 13	-20 38.9	PKS	0	10610		-160	317	
			0	10330	120		355	
4 53 18	2 9.0		0	4500		-78	475	
4 53 42	22 44.7	3CR132	0	64160	300	5	585	
4 54 30	- 4 56.0	NGC 1700	0	3918		-106	842	
			0	3976	40		13	
			0	800	50		1	DIS
4 54 54	- 0 56.0		0	13800		-91	475	
4 55 17	-42 52.6		0	3336		-216	680	
4 55 25	-42 52.6		0	3536		-216	680	
4 56 31	- 4 21.4	4C-04.17	0	35400		-106	600	
4 56 42	5 32.0		R	4695	10	-67	582	
			0	4734	50		582	
4 56 54	- 7 56.0	NGC 1720	0	4152	71	-119	789	
			0	4242	44		741	

R.A. (1950)	DEC. (1950)	NAME	OBS	HEL VEL (C*Z)	ERR	GAL CORR	REF	COMMENTS
4 56 59	-11 12.0	VV 699W	O	4569	50	-131	873	
4 56 59	-11 12.0	VV 699E	O	4507	50	-131	873	
4 57 18	- 7 49.8	NGC 1726	O	4072	24	-119	741	
			O	(4290)	200		555	
4 57 42	- 3 25.0	NGC 1729	R	3644	10	-103	752	
4 57 56	-26 5.8	NGC 1744	R	760	10	-179	216	
			R	740			171	
			O	676			13	
4 58 36	65 44.0	7 ZW 23	O	11735	200	166	820	
			O	9960			429	
4 59 0	3 30.0	2 ZW 28	R	8580		-77	745	
			O	8612			262	
			O	8580			591	
4 59 7	- 4 20.1	NGC 1741A/B VV 565 ARP 259	R O	4060 4000	25 80	-108	455 776	
4 59 7	- 4 20.1	NGC 1741B	O	4036	37	-108	148	
4 59 9	- 4 19.8	NGC 1741A VV 565SW	O O	3937 4110	10 100	-108	148 776	
4 59 17	- 4 22.2	IC 399	O	4015	20	-108	148	
4 59 54	25 12.2	3CR133	O	83190		11	946	
4 59 55	-62 8.8		O	8420		-240	896	
4 59 55	-62 8.3		O	9200		-240	896	
5 0 31	16 19.9	DDO 35	R	1390	15	-26	492	
5 1 0	1 30.0	NGC 1762	R	4633	15	-87	752	
5 2 8	-61 12.4	NGC 1796	O	987	15	-240	741	
			O	990			154	
5 2 12	1 45.0		O	3000		-87	475	
5 2 25	-69 38.3	NGC 1809	O	1310	30	-242	866	
5 2 32	-10 18.9	PKS	O	12300		-132	600	
5 3 6	-11 56.4	NGC 1784	R	2317		-138	744	
			R	2318	40		238	
			R	2313			283	
			R	2285			363	
			O	2314	25		283	
			O	2150	52		148	

R.A. (1950)	DEC. (1950)	NAME	OBS	HEL VEL (C*Z)	ERR	GAL CORR	REF	COMMENTS
5 3 30	-38 2.7	NGC 1792	R	1193	100	-212	544	
			R	1205	100		238	
			O	1140	100		844	
			O	1215			96	
			O	1254	64		148	
5 4 24	-17 37.0		R	4514	25	-158	582	
			O	4491	20		582	
5 4 33	-32 1.2	NGC 1800A/B	O	722	20	-199	741	
5 4 33	-32 1.2	NGC 1800A	O	725	50	-199	789	
5 4 33	-32 1.2	NGC 1800B	O	686	39	-199	789	
5 4 42	9 24.0	2 ZW 30	O	11710		-58	303	
5 5 12	0 36.0	2 ZW 31	O	15717		-93	262	
5 5 31	-16 21.7	DDO 36	R	2044	7	-155	492	
5 5 54	-37 34.7	NGC 1808	R	997		-212	310	
			O	960			173	
			O	1039	87		72	
			O	963			145	
			O	1031			155	
5 5 54	-37 34.7	NGC 1808A	O	1080		-212	235	KNOT
5 5 54	-37 34.7	NGC 1808C	O	937		-212	235	KNOT
5 5 54	-37 34.7	NGC 1808D	O	890		-212	235	KNOT
5 5 54	-37 34.7	NGC 1808E	O	880		-212	235	KNOT
5 5 54	-37 34.7	NGC 1808F	O	830		-212	235	KNOT
5 7 0	84 0.0		R	4138	15	211	752	
5 7 30	0 52.0		O	8400		-94	475	
5 7 42	- 0 47.0	2 ZW 32	O	8170		-101	303	
5 8 12	79 36.0	7 ZW 31	O	16090	200	201	820	
5 8 18	- 2 45.0	2 ZW 33	R	2832	19	-109	622	
			O	2850			389	
			O	2870	66		429	
			O	2838			262	
5 8 42	1 28.0	2 ZW 34	O	15778		-93	262	
			O	15593			197	

R.A. (1950)	DEC. (1950)	NAME	OBS	HEL VEL (C*Z)	ERR	GAL CORR	REF	COMMENTS
5 8 48	-31 40.0	DDO 230	R	980	10	-201	492	
5 9 6	5 8.0	NGC 1819 MRK 1194	0	4728		-78	921	
5 9 25	-14 51.0	A 509	0	2140	66	-153	741	
5 9 48	-15 44.8	NGC 1832	R	1936	15	-156	752	
			R	1912	20		544	
			R	1938			744	
			0	1950			174	
			0	2037	200		13	
5 10 6	-33 2.0	DDO 231	R	935	10	-205	492	
5 11 33	0 53.1	3CR135	0	38074		-97	160	
			0	38189	60		741	
			0	38160	60		159	
5 12 59	24 55.1	3CR136.1	0	19190	300	1	585	
5 13 36	- 0 12.0		0	9900		-103	475	SEYF
			0	9740			654	
5 13 53	-62 17.0		0	5100		-245	896	
5 14 24	0 52.0	2 ZW 35	0	7403		-99	262	
5 15 12	0 10.0	2 ZW 36	0	9015		-103	303	
5 15 48	66 11.0	7 ZW 35	0	13240	100	162	820	
5 16 24	2 34.0	2 ZW 37	0	16888		-94	262	
5 16 42	1 16.0	KDG 105A	0	8007	10	-99	908	
5 16 42	1 16.0	KDG 105B	0	8108	15	-99	908	
5 16 54	-21 35.8	DDO 37	R	(1811)		-179	492	
5 17 54	-32 12.0	NGC 1879 DDO 232	R	1247	10	-208	492	
5 18 24	-45 49.7	PKS	0	10500		-234	120	
			0	10490			628	PIC A
5 19 5	-61 46.5		0	4680		-247	896	
5 19 6	6 38.0	NGC 1875 ARP 327 VV 169	0	9167		-80	274	
5 19 18	3 26.0	KDG 107	0	4274	125	-93	908	

R.A. (1950)	DEC. (1950)	NAME	OBS	HEL VEL (C*Z)	ERR	GAL CORR	REF	COMMENTS
5 19 24	3 26.0	KDG 107	O	4238	133	-93	908	
5 19 54	43 30.0		R	6227	15	75	752	
5 20 15	-11 32.8	NGC 1888 ARP 123	O	2557		-150	13	
5 20 15	-11 32.6	NGC 1889 ARP 123	O O O	2471 2472 2557	20 40	-150	392 13 13	
5 21 12	76 37.0		R O	4177 4173	15 25	192	582 582	
5 21 13	-36 30.2	PKS	O O	16500 18300	500	-219	189 139	
5 22 19	4 27.4		R	521	10	-91	860	
5 24 0	-69 48.0		R O O O O	280 34 235 287 276	50	-247	10 483 252 1 11	LMC
5 26 0	-16 10.0		R O	2174 2181	10 50	-169	582 582	
5 26 28	-63 48.1	NGC 1947	O O O	1289 789 1179	46	-249	741 154 227	
5 26 53	55 47.3		R	2203	10	121	860	
5 27 16	73 41.2	DDO 38	R	1241	10	182	492	
5 30 7	-11 35.0		O	46400	300	-157	583	
5 30 30	-14 6.0	NGC 1954	R O	3126 3033	38	-165	744 789	
5 31 16	-21 58.9	NGC 1964	R R O O	1679 1690 1560 1849	50 50 58	-190	544 238 789 13	
5 32 6	77 17.0		O	4087	25	192	813	
5 36 34	69 21.3	NGC 1961 IC 2133 ARP 184	R R R O O O	3910 3941 3935 3936 3897 3870	100 10 10 15	166	544 838 744 838 838 13	

R.A. (1950)	DEC. (1950)	NAME	OBS	HEL VEL (C*Z)	ERR	GAL CORR	REF	COMMENTS
5 39 12	72 19.0		O	1142	31	176	813	
5 40 12	69 2.0	7 ZW 45	O	4290	200	164	820	
5 41 36	-64 19.2	NGC 2082	O	1374	30	-254	741	
			O	1114	47		373	
5 42 22	5 2.5		R	360	3	-103	796	
5 44 12	17 33.0		R	5582	15	-51	752	
5 45 15	-34 16.4	NGC 2090	R	1805		-229	363	
			O	914	100		789	DIS
			O	949	48		741	DIS
5 46 38	-25 29.5		O	11942	57	-210	741	
5 46 42	-25 29.4		O	11884	69	-210	741	
5 47 8	-47 46.5		O	15130	40	-250	680	
5 48 0	39 49.0		R	5194	10	45	752	
5 48 50	-32 8.6		O	12600	900	-227	439	
5 49 0	-31 45.0		O	12000		-226	439	
5 49 36	-32 15.5		O	12000		-227	439	
5 49 46	- 7 28.1	NGC 2110	O	2290	90	-156	686	
			O	2281	96		945	
5 49 53	75 18.3	DDO 39	R	818	10	183	492	
5 50 45	-17 53.0	IC 438	R	3114	10	-191	752	
5 51 6	46 26.0		R	6136	20	72	752	
5 51 6	78 30.0		R	4758	15	193	582	
			O	4821	60		582	
5 51 10	46 25.9		O	6150		72	669	SEYF
5 52 18	48 32.0	VV 601	O	5780	50	81	873	
5 53 6	3 23.1	2 ZW 40	R	760		-117	710	
			R	800			324	
			R	810	20		241	
			O	772	48		197	
			O	614			262	*
5 55 25	73 7.0		R	1085	10	175	860	

R.A. (1950)	DEC. (1950)	NAME	OBS	HEL VEL (C*Z)	ERR	GAL CORR	REF	COMMENTS
5 56 19	-69 34.1	NGC 2150	O	4440		-255	574	
5 57 43	-64 10.0		O	8065		-259	896	
5 58 37	-28 59.6	DDO 233	R	1395	15	-225	492	
5 58 48	36 7.0		R	1518	15	23	752	
5 59 4	-23 40.3	NGC 2139	R	1804	50	-212	544	
		IC 2154	R	1842	10		752	
			R	1800	26		238	
			O	1843			728	
			O	1913			13	
			O	1736	47		789	
6 0 24	7 50.0	2 ZW 42	O	5287		-103	262	
			O	5560			389	
6 1 18	8 40.0		R	5395	15	-100	752	
6 4 25	-20 21.8	PKS	O	49170		-207	851	
6 5 0	34 16.0	VV 596	O	2900	50	11	873	
6 5 0	80 29.0	KDG 108	O	3872	50	197	908	
6 5 23	-73 23.0	NGC 2199	O	4470		-253	896	
6 5 45	48 4.8	3CR153	O	82990		73	318	
6 5 47	-21 44.3	NGC 2179	R	2839	100	-211	885	
			O	2761	42		741	
6 6 0	80 28.0	KDG 108	O	4026	50	197	908	
6 7 0	81 10.0		R	4211	15	199	752	
6 7 21	-52 29.7	NGC 2191	O	4570		-262	574	
6 8 0	69 45.0		O	4062	15	161	813	
6 8 9	-75 21.5		O	10790		-251	896	
6 8 18	-34 5.0	NGC 2188	R	760	30	-241	216	
			O	678	16		148	
			O	730			95	
			O	745	20		555	
6 8 50	48 36.8		O	19990		74	255	
6 9 0	75 58.0		O	5110	49	182	813	
6 9 18	71 9.0		R	4061	10	165	582	
			R	4053	10		829	
			O	3960			582	KNOT

R.A. (1950)	DEC. (1950)	NAME	OBS	HEL VEL (C*Z)	ERR	GAL CORR	REF	COMMENTS
6 9 48	71 3.0	MRK 3	R	3952		165	707	SEYF
			O	3930			264	
			O	4107			887	
			O	4050			192	
			O	4080			280	
6 10 0	80 5.0	IC 440 KDG 109A	O	4353	59	195	908	
6 10 4	-21 47.8	NGC 2196	R	2315	30	-214	885	
			R	2330	20		775	
			O	2351	14		741	
			O	2289	48		148	
6 10 43	78 22.5	NGC 2146 4C 78.06 KDG 110	R	856		190	310	
			R	900			561	
			O	880			483	
			O	785	50		13	
			O	901	20		21	
			O	909	19		478	
			O	1110			748	
6 12 36	0 13.0		R	2340	10	-142	752	
6 13 18	-26 33.9	DDO 234	R	1799	20	-228	492	
6 14 0	-26 45.0	NGC 2206	R	6279	20	-229	752	
			O	6258			728	
6 14 14	-21 21.2	NGC 2207	O	2680	60	-215	13	
			O	2700	45		789	
6 14 32	-21 21.5	IC 2163	O	2798	53	-215	789	
6 14 49	-34 55.1	PKS	O	98630		-246	851	
6 15 36	42 39.0	KDG 111	O	7597	90	44	908	
6 15 53	78 33.3	NGC 2146A 4C 78.06 KDG 110	R	1494		190	744	
			R	1495			561	
			O	1567	58		789	
			O	1110			748	
6 16 0	42 35.0	KDG 111	O	7466	11	44	908	
6 17 6	59 8.0	KDG 112A	O	3163	15	116	812	
6 17 12	59 8.0	KDG 112B	O	3188	23	116	812	
6 18 18	-37 10.2	PKS	O	9640	60	-251	355	
			O	9770			317	
6 18 30	-16 2.0		R	2852	15	-202	582	
			O	2827	30		582	

R.A. (1950)	DFC. (1950)	NAME	OBS	HEL VEL (C*Z)	ERR	GAL CORR	REF	COMMENTS
6 19 26	-57 33.2	NGC 2221A	O	2532	35	-267	613	
6 19 26	-57 32.8	NGC 2221B	O	2357	35	-267	613	
6 19 26	-57 30.3	NGC 2222	O	2602	11	-267	613	
6 19 41	-27 12.4	NGC 2217	R	1620		-233	645	
			R	1413	100		544	
			R	1406	100		238	
			R	1630	35		775	
			O	1573			13	
			O	1585	150		13	
6 20 37	-52 40.0	PKS	O	15320	200	-267	350	
6 20 55	-64 26.3	NGC 2228	O	7260		-265	896	
6 21 28	74 20.0	MRK 4	O	4800		175	192	
6 21 43	-64 40.3		O	7430		-265	896	
6 22 30	59 7.0	IC 2166	R	2693	10	115	752	
			O	2671	30		813	
6 22 31	-22 48.7	NGC 2223	O	2608	30	-223	741	
			O	2745	100		789	
6 23 18	12 57.0	MRK 705	O	8545		-95	630	SEYF
6 23 48	-21 58.0	NGC 2227	R	2261	10	-222	752	
			O	2221	20		582	
6 25 0	74 28.0		R	5578	15	174	582	
			O	5568	25		582	
6 25 30	83 0.0	KDG 113	O	4704	30	203	908	
6 25 36	83 0.0	KDG 113	O	4461	67	203	908	
6 25 51	-54 32.7		O	15649		-269	526	
6 26 0	-47 9.0		O	11630	25	-266	719	
6 26 24	-54 28.6		O	15889		-269	526	
6 26 32	-63 19.5		O	11780		-267	896	
6 26 42	71 36.0		O	3502	30	164	813	
6 27 0	39 32.2		R	484	10	25	860	
6 27 16	57 15.9	MRK 619	O	13165		105	557	

R.A. (1950)	DEC. (1950)	NAME	OBS	HEL VEL (C*Z)	ERR	GAL CORR	REF	COMMENTS
6 27 33	-54 21.9		O	14930		-270	526	
6 28 26	-63 27.5		O	13970		-267	896	
6 28 47	-54 24.6		O	14810		-270	526	
6 29 2	-57 50.8		O	10010	120	-270	680	
6 30 30	21 4.0		R	5454	15	-62	752	
6 32 0	67 54.0	IC 445 KDG 114A	O	5210	102	148	908	
6 32 14	-67 51.3		O	4020		-265	896	
6 32 24	67 53.0	KDG 114B	O	12043	105	148	908	
6 32 29	26 19.1	4C 26.23	O O	12510 12060		-39	244 400	
6 34 23	-20 32.0	PKS	O O	16800 16800		-224	691 189	
6 34 47	49 18.1		R	451	10	67	860	
6 35 24	75 40.2	MRK 5	R O	790 870	20	177	368 192	
6 35 36	60 7.0		R	2105	10	115	752	
6 36 42	53 17.0		O	10275	100	85	521	
6 37 6	53 20.0		O	11130	100	85	521	
6 38 42	65 15.0		R	3552	20	136	752	
6 38 48	34 31.0	KDG 115	O	5029	70	-3	908	
6 38 54	34 30.0	KDG 115	O	5082	90	-3	908	
6 39 55	-66 0.3		O	12080		-268	896	
6 39 57	-58 28.6		O	2669	30	-273	613	
6 39 57	-58 23.0		O	2744	18	-273	613	
6 40 0	78 5.0	MRK 1195	O	11625		185	920	
6 41 30	57 57.0	7 ZW 93	O	12300	200	104	820	
6 41 44	86 38.4	KDG 116	O O	4287 4792	15	213	748 908	

6ʰ42ᵐ

R.A. (1950)	DEC. (1950)	NAME	OBS	HEL VEL (C*Z)	ERR	GAL CORR	REF	COMMENTS
6 42 3	60 23.7	NGC 2273B	R	2112		115	693	
6 42 9	86 37.5	KDG 116	O	4487		213	748	
			O	4738	30		908	
6 42 25	21 25.0	3CR166	O	73400		-67	946	
6 42 30	43 50.0	KDG 117	O	6209	70	39	908	
6 42 42	43 52.0	KDG 117	O	6163	20	39	908	
6 42 50	-27 35.2	NGC 2280	R	1900	20	-245	544	
			R	1902	10		238	
			O	2041	8		741	
			O	1919	71		789	
6 43 30	-74 14.0		O	6034		-258	442	
6 43 30	-74 13.0		O	6284		-258	442	
6 43 31	-74 15.0		O	5984		-258	442	
6 43 35	-74 16.0		O	5714		-258	442	
6 43 53	-63 39.7	NGC 2297	O	3450		-271	896	
6 44 0	33 38.0	NGC 2274 KDG 118A	O	4873	35	-10	812	
6 44 0	33 40.0	NGC 2275 KDG 118B	O	4927	20	-10	812	
6 44 9	-72 32.5		O	33100		-261	896	
6 45 12	-76 30.4		O	(3900)		-254	675	
6 45 36	60 54.0	NGC 2273 MRK 620	O	1950		116	438	
6 45 43	74 29.1	IC 450 MRK 6	O	5610	999	171	351	SEYF
			O	5400			192	
			O	5760			264	
6 46 3	-65 35.3		O	10700		-270	896	
6 46 8	-65 35.8		O	10370		-270	896	
6 46 48	25 41.0	KDG 119A	O	5030	39	-49	812	
6 46 54	25 41.0	KDG 119B	O	(5107)	50	-49	812	
6 47 25	-68 30.5		O	8960		-267	896	

R.A. (1950)	DEC. (1950)	NAME	OBS	HEL VEL (C★Z)	ERR	GAL CORR	REF	COMMENTS
6 47 35	45 13.0	3CR169.1	O O	189770 (189800)		43	850 946	
6 47 54	33 35.0	NGC 2294	O	(5090)	71	-12	789	
6 48 12	26 49.0		O	4740	100	-44	521	
6 48 12	57 14.0	KDG 120	O	5110	43	99	908	
6 48 21	-64 13.2	NGC 2305	O	3460		-272	574	
6 48 54	27 31.3	B2	O	12310		-41	485	
6 48 54	57 15.0	KDG 120	R O	1442 1418	10 30	99	752 908	
6 49 24	15 18.0	KDG 121	O	4751	30	-98	908	
6 49 30	15 19.0	KDG 121	O	4565	70	-98	908	
6 50 0	69 39.0		O	1283	13	152	813	
6 50 0	80 4.0		R R O	4962 4948 4986	15 20 30	191	582 752 582	
6 50 43	50 25.0	MRK 373	O	6000		67	307	
6 51 11	54 12.7	3CR171	O	71530		85	120	
6 52 8	39 49.8		R	1340	30	16	829	
6 52 12	39 9.3		R	398	10	13	860	
6 52 24	-40 48.0	NGC 2310	O O	(840) 1217	310 50	-270	555 741	
6 52 34	-64 51.5		O	10190		-272	896	
6 54 40	63 21.0		O	28600		125	401	
6 55 34	54 15.9	MRK 374	O O	13200 12955	180	83	307 562	SEYF
6 55 44	-62 46.8		O	9020		-274	896	
6 55 55	-70 41.5		O	(10760)		-265	896	
6 56 6	-62 29.8		O	8990		-275	896	
6 56 14	14 22.0		R	2336	10	-106	860	
6 56 32	-62 45.3		O	9470		-275	896	

6ʰ58ᵐ

R.A. (1950)	DEC. (1950)	NAME	OBS	HEL VEL (C*Z)	ERR	GAL CORR	REF	COMMENTS
6 58 18	63 20.0		O	28600		124	215	
6 58 24	1 59.0		R	1778	20	-160	752	
6 58 27	23 17.8	4C 23.18	O	27570		-65	400	
			O	27190			318	
6 59 35	64 5.7		R	4497	10	127	829	
6 59 36	39 18.0	MRK 1196	O	6499		11	920	
7 0 39	56 35.7	DDO 40	R	1382	7	93	492	
7 0 42	-28 37.4	NGC 2325	O	2014	75	-255	555	
			O	2248	33		741	
7 0 48	84 27.8	NGC 2268	R	2221		206	693	
			R	2231			744	
			O	2337			13	
7 1 46	17 39.3		R	1181	10	-93	860	
7 2 18	64 42.0	7 ZW118	O	23670	200	129	820	
7 2 22	67 45.6	MRK 375	O	3600		142	307	
7 2 48	74 54.3	3CR173.1	O	87540	300	171	585	
7 2 54	28 23.0	MRK 1197	O	4783		-43	920	
7 3 36	33 23.3	B2	O	26470		-19	595	
7 3 48	75 19.0	NGC 2314	O	3843	30	172	13	
			O	3951			13	
7 4 4	52 58.9		R	602	10	75	860	
7 4 18	50 45.0	NGC 2326	R	5985	15	64	752	
7 4 20	35 8.6	4C 35.16	O	23690	50	-11	13	
7 4 30	61 41.0		P	1796	5	115	752	
			R	1800	10		829	
7 4 48	31 45.0	KDG 123A	O	4300	95	-27	908	
			O	(4947)	84		812	
7 4 48	31 45.0	KDG 123B	O	4084	57	-27	908	
			O	4198	46		812	
7 5 0	35 4.0		O	23089	60	-11	13	
7 5 6	15 15.0		R	2202	15	-106	752	
			O	2182	15		936	

R.A. (1950)	DEC. (1950)	NAME	OBS	HEL VEL (C*Z)	ERR	GAL. CORR	REF	COMMENTS
7 5 6	53 31.6	DDO 41	R	3145	10	77	492	
7 5 19	48 41.9	NGC 2329	O	5760		54	255	
7 5 25	18 51.7	NGC 2339	R	2429	20	-89	544	
			R	2209			610	
			R	2261			744	
			R	2239			780	
			O	2361	40		13	
7 5 30	71 55.0		R	3120	30	159	481	GB 1
			R	3141	20		829	
			O	3170			144	
			O	3226	45		148	
7 5 41	44 27.7		R	426	10	34	860	
7 6 6	47 59.0	KDG 124	O	6092	50	51	908	
			O	6077	15		812	
7 6 6	48 0.0	KDG 124	O	6004	50	51	908	
			O	6227	26		812	
7 6 18	20 40.0	NGC 2341	O	5013	60	-81	567	
		KDG 125	O	5235	18		908	
			O	5581			748	
			O	5219			728	
7 6 24	20 43.0	NGC 2342	O	5581		-81	748	
		KDG 125	O	5132	10		908	
			O	5241	40		567	
			O	5256			728	
7 8 0	26 0.0	MRK 1198	O	7467		-57	920	
7 8 12	73 34.0	A 708A/B	R	2709		165	744	
		ARP 141A/B	O	2516	50		141	
		VV 123	O	2611	100		288	
7 8 12	73 34.0	ARP 141A	O	2735	141	165	789	
			O	2831			75	
7 8 12	73 34.0	ARP 141B	O	2728	100	165	789	
7 8 30	47 17.0	NGC 2344	R	962	25	47	238	
			R	967	50		544	
			O	914	38		148	
7 8 34	32 23.6	B2	O	20170	250	-26	696	
7 8 36	67 16.0	7 ZW129	O	4200	200	139	820	
7 9 12	28 40.0	KDG 126	O	4817	19	-44	908	

R.A. (1950)	DEC. (1950)	NAME	OBS	HEL VEL (C*Z)	ERR	GAL CORR	REF	COMMENTS
7 9 18	28 41.0	KDG 126	O	4862	21	-44	908	
7 9 44	75 49.7		R	1121	10	174	860	
7 10 31	85 50.9	NGC 2276	R	2418		210	693	
		KDG 127	R	2419	15		829	
		7 ZW134	R	2400	35		455	
		ARP 25	R	2410	10		622	
			O	2388	32		908	
			O	2391			13	
7 10 36	45 47.1	MRK 376	O	16800		39	306	SEYF
7 10 42	65 2.0	KDG 128A	O	4419	88	129	812	
7 11 16	64 47.9	NGC 2347	R	4424	10	128	829	
			O	4521			13	
7 11 18	64 49.0	KDG 128B	O	4471	53	128	812	
7 11 54	56 54.0	KDG 129	O	3228	51	92	908	
7 12 0	56 54.0	KDG 129	O	3042	31	92	908	
7 12 42	53 28.5	4C 53.16	O	19300		75	868	
7 13 5	12 11.9		R	2129	10	-123	860	
7 13 19	49 47.0	MRK 378	O	12563		57	437	
7 14 30	41 5.0	KDG 130A	O	6872	14	14	812	
7 14 36	41 5.0	KDG 130B	O	6869	10	14	812	
7 14 49	28 40.5	4C 28.18	O	24900		-46	400	
7 15 0	-57 15.0		O	1130		-282	719	
			O	(1412)			680	
7 15 18	-52 0.0		O	2505	20	-284	599	
7 15 24	77 53.9		R	2648	10	181	860	
7 16 0	-62 16.0	NGC 2369	O	3306	40	-279	373	
			O	3237	36		741	
			O	3300			574	
7 16 0	7 55.0	KDG 131A	O	5404		-143	728	
7 16 0	7 55.0	KDG 131B	O	5909		-143	728	
7 16 13	55 58.70		O	11710	141	86	651	

R.A. (1950)	DEC. (1950)	NAME	OBS	HEL VEL (C*Z)	ERR	GAL CORR	REF	COMMENTS
7 16 30	55 51.90		O	12737	158	86	651	
7 16 30	85 50.0	NGC 2300	O	2088		209	13	
		KDG 127	O	1946	30		13	
			O	2123	23		908	
7 16 48	55 48.90		O	13855	134	85	651	
7 16 52	55 49.35		O	11170	78	85	651	
7 16 57	55 45.55		O	11506	146	85	651	
7 17 0	71 42.0		R	2878	15	156	752	
7 17 15	55 53.60		O	11940	243	86	651	
7 17 15	55 54.35		O	11267	109	86	651	
7 17 24	55 48.37		O	11327	134	85	651	
7 17 24	55 51.40		O	11452	110	85	651	
7 17 26	55 51.14		O	12743	89	85	651	
7 17 29	55 59.15		O	13265	5	86	651	
7 17 35	55 47.15		O	19871		85	651	
7 17 39	55 47.60		O	9743	30	85	651	
7 17 59	55 45.90		O	10200	111	85	651	
7 18 1	55 58.24		O	12027	312	86	651	
			O	12020			255	
7 18 4	55 45.60		O	12382	20	85	651	
7 18 6	55 51.85		O	11433	104	85	651	
7 18 28	80 16.6	NGC 2336	R	2204	50	190	544	
			R	2202			693	
			R	2204			934	
			R	2208			744	
			R	2216	35		238	
			O	2252			13	
			O	1910			748	
7 18 56	-34 1.4	PKS	O	8900		-271	317	
7 19 6	65 10.2	4C 65.08	O	65500		128	864	
7 19 7	45 12.0		R	439	10	33	860	

R.A. (1950)	DEC. (1950)	NAME	OBS	HEL VEL (C*Z)	ERR	GAL CORR	REF	COMMENTS
7 19 12	-55 19.6	PKS	O	64750		-284	851	
7 19 17	-62 58.5	NGC 2381	O	3060		-279	896	
7 19 38	-57 58.1		O	1123	15	-283	680	
7 20 0	61 47.0		O	1727	30	112	813	
7 20 2	32 35.6	IC 2185	O	4529		-29	558	
7 20 30	33 31.0	MRK 1199	O	3894		-24	920	
			O	4094	100		556	
7 21 12	49 35.0		R	5949	5	54	752	
7 21 30	-68 54.0	NGC 2397	O	1363	48	-272	373	
			O	1307	50		741	
7 21 54	27 26.0	MRK 1200	O	7615		-55	920	
7 21 55	79 58.5	IC 467	O	2025	15	188	936	
			O	2247			748	
7 22 19	72 40.4	MRK 7	R	3090	75	160	368	
		7 ZW153	O	3045	60		261	
			O	3163			480	
			O	3100			280	
7 22 28	30 3.3	B2	O	5770		-42	485	
7 22 33	- 9 33.6	NGC 2377	R	2460		-213	705	*
		3C 178	O	2370			143	
7 22 36	30 4.0	MRK 1201	O	5412		-42	920	
7 23 0	47 12.0		R	3032	10	42	752	
7 23 18	- 0 48.9	PKS	O	38400		-181	600	
			O	38100			672	
7 23 22	33 55.5	NGC 2373	O	7523		-23	558	
7 23 37	69 19.1	NGC 2366	R	90	50	145	216	*
		MRK 71	R	110			171	
		DDO 42	R	102	5		492	
			R	100			744	
			R	100			693	
			R	96	4		489	
			R	107			934	
			O	194			13	
			O	145	18		148	
			O	93			483	
			O	122	2		650	HII REG

R.A. (1950)	DEC. (1950)	NAME	ORS	HEL VEL (C*Z)	ERR	GAL CORR	REF	COMMENTS
7 23 38	72 13.8	IC 2184A/B	R	3570		158	480	
		7 ZW156	O	3577	95		429	
		MRK 8	O	3630			480	
		VV 644	O	3494	50		873	
			O	3420			192	
			O	3240	60		386	
			O	3546	30		963	
7 23 38	72 13.8	KDG 135A	O	3617	20	158	908	
7 23 38	72 13.8	KDG 135B	O	3494	16	158	908	
7 23 42	69 10.0	NGC 2363	O	17	20	145	555	
7 24 12	33 55.0	NGC 2379	O	4030	65	-24	13	
7 24 24	19 44.0	KDG 136A	O	9898	90	-93	812	
7 24 28	-63 26.0		O	10190	.	-279	896	
7 24 30	19 43.0	KDG 136B	O	9882	110	-93	812	
7 24 50	40 52.2	DDO 43	R	355	7	10	492	
7 24 54	72 37.0	VV 141N	O	2592	50	159	873	
7 24 54	72 37.0	VV 141S	O	2546	50	159	873	
7 25 6	49 14.0		O	5887	53	51	813	
7 25 31	35 39.1		O	3915		-15	558	
7 25 38	56 57.2	MRK 72	O	13146		89	279	
			O	13096			341	
7 25 48	33 58.0	NGC 2389	R	3998		-24	744	★
			O	3816			13	
7 26 54	-75 5.0		O	4441	25	-262	599	
7 27 6	55 21.6	MRK 74	O	11062		81	279	
			O	10963			341	
7 27 7	63 20.9	MRK 73	O	4100		118	231	
			O	4400			233	
			O	4412			341	
7 27 42	-62 14.0	IC 2200/A	O	3089	73	-281	724	
7 27 42	-62 14.0	IC 2200	O	3160	100	-281	844	
			O	3170	28		613	
7 27 31	-62 15.5	IC 2200A	O	3280	25	-281	613	

R.A. (1950)	DEC. (1950)	NAME	OBS	HEL VEL (C★Z)	ERR	GAL CORR	REF	COMMENTS
7 28 29	55 18.2	MRK 75	O	8893		80	341	
			O	8919			279	
7 28 36	60 59.0	7 ZW162	O	9031	200	107	429	
7 28 49	0 9.9		R	1457	10	-180	860	
7 29 9	31 44.4		O	51600	300	-36	583	
			O	50999			444	
7 29 30	-62 9.0	NGC 2417	O	3179	20	-282	582	
7 30 24	74 34.0	VV 539N	O	3900	50	167	776	
7 30 24	74 34.0	VV 539S	O	3990		167	776	
7 30 24	74 37.0	NGC 2426	O	3708	27	167	812	
		KDG 137	O	3751	10		908	
7 30 30	74 37.0	NGC 2429	O	3811	20	167	812	
		KDG 137	O	3760	10		908	
7 32 0	65 43.0	NGC 2403I	O	205		128	650	HII REG
7 32 0	65 43.0	NGC 2403II	O	121		128	650	HII REG
7 32 0	65 43.0	NGC 2403III	O	89		128	650	HII REG
7 32 0	65 43.0	NGC 2403IV	O	153		128	650	HII REG
7 32 0	65 43.0	NGC 2403	R	128		128	121	
			R	140			934	
			R	128			340	
			R	137	5		74	
			R	138	5		275	
			R	137	5		93	
			R	128			156	
			R	128	8		404	
			O	125			243	
			O	70	40		13	
			O	133	10		218	
7 32 42	58 53.0	MRK 9	O	11690		96	192	SEYF
			O	11960			264	
7 33 21	-77 49.5		O	1380		-256	896	
7 33 39	35 21.3	NGC 2415	R	3786		-19	366	
			O	3619			558	
			R	3782	20		829	
			O	3808			163	
			O	3822	31		72	

R.A. (1950)	DEC. (1950)	NAME	OBS	HEL VEL (C*Z)	ERR	GAL CORR	REF	COMMENTS
7 34 0	-49 56.0		O	3003	30	-288	599	
7 34 6	42 3.0		R	5904	15	13	582	
			O	5856	100		582	
7 34 19	35 43.2		R	3990	10	-18	829	
7 34 24	71 14.8		R	2462	10	152	860	
7 34 28	80 33.5	3CR184.1	O	35320		190	657	
			O	35410	150		846	
			O	35410			318	
7 34 57	-66 14.8		O	7790		-278	896	
7 35 0	-69 10.0	NGC 2434	O	1453		-273	154	
			O	1327	43		741	
			O	1440			574	
7 35 1	-47 31.4	NGC 2427	R	967		-288	671	
			R	1000	50		666	
			O	(31)	23		741	DIS
7 35 50	49 28.3		O	6250		50	558	
7 36 0	48 51.0		R	6382	15	47	582	
			O	6358	25		582	
7 36 32	-69 25.0	NGC 2442	O	1429	9	-273	741	
			O	712			145	
7 37 16	39 21.0	NGC 2424	O	3307		0	910	
7 37 56	65 17.7	MRK 78	O	11156		125	341	SEYF
			O	10926	80		300	
			O	11300			233	
7 38 0	73 54.0		O	5100		163	768	*
7 38 1	40 13.8	DDO 46	R	364	10	3	492	
7 38 11	33 40.4	B2	O	109030		-29	899	
7 38 47	49 55.8	MRK 79	R	6643		52	707	SEYF
			O	6420			825	
			O	6500			231	
			O	6540			287	
			O	6637	80		300	
			O	6597			341	
7 39 3	16 55.1	DDO 47	R	265	10	-111	492	
7 39 24	52 27.0	NGC 2426 KDG 138	O	5858	35	64	908	

R.A. (1950)	DEC. (1950)	NAME	OBS	HEL VEL (C*Z)	ERR	GAL CORR	REF	COMMENTS
7 39 42	11 22.0	KDG 139	O	8770	21	-137	908	
7 39 42	67 23.0		O	4032	30	135	813	
7 39 54	9 29.9		O	18760		-145	255	
7 39 54	11 23.0	KDG 139	O	8801	33	-137	908	
7 39 54	52 29.0	NGC 2429 KDG 138	O	5451	60	64	908	
7 39 54	70 10.0		R O O	3882 3894 3926	10 30 42	147	582 582 813	
7 41 0	29 21.0	KDG 140A	O	4790	15	-51	908	
7 41 6	29 21.0	KDG 140B	O O	4672 4843	50 15	-52	567 908	
7 41 50	85 3.3		R	4365	25	206	829	
7 42 24	28 34.0	MRK 1202	O	4586		-56	920	
7 42 28	2 7.7	3CR187	O O	104930 104900		-177	851 946	
7 42 30	8 3.0	KDG 141	O O	4960 4976	15	-152	908 728	
7 42 38	62 30.3	MRK 82	O O	5637 5200		112	341 233	
7 42 48	8 3.0	KDG 141	O O	4896 4771	20	-152	908 728	
7 43 7	61 3.4	MRK 10	R O	8753 8690		105	707 192	SEYF
7 43 18	55 4.0	VV 654	O	5516	50	76	873	
7 43 31	39 8.4	C	O	4188		-3	19	* HII REG
7 43 31	39 9.4	NGC 2444 ARP 143 VV 117	O O	3649 3965	100	-3	789 19	
7 43 32	39 8.0	F	O	4257		-3	19	HII REG
7 43 32	39 8.4	NGC 2445 ARP 143 VV 117	R O	4020 3858	38	-3	744 789	

R.A. (1950)	DEC. (1950)	NAME	OBS	HEL VEL (C*Z)	ERR	GAL CORR	REF	COMMENTS
7 43 33	39 8.0	G	O	4227		-3	19	HII REG
7 43 33	39 8.4	D	O	4032		-3	19	HII REG
7 43 36	39 8.4	E	O	3970		-3	19	HII REG
7 43 42	59 8.0		R O	6494 6508	10 40	96	582 582	
7 44 0	28 4.0		R O	8260 8209	20 20	-59	582 582	
7 44 0	30 36.8	IC 475	O	4246		-46	558	
7 44 13	54 20.2	MRK 83	O O	14440 14600		72	341 233	
7 44 36	55 57.0	4C 56.16	O	10670		80	459	*
7 44 41	74 29.1	MRK 12	R R O O O	4130 3943 4003 4100 4025	75 15 60	165	368 829 261 280 480	
7 44 48	28 27.0	MRK 1203	O	7257		-57	920	
7 45 6	28 20.0	KDG 144A	O O	8450 8268	110 30	-58	567 908	
7 45 6	57 11.0	7 ZW181	O	13170	200	86	820	
7 45 9	-67 40.8		O	5100		-277	896	
7 45 12	28 19.0	KDG 144B	O O	8075 8165	140 15	-58	567 908	
7 45 18	-19 10.2	PKS	O	30820	150	-249	709	
7 45 36	-71 17.0	NGC 2466	O O	5750 5171	100 43	-271	844 373	
7 46 0	56 3.0	4C 56.16	O O O	10590 10500 10623	23	80	166 482 687	*
7 46 1	72 54.6		O	67600	300	158	529	
7 46 36	29 4.0	MRK 1204	O	4494		-54	920	
7 46 54	34 33.0		O	8520	100	-27	521	
7 47 0	30 51.0		R R O	4246 4308 4363	15 135	-46	779 582 582	

R.A. (1950)	DEC. (1950)	NAME	OBS	HEL VEL (C*Z)	ERR	GAL CORR	REF	COMMENTS
7 47 6	73 6.0	NGC 2441	O	3623		159	13	
7 47 36	50 21.0	KDG 145A	O	6639	62	52	812	
7 47 36	50 21.0	KDG 145B	O	6946	56	52	812	
7 47 54	24 1.0		O	2156	13	-80	813	
7 48 48	63 28.0	7 ZW185	O	11300	200	115	820	
7 49 6	78 8.8		R	2301	10	180	829	
7 51 5	55 50.1	MRK 84	O	6055		79	341	
			O	6200			233	
7 51 36	16 56.0	KDG 146A	O	13891	10	-115	908	
7 51 36	16 56.0	KDG 146B	O	13652	54	-115	908	
7 51 57	60 26.0	IC 2209	O	1560		101	192	
		MRK 13	O	1361	47		789	
7 52 3	39 19.1	MRK 382	O	10200		-4	306	SEYF
7 52 4	66 44.6		R	4080	15	130	829	
7 52 33	66 34.3		R	4913	10	129	829	
7 52 36	60 30.0	NGC 2460	R	1452		101	693	
			R	1450	10		829	
			R	1450	35		775	
			O	1442	50		13	
7 52 48	61 47.7		O	8293		107	558	
7 53 0	16 42.0	MRK 1205	O	13786		-116	920	
7 53 9	56 48.6	NGC 2463	O	8317		83	558	
7 53 18	49 42.0		O	3539	36	48	813	
7 54 0	73 12.0		O	4200		159	768	*
7 54 0	56 48.9	NGC 2469	O	3217		83	558	
7 54 8	52 59.6	NGC 2474/5	O	5019		64	13	
		KDG 147						
7 54 8	52 59.6	NGC 2474	O	5711		64	910	
			O	5527	108		908	
7 54 10	52 59.5	NGC 2475	O	5632	110	64	908	

R.A. (1950)	DEC. (1950)	NAME	OBS	HEL VEL (C*Z)	ERR	GAL CORR	REF	COMMENTS
7 54 24	18 48.0	MRK 1206	O	14137		-107	920	
7 54 42	59 7.0	7 ZW195	O	5700	200	94	820	
7 54 48	58 10.6	DDO 48	R	1090	10	89	492	
			R	1104	20		424	
7 54 54	25 17.0	NGC 2486 KDG 150A	O	4678	39	-75	812	
7 55 9	37 55.2	NGC 2484 4C 37.21	O	12990		-12	240	
			O	12390			485	
7 55 18	25 16.0	NGC 2487 KDG 150B	R	4826	15	-76	829	
			O	(4841)	97		812	
			O	4768	71		789	
			O	(4350)			571	
			O	5011			777	
7 55 54	60 25.0		R	5992	15	100	752	
7 56 30	16 33.0		R	4886	10	-118	582	
			O	4848	20		582	
7 57 11	39 58.5	NGC 2495 MRK 383	O	8400		-2	306	
7 57 17	77 57.4		R	2286	15	178	829	
7 58 8	50 52.6	NGC 2500	R	530	50	53	216	
			R	517			744	
			R	515			693	
			O	470			13	
7 58 18	61 31.0		R	1591	10	105	752	
7 58 21	-77 13.3	PKS	O	51560		-259	851	
8 0 0	15 57.0	NGC 2514	R	4856		-122	779	
			R	4854	15		582	
			O	4913	50		582	
			O	4871	71		789	
8 0 8	23 32.0	NGC 2512 MRK 384	R	4685		-85	779	
			O	4647	53		789	
			O	4800			306	
8 0 16	24 49.1	B2	O	13070		-79	485	
8 0 17	33 36.2		O	11735		-35	558	
8 0 18	73 28.0	KDG 152	O	5182	20	159	908	

121

R.A. (1950)	DEC. (1950)	NAME	OBS	HEL VEL (C*Z)	ERR	GAL CORR	REF	COMMENTS
8 0 27	25 14.6	MRK 385	O	8100		-77	306	
8 1 0	73 29.0	KDG 152	O	3590	20	159	908	
8 1 12	8 50.0	MRK 1208	O	4834		-154	920	
8 1 12	40 21.0		O	12202	75	-1	582	
8 1 18	10 9.0	MRK 1209	O	10008		-148	920	
8 1 30	5 15.0	MRK 1210	O	3890		-170	920	
8 2 36	24 18.5	3CR192	O	18000		-82	318	
			O	17970	39		134	
			O	17947	39		741	
8 2 43	- 1 2.6	PKS	O	26320		-195	511	
8 3 0	7 44.0	MRK 1211	O	15959		-159	920	
8 3 18	-11 17.0	NGC 2525	R	1569	20	-231	544	
			R	1585			744	
			O	2064			13	
8 3 24	12 41.0	IC 2226	O	10906	70	-138	582	
8 4 0	27 16.0	MRK 1212	O	12238		-68	920	
8 4 18	8 9.0	NGC 2526	O	4634		-158	728	
		KDG 154	O	4726	30		908	
8 4 21	39 9.0	MRK 622	O	7075		-7	438	SEYF
			O	6985			557	
			O	6901			874	
8 4 24	8 11.0	IC 2228	O	4624		-158	728	
		KDG 154	O	4654	32		908	
8 4 54	4 39.0		O	9125	100	-173	521	
8 5 0	76 35.0		R	1547	10	172	752	
8 5 10	30 55.5	4C 30.14	O	90200		-50	899	
8 5 22	72 57.0	MRK 14	O	3150		157	192	
8 6 38	42 36.9	3CR194	O	(93500)		9	946	
8 6 42	26 1.0	IC 495 KDG 155A	O	7535	72	-75	812	
8 6 42	26 1.0	IC 2229 KDG 155B	O	7396	50	-75	567	
			O	7493	28		812	

R.A. (1950)	DEC. (1950)	NAME	OBS	HEL VEL (C*Z)	ERR	GAL CORR	REF	COMMENTS
8 7 3	34 6.0	NGC 2532	R	5278	50	-34	544	
			R	5269	14		914	
			O	5153	50		13	
8 7 35	46 36.8	DDO 49	R	2254	10	29	492	
8 8 6	58 43.0	7 ZW217	O	7930	100	90	820	
8 8 13	25 21.3	NGC 2535/6 ARP 82	R	4410	25	-78	455	
8 8 13	25 21.3	NGC 2535 VV 9 KDG 156 HOL 94A ARP 82	R R O O O O O O O	4095 (4109) 4101 4073 4140 4080 4056 4243 4050	20 30 30 25 75 	-78	544 283 728 57 195 195 283 13 253	
8 8 17	25 19.7	NGC 2536	O	4170	30	-78	195	KNOT
8 8 18	25 19.8	NGC 2536 VV 9 KDG 156 HOL 94B ARP 82	O O O O	4046 4200 4149 4072	 60	-78	253 195 728 57	
8 8 19	25 19.8	NGC 2536	O	4080	30	-78	195	KNOT
8 8 57	55 49.4	NGC 2534	O	3517		76	341	
8 9 12	73 45.0	NGC 2523 ARP 9	O	3448		160	13	
8 9 42	36 24.4	NGC 2543 IC 2232 KDG 157	R O	2471 2415	9 27	-22	914 908	
8 9 42	46 9.0	NGC 2537 MRK 86 ARP 6 VV 138	R R O O O O	443 450 290 518 441 397	 29 20	27	693 744 13 72 279 13	KNOT
8 10 6	52 48.0		O	5654	59	60	813	
8 10 30	45 54.0	IC 2233	O	565	47	25	789	
8 10 42	7 43.2	4C 07.22	O	33880		-161	228	

R.A. (1950)	DEC. (1950)	NAME	OBS	HEL VEL (C*Z)	ERR	GAL CORR	REF	COMMENTS
8 11 2	49 12.9	NGC 2541	R	561		42	744	
			R	563			934	
			R	554			693	
			R	570	20		216	
			O	640	135		555	
			O	601	34		148	
8 11 18	18 36.0	KDG 159	O	5241	18	-112	908	
8 11 19	21 30.5	NGC 2545	R	3260	45	-98	885	
			R	(3381)	9		581	
			O	3192			321	
			O	3531	19		741	
8 11 24	18 36.0	KDG 159	O	5069	18	-112	908	
8 11 38	58 28.3		O	7880		89	255	
8 12 56	64 42.0		O	10965		119	714	
8 12 57	- 2 59.2	3CR196.1	O	59360	300	-205	585	
8 13 0	20 37.0		O	12800		-102	449	
			O	12800			229	
8 13 12	26 8.0	MRK 623	O	12650		-75	438	
8 13 43	70 52.3	A 814	R	165		147	171	HOL II
		DDO 50	R	158	10		492	
		ARP 268	R	155	5		552	
			R	159	12		93	
			R	160			934	
			R	156	14		489	
			R	150			156	
			R	161	13		216	
			R	158			693	
			O	220			13	
8 13 45	12 0.8		O	10500		-143	727	
8 14 30	21 50.0		R	3631	20	-97	939	
			O	3444	63		789	
			O	3617			777	
8 15 0	23 38.0	NGC 2554	O	4163		-88	321	
8 15 0	57 58.0	NGC 2549	O	1082	75	86	13	
8 15 4	68 4.1		O	(87100)	300	134	583	
8 15 19	67 35.6		O	7989		132	714	
8 15 9	-27 19.0	NGC 2559	O	1525		-275	728	

R.A. (1950)	DEC. (1950)	NAME	OBS	HEL VEL (C*Z)	ERR	GAL CORR	REF	COMMENTS
8 15 24	65 29.5		O	7453		122	714	
8 15 36	67 8.3		O	3943		130	714	
8 15 41	50 10.0	NGC 2552	R	517		46	693	
			R	511			363	
			O	780	210		555	
			O	508	25		741	
			O	388	43		789	
8 15 42	20 55.0		O	4801	100	-102	288	
8 15 42	68 47.0		O	4727	11	137	813	
8 15 55	74 8.9	NGC 2544	O	2851		161	278	*
		KDG 160A	O	2825	17		908	
		MRK 87	O	2736			341	
			O	2867	32		812	
8 16 10	32 12.3		O	5145		-45	558	
8 16 12	21 20.0		R	4092	20	-100	939	
8 16 12	74 9.0	KDG 160B	O	3599	22	161	908	
			O	3575	32		812	
8 16 18	21 36.0	NGC 2557	O	4944	100	-98	288	
8 16 24	20 40.0	NGC 2558	R	4979	20	-103	939	
8 16 24	21 13.0		R	4852	20	-100	939	
			O	4719			409	
8 16 30	21 9.7		O	3727		-100	409	
8 16 42	21 16.3		O	5505		-100	409	
8 16 48	-71 42.2		O	1458	20	-273	680	
8 16 48	21 11.7		O	28276		-100	409	
8 16 54	22 11.0	NGC 2565	R	3610		-96	779	
		MRK 386	R	3583	20		939	
			O	3422			321	
			O	3599			409	
8 17 0	21 8.0	NGC 2560	O	4903	100	-101	288	
8 17 6	21 13.5		O	4273		-100	409	
8 17 17	66 41.4		O	11091		128	714	
8 17 18	21 13.7		O	5180		-100	409	

8ʰ17ᵐ

R.A. (1950)	DEC. (1950)	NAME	OBS	HEL VEL (C∗Z)	ERR	GAL CORR	REF	COMMENTS
8 17 24	21 2.0		R	5030	20	-101	939	
8 17 29	21 17.5	NGC 2562	O	4963	50	-100	13	
			O	5100	100		2	
8 17 36	63 7.1		O	6229		111	714	
8 17 42	21 13.6	NGC 2563	O	4775	50	-100	13	
			O	4391			409	
8 17 42	21 14.3		O	4673		-100	409	
8 17 54	21 12.7		O	4542		-101	409	
8 18 1	47 12.2	3CR197.1	O	39000		31	240	
			O	38400			465	
8 18 7	67 7.8		O	4304		129	714	
8 18 30	21 5.0	NGC 2570	R	6541	20	-101	939	
8 18 48	73 35.0	NGC 2551	O	2296		159	13	
8 18 51	3 19.7	IC 2327	R	2685	10	-181	829	
			R	2683	10		752	
8 19 15	64 30.5		O	11287		117	714	
8 19 15	74 35.7	DDO 51	R	3523	20	163	492	
8 19 33	3 25.7	IC 503	R	4131	15	-181	829	
8 19 48	22 43.0		O	2123	100	-93	288	
8 19 52	6 6.7	3CR198	O	24450		-170	120	
8 19 52	65 46.0		O	13827		123	714	
8 19 53	54 18.7		O	92000		67	717	
8 20 12	22 49.0		R	4374	20	-93	939	
8 20 12	67 1.9		O	4217		129	714	
8 20 42	21 29.0	IC 2338/9 ARP 247	R	5400	50	-100	455	
8 20 42	21 29.0	IC 2338	O	5418	19	-100	908	
		KDG 161A	O	5230	100		288	
8 20 48	21 30.0	IC 2339	R	5348	30	-100	939	
		KDG 161B	O	5194	100		288	
			O	5391	20		908	

R.A. (1950)	DEC. (1950)	NAME	OBS	HEL VEL (C*Z)	EPR	GAL CORR	REF	COMMENTS
8 20 48	21 38.0		O	4846		-99	321	*
8 20 56	- 4 49.3		O	6900	150	-213	795	
8 20 57	- 4 48.0		O	6540	150	-213	795	
8 21 0	- 4 51.0	NGC 2583	O	5910	150	-213	795	
8 21 0	11 43.0	KDG 162	O	(9502)	76	-146	908	
8 21 3	- 4 49.0		O	6780	150	-213	795	
8 21 6	- 4 50.0		O	(7200)	150	-213	795	
8 21 6	21 12.0		R	4649	20	-101	939	
8 21 12	25 51.0	MRK 624	O	8455		-78	438	
8 21 18	- 4 47.0		O	(7200)	150	-213	795	
8 21 18	11 40.0	KDG 162	O	9423	20	-146	908	
8 21 24	17 29.7	MRK 387	O	11070		-119	436	
8 21 54	56 10.6		O	57860		76	726	
8 21 57	56 8.1		O	58190		76	726	
8 22 12	47 17.0		O	38658		31	444	
			O	38964			444	
			O	38963			451	
			O	37300	300		583	
8 22 21	67 56.9		O	115100		133	717	
8 22 24	20 30.0	NGC 2582	O	4514	75	-105	582	
			O	4383			321	
8 22 28	56 6.4		O	13040		76	726	
8 22 28	65 24.6		O	14008		121	714	
8 22 30	20 41.0	NGC 2590	R	4998	25	-104	752	
			O	4990	25		740	
8 22 36	46 8.0		O	2125	23	25	813	
8 22 54	19 36.0	IC 2363	O	7616		-109	777	
8 23 0	73 55.0		R	3641	15	160	752	
8 23 12	21 37.0		R	5344	20	-99	872	
			O	(4500)			54	DIS

R.A. (1950)	DEC. (1950)	NAME	OBS	HEL VEL (C*Z)	ERR	GAL CORR	REF	COMMENTS
8 23 16	64 23.6		O	11261		116	714	
8 23 31	63 26.6		O	6825		112	714	
8 23 48	68 38.0	ARP 300 VV 106	O	3800		136	535	
8 23 54	68 39.0	ARP 300 VV 106	O	3900		136	535	
8 24 0	20 32.0	IC 2373	R	7507	20	-105	939	
8 24 15	55 15.9		O	11329		71	558	
8 24 18	55 52.0	MRK 88 1 ZW 14	O	9200		74	233	
			O	9175	100		521	
			O	9160			262	
			O	9138			341	
8 24 21	29 28.7	3CR200	O	137300		-60	946	
			O	137100			718	
8 24 39	67 20.9		O	11119		130	714	
8 24 48	21 38.8	NGC 2595 3 ZW 59	R	4300		-100	779	
			R	4312			389	
			R	4332	20		939	
			O	4485			303	
			O	4367			321	
			O	4585			197	
			O	4111			262	
8 24 51	63 30.3		O	6918		112	714	
8 25 7	42 1.2	DDO 52	R	392	10	3	492	
8 25 12	-67 57.0	NGC 2601	O	3890	100	-280	844	
			O	3244	49		373	
8 25 18	55 40.0	KDG 163	O	7704	44	73	908	
8 25 24	30 36.2	IC 2378	O	14960		-55	255	
8 25 24	55 40.0	KDG 163	O	7755	44	73	908	
8 25 29	67 11.8		O	4127		129	714	
8 25 38	25 30.0	MRK 388	O	750		-81	305	
			O	0			306	
8 25 42	-77 37.0		O	5320	36	-260	599	
8 25 56	52 14.6	MRK 89	R	1715		56	368	
			O	1704			341	
			O	1400			233	
			O	1400			231	

R.A. (1950)	DEC. (1950)	NAME		OBS	HEL VEL (C*Z)	ERR	GAL CORR	REF	COMMENTS
8 26 16	52 51.9	MRK	90	O	4308		59	279	
				O	4231			341	
8 26 18	66 4.0		1	O	53472	100	124	633	★
8 26 18	66 4.0		2	O	54920		124	401	
				O	54723	100		633	
8 26 18	66 4.0		4	O	54647	100	124	633	
8 26 35	- 3 40.3			O	3060		-210	896	
8 27 0	52 28.0			R	5086	20	57	582	
				O	5090	50		582	
8 28 21	32 29.6	4C 32.25		O	15250	250	-46	696	
				O	15373	69		741	
8 28 45	52 46.6	MRK	91	O	5111		58	279	
				O	5102			341	
8 28 49	75 18.6	MRK	15	O	6660		166	280	
				O	6359	60		261	
8 29 6	19 23.0	ARP	58	O	11180	150	-111	195	
				O	11120	60		195	
				O	11210	150		195	
8 29 6	24 11.0	IC	509	R	5484	10	-88	582	
				O	5602	16		813	
				O	5494	40		582	
8 29 12	19 22.0	ARP	58B	O	11240	90	-111	195	KNOT
				O	11270	150		195	KNOT
8 29 12	22 44.0	NGC 2599		R	4745		-95	779	
		MRK	389	R	4732	20		939	
				O	5632	60		562	DIS
				O	4690			321	
8 29 33	66 21.1	DDO	53	R	19	7	125	492	
		7 ZW238		O	6			714	
8 29 54	57 43.0	KDG	164A	O	8111	120	83	567	
				O	7957	26		908	
8 30 6	57 45.0	KDG	164B	O	5387	27	83	908	★
8 30 29	-59 37.0			O	6330		-292	896	
8 31 4	55 44.7	4C 55.16		O	72550		73	850	
8 31 11	-22 48.1	NGC 2613		R	1831	80	-268	238	
				O	1555			13	
				O	1710	40		13	

8ʰ31ᵐ

R.A. (1950)	DEC. (1950)	NAME		OBS	HEL VEL (C*Z)	ERR	GAL CORR	REF	COMMENTS
8 31 54	1 50.0	KDG	165	O	4142	21	-190	908	
8 32 0	1 50.0	KDG	165	O	4310	10	-190	908	
8 32 10	66 24.3	MRK	93	O	4700		125	231	
				O	5288			279	
				O	5171			714	
				O	5226			341	
8 32 11	46 39.9	MRK	92	O	4308		26	341	
				O	3900			231	
8 32 11	46 40.0		A	O	4510		26	926	*
8 32 12	46 40.0		B	O	4445		26	926	*
8 32 12	28 39.0	NGC	2608	R	2119		-66	779	
		ARP	12	R	2126			610	
				O	2125	15		936	
				O	2119	100		13	
8 32 14	69 57.2			R	156	10	142	860	
8 32 28	30 42.3	MRK	390	O	7200		-55	306	
8 32 36	- 2 56.0			O	6600		-208	491	
8 33 27	66 18.7			O	7926		125	714	
8 33 34	66 42.2			O	9014		126	714	
8 33 57	65 17.8			O	5608		120	714	
8 34 4	51 48.9	MRK	94	O	761		53	279	HII REG
				O	750			421	
				O	706			341	
8 34 4	51 49.2			O	900		53	421	
8 34 7	-26 14.1			R	880	20	-276	532	HE2 -10
				O	870			532	
				O	900			604	
8 34 36	28 52.0	NGC	2619	R	3471	10	-65	752	
8 34 48	20 41.0			R	4678	10	-106	752	
8 35 18	25 5.0	NGC	2622	O	9404		-84	920	SEYF
		MRK	1218						
8 35 24	- 2 17.0			O	1800		-206	491	
				O	1940	100		521	

R.A. (1950)	DEC. (1950)	NAME	OBS	HEL VEL (C*Z)	ERR	GAL CORR	REF	COMMENTS
8 35 24	25 56.0	NGC 2623 VV 79 ARP 243	R O	5400 5435	40	-86	610 13	
8 35 31	19 53.5	NGC 2625 MRK 625	O	4525		-110	438	
8 35 36	17 48.0	KDG 166	O	7988	75	-120	908	
8 35 43	-72 25.5		O	15830		-272	896	
8 35 48	17 48.0	KDG 166	O	(10098)	100	-120	908	
8 36 13	29 1.3	B2	O	23750	250	-64	696	
8 36 43	67 2.3		O	3864		128	714	
8 36 59	29 59.8	4C 29.30	O O	19302 19560	35	-59	741 485	
8 37 1	24 13.8	4C 24.18	O	12780		-89	240	
8 37 4	66 37.5		O	11603		126	714	
8 37 12	73 10.0	NGC 2614	R	3458	15	156	752	
8 38 7	32 35.7	B2 N	O O	20430 20800	250 179	-46	696 741	
8 38 7	32 35.7	B2 S	O	20893	69	-46	741	
8 38 10	77 5.8		R	721	10	173	860	
8 38 18	- 3 57.0	NGC 2642	O	4439		-213	13	
8 38 36	64 47.5		O	11383		117	714	
8 38 36	77 5.0	7 ZW244	O	39070	200	173	820	
8 38 54	5 9.0	NGC 2644	O	1990	20	-177	813	
8 39 16	64 26.3		O	11601		115	714	
8 39 17	65 27.0		O	7383		120	714	
8 39 30	65 9.0		O	6948		119	714	
8 39 33	68 8.5		O	3897		133	714	
8 39 40	65 9.3		O	6955		119	714	
8 39 54	14 28.0	NGC 2648 KDG 168	R O	1925 1918	11 60	-136	914 908	

131

R.A. (1950)	DEC. (1950)	NAME	OBS	HEL VEL (C*Z)	ERR	GAL CORR	REF	COMMENTS
8 40 0	14 27.0	KDG 168	O	2126	60	-136	908	
8 40 3	50 23.2	NGC 2639	O	3314	75	45	13	
8 40 4	50 22.8		O	1680		45	877	
8 40 54	34 54.0	NGC 2649	O	4075	39	-35	813	
8 41 43	64 18.2		O	10586		115	714	
8 42 24	37 7.0	MRK 626	O	3934		-23	438	
			O	3923			558	
8 42 34	74 17.0	NGC 2633	O	2228		161	13	
		VV 519	O	2221			748	
		KDG 169	O	2164	30		908	
		ARP 80	O	2164	50		873	
8 42 36	9 50.0	NGC 2657	R	4141	15	-157	752	
8 42 36	73 44.0	IC 2389	O	2632		158	13	
8 43 6	12 58.0	KDG 171	O	8976	15	-143	908	
8 43 9	-33 37.1	NGC 2663	O	2280	200	-289	350	
		PKS	O	2150	60		355	
			O	2190			317	
8 43 12	12 49.0	NGC 2661	R	4120	15	-144	752	
8 43 18	12 59.0	KDG 171	O	8931	15	-143	908	
8 43 24	36 37.4	MRK 631	O	3215		-26	438	
8 43 38	31 37.2	B2	O	19990	250	-52	696	
8 43 42	49 44.0		O	3060		41	220	
8 44 7	70 17.7	KDG 172A/B	R	3626	15	143	829	
8 44 7	70 17.7	KDG 172A	O	3659	15	143	908	
8 44 7	70 17.7	KDG 172B	O	3467	15	143	908	
8 44 11	54 3.7	NGC 2656	O	13430	200	64	350	
		4C 54.17	O	13580			471	
8 44 33	70 20.9	MRK 95	O	3500		143	231	
8 44 36	73 40.0	NGC 2646	O	3546		158	13	
8 44 54	31 58.8	IC 2402	O	20400	600	-50	824	
		4C 31.32	O	20250			400	
			O	20217	77		741	

R.A. (1950)	DEC. (1950)	NAME	OBS	HEL VEL (C*Z)	ERR	GAL CORR	REF	COMMENTS
8 44 56	47 28.5		O	8970		30	558	
8 45 4	70 29.1	NGC 2650	R	3826	15	144	829	
8 45 12	60 25.0	NGC 2654	R	1295		95	693	
			R	1369	27		885	
			O	1360	65		13	
8 45 34	46 26.1	MRK 96	O	6600		24	233	
			O	6600			231	
			O	7070			341	
8 45 48	- 2 50.0	KDG 173	O	3899	15	-209	908	
8 45 48	- 2 47.0	KDG 173	O	3941	25	-209	908	
8 45 48	1 15.0	KDG 174	O	(8553)	90	-194	908	
8 46 6	1 13.0	KDG 174	O	8738	40	-194	908	
8 46 30	19 15.6	NGC 2672/3 ARP 167 KDG 175	R	3964		-114	610	
8 46 30	19 15.6	NGC 2672 HOL 99A	O	4223	100	-114	13	
8 46 34	19 15.6	NGC 2673 HOL 99B	O	3792	65	-114	13	
8 46 34	65 49.5	MRK 97	O	6970		122	436	
			O	7029			341	
			O	7200			308	
8 46 53	72 0.1	MRK 98	O	3100		150	231	
			O	3455			341	
			O	3100			233	
8 47 25	61 12.6	MRK 99	O	3800		99	233	
			O	3500			231	
8 47 49	57 18.0	MRK 17	O	6830		80	192	
8 47 54	29 25.0	MRK 628	O	8100		-63	438	
			O	8050	100		521	
8 47 58	73 22.7	MRK 16	O	2500		157	280	
			O	2343	60		261	
8 48 0	47 44.0	NGC 2676	O	6010		31	220	
8 48 4	24 30.1		R	2739	10	-88	860	

R.A. (1950)	DEC. (1950)	NAME	OBS	HEL VEL (C*Z)	ERR	GAL CORR	REF	COMMENTS
8 48 30	31 4.0	KDG 176A	O	1983	77	-55	908	
8 49 6	- 2 10.0	KDG 177	O	3297	18	-207	908	
8 49 6	- 2 9.0	KDG 177	O	3316	14	-207	908	
8 49 9	78 24.9	NGC 2655	R	1389	33	178	390	
		ARP 225	R	(1450)			645	
			O	1299	65		13	
			O	1517			106	
8 49 35	33 36.5	NGC 2683	R	260	25	-42	183	
			R	413			934	
			O	400	50		1	
			O	564			149	
			O	336	65		13	
			O	310	25		283	
			O	335			13	
			O	285	35		72	
8 49 56	51 30.2	NGC 2681	R	722		50	310	
			O	736			13	
			O	703	30		13	
8 50 12	45 31.6		R	1881	10	19	860	
8 51 1	65 19.2		O	84700	300	119	583	
8 51 12	47 17.0		O	1732	38	28	813	
8 51 16	34 44.9		R	2200	10	-36	860	
8 51 18	32 49.0	KDG 178B	O	4322	25	-46	908	
8 51 18	32 52.0	IC 2421	R	4385	10	-46	582	
		KDG 178A	O	4395	25		582	
			O	4383	30		908	
8 51 32	39 43.7	NGC 2691	R	3990		-10	480	SEYF
		MRK 391	R	3992			707	
			O	3900			306	
			O	3936			480	
8 51 41	58 55.5	NGC 2685	R	870		88	682	
		ARP 336	R	883			310	
			R	870			942	
			R	930			645	
			R	875	15		455	
			R	880	35		775	
			O	830			106	
			O	910	30		110	
			O	867	40		72	
			O	884	40		13	

R.A. (1950)	DEC. (1950)	NAME	OBS	HEL VEL (C*Z)	ERR	GAL CORR	REF	COMMENTS
8 51 48	17 53.0	MRK 1220	0	19461		-121	920	
8 51 54	57 45.0	VV 761N	0	11932	50	82	873	
8 51 54	57 45.0	VV 761E	0	11817	50	82	873	
8 52 18	-59 1.2	NGC 2714	0	2715		-294	574	
8 53 6	52 18.0	KDG 179A	0	4266	60	54	567	
			0	3661			594	
8 53 20	52 15.5	NGC 2692 KDG 179B	0	3749		54	594	
8 53 30	51 32.0	NGC 2694	0	5123	75	50	13	
8 53 30	51 33.0	NGC 2693	0	4956	50	50	13	
8 53 36	- 3 10.0	NGC 2708	R	2021		-211	744	
			0	2060	47		789	
8 53 49	- 3 21.0		0	5940		-212	896	
8 54 30	66 39.8	MRK 100	0	3586		125	341	
			0	4400			233	DIS
8 54 42	3 6.0	NGC 2713	0	3969	85	-187	741	
			0	3917			777	
			0	3997			180	
8 55 0	3 17.0	NGC 2716	0	3537	50	-186	13	
8 55 6	3 23.0	A	0	61241	100	-186	13	*
8 55 6	3 23.0	B	0	60964	50	-186	13	
8 55 18	3 22.0		0	60959	150	-186	13	
8 55 18	3 23.0		0	30403	50	-186	13	
8 55 18	37 16.0		0	12870		-23	573	II- HZ4
			0	12885			591	
8 55 18	37 16.0	B	0	12820	40	-23	573	*
8 55 24	3 21.0		0	20575	50	-186	13	
8 55 28	53 58.3	NGC 2701	R	2328	10	63	752	
			R	2338	20		544	
			0	2329	59		741	
			0	2421	30		741	KNOT
8 55 48	6 31.0		R	3546		-173	389	
			R	3555	30		622	
			0	3556			389	

8ʰ55ᵐ

R.A. (1950)	DEC. (1950)	NAME	OBS	HFL VEL (C*Z)	ERR	GAL CORR	REF	COMMENTS
8 55 50	-67 43.0		O	6330		-282	896	
8 56 6	55 54.0	NGC 2710	R	2538	15	72	752	
8 56 12	6 30.0	NGC 2718	R	3831	10	-173	752	
			R	3848	13		622	
8 56 12	45 6.0	NGC 2712	O	1840	200	17	13	
			O	1832			641	
8 57 7	35 55.0	NGC 2719A	O	3053		-30	253	
		HOL 105B	O	3318			57	
		ARP 202						
		KDG 181B						
8 57 7	35 55.5	NGC 2719	O	3190		-30	253	
		HOL 105A	O	3181			57	
		ARP 202	O (	3073)			594	
		KDG 181A						
8 57 11	5 15.4		O	3778		-178	558	
8 57 42	3 23.0	NGC 2723	O	3725	65	-186	13	
8 58 2	60 20.9	MRK 18	O	3250		95	279	*
8 58 5	29 13.6	3CR213.1	O	58160	300	-65	585	
8 58 26	-73 8.0		O	11510		-271	896	
			O (	946)	140		680	DIS
8 59 42	26 8.0	NGC 2735	R	2431		-81	779	
		ARP 287	O (	7900)			535	DIS
		VV 40						
9 0 12	30 47.0	IC 2428	O	4277	15	-57	813	
9 0 24	18 27.0	MRK 1221	O	3269		-119	920	
9 0 36	20 52.0	MRK 1222	O	9457		-107	920	
9 0 48	3 34.0	KDG 183	O	7935	60	-185	908	
9 1 0	3 34.0	KDG 183	O	3694	60	-185	908	
9 1 0	55 43.0	1 ZW 17	O	14123		71	262	
9 1 1	51 48.8	MRK 101	R	4750		51	368	
			O	4757			341	
			O	4800			278	
9 1 3	60 8.8	NGC 2726	O	1401	60	94	261	

R.A. (1950)	DEC. (1950)	NAME	OBS	HEL VEL (C*Z)	ERR	GAL CORR	REF	COMMENTS
9 1 8	22 10.0	NGC 2738	R	3102	15	-101	829	
9 1 42	14 42.0	IC 2431	O	14856		-136	920	
		VV 645	O	14860	150		776	
		MRK 1224	O	15040	100		776	
9 1 48	13 45.0	KDG 184	O	8313	53	-141	908	
9 1 50	18 39.6	NGC 2744	R	3431	10	-118	829	
		VV 612	O	3330	100		776	
			O	3450	50		13	
9 2 0	78 16.0	NGC 2715	O	1158		178	13	
			O	1308	10		936	
9 2 6	13 46.0	KDG 184	O	8479	97	-141	908	
9 2 15	45 31.2		R	2017	10	19	860	
9 2 30	18 31.0	NGC 2749	O	4203	40	-118	13	
			O	4270			106	
9 2 36	-73 47.0		O	5694	24	-270	599	
9 2 48	25 38.0	NGC 2750A/B	R	2685	10	-83	752	
		KDG 186						
9 2 48	25 38.0	NGC 2750A	O	2710	13	-83	908	
		VV 541W	O	2689	50		873	
9 2 54	25 38.0	NGC 2750B	O	2636	13	-83	908	
		VV 541E	O	2614	50		873	
9 2 54	18 32.0	NGC 2752	O	4022		-118	321	
9 3 18	34 49.0		O	2156	30	-36	813	
9 3 18	79 33.4		R	2073	10	183	860	
9 3 36	60 40.0	NGC 2742	R	1296		96	693	
			O	1267	10		936	
			O	1291	57		148	
			O	1276			688	*
9 4 12	25 32.0	NGC 2753A	O	2791	31	-84	908	
		KDG 187						
9 4 30	-15 18.0	NGC 2763	R	1889		-252	744	
			O	1818	58		789	
			O	1632	120		555	
			O	1998	37		741	
9 4 30	16 56.4		O	15289		-126	726	

9ʰ4ᵐ

R.A. (1950)	DEC. (1950)	NAME	OBS	HEL VEL (C∗Z)	ERR	GAL CORR	REF	COMMENTS
9 4 33	16 51.2		O	22660		-126	726	
9 4 41	16 50.8		O	23680		-126	726	
9 5 0	3 35.0	NGC 2765	O	3827	30	-185	813	
9 5 15	64 12.3		O	64500		113	717	
9 5 18	35 50.0	KDG 188	O	8204	15	-31	908	
9 5 24	35 49.0	KDG 188	O	7174	15	-31	908	
9 5 26	21 38.8	NGC 2764	R	2714	10	-103	829	
			O	2627	20		741	
9 5 31	6 7.9	DDO 54	R	1305	10	-175	492	
9 5 42	64 37.0	7 ZW261	O	5080	200	115	820	
9 5 51	- 9 31.40		O	(18168)	100	-234	633	
9 5 59	- 9 21.4	MC	O	48120	150	-233	721	
9 6 6	- 9 25.49		O	16510	100	-234	633	
			O	16370	150		721	
			O	16340			255	
			O	16303			444	
9 6 10	11 13.9		O	48900		-152	717	
9 6 26	- 9 19.51		O	15769	100	-233	633	
9 6 30	- 9 28.43		O	14511	100	-234	633	
			O	14330	150		721	
9 6 30	33 20.0	NGC 2770 HOL 111A	O	1955	71	-44	789	
9 6 32	- 9 23.65		O	15592	100	-234	633	
9 6 53	- 9 29.67		O	16247	100	-234	633	
9 7 0	-75 36.0		O	4746	11	-266	599	
9 7 0	- 9 10.55		O	17536	100	-233	633	
			O	17750	150		721	
9 7 0	50 39.0	NGC 2769	O	4703	56	45	908	
9 7 1	- 9 25.57		O	16815	100	-234	633	
9 7 6	50 36.0	NGC 2771	O	4992	70	45	908	

R.A. (1950)	DEC. (1950)	NAME	OBS	HEL VEL (C∗Z)	ERR	GAL CORR	REF	COMMENTS
9 7 13	- 9 31.07		O	16254	100	-234	633	
9 7 18	79 24.0	NGC 2732	O	2121		182	13	
9 7 31	-22 48.2	DDO 56	R	727	10	-271	492	
9 7 33	- 9 26.01		O	16524	100	-234	633	
9 7 42	7 15.0	NGC 2775	O	1135	75	-170	13	
9 7 48	60 15.0	NGC 2768	O	1408	175	94	13	
9 7 51	- 9 24.73		O	16564	100	-234	633	
			O	16550	150		721	
9 8 12	76 41.0	NGC 2748	O	1489		171	13	
9 8 18	-67 44.0	NGC 2788	O	1650	100	-282	844	
			O	1548	47		373	
9 8 18	46 50.5	MRK 102	O	4239		25	341	
			O	4302			278	
9 8 45	37 36.5		O	31200	250	-22	696	
9 8 54	45 11.0	NGC 2776	R	2630	10	17	752	
			R	2629	20		544	
			R	2620	10		251	
			O	2595			784	
			O	2673			13	
9 8 58	-14 50.8	DDO 57	R	2054	25	-250	492	
9 9 3	35 44.1	DDO 55	R	1880	10	-32	492	
9 9 6	-14 36.0	NGC 2781	O	2195	24	-250	741	
			O	1770	70		555	
9 9 12	35 13.0	NGC 2778	O	2051		-34	594	
9 9 36	35 7.0	NGC 2780	O	2208		-35	594	
9 10 6	-23 58.0	NGC 2784	O	708		-273	13	
9 10 30	17 51.0		O	7710	100	-122	521	
9 10 33	30 12.4	KDG 192A	O	6760		-60	748	
			O	6542	52		908	
9 10 40	30 12.0	NGC 2783	O	6793		-60	842	
		KDG 192B	O	6517	40		908	
			O	6660			748	

139

R.A. (1950)	DEC. (1950)	NAME	OBS	HEL VEL (C*Z)	ERR	GAL CORR	REF	COMMENTS
9 10 40	35 2.2	B2	O	7240		-35	485	
9 10 54	40 19.3	NGC 2782	R	2562		-8	707	
		ARP 215	R	2561			310	
			R	2570	35		775	
			O	2755	5		413	
			O	2517	20		13	
			O	2538	29		112	
			O	2586			665	
			O	2579	15		112	
			O	2536			399	
			O	2358	35		72	
			O	2566	15		50	
			O	2526	12		50	
			O	2536	12		35	
9 11 18	36 18.0	KDG 193	O	6451	75	-29	908	
9 11 24	16 57.0		R	8367	15	-126	582	
			O	8263	100		582	
9 11 30	36 18.0	KDG 193	O	7474	15	-29	908	
9 11 36	47 7.0		O	4210		27	449	
9 11 42	67 58.0	MRK 103	O	9434		131	279	
			O	9391			341	
9 12 9	-60 35.3		O	2940		-292	896	
9 12 13	19 54.3	NGC 2790	O	7800		-112	491	
		MRK 1228						
9 12 26	-63 27.5		O	(6830)		-289	896	
9 12 33	44 31.0	ARP 55	O	11773	50	13	141	
9 12 36	12 6.0	IC 530	O	4899	64	-148	813	
9 12 54	59 58.9	MRK 19	O	4230		93	192	
9 13 0	21 9.0	IC 2453	O	8790		-106	920	
		MRK 1229						
9 13 13	53 39.1	MRK 104A/B	O	2199		61	341	
			O	2000			231	
9 13 13	53 39.1	MRK 104B	O	2143	20	61	929	*
9 13 13	53 39.1	MRK 104A	O	2163	30	61	929	*
9 13 27	73 58.2	IC 529	R	2264	10	159	829	
			O	2091	100		789	

R.A. (1950)	DEC. (1950)	NAME	OBS	HEL VEL (C*Z)	ERR	GAL CORR	REF	COMMENTS
9 13 39	38 30.7	B2	O	21330	250	-17	696	
9 13 42	34 38.5	NGC 2793	R	1681	3	-37	745	
			O	1669	22		148	
			O	1691			708	
9 13 51	20 24.0	NGC 2804	O	8070		-109	950	
			O	8150	200		843	
9 13 54	-16 6.0	NGC 2811	O	2514	75	-254	13	
9 13 54	19 10.0	NGC 2802	O	8634		-115	594	
9 13 54	19 10.0	NGC 2803	O	8833		-115	594	
9 14 3	20 15.0	NGC 2806	O	8170		-110	950	
9 14 5	-23 25.5	NGC 2815	O	2590	46	-272	741	
9 14 6	25 38.0	MRK 1230	O	1433		-83	920	
9 14 10	42 12.6	NGC 2798/9 ARP 283	R	1740	40	1	455	
9 14 10	42 12.6	NGC 2798	R	1726	12	1	885	
		VV 50	O	1719	58		789	
		KDG 195A	O	1769	15		908	
			O	1708	75		13	
			O	1899			748	
9 14 18	42 12.3	NGC 2799	O	1813	45	1	789	
		VV 50	O	1999			748	
		KDG 195B	O	1877	15		908	
9 14 29	-62 54.2	NGC 2842	O	2780		-290	574	
			O	2970			896	
9 14 54	45 52.0		O	8096	25	20	582	
9 14 54	69 25.0	NGC 2787	O	639	40	138	13	
			O	551			13	
9 15 30	34 46.0		O	6571		-37	708	
9 15 36	16 31.0	MRK 704	O	8958		-128	630	SEYF
9 15 37	-22 8.8	NGC 2835	R	870		-269	171	
			R	880	20		216	
			O	909			13	
9 15 41	-11 53.0	3C 218	O	16160		-241	846	
			O	16100			846	HYA A
			O	16160	60		13	

R.A. (1950)	DEC. (1950)	NAME	OBS	HEL VEL (C*Z)	ERR	GAL CORR	REF	COMMENTS
9 15 43	71 37.0	MRK 105	O	3550		148	436	
			O	3506			341	
			O	3100			231	
			O	3700			233	
9 15 57	32 3.9	B2	O	18650		-51	485	
9 16 7	26 28.8	NGC 2824 MRK 394	O	9300		-79	306	
9 16 12	34 4.0	NGC 2827	O	6826		-40	708	
9 16 13	34 13.3		O	5065	31	-39	741	
9 16 14	34 13.1	NGC 2823 B2	O O	7000 7151		-39	485 708	
9 16 18	33 57.0	NGC 2825	O	7847		-41	708	
9 16 18	55 34.3	MRK 106	O O	37011 36500		71	341 231	SEYF
9 16 18	64 19.1	NGC 2805 HOL 124B	R R R R R O O	1737 1726 1733 1734 1742 1688 1916	13 5 25	114	744 834 693 883 283 283 13	
9 16 24	33 50.0	NGC 2826	O	6263		-41	708	
9 16 36	33 38.0		O	7089		-42	708	
9 16 42	33 57.0	NGC 2830 ARP 315 HOL 123A	O	6237		-41	708	
9 16 43	33 57.3	NGC 2831 ARP 315 HOL 123B	O O O	5046 4977 5155	65	-41	594 708 13	
9 16 44	33 57.8	NGC 2832 ARP 315 HOL 123C	O O O O	6946 6793 6803 6964	50 25	-41	13 594 708 741	
9 16 50	33 7.9	B2	O O	14902 15050		-45	708 485	
9 16 55	75 50.9		R	659	10	167	860	
9 17 0	71 45.3	MRK 20 MRK 107	O O O O	3620 8576 3491 3488	60	149	280 341 261 279	DIS

142

R.A. (1950)	DEC. (1950)	NAME	OBS	HEL VEL (C*Z)	ERR	GAL CORR	REF	COMMENTS
9 17 9	64 27.8	NGC 2814	R	1634	20	115	883	
		HOL 124C	O	1662			253	
			O	1671			57	
			O	1682			253	
			O	1588	100		8	
9 17 24	-12 2.2	DDO 60	R	1945	10	-242	492	
9 17 27	64 27.1	IC 2458	R	1480		115	834	
		MRK 108	O	1100			233	
		7 ZW276	O	1424			341	
		HOL 124D	O	1404	80		300	
			O	1391			57	
			O	1470			253	
			O	1485			253	
9 17 30	1 15.0	1	O	5195		-195	750	★
			O	4938	150		843	
9 17 30	1 15.0	2	O	5355		-195	750	
9 17 30	1 15.0	3	O	5145		-195	750	
9 17 36	33 17.0		O	6580		-44	708	
9 17 36	33 52.0	NGC 2839	O	7942		-41	708	
9 17 44	64 28.3	NGC 2820	R	1692		115	834	
		HOL 124A	R	1574	10		883	
		7 ZW276	O	1667			253	
			O	1602			57	
			O	1673			57	
			O	1581	100		8	
			O	1687			253	
			O	1702			253	
9 17 49	-16 18.8	NGC 2848	O	2096	47	-254	789	
			O	2255	92		741	
9 17 51	45 51.7	3CR219	O	52320		20	120	
9 17 54	29 5.0	KDG 196	O	6541	28	-66	908	
9 18 6	29 9.0	KDG 196	O	10565	20	-65	908	
9 18 15	-12 22.0	DDO 61	R	1906	20	-243	492	
9 18 35	51 11.3	NGC 2841	R	661	25	48	183	
			R	638			934	
			R	635			693	
			R	641	100		238	
			R	665	100		544	
			O	600	50		1	
			O	740			13	
			O	584	40		13	

R.A. (1950)	DEC. (1950)	NAME	OBS	HEL VEL (C*Z)	ERR	GAI CORR	REF	COMMENTS
9 18 37	40 22.0	NGC 2844	O O	1485 1491	10	-7	741 910	
9 18 42	76 45.0	7 ZW277	O	1490	100	171	820	
9 18 48	33 37.0		O	7160		-42	708	
9 19 5	47 27.5	MRK 109 KDG 198A	O O O	9096 9116 9043	20	29	341 279 908	
9 19 6	-11 41.0	NGC 2855	O O	1908 1913	50	-241	13 106	
9 19 6	34 3.0		O	7095		-40	708	
9 19 6	47 28.0	KDG 198B	O	8938	50	29	908	
9 19 11	-22 17.5	DDO 62	R	849	10	-269	492	
9 20 0	40 23.0	NGC 2852 KDG 199A	O	1888		-7	594	
9 20 6	40 25.0	NGC 2853 KDG 199B	O	1800		-7	594	
9 20 12	34 56.0		R O	1630 1607	100 80	-36	943 879	
9 21 0	2 20.0	NGC 2861	R	5134	15	-190	752	
9 21 0	17 22.0		O	12930	100	-124	521	
9 21 0	64 47.0	KDG 200	O	5021	30	116	908	
9 21 6	64 45.0	KDG 200	O	5369	31	116	908	
9 21 12	-22 58.0	NGC 2865	O	2714	75	-271	13	
9 21 12	49 34.0	NGC 2857 ARP 1	R O	4888 4864	10 15	40	582 582	
9 21 18	34 44.0	NGC 2859	R O O O	1587 1607 1694 1500	40 100 100	-37	610 879 13 2	
9 21 21	14 23.6		O	40900	150	-138	529	
9 21 22	-21 22.9	PKS	O	15600		-267	828	
9 21 29	18 1.4	MRK 397	O	23100		-121	306	

R.A. (1950)	DEC. (1950)	NAME	OBS	HEL VEL (C*Z)	ERR	GAL CORR	REF	COMMENTS
9 21 30	34 52.0		R	1850	50	-36	943	
			O	1890	80		879	
9 21 36	6 9.0	NGC 2864	O	3546		-174	641	
9 21 44	52 30.2	MRK 110	O	10700		55	233	SEYF
9 21 52	17 52.6	MRK 398	R	4021		-121	779	
			O	4200			306	
9 21 54	26 59.0	NGC 2862	O	4156	30	-76	813	
9 22 14	-63 35.5	NGC 2887	O	2850		-289	574	
9 22 24	-24 53.0		O	2413	25	-275	582	
9 22 34	36 40.1	4C 36.14	O	33510		-27	400	
			O	33760	250		696	
9 22 42	34 30.0		R	1695	100	-38	943	
			O	1758	80		879	
9 23 0	34 52.0	SE	O	7442	80	-36	879	
9 23 0	34 52.0	NW	R	1630	35	-36	943	
			O	1682	80		879	
9 23 1	11 38.9	NGC 2872/4 ARP 307	R	3710		-150	610	
9 23 1	11 38.9	NGC 2872 KDG 202A	O	2954		-150	253	
			O	3035	42		908	
			O	2955	50		8	
9 23 5	35 6.8	MRK 399	O	4800		-35	306	
9 23 6	11 38.5	NGC 2874	R	3775	50	-150	829	
		NGC 2875	O	3595	50		8	
		KDG 202B	O	3725	50		908	
			O	3591			253	
9 23 12	2 26.0	NGC 2877	O	6900		-190	491	
9 23 12	19 36.1	MRK 400	R	2533		-113	779	
			O	2400			306	
9 23 12	33 0.5	4C 32.29	O	41884	50	-45	741	
9 23 18	12 57.0		O	7800		-144	491	SEYF
9 23 27	78 30.5		O	67200		179	717	
			O	67600	300		583	

R.A. (1950)	DEC. (1950)	NAME	OBS	HEL VEL (C*Z)	ERR	GAL CORR	REF	COMMENTS
9 23 30	68 37.7	7 ZW280A	0	3924	36	135	500	
		VV 106B	0	3638			341	
		ARP 300A	0	3500			231	
		MRK 111A	0	3616	48		830	
9 23 30	68 37.7	MRK 111B	0	3713	72	135	830	
9 23 48	68 40.0	7 ZW280B	0	3860	62	135	500	
		ARP 300B						
		VV 106A						
9 23 48	74 48.0	A	0	40807	105	163	444	*
9 23 48	74 48.0	B	0	(40008)		163	444	
9 24 9	-27 49.1	NGC 2888	0	2233	67	-281	741	
9 24 12	57 36.0	NGC 2870	0	3287	20	81	813	
9 24 32	-11 25.3	NGC 2889	0	3417	76	-240	741	
9 24 55	30 12.3	B2	0	8040		-60	485	
9 25 0	12 30.0	KDG 204A	0	8642	12	-146	908	
9 25 0	12 30.0	KDG 204B	0	8692	10	-146	908	
9 25 24	50 59.0	VV 716	0	7486	50	47	873	
9 25 30	76 42.0	VV 58	0	2286	50	171	873	
9 25 42	20 45.0		0	57612	100	-107	13	
9 25 42	62 44.0	NGC 2880	0	1514	50	107	13	
9 26 0	74 0.0		0	3300		159	768	*
9 26 30	-76 24.5	NGC 2915	R	703		-264	660	
9 26 37	56 4.3	MRK 114	0	7537		73	279	
			0	7469			341	
9 26 39	-76 24.5	NGC 2915	0	441	7	-264	599	HII REG
			0	414	32		373	
			0	502			660	
			0	392			660	
9 27 19	29 45.5	NGC 2893	R	1711		-62	480	
		MRK 401	0	1800			306	
			0	1680			480	
9 27 26	49 28.7	MRK 115	0	7800		40	233	
			0	7757			341	

R.A. (1950)	DEC. (1950)	NAME	OBS	HEL VEL (C*Z)	ERR	GAL CORR	REF	COMMENTS
9 27 36	16 34.0	KDG 205	O	8589	25	-127	908	
9 27 36	16 35.0	KDG 205	O	8613	20	-127	908	
9 27 42	4 22.0	NGC 2900	O	5354	43	-182	813	
9 28 30	-14 31.0	NGC 2902	O	2065	39	-249	741	
9 29 20	-16 30.9	NGC 2907	O	2065	33	-254	741	
9 29 20	21 43.2	NGC 2903	R	560		-102	171	
			R	561	25		183	
			R	580	20		216	
			R	590			294	
			O	621	24		732	
			O	589			33	*
			O	540			264	
			O	642	65		13	
			O	645			13	
			O	486			513	
9 29 58	21 44.9		R	448	10	-102	860	
9 30 24	23 21.0	VV 553	O	7800		-94	535	
9 30 30	55 28.0	1 ZW 18A/B	R	759		70	918	
			R	767	30		622	
			O	300			233	
9 30 30	55 28.0	1 ZW 18A MRK 116A	R	760	60	70	241	
			R	760			389	
			O	778			389	
			O	764			123	
9 30 31	55 27.8	1 ZW 18B MRK 116B	O	914	6	70	123	
9 31 0	10 22.0	NGC 2911 ARP 232	O	3032		-155	281	
			O	3140	75		13	
			O	3225			106	
9 31 24	10 20.0	NGC 2914 ARP 137	R	3151	100	-156	885	
			O	3370	100		13	
9 31 24	11 14.0		O	2510	100	-152	521	
9 31 36	0 29.0	KDG 206	O	4813	20	-197	908	
9 31 36	0 30.0	KDG 206	O	4715	28	-197	908	
9 32 6	21 56.0	NGC 2916	O	3550	71	-101	789	
9 32 22	30 38.0	MRK 402	O	7327		-57	438	
			O	7200			359	

R.A. (1950)	DEC. (1950)	NAME	OBS	HEL VEL (C*Z)	ERR	GAL CORR	REF	COMMENTS
9 32 39	25 24.1	4C 25.26	O	84850		-84	400	
9 32 48	59 37.0	7 ZW285	O	12100	100	91	820	
9 32 49	-16 10.6	NGC 2924	O	4615	14	-253	741	
9 33 0	25 27.0	IC 544 KDG 207	O	7465	15	-83	908	
9 33 6	33 26.7		O	22784		-43	595	
9 33 12	25 10.0	IC 545 KDG 207	O O	6000 6208	29	-85	491 908	
9 34 18	48 51.0	1 ZW 19	O	10133		37	262	
9 34 27	-20 54.2	NGC 2935	O	2244	49	-265	741	
9 34 30	1 20.0		O	15150	100	-193	521	7W SEY
9 34 36	- 4 49.0		O	15428		-217	322	
9 34 54	-78 0.8		O	5820		-259	896	
9 35 6	2 58.0	NGC 2936/7 ARP 142	O O	7393 6705	100	-187	321 288	
9 35 6	2 58.0	NGC 2936/7	O	4845		-187	141	KNOT
9 35 6	2 58.0	NGC 2936 VV 316A	O O O	7039 6867 6927	50	-187	141 440 321	
9 35 6	2 58.0	NGC 2937 VV 316B	O O	6899 7167	50	-187	141 440	
9 35 24	9 45.0	NGC 2939	O	3367		-158	594	
9 35 24	9 50.0	NGC 2940	O	2986		-157	594	
9 36 0	71 24.9	A 936 DDO 63	R R R R R	141 133 141 140 140	4 5 2 2	148	693 390 492 754 769	HOL I
9 36 6	34 14.0	NGC 2942	P O O	4412 4751 4478	20 116 100	-38	582 741 582	
9 36 12	36 47.0	KDG 208	O	5976	14	-25	908	
9 36 18	36 48.0	KDG 208	O	5948	11	-25	908	

R.A. (1950)	DEC. (1950)	NAME		OBS	HEL VEL (C*Z)	ERR	GAL CORR	REF	COMMENTS
9 36 20	32 32.3	NGC	2944	R	6831	10	-47	622	
		ARP	63	0	6983	160		429	
		VV	82						
9 36 23	- 4 38.0	ARP	321	0	6737	100	-216	288	
		VV	116C	0	6852	30		43	
9 36 23	- 4 37.3	ARP	321	0	6666	100	-216	288	
		VV	116A	0	6577	40		43	
9 36 24	32 36.0	ARP	129E/W	0	7000		-47	535	
		3 ZW 60		0	6800			389	
		VV	83	0	6801			262	
		KDG	209	0	6517			197	
9 36 24	32 36.0	ARP	129W	0	6580	100	-47	288	
		KDG	209A	0	6811	33		908	
9 36 24	32 36.0	ARP	129E	0	6377	150	-47	288	
		KDG	209B	0	6631	50		908	
9 36 24	32 36.0			0	6407		-47	347	KNOT
9 36 25	- 4 38.4	ARP	321	0	6816	150	-216	288	
		VV	116B	0	6797	40		43	
9 36 25	- 4 37.8	ARP	321	0	6696	150	-216	288	
		VV	116E	0	6600			360	
9 36 25	- 4 36.6	ARP	321	0	6422	100	-216	288	
		VV	116D	0	6451	30		43	
9 36 29	34 4.5			0	5739		-39	558	
9 36 51	36 7.5	3CR223		0	41130		-28	143	
				0	41040	30		134	
				0	41050	30		741	
9 37 10	48 33.8			0	415		35	558	
9 37 42	15 9.0	NGC	2954	0	3733	30	-133	813	
9 37 56	21 27.4	MRK	403	0	7270		-103	374	
				0	7225			874	
				0	7200			359	
				0	7260			557	
				0	7360			438	
9 38 0	3 48.0	NGC	2960	0	4964	33	-183	813	
9 38 15	36 6.6	NGC	2955	0	7056	34	-28	741	
				0	1750	190		555	DIS

149

R.A. (1950)	DEC. (1950)	NAME	OBS	HEL VEL (C*Z)	ERR	GAL CORR	REF	COMMENTS
9 38 17	5 23.6	NGC 2962	R	2116		-176	610	
			O	1970			321	
			O	2039	47		741	
9 38 19	39 58.7	3CR223.1	O	32250	90	-8	134	
			O	32244	64		741	
9 38 48	75 5.0	NGC 2977	O	3072	14	164	813	
9 38 50	- 1 29.6	PKS	O	114830		-204	851	
9 39 0	59 5.0	NGC 2950	O	1339		89	13	
			O	1500	75		2	
			O	1359			910	
			O	1430	50		13	
9 39 6	76 34.8	MRK 118	O	2383		171	279	
			O	2308			341	
9 39 21	9 12.0		O	62000	300	-160	529	
9 39 30	0 34.0	NGC 2967	O	2245		-196	13	
9 39 36	4 54.0	NGC 2966	R	2048	50	-178	752	
		MRK 708	O	1898			929	
9 39 56	32 4.6	NGC 2964	R	1319	20	-49	829	
		KDG 210A	R	1311	19		914	
			O	1601			748	
			O	1251	40		908	
			O	1340	50		13	
9 39 56	32 4.6	MRK 404	O	1245		-49	438	HII REG
			O	1200			359	
9 40 0	- 3 28.1	NGC 2974	R	2072	50	-211	782	
			O	2008			106	
			O	2000			360	
			O	2013	50		13	
9 40 10	66 12.4	MRK 119	O	2900		124	233	
9 40 15	32 9.6	NGC 2968	O	1600	48	-49	741	
		KDG 210B	O	1579	80		908	
			O	1333			748	
			O	1608	73		148	
9 40 30	- 2 1.0		O	4500		-206	491	
9 40 35	32 12.5	NGC 2970	O	1678	40	-48	741	
		MRK 405	O	12785			557	DIS
9 40 54	- 5 4.0	VV 52 ARP 253	O	1946	52	-217	148	

R.A. (1950)	DEC. (1950)	NAME	OBS	HEL VEL (C*Z)	ERR	GAL CORR	REF	COMMENTS
9 41 0	68 49.0	NGC 2959 KDG 211	O	4524	50	136	908	
9 41 8	29 50.1	MRK 406	R O O O O O	5093 4968 5010 5100 5160 5100		-61	779 874 374 438 557 359	
9 41 12	68 50.0	NGC 2961 KDG 211	O	(4620)	63	136	908	
9 41 18	-20 14.0	NGC 2983	O	2015	100	-263	13	
9 41 30	- 0 25.0	KDG 212	O	1402	20	-200	908	
9 41 36	- 0 26.0	KDG 212	O	(1495)	60	-200	908	
9 41 48	-21 3.0	NGC 2986	O	2397	100	-265	13	
9 42 12	27 30.0	IC 2505 KDG 213	O	9899	113	-72	908	
9 42 14	72 40.9	MRK 120	O	2100		154	308	
9 42 18	27 29.0	IC 2506 KDG 213	O	10142	88	-72	908	
9 42 32	23 19.9		R	2132	10	-93	860	
9 42 36	-31 36.0	DDO 235	R	1256	15	-285	492	
9 42 36	9 20.0		O	5626		-159	777	
9 42 43	73 13.2	NGC 2957A MRK 121A	O O O	6480 6895 6724	98	156	341 830 278	★
9 42 47	73 13.1	NGC 2957B MRK 121B	O	6701	147	156	830	
9 42 58	- 8 25.3		O	46200		-228	717	
9 43 4	-18 8.6	NGC 2989	O	4166	22	-257	741	
9 43 10	68 8.9	NGC 2976	O O	42 -18		133	13 888	
9 43 14	73 11.8	NGC 2963 MRK 122	O	6553		156	278	
9 43 18	5 56.6	NGC 2990	O O	3117 3155	36	-173	741 910	

R.A. (1950)	DEC. (1950)	NAME	OBS	HEL VEL (C★Z)	ERR	GAL CORR	REF	COMMENTS
9 43 18	45 59.0	1 ZW 21	O	4870		22	389	
			O	4936			262	
9 43 23	-14 5.7	NGC 2992	O	2313	21	-246	741	
		ARP 245	O	2446	60		761	
			O	2473	90		965	
			O	2254	75		945	
			O	2100			319	
9 43 24	-14 8.2	NGC 2993	O	2050		-246	319	
		ARP 245	O	2373	44		741	
9 43 28	-30 57.6	NGC 2997	R	1089	10	-284	752	
			R	1084			671	
			R	1049	22		238	
			R	1091	50		544	
			O	1171	27		732	
			O	1090	20		919	
			O	1085	44		741	
			O	1030			145	
			O	1085			155	
			O	(1050)	210		555	
9 43 36	5 56.0	NGC 2990	O	3198	43	-173	789	
			O	3300			491	
9 43 42	56 20.2	MRK 123	O	7629		76	341	
			O	7700			279	
9 43 44	3 16.6	IC 563	O	6093		-185	253	
		HOL 143A	O	6104	50		8	
		ARP 303						
9 43 46	3 18.1	IC 564	O	6026		-184	253	
		HOL 143B	O	6024	50		8	
		ARP 303						
9 43 56	42 45.2	HOL 142	O	5394		6	558	
9 44 0	- 4 38.0		O	3850		-215	360	★
9 44 6	21 57.8		R	735	10	-100	860	
9 44 7	-30 12.4	NGC 3001	O	2250	72	-283	741	
9 44 30	1 11.7		R	1855	10	-193	860	
9 44 33	73 28.6		O	17770		157	244	
9 44 43	39 19.0	MRK 407	O	1500		-11	359	
			O	1661	39		562	
			O	1650			438	

R.A. (1950)	DEC. (1950)	NAME	OBS	HEL VEL (C*Z)	ERR	GAL CORR	REF	COMMENTS
9 44 46	39 45.8	B2	O	12340		-9	485	
9 44 57	58 12.2	MRK 21	O	8540		85	280	
			O	8503	60		261	
9 45 7	7 39.3	3CR227	O	25830	30	-166	741	SEYF
			O	25632			160	
			O	25810	60		159	
			O	26310			575	
9 45 8	33 6.9	MRK 408	O	1470		-43	438	
			O	1500			359	
9 45 18	47 36.0	KDG 215	O	9779	20	31	908	
9 45 24	-30 43.0		O	2466	12	-284	692	
9 45 24	50 43.3	MRK 124	O	16931		47	341	SEYF
			O	16400			231	
9 45 29	-30 42.9		O	2498	15	-284	743	
9 45 30	47 35.0	KDG 215	O	9739	10	31	908	
9 45 35	44 19.0	NGC 2998	R	4777	10	14	752	
		HOL 144A	O	4720	41		741	
			O	4767	15		936	
9 45 36	33 39.0	NGC 3003	R	1481		-40	744	
			R	1482	9		914	
			R	1479			693	
			R	1482	10		829	
			O	1476	60		13	
9 46 0	72 31.0	NGC 2985	O	1277	50	153	13	
9 46 3	55 49.0	MRK 22	O	1500		73	192	
9 46 24	- 7 49.0		R	6580	15	-226	582	
			O	6526	45		582	
9 46 42	32 27.0	NGC 3011	O	1440		-47	438	
		MRK 409	O	1500			359	
9 46 48	1 22.0	NGC 3015	O	7500		-192	491	
9 47 3	46 11.6	MRK 125	O	7400	80	24	300	
			O	7420			279	
9 47 6	0 51.0	VV 620W	O	1827	50	-194	873	
9 47 6	0 51.0	VV 620E/A	O	1849	50	-194	873	

R.A. (1950)	DEC. (1950)	NAME	OBS	HEL VEL (C*Z)	ERR	GAL CORR	REF	COMMENTS
9 47 6	0 51.0	VV 620E/B	O	1909	50	-194	873	
9 47 6	12 56.0	NGC 3016	O	8843	48	-142	644	
9 47 7	0 51.4	NGC 3018 KDG 216	O	1847	14	-194	908	
9 47 12	34 39.0		O	6570		-35	449	
9 47 18	0 51.0	MRK 1236	O	1814		-194	920	*
9 47 18	0 51.3	VV 620SE	O	1810	50	-194	776	* KNOT
9 47 18	0 51.3	N 3023 VV 620 KDG 216	O O	1877 1845	14 50	-194	908 776	
9 47 24	13 3.0	NGC 3020 HOL 147A	R O	1430 1424	10 22	-142	752 644	
9 47 24	31 43.3	DDO 64	R	520	15	-50	492	
9 47 24	31 43.3	DDO 64S	O	430		-50	315	KNOT
9 47 24	31 43.3	DDO 64N	O	423		-50	315	KNOT
9 47 24	44 34.0	MRK 1237	O	4425		15	920	
9 47 42	- 1 22.0	MRK 1239	O	5962		-202	920	SEYF
9 47 48	13 0.0	NGC 3024	O	1506	25	-142	644	
9 48 0	33 47.3	NGC 3021	O R R O	1495 1541 1537 1529	63 20 16 32	-39	741 829 914 148	
9 48 1	28 47.0	NGC 3026	R O	1492 1468	10 20	-65	829 813	
9 48 37	71 32.5		O	54600		149	717	
9 48 42	8 4.0	KDG 217A	O	451	29	-164	908	
9 48 42	8 4.0	KDG 217B	O	493	14	-164	908	
9 49 5	-32 31.1	NGC 3038	O O	2720 2660	 53	-286	574 741	
9 49 10	1 41.0	DDO 65	R	1853	10	-190	492	
9 49 14	29 28.6	NGC 3032	R R O	1561 1501 1568	20 150	-61	829 610 13	

R.A. (1950)	DEC. (1950)	NAME	OBS	HEL VEL (C∗Z)	ERR	GAL CORR	REF	COMMENTS
9 49 17	52 27.6	MRK 126	O	11673	80	56	300	
			O	11681			278	
9 49 41	-73 41.0	NGC 3059	O	1209		-269	154	
			O	1270	20		741	
			O	1330	100		844	
			O	1302	33		599	
			O	1232	34		373	
9 49 48	43 5.1		O	4816	50	8	741	
9 50 6	37 59.0	MRK 410	O	6900		-17	359	
			O	6969			438	
9 50 18	36 18.0		O	5482		-26	777	
9 50 22	16 54.9	NGC 3041	R	1417	20	-123	544	
			O	1305	58		789	
			O	1343	28		741	
9 51 4	51 28.8	MRK 127	O	11000		52	233	
			O	10847			341	
9 51 6	1 49.0	NGC 3044	R	1292	6	-189	914	
			O	1326	25		148	
			O	1359			149	
9 51 18	72 26.0	NGC 3027	R	1059		153	744	
			R	1072	20		544	
			R	1061	20		238	
			O	1079			13	
9 51 27	69 17.8	NGC 3031	R	-31	6	139	493	M 81
		KDG 218A	R	-38			693	
			R	(-40)			934	
			R	-40			285	
			O	-45			25	
			R	-45			27	
			R	-40			93	
			R	29	168		934	∗
			R	(-40)			457	
			O	-186	72		149	
			O	-30	50		1	
			O	-30			150	
			O	-60			193	
			O	-129	28		150	
			O	-46	15		441	
			O	-53	30		555	
			O	-55	20		13	
			O	-64			13	
			O	-51			748	
9 51 43	69 55.0	NGC 3034	R	248		142	249	M 82
		ARP 337	R	70	130		216	∗
		KDG 218B	R	190			58	
		3CR231	R	135			171	

R.A. (1950)	DEC. (1950)	NAME	OBS	HEL VEL (C*Z)	ERR	GAL CORR	REF	COMMENTS
9 51 43	69 55.0	NGC 3034	R	240	5	142	301	
			R	230			934	
			R	184	15		93	
			R	238	10		621	★
			R	240	10		470	★
			O	216	45		945	
			O	256	63		149	
			O	275			13	
			O	250			604	
			O	440			604	
			O	290	50		1	
			O	180			748	
			O	305			91	
			O	237	3		325	
			O	239	50		199	
			O	263	75		13	
9 52 6	-25 28.0	NGC 3054	O	2194	42	-273	148	
9 52 7	-18 24.3	NGC 3052	O	3616	45	-257	741	
9 52 12	9 30.0	NGC 3049 MRK 710A/B	O	1567	10	-157	813	
9 52 12	9 30.0	MRK 710A	O	1415		-157	928	★
9 52 12	9 30.0	MRK 710B	O	1463		-157	928	★
9 52 18	-28 3.8	NGC 3056	O	1047	48	-278	741	
9 52 24	14 32.0	KDG 219	O	7194	27	-134	908	
9 52 30	14 33.0	KDG 219	O	7182	11	-134	908	
9 52 42	4 31.0	NGC 3055	O	1913		-178	13	
9 52 43	53 32.8	NGC 3043	O	2935	51	62	741	
			O	2973	15		813	
9 52 48	8 37.0		O	1283	100	-161	13	
9 53 6	16 39.0		R	1105	10	-124	860	
9 53 16	60 19.5	MRK 128	O	9300		96	308	
9 53 24	10 44.0	IC 577 KDG 220	O	9016	45	-151	908	
9 53 27	60 12.3	7 ZW301 MRK 23	O O	9140 9740	55	96	430 192	
9 53 33	46 41.8	MRK 129	O O	4695 4597		27	279 341	

R.A. (1950)	DEC. (1950)	NAME	OBS	HEL VEL (C*Z)	ERR	GAL CORR	REF	COMMENTS
9 53 36	10 43.0	IC 578 KDG 220	0	8954	30	-151	908	
9 53 52	29 3.8	DDO 68	R	504	10	-63	492	
9 54 0	15 53.0	MRK 712A	0	4562		-127	928	*
9 54 0	15 53.0	MRK 712B	0	4531		-127	928	*
9 54 36	33 51.0	IC 2524 MRK 411	0 0	1498 1500		-38	438 359	
9 54 42	7 26.0		R 0	6518 6600		-165	707 491	SEYF
9 55 4	32 28.0	MRK 412	0 0	4554 4500		-45	438 359	
9 55 26	32 36.5	NGC 3067	R 0 0 0 0	1484 1399 1506 1550 1396	14 50 53	-45	914 323 13 272 789	
9 55 46	29 6.6	NGC 3068B ARP 174B	0	6236	32	-62	38	
9 55 47	29 7.0	NGC 3068A ARP 174A	0	6409	27	-62	38	
9 55 54	13 29.0		0	2700		-138	491	
9 56 0	52 29.0	1 ZW 23	0 0	12261 12143		57	262 774	
9 56 0	54 45.1	MRK 24	0	13618	60	69	261	
9 56 12	-26 41.0	NGC 3078	0 0 0	2540 2481 2468	 50	-275	244 13 281	
9 56 21	31 56.3	MRK 413	0 0	11400 11590		-48	359 438	
9 56 24	10 36.0	IC 584	0	5400		-151	491	
9 56 29	30 59.1	DDO 69	R R	26 23	10 3	-53	492 677	LEO A
9 56 42	35 38.0	NGC 3074	R 0 0	5150 5161 5000	15 20	-29	582 582 449	

9ʰ56ᵐ

R.A. (1950)	DEC. (1950)	NAME	OBS	HEL VEL (C*Z)	ERR	GAL CORR	REF	COMMENTS
9 56 45	47 32.7	MRK 130	O	25650		32	436	
9 56 59	-33 59.0	NGC 3087	O	2662	26	-287	741	
9 57 6	-27 53.0		O	1477		-277	683	
9 57 11	-22 35.1	NGC 3081	O	2413	13	-266	741	
			O	2403			910	
9 57 18	13 17.0	NGC 3080 MRK 1243	O	11049		-139	920	
9 57 22	-28 5.1	NGC 3089	O	2653	23	-278	741	
9 57 23	5 34.1	DDO 70	R	295	10	-173	492	SEX B
9 57 24	- 2 37.0		O	6376		-206	750	
9 57 29	55 51.7	NGC 3073 MRK 131	O	1176		74	565	
9 57 42	- 2 43.0		O	9766		-206	750	
9 57 42	72 25.0	NGC 3065	O	2051		153	13	
			O	1984			106	
9 57 52	72 21.9	NGC 3066 MRK 133	O	2132		153	13	*
			O	2070			287	
			O	1600			233	
			O	1847			774	
			O	2033	80		300	
9 57 53	-19 23.8	NGC 3091	O	3882	37	-259	741	
9 57 55	-33 59.8	IC 2532	O	2900		-287	574	
9 58 0	- 2 43.0	NGC 3090	O	6236		-206	750	
			O	6046	65		843	
9 58 12	-14 43.0		R	9084	15	-246	582	
			O	9011	40		582	
9 58 35	0 9.7		O	13790		-195	726	
9 58 35	55 55.4	NGC 3079 HOL 156A 4C 55.19	R	1131		75	744	
			R	1120			216	
			R	1170			171	
			O	1171			13	
			O	1065	53		789	
			O	1168	71		72	
			O	939	48		72	
			O	1150			618	

R.A. (1950)	DEC. (1950)	NAME	OBS	HEL VEL (C*Z)	ERR	GAL COPR	REF	COMMENTS
9 58 38	0 6.4		O	26800		-195	726	
9 58 57	29 1.6	3CR234	O	55440		-62	120	
9 59 21	68 58.5	NGC 3077	R	15	5	138	551	
			R	10			934	
			R	10	3		390	
			O	-10			425	
			O	-41			211	
			O	-158			13	
9 59 27	24 57.2	NGC 3098	O	1340	30	-83	741	
			O	1287	30		813	
9 59 44	19 16.7		R	2010	10	-110	860	
9 59 44	43 25.7	MRK 134	O	5405		11	436	
			O	5400			308	
9 59 51	80 31.7	NGC 3057	R	1529	15	188	492	
9 59 59	-44 23.6	PKS	O	6300		-296	870	
10 0 11	20 6.5	4C 20.20	O	50250		-106	400	
10 0 22	59 40.7	7 ZW308	R	(2602)	20	94	622	
		MRK 25	O	2730			192	
			O	3000			264	
			O	3206			774	
10 0 47	-25 54.8	NGC 3109	R	404		-273	671	
		DDO 236	R	403	10		492	
			R	404	5		905	
			R	403	10		93	
			R	403	2		489	
			R	408			156	
			R	403	2		383	
			R	408			121	
			R	405			126	
			R	401	5		137	
			O	441			13	
10 0 48	40 58.0	NGC 3104	R	630		-1	363	
		ARP 264	O	725	64		148	KNOT
		VV 119						
10 1 16	14 27.8		O	8991	59	-133	741	
10 1 42	13 51.7		R	2806		-136	780	
			O	2743	41		741	
10 2 1	51 50.4		O	14067	50	55	741	

10ʰ2ᵐ

R.A. (1950)	DEC. (1950)	NAME	OBS	HEL VEL (C*Z)	ERR	GAL CORR	REF	COMMENTS
10 2 36	19 30.0	KDG 222	O	3829	17	-109	908	
10 2 36	19 31.0	KDG 222	O	3683	10	-109	908	
10 2 44	- 7 28.5	NGC 3115	R	680		-222	216	
			R	658	24		390	
			O	642			321	
			O	600	50		1	
			O	629			36	
			O	728	9		530	
			O	648	12		13	
			O	668	105		555	
			O	662			937	
			O	743	7		149	
			O	695			503	
			O	591	40		72	
			O	719	38		392	
10 2 46	53 57.5	MRK 135	O	13246	80	65	300	
			O	12800			233	
10 3 5	35 8.8	3CR236	O	29400	600	-30	824	
			O	29680	250		696	
			O	29650	30		159	
			O	29667	35		741	
10 3 18	0 53.0	IC 590A	O	6216	31	-191	594	
10 3 18	0 53.0	IC 590B	O	6383	35	-191	594	
10 3 24	29 12.0		O	1402	57	-61	246	
10 3 42	77 9.1	MRK 136	O	10073		174	278	
10 3 48	-43 59.0		O	3350	20	-295	719	
10 3 49	26 9.4	B2	O	35010	250	-76	696	
10 4 10	14 37.0	NGC 3121B KDG 224B	O	9232		-132	748	
10 4 10	14 37.2	NGC 3121A KDG 224A	O	9532		-132	748	
10 4 17	-18 58.3	NGC 3124	R	3570	15	-256	752	
			O	3381	67		741	
10 4 18	-29 40.0		O	1110	30	-279	655	
10 4 24	38 7.0		O	15635		-15	347	
10 4 24	38 7.0		O	15665		-15	347	KNOT

R.A. (1950)			DEC. (1950)		NAME		OBS	HEL VEL (C*Z)	ERR	GAL CORR	REF	COMMENTS
10	4	30	-67	8.0	NGC	3136	O	1821	50	-281	113	
							O	1647	44		741	
							O	(1091)			154	
10	4	48	12	31.0	IC	591	O	2700		-141	491	
10	4	55	17	20.7			O	7919		-119	558	
10	4	55	52	5.4			R	1092	10	56	752	
10	5	6	28	16.5	B2		O	44330	250	-65	696	
10	5	24	32	6.0	NGC	3126	O	5019	36	-46	813	
10	6	39	30	23.8	DDO	73	R	1371	15	-54	492	
10	6	50	-38	8.0			O	4845	32	-290	790	
10	6	50	-38	7.5			O	4512		-290	790	
10	6	54	54	45.0	VV	533E/W	O	(1100)		70	776	
							O	1150			774	
10	6	54	54	45.0	VV	533W	O	1098	50	70	873	
10	7	6	-16	23.0	NGC	3140	O	8458	50	-249	582	
10	7	13	14	16.7	4C	14.36	O	(64500)		-133	600	
10	7	24	17	56.0	KDG	225	O	10369	20	-115	908	
10	7	24	23	20.0	MRK	716	O	17189		-89	630	SEYF
10	7	30	-66	48.0	IC	2554	O	1850	100	-281	844	
							O	1377	35		373	
							O	1243	40		719	*
10	7	30	17	58.0	KDG	225	O	10157	20	-115	908	
10	7	30	77	58.0	MRK	1246	O	7912		178	920	
10	7	42	-12	11.3	NGC	3145	R	3648	25	-236	752	
							O	3650	15		740	
							O	3855			13	
10	7	55	16	55.9	MRK	1247	O	4920		-120	558	
10	8	16	-25	34.7	DDO	237	R	2516	10	-271	492	
10	8	26	59	8.3	MRK	26	O	9132		92	278	
10	8	32	58	58.9	MRK	27	O	2139	60	91	261	
							O	2200			280	

R.A. (1950)	DEC. (1950)	NAME	OBS	HEL VEL (C*Z)	ERR	GAL CORR	REF	COMMENTS
10 8 34	- 4 27.7	A 1009	R	330	13	-210	216	SEX A
		DDO 75	R	321	10		492	
			R	325	8		93	
			R	320			156	
			O	371	30		13	
			O	436			13	
			O	370	8		11	
			O	369	30		13	KNOT
10 8 44	0 41.5		R	3636	10	-191	752	
10 9 0	58 38.7	MRK 28	O	9280		89	280	
			O	9047	60		261	
10 9 4	-38 22.9		O	18900		-290	675	
10 9 18	67 39.5	MRK 138	O	4500		132	308	
10 10 6	3 22.7	NGC 3156	R	(1135)		-179	815	
			O	1164	23		741	
10 10 11	35 31.6	MRK 414	O	11330		-27	374	
		KDG 226A	O	11326	44		908	
			O	11430			557	
			O	11400	60		567	
			O	11444			438	
			O	11700			359	
10 10 12	12 55.0	NGC 3153	R	2806	15	-138	752	
			O	2827	63		789	
10 10 12	35 33.0	KDG 226B	O	11533	45	-27	908	
			O	11530			557	
10 10 36	38 52.0	NGC 3151	O	7140	59	-10	72	
10 10 42	13 52.0		O	2400		-134	491	
10 10 42	39 6.0	NGC 3152	O	6471	96	-9	72	
10 10 48	22 59.0	NGC 3162	R	1302		-90	693	
			R	1290			780	
			O	1456	65		13	
10 10 48	68 2.0	7 ZW313	O	27500	200	134	820	
10 10 54	39 1.0	NGC 3158	O	7024	50	-9	13	
10 10 55	38 54.1	NGC 3159	O	6950		-10	13	
		HOL 172C						
10 10 56	34 57.7		O	42810		-30	726	

R.A. (1950)	DEC. (1950)	NAME	OBS	HEL VEL (C*Z)	ERR	GAL CORR	REF	COMMENTS
10 11 0	18 22.0	KDG 227	O	3635	50	-113	908	
10 11 2	38 54.3	NGC 3161	O	6204		-10	13	
10 11 5	34 56.0		O	13700		-30	726	
10 11 6	18 22.0	KDG 227	O	3486	50	-113	908	
10 11 9	3 40.5	NGC 3166	O	1378		-178	748	
		HOL 173A	O	1200	31		908	
		KDG 228A	O	1381	50		13	
10 11 12	38 53.0	NGC 3163	O	6245		-10	13	
		HOL 172B						
10 11 12	39 41.9		R	2056	10	-5	860	
10 11 22	65 23.3		R	3315	10	122	829	
10 11 38	3 43.2	NGC 3169	R	1093	80	-178	238	
		HOL 173B	O	1281	60		13	
		KDG 228B	R	1240	20		775	
			O	1294			748	
			O	1314	31		908	
			O	1205			910	
			O	1312			13	
10 12 24	-28 37.2	NGC 3175	O	1125	30	-276	741	
10 12 26	55 55.0		O	7254	32	76	741	
10 12 30	21 21.0	2 ZW 44	R	6028		-98	389	
			R	6206	50		622	
			O	6150			389	
			O	6147			262	
10 12 46	44 2.2	MRK 139	O	5183		16	278	
10 12 48	73 39.0	NGC 3147	O	2721	80	160	13	
10 13 0	60 29.0	NGC 3168A	O	9283	37	99	594	
10 13 0	60 29.0	NGC 3168B	O	9019	43	99	594	
10 13 18	2 56.1	IC 600	R	1271	10	-181	860	
10 13 24	5 12.0		O	9300		-171	491	
10 13 25	45 34.3	MRK 140	O	1630	80	24	300	
			O	1630			341	
			O	1500			233	
10 13 45	5 4.2		O	13786	110	-172	741	

163

R.A. (1950)	DEC. (1950)	NAME	OBS	HEL VEL (C*Z)	ERR	GAL CORR	REF	COMMENTS
10 13 48	21 23.0	NGC 3177	R	1299	14	-97	914	
			R	1317			780	
			O	1220	65		13	
10 13 54	12 54.0	KDG 229	O	9524	30	-137	908	
10 14 0	12 50.0	KDG 229	O	9403	65	-138	908	
10 14 2	53 42.7		O	13622	34	66	741	
10 14 6	39 17.6		O	61800	300	-7	529	
10 14 19	39 46.0	4C 39.30	O	17880		-5	400	
10 14 28	60 18.5	MRK 29	O	12780	60	98	261	
			O	9210			279	DIS
10 14 36	4 34.9		R	1349	10	-173	860	
10 14 36	15 45.0	MRK 629	O	9725		-124	438	
10 14 54	21 56.0	NGC 3185	R	1218	14	-94	914	
			O	1241	65		13	
10 15 0	22 8.0	NGC 3187	O	1589	22	-93	148	
		ARP 316						
		VV 307						
10 15 18	41 40.0	NGC 3184	R	593		4	693	
		NGC 3180	R	595			744	
			R	588	12		390	
			O	443	100		13	
			O	395			13	
10 15 24	22 5.0	NGC 3190	O	1319	60	-94	13	
		VV 307	O	1380			13	
		ARP 310						
10 15 36	7 17.0	IC 601	O	3642	36	-162	594	
10 15 39	64 13.2	MRK 141	O	12903	80	117	300	SEYF
			O	11600			231	
			O	11600			233	
10 15 42	7 18.0	IC 602	O	3600		-162	491	
		KDG 230	O	3608	31		594	
10 15 42	22 9.0	NGC 3193	O	1371	50	-93	13	
		ARP 316	O	1398	14		938	
		HOL 175B	O	1300	100		2	
10 16 1	46 42.3	NGC 3191	O	9145	28	30	741	

R.A. (1950)	DEC. (1950)	NAME	OBS	HEL VEL (C*Z)	ERR	GAL CORR	REF	COMMENTS
10 16 12	-17 43.9	NGC 3200	O	3567	19	-250	741	
10 16 19	57 40.2	NGC 3188A	O	8020		86	280	
		MRK 30	O	8095	60		261	
10 16 24	57 40.3	NGC 3188	O	7890		86	280	
		MRK 31	O	7748	60		261	
10 16 32	52 18.9		R	659	10	59	860	
10 16 52	45 48.0	NGC 3198	R	680		26	171	
			R	662			934	
			R	677	25		183	
			R	633			693	
			R	645			216	
			O	670			483	
			O	649			13	
10 17 8	65 25.4		R	3296	15	123	829	
10 17 14	-26 26.9	NGC 3203	O	2424	15	-271	741	
10 17 24	-25 36.0	NGC 3208	R	3006	15	-269	752	
			O	3007	20		582	
10 17 32	43 16.3	NGC 3202	O	6745	59	13	741	
10 17 50	48 45.6	4C 48.29	O	15600	600	41	824	
10 18 6	25 37.0		O	1300		-76	491	
10 18 30	28 11.0	1 ZW 24	O	15007		-63	262	
10 18 30	57 11.0	NGC 3206	R	1158		83	744	
			O	1192	22		148	
10 18 36	-37 32.0		O	7633	50	-287	582	
10 19 1	79 2.0		O	2565		182	591	
10 19 20	-34 0.8	NGC 3223	O	2733		-283	154	
		IC 2571	O	2911	55		741	
10 19 22	- 4 37.5		O	16130		-208	255	
10 19 24	36 50.0	IC 2566 KDG 231	O	7753	36	-19	908	
10 19 30	78 52.0	KDG 232	O	2832	21	182	908	
10 19 32	19 7.3		O	26530		-107	726	
10 19 36	21 29.0		O	16120		-96	449	

10ʰ19ᵐ

R.A. (1950)	DEC. (1950)	NAME	OBS	HEL VEL (C*Z)	ERR	GAL CORR	REF	COMMENTS
10 19 36	21 50.0		0	4103		-94	777	
10 19 36	36 51.0	IC 2568 KDG 231	0	8080	60	-19	908	
10 19 36	46 29.0		R 0	5062 4983	10 40	30	582 582	
10 19 40	19 9.4		0	26500		-107	726	
10 19 48	20 8.0	NGC 3222	0	5577	40	-102	13	
10 19 48	78 51.0	KDG 232	0	2754	22	182	908	
10 20 11	71 7.9	DDO 77	R	1011	10	149	492	
10 20 30	18 12.0	MRK 630	0	3560		-111	438	
10 20 36	53 21.0	VV 312A	0	9675	14	65	594	
10 20 36	53 21.0	VV 312B	0	9351	26	65	594	
10 20 43	20 9.2	NGC 3226 ARP 94 VV 209 KDG 234A HOL 187B	R 0 0 0 0	1138 1338 1543 1349 1444	15 15 42	-102	497 13 594 185 106	
10 20 48	11 13.0	MRK 721A IC 606	0	9463		-143	928	*
10 20 48	11 13.0	MRK 721B IC 606	0	9863		-143	928	*
10 20 48	20 6.0	NGC 3227 ARP 94 VV 209 KDG 234B HOL 187A	R R R R R R 0 0 0 0 0 0 0	1106 1152 1284 1165 1199 1260 1080 1140 1010 1250 1150 1111 1175	20 25 32 40 60 60 42 38 30 20	-102	707 779 693 829 390 267 816 816 112 594 2 13 185	SEYF
10 21 24	-28 59.0		0	17660	90	-275	655	
10 21 59	15 0.6	DDO 79	R R	1390 1370	10 20	-126	492 424	

166

R.A. (1950)	DEC. (1950)	NAME	OBS	HEL VEL (C*Z)	ERR	GAL CORR	REF	COMMENTS
10 22 1	-32 13.7	NGC 3241	O	2874	59	-280	741	
10 22 11	55 47.0		O	7621	30	77	741	
10 22 23	51 55.7	MRK 142	O	13400		58	233	SEYF
10 22 24	17 25.0	NGC 3239 VV 95 KDG 236 ARP 263	R R O	755 751 880		-114	744 693 13	
10 23 0	11 59.0		O	2342	35	-139	813	
10 23 0	80 3.0	NGC 3212 KDG 237	O	9767	26	187	908	
10 23 18	-39 35.0	NGC 3244	O	(2640)	100	-288	844	
10 23 28	62 35.4	MRK 143	O O	9650 9300		110	436 308	
10 23 48	56 31.5	MRK 32	O O	851 880	60	81	261 280	
10 23 54	44 15.7	MRK 144	O O	8274 8232		19	341 278	
10 24 0	80 2.0	NGC 3215 KDG 237	O	9428	90	187	908	
10 24 12	28 54.0	NGC 3245A	O	1486	100	-58	789	
10 24 21	-39 41.4	NGC 3250	O	2871	49	-288	741	
10 24 24	- 3 12.0	IC 614	O	2252	250	-202	556	
10 24 24	10 39.0		O O	19636 19700	50 300	-145	13 2	
10 24 30	20 42.0		O	5760		-98	449	
10 24 30	28 46.0	NGC 3245	O	1261	30	-58	13	
10 24 36	- 3 4.0		O	11442	64	-201	843	
10 24 48	68 40.3	IC 2574 DDO 81 7 ZW330	R R R R R R O	52 45 50 55 47 38 28	10 5 10 10	139	489 156 402 934 93 492 13	

10ʰ25ᵐ

R.A. (1950)			DEC. (1950)		NAME		OBS	HEL VEL (C*Z)	ERR	GAL CORR	REF	COMMENTS
10	25	6	-28	32.0			0	8973		-273	683	
10	25	6	22	0.0	2 ZW	46	0	12617		-92	262	
10	25	12	19	45.0	2 ZW	47	0	12372		-103	262	
10	25	18	67	3.7			R	1120	10	131	860	
10	25	42	-43	38.9	NGC 3256		0	2758	30	-291	49	
					VV	65	0	2954			145	
							0	2606	120		719	
10	25	47	40	6.0	MRK	415	0	9000		-1	359	
10	25	48	12	58.0	NGC 3253		R	9682	20	-134	582	
							0	9711	20		582	
10	25	48	19	48.4			R	1111	10	-102	860	
10	25	48	40	6.0	MRK	415	0	8790		-1	438	
10	26	0	-31	15.0	IC 2580		R	3132	10	-277	752	
							0	3137	40		582	
10	26	12	4	28.0			0	2180	100	-171	820	
10	26	30	29	45.0	NGC 3254		R	(1366)		-53	693	
							0	1223			641	
							0	1228	60		13	
10	26	31	-35	24.2	NGC 3257		0	3063	67	-283	741	
10	26	37	70	18.4	DDO	80	R	1916	15	146	492	
					VV	294						
10	26	38	-35	21.0	NGC 3258		0	2848	31	-283	741	
10	26	51	-35	20.4	NGC 3260		0	2453	54	-283	741	
10	26	54	-44	24.0	NGC 3261		0	2612	50	-291	741	
10	27	8	16	26.4	MRK	631	0	3215		-118	438	
10	27	16	-35	0.2			0	1892	50	-283	741	
10	27	34	-35	4.0	NGC 3267		0	3760		-283	574	
							0	3749	45		741	
10	27	36	- 2	55.0			0	11460		-200	750	
							0	8960	66		843	
10	27	38	-35	7.2			0	1781	32	-283	741	

R.A. (1950)	DEC. (1950)	NAME	OBS	HEL VEL (C*Z)	ERR	GAL CORR	REF	COMMENTS
10 27 42	-34 58.1	NGC 3269	O	3820		-282	574	
			O	3794	44		741	
10 27 46	-35 4.2	NGC 3268	O	2801	44	-282	741	
10 28 6	-30 8.0		O	4170	60	-275	655	
10 28 12	-35 6.2	NGC 3271 IC 2585	O	3824	21	-282	741	
10 28 14	-35 21.4	NGC 3273	O	2459	70	-283	741	
10 28 19	34 45.6		R	1493	10	-27	860	
10 28 24	26 18.7	IC 2583	O	8470		-70	558	
10 28 36	4 44.0		O	1307	15	-169	813	
10 28 39	-36 29.0	NGC 3275	O	3241	40	-284	741	
10 28 42	25 7.0	NGC 3270	O	6262	60	-75	813	
10 28 42	25 7.0	NGC 3270/A	O	(275)		-75	154	
10 28 48	29 6.0	A	O	17865	30	-56	396	SEYF
10 28 48	29 6.0	B	O	18015	30	-56	396	SEYF
10 28 58	25 33.9		R	1277	10	-73	860	
10 29 5	65 18.1	NGC 3259	O	1703		124	910	
			O	1866			13	
10 29 22	54 39.4	MRK 33 ARP 233	R	1620		73	368	
			O	1410			264	
			O	1374	20		178	
			O	1620			224	
10 29 30	27 55.6	NGC 3274	O	536	27	-61	741	
			O	530	41		789	
10 29 40	-34 35.3	NGC 3281	O	3435	26	-281	741	
			O	3460			574	
10 30 18	28 47.0	NGC 3277	O	1460	75	-57	13	
10 30 19	58 30.1	3CR244.1	O	128300		92	946	
10 30 52	60 17.3	MRK 34	O	15450		100	280	SEYF
			O	15200			224	
10 30 54	61 53.0		O	(6192)		108	774	

10ʰ31ᵐ

R.A. (1950)	DEC. (1950)	NAME	OBS	HEL VEL (C*Z)	ERR	GAL CORR	REF	COMMENTS
10 31 36	39 53.0	KDG 238A	O	12862	98	0	908	
10 31 39	35 30.9		R	1516	10	-23	829	
10 31 42	39 54.0	KDG 238B	O	13177	50	0	567	
			O	12951	30		908	
10 31 44	64 42.4	MRK 145	O	10019		121	341	
			O	10800			308	
10 32 4	21 54.6	NGC 3287	R	1307	10	-90	829	
			R	1305	10		914	
			O	959	100		555	
			O	1302	50		741	
10 32 5	46 49.1	MRK 146	O	3302		34	341	
			O	3600			233	
10 32 18	-28 19.0		O	3630	60	-271	655	
10 32 26	63 47.7	MRK 147	O	7033		117	279	
			O	7012			341	
10 32 36	-16 54.0	NGC 3290 ARP 53	O	10616		-244	321	
10 32 37	44 34.4	MRK 148	O	7200		23	308	
10 32 41	-72 59.0	IC 2596	O	3390		-248	896	
10 33 0	-24 29.8	DDO 238	R	1048	10	-263	492	
10 33 6	58 51.0	NGC 3286 KDG 239A	O	7584	61	94	594	
10 33 12	58 49.0	NGC 3288 KDG 239B	O	7588	52	94	594	
10 33 24	37 35.0	NGC 3294 HOL 202A	R	1569	50	-12	544	
			R	1592			744	
			O	1469			13	
10 33 43	-27 14.99		O	2389	96	-268	527	
			O	4147	50		741	DIS
10 33 50	-27 11.70		O	4829	25	-268	527	
10 33 50	-26 54.20	NGC 3305	O	3963		-267	633	
			O	3971	75		527	
10 33 54	31 48.4	DDO 83 ARP 267	R	584	10	-41	492	

R.A. (1950)	DEC. (1950)	NAME	OBS	HEL VEL (C*Z)	ERR	G/L CORR	REF	COMMENTS
10 33 58	14 25.8	NGC 3300	O	2992	31	-125	741	
			O	3094			910	
10 34 1	-27 10.71	NGC 3308	O	3578		-268	633	
			O	3513	23		527	
			O	3729	20		741	
10 34 3	-26 44.60		O	4158		-267	633	
10 34 4	-27 19.35		O	2742	56	-268	527	
10 34 6	-27 3.57		O	3292		-268	633	
10 34 8	-27 13.48		O	4927	133	-268	527	
10 34 12	22 9.0	NGC 3301	O	1333	75	-88	13	
10 34 14	-27 15.49	NGC 3309	O	4052		-268	633	
			O	3710			255	
			O	4086	38		741	
			O	4118	98		527	
10 34 18	18 23.0	ARP 192S	O	6060	100	-106	288	
10 34 18	18 23.0	ARP 192N	O	6317	100	-106	288	
10 34 20	-27 18.2		O	4736	40	-268	741	
10 34 21	-27 16.07	NGC 3311	O	3743		-268	633	
			O	3622	21		741	
			O	3835	103		527	
10 34 24	-27 12.58		O	2772	59	-268	527	
			O	2280			792	
10 34 24	18 24.0	NGC 3303A ARP 192 VV 71 KDG 240A	O	6470	50	-106	141	
10 34 24	18 24.0	NGC 3303B ARP 192 VV 71 KDG 240B	O	6397		-106	141	
10 34 31	12 54.8	NGC 3306	R	2884	15	-132	829	
10 34 32	-41 38.4		O	25500		-238	675	
10 34 33	-27 33.70		O	2753	16	-269	527	
10 34 39	64 31.5	MRK 149	O	1619	80	121	300	
			O	500			233	DIS

R.A. (1950)	DEC. (1950)	NAME	OBS	HEL VEL (C*Z)	ERR	GAL CORR	REF	COMMENTS
10 34 41	-27 18.30	NGC 3312	O	2780	50	-268	527	
		IC 629	O	2872			633	
			O	2811	23		741	
			O	2533	75		72	
10 34 51	-27 25.43	NGC 3314	O	2916	120	-268	527	
10 34 56	-27 12.54		O	4858		-268	633	
			O	4835	13		741	
10 34 58	-26 55.94		O	3840		-267	633	
10 34 59	5 53.0	KDG 241A	O	8325	30	-162	908	
10 35 0	5 51.0	IC 628 KDG 241B	O	8718	60	-162	908	
10 35 6	-41 22.0	NGC 3318	O	2610	100	-287	844	
			O	2887			154	
10 35 16	-27 20.02	NGC 3316	O	3887		-268	633	
			O	3609	91		72	
			O	3960	54		527	
10 35 22	-39 0.7		O	15000		-285	675	
10 35 26	-26 49.30		O	3012		-267	633	
			O	3030	103		527	
10 35 40	44 46.9	MRK 150	O	3660		24	341	
			O	3900			233	
			O	3720	80		300	
10 35 42	53 46.0	NGC 3310	R	981	25	69	238	
		ARP 217	R	1010	50		544	
			R	970	15		455	
			O	1090	13		112	
			O	1041	19		112	
			O	1050	23		103	
			O	1021			641	
			O	1019	9		596	
			O	986	5		164	
			O	1039	30		13	
			O	998			13	
10 35 46	54 14.0		O	21177	87	72	741	
10 35 54	- 6 54.0	IC 630 MRK 1259	O	2161		-211	920	
10 36 14	41 56.8	NGC 3319	R	743		10	693	
			R	744			744	
			R	742	25		183	
			R	760	40		216	
			O	749			641	
			O	790			483	

R.A. (1950)	DEC. (1950)	NAME	OBS	HEL VEL (C*Z)	EPR	GAL CORR	REF	COMMENTS
10 36 14	41 56.8	NGC 3319	O	826		10	13	
10 36 30	48 12.0	KDG 242	O	842	10	42	908	
10 36 36	47 39.0	NGC 3320	R	2331		39	744	
10 36 48	- 0 8.0	IC 633	O	5400		-186	491	
10 36 48	48 11.0	KDG 242	O	1476	12	42	908	
10 36 54	5 22.0	NGC 3326	O	7800		-164	491	
10 37 42	39 19.0	KDG 243	O	9017	13	-2	908	
10 37 43	30 13.5	4C 30.19	O	27270	250	-48	696	
			O	27800			600	
			O	27314	40		741	
10 37 48	39 20.0	KDG 243	O	8916	10	-2	908	
10 37 56	-27 30.98		O	3970		-267	527	
10 38 7	-46 3.9		O	6199	55	-289	680	
10 38 8	-46 3.9		O	5429	70	-289	680	
10 38 10	-46 3.7		O	6159	30	-289	680	
10 38 48	63 16.0	7 ZW342	O	12040	200	116	820	
10 39 2	-23 7.6	DDO 85	R	1199	15	-258	492	
10 39 15	48 1.7	MRK 151	O	1515		42	341	
			O	1500			233	
10 39 30	14 0.0	NGC 3338	R	1299		-125	744	
			R	1306			934	
			R	1298			693	
			O	1330			13	
10 39 48	16 1.0	KDG 245	O	14989	95	-116	908	
10 39 51	34 42.8	DDO 84	R	633	10	-24	492	
			O	599	40		964	
		VV 794	O	599	50		873	
10 39 54	16 0.0	KDG 245	O	14608	103	-116	908	
10 40 10	13 43.2	VV 112	R	1210	3	-126	745	
10 40 13	13 55.0		O	1070		-125	591	
10 40 24	20 40.9	MRK 416	R	1308		-94	779	
			O	1320			438	
			O	1200			359	

10ʰ40ᵐ

R.A. (1950)	DEC. (1950)	NAME	OBS	HEL VEL (C*Z)	ERR	GAL CORR	REF	COMMENTS
10 40 30	-36 5.3	NGC 3347	O	2923	25	-281	741	
10 40 31	31 46.8		O	10840		-39	485	
			O	10541	50		741	
10 40 31	77 4.3	NGC 3329	O	1689		176	321	
			O	1948	25		741	
10 40 47	25 11.1	NGC 3344	R	728	5	-72	390	
			R	590			934	
			R	589	10		216	
			R	581	25		183	
			O	576			641	
			O	579	150		13	
10 40 54	12 22.0	KDG 246	O	7922	31	-132	908	
10 40 59	15 8.1	NGC 3346	R	1266	10	-120	752	
			R	1258	4		914	
			O	1163	30		813	
			O	1100	50		741	
10 41 0	12 21.0	KDG 246	O	7893	30	-132	908	
10 41 8	16 9.2	MRK 632	O	11900		-115	438	
			O	12075			557	
10 41 16	-36 8.8	NGC 3358	O	2910	85	-280	741	
10 41 16	56 41.1		R	815	10	85	860	
10 41 17	60 37.8	DDO 86	R	1018	10	104	492	
10 41 18	11 58.0	NGC 3351	R	779	3	-134	578	M 95
			R	776			693	
			R	779	10		752	
			O	779	9		578	
			O	688	200		13	
			O	913	8		507	
			O	794	53		72	
			O	796	25		283	
10 41 24	- 1 1.0		R	7808		-188	707	SEYF
			O	7800			491	
10 41 36	7 1.3	NGC 3356 VV 529	R O	6184 5800	10	-155	829 535	
10 42 16	56 13.4	NGC 3353 MRK 35	O O	970 1020	12	83	741 224	
			O	862	10		178	
10 42 17	6 51.8	NGC 3362	O	8360		-156	895	

174

R.A. (1950)	DEC. (1950)	NAME	OBS	HEL VEL (C*Z)	ERR	GAL CORR	REF	COMMENTS
10 42 38	29 28.9		O	6350		−50	558	
10 42 46	9 40.3		O	16250		−144	895	
10 43 20	63 29.2	NGC 3359	R	1008	5	117	512	
			R	1018			934	
			R	1003	25		183	
			P	1010			156	
			O	1010			483	
			O	1008			13	
10 43 24	73 6.0	NGC 3348	O	2855	75	160	13	
10 44 0	− 2 27.0		O	6300		−193	491	
10 44 0	14 1.0	NGC 3367	R	3040	20	−124	544	
		4C 14.37	O	2879	100		13	
10 44 8	12 5.1	NGC 3368	R	966	32	−132	390	M 96
			R	900			934	
			R	831			310	
			R	891			693	
			O	940	50		1	
			O	924	46		72	
			O	927	40		13	
10 44 11	9 18.6		O	68000	300	−145	583	
10 44 16	54 18.1		R	767	10	74	860	
10 44 18	9 18.9		O	27400	300	−145	583	
10 44 18	26 48.0	VV 727	O	(6200)		−63	776	
10 44 30	17 32.0	NGC 3370	R	12 /	4	−107	914	
			O	1400			13	
10 44 44	14 19.9	NGC 3377A DDO 88	R	571	10	−122	492	
10 45 6	14 15.0	NGC 3377	O	718	40	−122	13	
10 45 6	66 38.0	KDG 248	O	6868	18	132	908	
10 45 12	12 51.0	NGC 3379 HOL 212A	O	963	40	−129	72	M 105
			O	754	155		555	
			O	780	50		1	
			O	845	50		1	
			O	905	16		742	
			O	909	13		938	
			O	862	30		13	
10 45 12	66 37.0	KDG 248	O	6594	25	132	908	

R.A. (1950)	DEC. (1950)	NAME	OBS	HEL VEL (C*Z)	ERR	GAL CORR	REF	COMMENTS
10 45 36	34 58.0	NGC 3381	R	1627	9	-22	914	
10 45 42	12 54.0	NGC 3384	O	699	40	-128	72	
		HOL 212B	O	781	30		13	
10 45 42	26 51.0		R	6296	10	-62	582	
			O	6295	20		582	
10 45 44	-31 16.2	NGC 3390	O	2850	40	-272	741	
10 45 50	12 47.9	NGC 3389	O	1257		-129	157	
		HOL 212C	O	1334			13	
10 45 52	-72 9.8		O	7700		-268	896	
10 45 54	50 18.2	MRK 152	O	6975		54	341	
			O	6883			278	
10 46 0	26 19.5		R	6432		-65	366	
			O	6447	67		246	
10 46 4	52 35.8	MRK 153	O	2300		66	233	
10 46 17	65 47.7	DDO 87	R	336	15	128	492	
10 46 42	33 1.0	VV 538	O	1680	50	-31	873	
10 46 48	23 13.8	MRK 417	O	9850		-79	374	
			O	9800			438	
10 47 2	33 14.7	NGC 3395/6	R	1605		-30	283	
		ARP 270	R (	1625)			914	
		VV 246	O	1622	25		283	
10 47 2	33 14.7	NGC 3395	R	1681		-30	744	★
		HOL 215A	P (	1621)	9		581	
		KDG 249A	O	1644			253	
			O	1751			13	
			O	1633	30		8	
			O	1530			558	
			O	1622			242	
10 47 8	33 15.3	NGC 3396	O	1651		-30	253	
		HOL 215B	O	1643			13	
		KDG 249B	O	1712	29		242	★
			O	1637	30		8	
10 47 28	9 20.8		O	9830		-143	895	
10 47 42	- 1 2.0		R	4544	15	-186	582	
			O	4531	20		582	
10 47 49	19 54.5	DDO 89	R	1255	10	-95	492	

R.A. (1950)	DEC. (1950)	NAME	OBS	HEL VEL (C*Z)	ERR	GAL CORR	REF	COMMENTS
10 47 49	50 26.0	MRK 154	O	12900		56	308	
10 48 18	1 27.0	IC 649A KDG 252A	O	11563	12	-176	908	
10 48 18	1 27.0	IC 649B KDG 252B	O	11614	10	-176	908	
10 48 18	13 41.0	NGC 3412	O	861	75	-124	13	
10 48 24	44 50.1	MRK 155	O	1800		28	233	
10 48 30	- 1 52.0	IC 651	O	4200		-189	491	
10 48 36	6 6.3	NGC 3423	R O	1013 865	55	-157	744 741	
10 48 36	28 14.5	NGC 3414 ARP 162	R O	1414 1449	100	-54	610 13	
10 48 42	14 13.0	NGC 3419	R O	3021 2982		-121	610 13	
10 48 42	14 17.0	NGC 3419A	O	3013	50	-121	789	
10 48 50	43 58.7	NGC 3415	R O	3303 3207	20 70	24	829 741	
10 48 56	44 1.8	NGC 3416	O	3276		24	558	
10 49 0	-34 9.0		O	1410	90	-275	655	
10 49 0	33 10.0	NGC 3424 HOL 218A	O	1421	47	-30	789	
10 49 6	58 42.1	NGC 3408	O	9604	32	96	741	
10 49 6	61 39.0	NGC 3407	O	4994		110	471	
10 49 18	59 57.1		O	8417	91	102	741	
10 49 24	33 13.0	NGC 3430 IC 2613 HOL 218B	R R R O O O	1583 1594 1583 1742 1604 1504	10 71 25	-29	283 693 752 13 789 283	*
10 49 26	10 24.8	NGC 3433 ARP 206 VV 11	R R O	2719 2720 2621	10 10 48	-138	829 752 741	
10 49 30	30 20.0	KDG 254A	O O	10808 10416	70 30	-44	567 908	

177

R.A. (1950)	DEC. (1950)	NAME	OBS	HEL VEL (C*Z)	ERR	GAL CORR	REF	COMMENTS
10 49 35	6 25.8		0	15410		-155	895	
10 49 36	30 21.0	KDG 254B	0	10532	82	-43	908	
10 49 42	36 53.1	NGC 3432	R	641	25	-11	183	
		ARP 206	R	615			693	
		VV 11	0	670			125	
			0	609			13	
10 49 54	7 30.0	NGC 3441	0	6600		-150	491	
10 49 54	23 12.0	NGC 3437	R	1275	7	-78	914	
			0	1198	21		813	
			0	1149	72		741	
10 50 6	73 57.0	NGC 3403	0	1244		164	13	
10 50 11	50 33.0	MRK 156	0	1223	80	57	300	
			0	1400			233	
10 50 21	34 10.6	NGC 3442	0	1710		-24	438	
		MRK 418	0	1780	45		562	
			0	1800			359	
10 50 24	4 54.0		0	5700		-161	491	
10 50 34	-32 39.7	NGC 3449	0	3267	48	-273	741	
10 50 42	17 2.0	NGC 3447	R	1062		-107	283	
		VV 252	R	1071	9		581	
		KDG 255	0	1067	25		283	
			0	965	47		72	
10 50 48	17 2.0	NGC 3447A	0	1014	47	-107	72	
		VV 252						
		KDG 255						
10 51 12	54 34.0		R	1480	20	76	684	★
			R	1500	10		833	
10 51 33	56 13.9		0	14520	50	85	741	
10 51 34	57 15.3	NGC 3445	0	1984	64	89	148	
		KDG 256A	0	1969	30		908	
		ARP 24						
		VV 14						
10 51 36	57 14.0	KDG 256B	0	1915	25	89	908	
		ARP 24						
		VV 14						
10 51 40	54 34.5	NGC 3448	R	1370	15	77	833	★
		ARP 205	P	1380	20		455	
			R	1360	10		497	
			R	1350	20		684	
			R	1321	24		390	

R.A. (1950)	DEC. (1950)	NAME	OBS	HEL VEL (C*Z)	EPR	GAL CORR	REF	COMMENTS
10 51 40	54 34.5	NGC 3448	O	1399		77	641	
			O	1333	32		149	
			O	1404	12		148	
10 51 40	54 34.5	NGC 3448W	O	1248	55	77	684	
10 51 40	54 34.5	NGC 3448E	O	1417	55	77	684	
10 51 41	61 33.3	NGC 3435	R	5158	10	110	829	
			O	5141	61		741	
10 51 49	17 36.7	NGC 3454	R	1111		-104	893	
		KDG 257	O	1157	15		908	
		HOL 221B	O	1149			253	
			O	1194			57	
10 51 52	17 33.1	NGC 3455	R	1102	10	-105	829	
		KDG 257	R	1109			893	
		HOL 221A	O	1149	15		908	
			O	1140			57	
			O	1104			253	
10 52 6	49 59.6	MRK 157	R	1353	20	54	829	
			O	1400			233	
			O	1365	80		300	
10 52 14	40 43.0		O	34990	80	8	395	
			O	34950	92		911	
10 52 15	-20 48.1	NGC 3464	O	3836		-248	741	
10 52 15	40 43.2		O	34560	125	8	911	
10 52 16	40 43.2		O	35020	80	8	395	
			O	35190	38		911	
10 52 16	40 43.5		O	35140	80	8	395	*
			O	34680	53		911	
10 52 16	40 43.5		O	34960	80	8	395	
			O	35010	60		911	
10 52 18	40 43.6		O	(35050)	80	8	395	
			O	33120	115		911	
10 52 18	40 43.8		O	33180	210	8	911	
10 52 24	58 9.0	VV 628	O	6924	50	94	873	
			O	7085	100		776	
10 52 48	17 24.6		R	1069	10	-105	860	
10 52 58	57 23.0	NGC 3458	O	1800	32	90	741	
			O	1846			910	

R.A. (1950)	DEC. (1950)	NAME	OBS	HEL VEL (C*Z)	ERR	GAL CORR	REF	COMMENTS
10 53 30	6 26.0		O	1000		-154	491	
10 53 42	6 26.0		O	1354		-154	683	
10 54 30	37 50.0	MRK 633	O	10730		-5	438	
10 55 0	1 52.9		O	11450		-172	255	
10 55 6	57 2.0		O	39914	100	89	13	
10 55 21	20 45.0	MRK 634	O	19828		-88	874	
			O	19020			557	
10 55 24	57 3.0		O	19150	100	89	13	
10 55 24	72 53.0	KDG 258B	O	8019	15	160	908	
10 55 25	72 54.7	MRK 159	O	7900		160	231	
		KDG 258A	O	7900			233	
			O	8057	15		908	
10 55 40	59 46.8	NGC 3470	O	6669	43	102	741	
		KDG 259A	O	6635	43		908	
10 55 42	57 2.0		O	41631	300	89	13	
10 55 48	24 28.0	MRK 419	O	6765		-70	438	
10 55 48	59 45.0	KDG 259B	O	6696	20	102	908	
10 56 2	61 47.7	NGC 3471	R	2254	100	112	885	
		MRK 158	O	2047			341	
			O	2067	80		300	
			O	2000			233	
10 56 35	46 23.4	NGC 3478	O	6688	42	38	741	
10 56 54	50 17.0	KDG 260	O	7377	42	57	908	
10 57 0	50 19.0	KDG 260	O	7178	44	57	908	
10 57 6	8 17.3		O	8390		-144	895	
10 57 6	29 55.0		O	625		-43	347	
10 57 6	74 25.3	4C 74.17	O	32100	600	167	824	
10 57 21	-66 3.8		O	1470		-276	896	
10 57 24	15 6.6	NGC 3485	R	1436	4	-114	914	
			O	1513			741	
10 57 41	14 10.2	NGC 3489	R	695		-118	780	
			R	588			610	
			O	711			106	
			O	692	65		13	

R.A. (1950)	DEC. (1950)	NAME	OBS	HEL VEL (C*Z)	ERR	GAL CORR	REF	COMMENTS
10 57 41	14 10.2	NGC 3489	O	688		-118	641	
			O	(600)			1	
10 57 42	29 14.6	NGC 3486	R	708		-46	363	
			R	685			934	
			R	681			744	
			O	1116	100		13	
10 58 8	10 49.3		O	10220		-133	726	
10 58 12	11 0.0		R	10805	20	-132	582	
			O	10793	20		582	
10 58 24	11 19.0	MRK 728	O	10430		-131	874	
10 58 38	10 45.0		O	12470		-133	726	
10 58 41	3 53.8	NGC 3495	R	1146		-162	744	
			O	1114	15		936	
			O	994	64		148	
10 58 47	-22 27.6		O	41400	300	-250	529	
			O	(41949)			444	
10 59 6	28 57.3		R	699	10	-47	860	
10 59 7	45 29.8	MRK 161	R	5990		34	368	
			O	6000			233	
			O	5912			341	
10 59 21	10 33.8		O	10334		-134	683	SEYF
10 59 54	2 54.0	KDG 261	O	11937	21	-166	908	
10 59 54	52 22.9		R	947	10	68	860	
10 59 57	17 0.2		R	1036	10	-104	829	
11 0 0	2 53.0	KDG 261	O	11977	32	-166	908	
11 0 0	50 56.0	KDG 262	O	6039	91	61	908	
11 0 0	50 57.0	KDG 262	O	6299	63	61	908	
11 0 12	18 15.0	NGC 3501 KDG 263A	O	1083	20	-98	908	
11 0 29	28 14.5	NGC 3504	R	1531	25	-50	885	
			O	1539			32	
			O	1513	50		13	
11 0 36	11 21.0	NGC 3506	R	6409	12	-130	914	
			O	6300			491	
			O	6489			741	

11ʰ0ᵐ

R.A. (1950)	DEC. (1950)	NAME	OBS	HEL VEL (C*Z)	ERR	GAL CORR	REF	COMMENTS
11 0 48	-22 50.0	NGC 3511	O	1221	31	-250	148	
11 0 48	3 36.0	A	O	7454	22	-163	594	
11 0 48	3 36.0	B	O	7293	69	-163	594	
11 0 48	18 24.0	NGC 3507	R	980	10	-97	752	
		KDG 263B	O	940	10		908	
11 0 48	70 42.0		O	25120		152	401	
11 0 53	38 11.9	MRK 420	O	12815		-1	438	
			O	12900			359	
11 1 1	29 9.3	NGC 3510	R	709		-46	366	
			O	717			163	
			O	719			13	
11 1 5	43 45.3	KDG 264	O	14927	20	26	908	
			O	15074			748	
11 1 6	41 7.2	ARP 148A	O	10346	60	13	13	MYI NEB
		VV 32	O	10363			90	
11 1 6	41 7.2	ARP 148B	O	10601		13	90	
		VV 32						
11 1 7	43 45.2	KDG 264	O	14851	19	26	908	
			O	14874			748	
11 1 19	-22 58.5	NGC 3513	O	1154	63	-250	789	
			O	1244	20		741	
11 1 20	28 18.3	NGC 3512	R	1388		-50	693	
			R	1370	7		914	
			O	1502			13	
11 1 20	38 30.5		O	9350		0	757	
11 1 40	38 28.6	B2	O	9000		0	485	
		MRK 421	O	(300)			374	DIS
			O	9000			757	
11 1 41	38 28.6		O	9500		0	757	
11 1 48	5 6.0	NGC 3509A/B	O	7600		-156	77	
		ARP 335						
		VV 75						
		KDG 265						
11 1 48	5 6.0	NGC 3509A	O	7670	32	-156	908	
11 1 48	5 5.0	NGC 3509B	O	7644	15	-156	908	

182

R.A. (1950)			DEC. (1950)		NAME	OBS	HEL VEL (C★Z)	ERR	GAL CORR	REF	COMMENTS
11	2	0	38	28.3		O	29800		0	757	
11	2	16	29	24.6	MRK 36	O	630		-44	280	
						O	660			224	
11	2	16	38	24.0		O	9350		0	757	
11	2	18	45	1.0	MRK 162	O	6300		33	233	
						O	6433	80		300	
11	2	26	38	17.6		O	9350		0	757	
11	2	26	38	30.0		O	8550		0	757	
11	2	36	56	47.0	NGC 3517A KDG 266	O	8214	30	90	908	
11	2	36	56	48.0	NGC 3517B KDG 266	O	(8386)	30	90	908	
11	2	40	30	25.9	B2	O	21640		-39	485	
11	3	16	20	5.7	DDO 91	R	1330	7	-88	492	
11	3	18	0	14.2	NGC 3521	R	810		-175	171	
						O	730	50		1	
						R	809			934	
						R	845	20		216	
						O	815			88	
						O	789	30		13	
11	3	23	72	50.4	NGC 3516	R	2503		161	310	SEYF
						O	2650			959	
						O	2505	100		288	
						O	2700			167	
						O	2593	17		789	
						O	2614	70		555	
						O	2614	50		13	
						O	2632			13	
11	3	35	48	54.3	MRK 163	O	7500		52	308	
11	3	48	11	41.0	KDG 267A	O	9370	46	-127	908	
11	3	53	57	57.6		O	9820	138	95	741	
11	3	54	11	40.0	NGC 3524 KDG 267B	O	1321	40	-127	908	
11	4	0	51	30.0		O	2097	33	65	813	
11	4	22	18	42.1		O	53400		-95	868	

11ʰ4ᵐ

R.A. (1950)	DEC. (1950)	NAME	OBS	HEL VEL (C*Z)	ERR	GAL CORR	REF	COMMENTS
11 4 24	7 26.0	NGC 3526	0	1164	28	−145	813	
11 4 39	18 42.0	VV 239B ARP 191 KDG 268	0	7982	80	−94	38	
11 4 41	18 42.1	VV 239A ARP 191 KDG 268	0	8211	77	−94	38	
11 5 6	19 49.0		R	1165	10	−89	860	
11 6 6	0 33.0	KDG 269	0	7613	18	−173	908	
11 6 6	0 34.0	KDG 269	0	(7531)	12	−173	908	
11 7 4	37 54.9		0	103700		0	400	
11 7 7	46 22.0		R	1391	10	41	860	
11 7 10	−37 4.8		0	2885	72	−274	741	
11 7 11	24 32.0	ARP 301A HOL 231A VV 229	0 0 0	6173 6301 6155	61	−66	594 253 57	
11 7 12	13 2.0	KDG 272	0	12716	20	−119	908	
11 7 12	13 3.0	KDG 272	0	12846	45	−119	908	
11 7 14	24 31.7	ARP 301B HOL 231B VV 229	0 0 0	6149 5983 5837	46	−66	594 253 57	
11 7 19	10 59.5	NGC 3547	R 0	1614 1543	23 28	−128	914 741	
11 7 24	37 14.0	NGC 3545A KDG 273A	0	8442	50	−3	908	
11 7 24	37 14.0	NGC 3545B KDG 273B	0	8771	50	−3	908	
11 7 36	−37 16.0	NGC 3557	0	3151	32	−274	741	
11 7 48	57 55.0	7 ZW377	0	9260	200	96	820	
11 7 55	29 2.3	NGC 3550A KDG 274A	0 0 0	10474 10520 10160	60	−44	347 255 908	
11 7 55	29 2.3	NGC 3550B KDG 274B	0	10196	113	−44	908	

R.A. (1950)	DEC. (1950)	NAME	OBS	HEL VEL (C*Z)	ERR	GAL CORR	REF	COMMENTS
11 7 59	28 35.0		O	10386		-46	347	
11 7 59	28 36.0		O	9576		-46	347	KNOT
11 8 4	53 39.4	NGC 3549	O	2847	22	95	741	
11 8 5	29 2.8		O	8731		-43	59	
11 8 13	-37 16.7	NGC 3564	O	2811	44	-274	741	
			O	2860			574	
11 8 14	23 44.3		O	(113320)		-69	585	
11 8 14	28 49.0	NGC 3558 MRK 422	O	9000		-45	359	
11 8 18	-48 49.9		O	2792	40	-282	680	
11 8 18	19 27.0	KDG 275	O	6233	14	-89	908	
11 8 24	19 28.0	KDG 275	O	6247	25	-89	908	
11 8 26	-37 10.7	NGC 3568	O	2476	6	-273	741	
11 8 30	9 54.3		O	1650		-133	895	
11 8 31	28 57.8		O	8939		-44	347	AMB KNT
			O	8839			59	
			O	8900			123	
11 8 31	28 58.1	NGC 3561B ARP 105 VV 237	O	8579		-44	347	
			O	8550	53		59	
11 8 31	28 59.2	NGC 3561A ARP 105 VV 237	O	8846		-44	347	
			O	8803	10		59	
11 8 37	55 56.7	NGC 3556	R	698	3	87	769	M 108
			R	695			171	
			R	701			934	
			R	699			744	
			R	700	65		216	
			R	694	25		183	
			R	670			156	
			O	636	75		13	
			O	719	47		789	
			O	680			483	
			O	735			30	
			O	650			13	
11 8 43	27 14.0	NGC 3563A	O	9980		-52	485	

11ʰ8ᵐ

R.A. (1950)			DEC. (1950)		NAME		OBS	HEL VEL (C*Z)	ERR	GAL CORR	REF	COMMENTS
11	8	43	27	14.0	B2		O	10952		-52	748	
11	8	44	27	14.1	NGC	3563B	O	10052		-52	748	
11	8	48	- 9	42.0			R	7780	10	-209	582	
							O	7772	20		582	
11	8	54	41	6.8	4C 41.23		O	22100		15	595	
11	9	2	-18	1.0	NGC	3571	O	3377	24	-234	741	
11	9	12	- 2	10.0			O	5400		-182	491	
11	9	34	51	54.4	MRK	164	O	3000		68	308	
11	10	6	9	20.0	MRK 731A IC 676		O	1398		-134	928	*
11	10	6	9	20.0	MRK 731B IC 676		O	1418		-134	928	*
11	10	15	10	28.6			R	1297	10	-129	860	
11	10	32	53	52.0	DDO	92	R	928	10	78	492	
11	10	36	26	8.0			R	6442	20	-57	752	
11	10	52	47	51.0			O	5351		49	558	
11	10	54	-26	29.0	NGC	3585	O	1491	75	-254	13	
11	10	54	48	33.0	NGC	3577	O	5219	78	52	594	
11	11	16	57	3.8			O	10015	20	93	741	
11	11	23	48	35.5	NGC	3583	R	2130	10	53	752	
							O	2112	35		741	
							O	2160	120		350	
							O	4834	126		594	DIS
11	11	55	56	51.1			O	10369	31	92	741	
11	11	59	13	5.5	NGC	3593	R	627	9	-117	581	
							R	621	6		818	
							R	617			610	
							R	550	35		775	
							O	668			106	
							O	547	75		13	
							O	630			210	
11	12	24	29	48.0			O	15008	65	-38	902	
11	12	29	15	3.6	NGC	3596	R	1202		-108	693	
							R	1160			363	
							O	1206	20		741	

186

R.A. (1950)	DEC. (1950)	NAME	OBS	HEL VEL (C*Z)	ERR	GAL CORR	REF	COMMENTS
11 12 30	29 48.0		0	14195	65	-38	902	
11 12 54	5 23.0	NGC 3601	0	7800		-150	491	
11 13 24	24 57.4	B2	0	30680	250	-61	696	
11 13 24	29 43.0		0	8334	65	-38	902	
11 13 30	-33 41.0		0	3005	75	-267	582	
11 13 30	29 30.0		0	13185	65	-39	902	
11 13 30	29 40.0		0	8660		-38	255	
			0	8967	65		902	
11 13 36	29 31.0		0	13665	65	-39	902	
11 13 42	29 30.0		0	14402	65	-39	902	
11 13 42	29 34.0		0	14267	65	-39	902	
11 13 42	29 36.0		0	13895	65	-39	902	
11 13 51	29 3.0	MRK 37	0	6965		-41	436	
11 13 53	29 31.0		0	14969		-39	347	
			0	14840			318	
			0	15023	34		741	
11 13 53	29 31.2	4C 29.41	0	14700		-39	485	*
			0	14450			318	
			0	14550			400	
			0	14610	68		741	
11 13 55	29 31.0		0	12989		-39	347	KNOT
11 14 12	18 17.0	NGC 3605 HOL 240C	0	693	65	-92	13	
11 14 17	18 19.7	NGC 3607 HOL 240A KDG 278A	0	951	40	-92	13	
11 14 24	18 26.0	NGC 3608 KDG 278B HOL 240B	0	1210	50	-92	13	
11 14 54	4 50.0	NGC 3611	R	1567	61	-151	818	
			R	1567	61		885	
			0	1754	75		13	
11 15 12	26 54.0	NGC 3609	0	5673	90	-51	594	

11ʰ15ᵐ

R.A. (1950)	DEC. (1950)	NAME	OBS	HEL VEL (C*Z)	ERR	GAL CORR	REF	COMMENTS
11 15 12	29 35.0		0	7132	65	-38	902	
11 15 24	- 1 49.0		R	7404	10	-178	582	
			0	7446	30		582	
11 15 26	54 1.3	MRK 38	0	11090		80	280	
			0	10664	60		261	
11 15 30	54 1.4	MRK 39	0	10893	60	80	261	
11 15 34	46 1.2	NGC 3614	0	2293		41	741	
11 15 36	26 54.0	NGC 3612	0	5521	150	-51	594	
11 15 36	28 33.0		0	9810		-43	449	
11 15 36	59 4.0	NGC 3610	0	1765	50	103	13	
			0	1850	75		2	
11 15 37	63 33.1	MRK 165	0	3237	80	123	300	
			0	3800			233	
11 15 40	19 7.3		0	1138		-88	558	
11 15 42	58 17.0	NGC 3613	0	2054	75	100	13	
11 15 54	-32 32.4	NGC 3621	R	731		-264	671	
			R	720	10		216	
			0	670	70		113	
			0	990	100		844	
			0	587	50		789	
11 16 18	13 22.0	NGC 3623	R	818		-114	693	M 65
		ARP 317	P	804	21		818	LEO TRI
		HOL 246B	R	813			804	
			0	800	50		1	
			0	814			47	
			0	705	50		13	
			0	620	43		72	
11 16 19	28 10.5	B2	0	20040	250	-45	696	
11 16 29	62 45.4	MRK 166	0	3140		120	341	
			0	3500			231	
			0	3500			233	
11 16 30	58 2.0	NGC 3619	0	1649	75	99	13	
11 16 45	51 46.1		0	1326	15	69	101	
			0	1389	12		477	
11 16 52	- 2 47.8		0	7290		-181	143	

R.A. (1950)	DEC. (1950)	NAME	OBS	HEL VEL (C*Z)	ERR	GAL CORR	REF	COMMENTS
11 17 30	18 38.0	NGC 3626	R	1473	12	-89	818	
		NGC 3622	P (	1537)	9		581	
			R (	1562)			645	
			O	1452	100		13	
11 17 36	- 2 47.0		O	7692	25	-180	582	
11 17 38	13 15.8	NGC 3627	R	716		-114	693	M 66
		ARP 16	R	740			171	
		HOL 246A	P	720	40		216	
			R	736			804	
			R	735			934	
			R	740			736	
			R	722			817	
			O	744	50		13	
			O	808	20		741	
			O	650	50		1	
			O	638	36		72	
11 17 40	13 52.1	NGC 3628	R	847		-111	693	
		ARP 317	R	842	25		183	
		VV 308	R	855			736	
		HOL 246C	R	847			934	
			P	849			804	
			O	842			13	
11 17 41	2 47.9	DDO 94	R	1609	15	-158	492	
11 17 42	3 14.3	NGC 3630	O	1514	22	-157	741	
			O	1537			321	
11 17 52	27 14.2	NGC 3629	R	1508		-48	744	
		HOL 247A	R	1523	28		914	
			O	1588	25		741	
11 18 4	23 44.3	3CR256	O	113320	300	-65	585	*
11 18 8	- 9 59.0	NGC 3637	O	1855	36	-206	741	
11 18 13	53 26.7	NGC 3631	R	1165	25	78	183	
		ARP 27	O	1087			13	
			O	1180	25		283	
11 18 18	21 37.0	KDG 280	O	6244	16	-75	908	
11 18 24	21 38.0	KDG 280	O	6156	30	-75	908	
11 18 30	3 31.0	NGC 3640	O	1354	40	-155	13	
11 18 38	34 37.9	IC 2738	O	10340		-12	255	
			O	13020			220	
11 19 5	20 26.7	NGC 3646	O	4524	200	-80	567	
		KDG 281	O	4185	16		45	
			O	4307	10		908	
			O	4278			748	

11ʰ19ᵐ

R.A. (1950)	DEC. (1950)	NAME	OBS	HEL VEL (C*Z)	ERR	GAL CORR	REF	COMMENTS
11 19 5	20 26.7	NGC 3646	O	4425		-80	13	
11 19 12	12 1.0	MRK 734	O	15009		-119	630	SEYF
11 19 21	18 8.8		O	59860		-91	726	
11 19 29	18 1.7		O	25930		-91	726	
11 19 30	59 21.0	NGC 3642	O	1623	50	105	13	
11 19 37	18 6.8		O	50250		-91	726	
11 19 37	20 29.0	NGC 3649	O	4689	130	-80	567	
		KDG 281	O	4524			748	
			O	4442			910	
			O	4402	24		908	
11 19 57	38 2.5	NGC 3652	O	2096		4	558	
11 20 17	16 51.7	NGC 3655	O	1492	27	-96	741	
11 20 32	30 12.9	MRK 635	O	7110		-33	557	
11 20 48	54 7.0	NGC 3656	O	2803	100	82	288	
		ARP 155	O	3003			777	
		VV 22						
		KDG 282						
11 20 48	54 7.0	KDG 282A	O	2878	30	82	908	
11 21 6	18 5.0	NGC 3659	R	1276	10	-90	914	
			O	1152	50		789	
			O	1354			741	
11 21 6	53 11.8	NGC 3657	O	1227		77	910	
11 21 21	21 45.3		O	64800		-73	717	
11 21 36	3 35.0	KDG 283A	O	1321	10	-153	908	
11 21 50	3 36.3	NGC 3664	R	1379	25	-153	492	
		ARP 5	O	1406	28		148	
		KDG 283B	O	1353	17		908	
		DDO 95						
		VV 251						
11 21 50	11 37.1	NGC 3666	O	1077	31	-119	741	
11 22 1	39 2.2	NGC 3665 B2	O	2002	50	10	13	
11 22 6	19 36.0	3CR258	O	49770	300	-83	585	
			O	49500			946	

R.A. (1950)	DEC. (1950)	NAME	OBS	HEL VEL (C*Z)	ERR	GAL CORR	REF	COMMENTS
11 22 24	-13 18.0		R	5384	15	-214	582	
			O	5392	30		582	
11 22 24	64 0.0		R	3726	5	127	582	
			O	3724	25		582	
11 22 30	- 9 31.2	NGC 3672	R	1867	10	-202	752	
			R	(1855)			693	
			O	1857			656	
			O	(2045)			13	
			O	1857	10		740	
11 22 30	63 43.0	NGC 3668	O	3625	50	125	810	
11 22 42	54 39.4	1 ZW 26	R	6080		85	389	SEYF
		MRK 40	R	6323	50		622	
		ARP 151	O	6250			122	
		VV 144	O	6162	17		86	
			O	6280			274	
			O	6169			215	
			O	6300			186	
			O	6250			389	
			O	6171			262	
			O	6060			224	
11 22 44	-26 27.8	NGC 3673	R	1933	15	-249	698	
			O	2396W			741	
11 22 54	23 5.0		O	6563	75	-66	582	
11 23 1	47 16.4	MRK 168	O	10219		50	341	
			O	10200			308	
11 23 22	20 22.4	4C 20.25	O	39480		-79	400	
11 23 24	-35 7.3		O	9860	120	-265	350	
11 23 24	43 52.0	NGC 3675	R	767		34	693	
			O	756			524	
			O	742	59		72	
			O	688	40		13	
11 23 28	-35 6.8	PKS	O	9690	60	-265	355	
			O	10070			317	
			O	9786	41		741	
11 23 33	47 15.0	NGC 3677 KDG 284A	O	7475	43	50	908	
11 23 34	35 36.7		O	10190		-5	255	
11 23 48	47 17.0	KDG 284B	O	7408	44	50	908	

11ʰ23ᵐ

R.A. (1950)	DEC. (1950)	NAME	OBS	HEL VEL (C*Z)	ERR	GAL CORR	REF	COMMENTS
11 23 53	59 25.8	MRK 169	O	1100		107	233	
		IC 691	O	1230			287	
			O	1461	80		300	
11 23 54	17 8.0	NGC 3681	R	1244		-93	744	
			O	1314	65		13	
			O	1146	63		882	
11 23 56	64 24.8	MRK 170	O	953		129	341	
			O	1050			308	
11 24 0	59 25.0	IC 691	O	1293		107	774	
11 24 0	63 42.3		O	3304	120	126	810	
11 24 8	35 31.0	MRK 423	O	9560		-6	438	
			O	9632			874	
			O	9600			359	
			O	9660			374	
			O	9600			557	
11 24 34	17 18.3	NGC 3684	R	1173		-92	693	
			R	1158	7		914	
			R	1171			744	
			O	1098	31		882	
			O	1422	75		13	
11 24 36	79 16.0	7 ZW403	O	-102	100	188	820	
			O	-110	50		430	
11 24 42	66 52.0	NGC 3682	O	1608	16	139	813	
11 24 48	9 0.5	IC 2828	O	1020		-129	895	
11 24 52	54 11.3		O	2895	32	83	477	
11 24 52	57 9.2	NGC 3683	O	1686	18	97	741	
11 24 54	-28 58.0		O	7454		-254	683	
11 25 0	8 31.0		O	5831		-131	683	
11 25 6	17 30.0	NGC 3686	R	1142		-91	693	
			R	1157			744	
			O	1148	22		882	
			O	1022	60		13	
11 25 18	- 0 57.0	KDG 286	O	(12531)	122	-170	908	
11 25 23	29 47.0	NGC 3687	O	2407	28	-33	741	
11 25 24	- 0 56.0	KDG 286	O	(12657)	83	-170	908	

R.A. (1950)	DEC. (1950)	NAME	OBS	HEL VEL (C*Z)	ERR	GAL CORR	REF	COMMENTS
11 25 30	22 16.0	VV 594	O	6200	100	-69	776	
11 25 33	17 11.8	NGC 3691	O	997	38	-92	882	
11 25 33	25 56.2	NGC 3689	O	2758	26	-51	741	
		4C 25.35	O	2690			166	
11 25 36	-36 15.0		R	3023	10	-266	752	
			O	2976	25		582	
11 25 42	58 50.0	MRK 171A/B	O	3000		105	287	
11 25 42	58 50.0	NGC 3690	O	2948	80	105	300	
		VV 118B	O	3051	22		908	
		KDG 288	O	2996	14		148	
		MRK 171B	O	1400			233	DIS
		ARP 299	O	3040			604	
			O	2973			341	
11 25 43	58 50.4	IC 694	O	3111	14	105	148	
		VV 118A	O	2998	80		300	
		KDG 288	O	3091	22		908	
		MRK 171A	O	1400			233	DIS
		ARP 299	O	3160			604	
11 25 48	9 41.0	NGC 3692	R	1740	19	-126	914	
11 26 14	35 41.4		O	2254		-4	558	
11 26 42	- 4 8.0	MRK 1298	O	18081		-181	920	
11 26 42	20 50.0	VV 498	O	1369	50	-75	873	
			O	1394	50		873	KNOT
11 26 48	58 44.0		O	10160		104	282	
11 26 51	22 3.5	MRK 172	O	10089		-69	437	
11 27 6	22 24.0		R	6502	10	-68	582	
			R	6500			858	
			O	6547	100		582	
11 27 17	-36 6.7	NGC 3706	O	3045	43	-265	741	
11 27 32	9 33.2	NGC 3705	O	1084	34	-126	741	
		HOL 259A						
11 27 33	20 47.9		O	38700		-75	717	
11 27 45	48 23.0	MRK 173	O	8400		57	308	
11 27 47	37 0.7	MRK 424	O	1980		2	438	
			O	2100			359	

R.A. (1950)	DEC. (1950)	NAME	OBS	HEL VEL (C*Z)	ERR	GAL CORR	REF	COMMENTS
11 28 12	20 45.0	IC 701 ARP 197 VV 3	O	6182	50	-75	141	
11 28 32	56 24.7	MRK 174	O	15300		94	308	
11 28 58	- 2 2.0		R	4745	5	-172	752	
11 29 4	-30 1.9	NGC 3717	O	1778	42	-254	741	
11 29 6	74 54.0	NGC 3752	O	1961	14	172	813	
11 29 7	71 5.0	ARP 329GRP	O	15580		157	77	*
11 29 7	71 5.1	VV 172E ARP 329	O	15480	60	157	188	
11 29 7	71 5.3	VV 172D ARP 329	O	15690	60	157	188	
11 29 8	71 5.6	VV 172C ARP 329	O	15820	60	157	188	
11 29 8	71 5.8	VV 172B ARP 329	O	36880	60	157	188	
11 29 8	71 6.0	VV 172A ARP 329	O	16070	60	158	188	
11 29 12	28 38.0	NGC 3714	O	6900		-37	491	
11 29 26	49 16.2		O	9863		61	550	
11 29 26	49 17.2		O	9363		62	550	
11 29 38	62 47.0	MRK 175	O	3648		123	341	
11 29 40	1 5.7	NGC 3719 KDG 289	O	5883	15	-160	908	
11 29 47	53 13.0	VV 150D	O O	8300 7500	 600	80	360 816	 DIS
11 29 48	1 4.8	NGC 3720 KDG 289	O O	6016 5912	28 10	-160	741 908	
11 29 50	53 13.5	VV 150B	O O O	8000 6700 8100	 300	80	360 40 816	 DIS
11 29 50	53 13.5	VV 150C	O	8700	600	80	816	
11 29 50	53 20.6	NGC 3718 ARP 214 KDG 290A	R R R R	990 1074 1017 990	30 25 10 10	81	770 183 497 455	

R.A. (1950)	DEC. (1950)	NAME	OBS	HEL VEL (C*Z)	ERR	GAL CORR	REF	COMMENTS
11 29 50	53 20.6	NGC 3718	R	990	50	81	544	
			R	1004	32		390	
			R	962	40		238	
			R	992			693	
			O	1012	10		908	
			O	1050	100		13	
11 29 51	49 23.1		O	10097		62	550	
11 29 54	53 13.5	VV 150GRP	O	8013		80	341	
		1 ZW 27	O	7960			262	
		ARP 322						
11 29 54	53 13.5	MRK 176	R	8208		80	707	SEYF
		VV 150A	O	7800			308	*
			O	7740	60		816	
			O	7800	60		816	
			O	7830	120		816	
			O	8070	90		816	
			O	7900			40	
11 29 55	62 6.2		R	3251	15	120	829	
11 30 0	18 25.2		O	12385		-85	558	
11 30 6	49 22.6		O	11251		62	550	
11 30 11	43 47.5		O	6864		36	558	
11 30 31	- 3 44.5	4C-03.43	O	14610		-178	318	
11 30 34	49 33.6		O	3180		63	550	
11 30 36	47 18.0	NGC 3726	R	863		53	693	
			R	(762)			183	
			R	869			744	
			O	948	75		13	
11 30 37	55 20.9	MRK 177	O	1800		90	308	
11 30 42	32 52.0	KDG 291	O	6253	22	-16	908	
11 30 45	49 30.7	MRK 178	R	200		63	368	
			O	218			341	
			O	0			233	
			O	218	80		300	
			O	318			550	
11 30 51	62 9.9	NGC 3725	O	3180	80	121	300	
		MRK 179	O	4400			233	DIS
			O	3284	44		741	
11 31 0	21 39.0	IC 707	O	6600		-69	491	

11ʰ31ᵐ

R.A. (1950)	DEC. (1950)	NAME	OBS	HEL VEL (C*Z)	ERR	GAL CORR	REF	COMMENTS
11 31 0	32 54.0	KDG 291	O	6419	90	-16	908	
11 31 4	53 24.4	NGC 3729	R	1000	50	81	770	
		KDG 290B	O	1035	11		741	
			O	998	29		908	
			O	1000	200		770	
			O	1020	40		770	
11 31 16	49 20.3	IC 708	O	9613		62	550	
11 31 21	21 22.1	4C 31.32	O	20000		-70	189	
11 31 31	49 19.1	IC 709	O	9549		62	550	
11 31 41	- 9 34.2	NGC 3732	O	1712	8	-198	741	
11 32 3	49 13.9	IC 711	O	9724		62	550	
11 32 6	49 21.2	IC 712	O	10054		63	550	
			O	9980			255	
11 32 43	-45 6.7		O	4948	85	-273	680	
11 32 47	49 23.0		O	11307		63	550	
11 32 52	-45 26.7		O	26400		-273	675	
11 32 56	55 13.4	NGC 3737	O	5580		90	255	
		HOL 266A						
11 32 59	49 4.8		O	11179		62	550	
11 33 0	16 15.0	MRK 636	O	5245		-93	438	
11 33 1	49 26.6		O	10454		63	550	
11 33 1	50 0.9		O	9727		66	550	
11 33 4	49 25.6		O	9870		63	550	
11 33 4	54 47.9	NGC 3738	O	176	29	88	741	
		ARP 234	O	163	43		789	
11 33 6	70 48.6	NGC 3735	O	2738	39	157	789	
			O	2627	42		741	
11 33 22	49 43.6		O	9467		65	550	
11 33 36	20 48.0		O	6530		-72	449	
11 33 48	21 52.0	NGC 3758A/B	O	9067		-67	630	SEYF
11 33 48	21 52.0	NGC 3758A	O	8854	40	-67	831	
		MRK 739A						

R.A. (1950)		DEC. (1950)		NAME		OBS	HEL VEL (C*Z)	ERR	GAL CORR	REF	COMMENTS
11 33 48		21 52.0		NGC 3758B MRK 739B		O	8940	39	-67	831	
11 33 49		49 24.5				O	11223		64	550	
11 33 54		36 41.3		NGC 3755		R	1565	15	3	752	
11 33 54		49 20.4				O	9641		63	550	
11 33 58		55 7.4		MRK 41 IC 2943		O	5795		90	278	
11 34 5		54 34.3		NGC 3756		O	1101	48	88	741	
11 34 11		55 6.0		NGC 3759		O	5530		90	910	
11 34 18		20 12.2		MRK 182		O	6304		-74	437	
11 34 18		20 14.0		MRK 181		R O O O	6201 6300 6200 6192	35	-74	779 308 644 341	
11 34 30		3 7.0		KDG 292		O	8745	21	-149	908	
11 34 36		3 6.0		KDG 292		O	8734	12	-149	908	
11 34 36		18 7.0		NGC 3768		O O	3301 3372	22	-84	644 703	
11 34 36		39 32.0		KDG 293		O	9391	17	17	908	
11 34 36		39 33.0		KDG 293		O	9178	12	17	908	
11 34 42		62 2.3				O	3549	100	121	810	
11 35 2		48 10.3		NGC 3769 ARP 280 KDG 294A HOL 270A		O O	737 740	50	58	253 8	
11 35 6		24 24.4				R	1572	10	-54	860	
11 35 6		48 9.0		NGC 3769A ARP 280 KDG 294B HOL 270B		O	709		58	253	
11 35 7		22 17.2		NGC 3746 ARP 320C VV 282		O	9225		-64	274	COP SEP
11 35 8		22 18.0		NGC 3745 ARP 320B VV 282		O	9401		-64	274	

11ʰ35ᵐ

R.A. (1950)	DEC. (1950)	NAME	OBS	HEL VEL (C*Z)	ERR	GAL CORR	REF	COMMENTS
11 35 12	- 7 0.0		R	9480	20	-187	582	
11 35 13	22 18.2	NGC 3748 ARP 320A VV 282	O	8977		-64	274	
11 35 15	22 15.1	NGC 3750 ARP 320G VV 282	O	8999		-64	274	
11 35 17	22 12.9	NGC 3751 ARP 320H VV 282	O	9580		-65	274	
11 35 18	22 15.6	NGC 3753 ARP 320F VV 282	O	8672		-64	274	
11 35 18	22 15.7	NGC 3754 ARP 320E VV 282	O	9112		-64	274	
11 35 21	20 8.2		O	5340	41	-74	564	
11 35 29	-44 48.0		O	(25500)		-271	675	
11 35 38	12 23.4	NGC 3773	R O O	979 1014 973	21	-109	815 741 910	
11 35 54	58 10.0	7 ZW415	O	1020	200	104	820	
11 36 0	68 49.2	MRK 183	O O	12295 12275		150	341 278	
11 36 15	21 15.5	MRK 637	O	19560		-69	557	
11 36 18	45 53.0	2 ZW 53	O	17313		48	262	
11 36 24	3 51.0	MRK 1302	O	(5575)		-145	920	
11 36 30	-37 28.0	NGC 3783	O O O O	2855 3063 3003 2550	30	-263	574 154 454 655	SEYF
11 36 40	56 32.9	NGC 3780	R O	2394 2391	10 85	97	829 741	
11 36 42	46 47.0	NGC 3782	O	699	47	52	789	
11 37 1	-45 18.8		O	20400		-271	675	

R.A. (1950)	DEC. (1950)	NAME	OBS	HEL VEL (C*Z)	ERR	GAL CORR	REF	COMMENTS
11 37 4	32 11.1	NGC 3786	O	2741		-16	253	
		KDG 295	O	2687	12		908	
		HOL 272B	O	2759	80		8	
		ARP 294						
		VV 228						
11 37 6	32 12.6	NGC 3788	O	2326		-16	253	
		KDG 295	O	2627	20		908	
		HOL 272A	O	2348	80		8	
		ARP 294						
		VV 228						
11 37 21	17 13.9	MRK 745	O	3387		-87	558	
11 37 33	15 36.2	NGC 3799	O	3545		-94	253	
		KDG 296	O	3516	50		8	
		ARP 83						
		VV 350						
11 37 36	24 59.0	NGC 3798	O	3509		-50	703	
11 37 38	15 37.2	NGC 3800	O	3528		-94	253	
		KDG 296	O	3560	50		8	
		ARP 83						
		VV 350						
11 37 42	9 17.0	IC 719	O	1600		-122	491	
11 37 42	18 0.0	NGC 3801	O	3230		-83	166	
		4C 17.52	O	3400			868	
			O	3439			703	
			O	3254	70		644	
11 37 48	28 40.0		O	2970		-33	178	
11 37 58	16 56.3		O	61200		-88	868	
11 38 7	22 42.3	NGC 3808	O	7020		-61	195	
		ARP 87	O	7080			195	
		VV 300						
11 38 8	22 43.5	NGC 3808A	O	7250		-61	195	
		ARP 87	O	7200			195	
		VV 300						
11 38 10	35 29.0	MRK 426	O	1500		0	359	
11 38 12	20 38.0	NGC 3805	O	6472	35	-71	644	
			O	6697	178		559	
11 38 12	35 28.0	MRK 426	O	1500		0	438	
11 38 12	56 28.8	NGC 3804	R	1385	10	97	829	

11ʰ38ᵐ

R.A. (1950)	DEC. (1950)	NAME	OBS	HEL VEL (C*Z)	ERR	GAL CORR	REF	COMMENTS
11 38 15	11 14.3	MC 2	O	24280		-113	661	
11 38 24	11 44.9	NGC 3810	R	995		-111	693	
			O	900			465	
			O	1005			13	
			O	972	65		13	
11 38 30	- 6 13.0	VV 544	O	2070	50	-183	873	
11 38 30	25 7.0	NGC 3812	O	3578		-49	703	
11 38 35	20 6.2		O	5439	20	-73	564	
11 38 36	47 58.2	NGC 3811	O	3041		58	341	
		MRK 185	O	3000			308	
11 38 40	36 49.4	NGC 3813	R	1464		6	693	
			O	1386			321	
			O	1514	17		741	
11 38 42	20 14.0		O	6554	210	-72	559	
11 38 58	15 43.3		O	34800		-93	600	
11 39 0	38 16.2	MRK 638	O	6610		13	557	
11 39 12	20 23.0	NGC 3816	O	5547	49	-71	644	
			O	5742	127		559	
11 39 15	16 15.0	MRK 747	O	840		-90	558	
		HOL 275						
11 39 24	5 53.0	NGC 3818	O	1498	65	-135	13	
11 39 36	20 15.0		O	7752		-72	753	
11 39 36	20 36.0	NGC 3821	O	5535	44	-70	644	
			O	5807	103		559	
11 39 42	0 37.0		O	5400		-156	491	
11 39 42	15 16.5		R	1010	10	-94	860	
11 39 48	10 33.0	NGC 3825	O	6143	200	-115	843	
11 39 48	20 24.0		O	5972	60	-71	559	
11 40 0	19 7.0	NGC 3827	O	3268	20	-77	644	
			O	3143	27		559	
11 40 6	20 18.0		O	6314	18	-71	559	
11 40 18	20 14.0		O	7282	33	-71	559	

R.A. (1950)	DEC. (1950)	NAME	OBS	HEL VEL (C*Z)	ERR	GAL CORR	REF	COMMENTS
11 40 25	59 23.0	DDO 96	R	1318	15	111	492	
11 40 36	19 39.0		O	6531		-74	753	
11 40 36	20 1.0		O	7646	23	-72	564	*
			O	7526	101		559	
11 40 36	20 17.0		O	7016	50	-71	559	
11 40 44	31 43.9		R	1789	10	-17	860	
11 40 45	24 10.7	MRK 639	O	10005		-53	557	
11 40 48	20 1.0	IC 2951	O	6172		-72	753	
11 40 48	36 23.6	MRK 427	O	6355		4	438	
			O	6600			359	
11 40 54	23 0.0		O	6906	6	-58	559	
11 41 0	19 53.0		O	6564		-73	753	
11 41 6	20 18.0		O	6399		-71	753	
11 41 12	20 15.0		P	6746		-71	967	
			O	6698	32		559	
11 41 18	20 3.0		O	5624		-72	753	
11 41 18	20 10.0	NGC 3837	O	6226	57	-71	559	
			O	6353	22		564	
11 41 18	20 13.0		O	6040		-71	753	
11 41 18	20 15.0		O	6589	15	-71	564	
11 41 18	20 21.0	NGC 3840	O	7352	21	-70	559	
			O	7417	17		564	
11 41 18	20 28.0		O	6497	16	-70	559	
11 41 23	20 13.7	NGC 3842/41	O	6180		-71	255	
11 41 23	20 13.7	NGC 3842	O	6230	30	-71	564	
			O	6258	16		564	
			O	6111	36		644	
11 41 24	20 3.0		O	4937		-72	753	
11 41 24	20 5.0		O	8004		-71	753	
11 41 24	20 15.0	NGC 3841	O	6216		-71	753	

R.A. (1950)	DEC. (1950)	NAME	OBS	HEL VEL (C*Z)	ERR	GAL CORR	REF	COMMENTS
11 41 24	20 18.0	NGC 3844	O	6824	15	-70	564	
			O	6868	92		559	
11 41 30	20 0.0		O	7733		-72	753	
11 41 30	20 16.0	NGC 3845	O	5643	121	-71	559	
11 41 31	5 31.0		O	3510		-136	895	
11 41 33	37 27.8	MRK 428	O	12515		10	438	
			O	12300			359	
11 41 36	20 7.0		O	6453		-71	753	
11 41 42	20 15.0	NGC 3851	O	6469		-71	753	
11 41 43	70 0.6		R	2702	10	155	829	
			O	2545			558	
11 41 44	49 6.9		R	896	10	65	860	
11 41 48	20 0.0		O	6823		-72	753	
11 41 48	20 6.0		O	4527		-71	753	
			O	5439	20		564	DIS
11 41 50	37 25.3	B2	O	34330	150	10	958	*
11 41 50	37 25.3	B2	O	34630	999	10	958	
11 41 54	20 20.0		O	6761	88	-70	559	
11 42 0	57 47.0	1	O	34200	120	104	504	*
11 42 0	57 47.0	4	O	34500	120	104	504	
11 42 0	57 47.0	5	O	34500	120	104	504	
11 42 3	11 16.9	MC 2	O	57260		-111	661	
11 42 6	20 3.0		O	6419		-71	753	
			O	8522	67		559	DIS
11 42 6	20 9.0		O	7792	29	-71	564	
11 42 12	19 49.0	NGC 3857	O	6255		-72	753	
11 42 12	19 53.0		O	6556		-72	753	
11 42 12	19 58.0		O	4980		-72	753	
11 42 12	20 5.0	NGC 3860	O	5461	21	-71	564	

R.A. (1950)	DEC. (1950)	NAME	OBS	HEL VEL (C*Z)	ERR	GAL CORR	REF	COMMENTS
11 42 12	20 24.0		O	6724	78	-70	559	
			O	6590	16		564	
11 42 12	20 58.0	NGC 3883	O	7406	136	-67	559	
11 42 18	19 44.0	NGC 3859	O	5508	56	-73	559	
11 42 18	19 46.0		O	6457		-72	753	
11 42 18	20 3.0		O	8248		-71	753	
11 42 19	- 8 57.6	NGC 3865	O	5714	70	-190	741	
11 42 24	20 8.0		O	5340	41	-71	564	
11 42 29	19 53.0	NGC 3862	O	6705	25	-72	644	
		3CR264	O	6240			120	
			O	6592	111		559	
			O	6468	20		564	
11 42 30	19 54.0	IC 2955	O	6345	.	-72	753	
11 42 30	20 15.0	NGC 3861	R	5076		-70	967	
		KDG 299A	O	5180	30		908	
			O	5015	18		564	
			O	5040	126		559	
11 42 36	20 15.0	KDG 299B	O	6079	41	-70	908	
11 42 36	19 40.0	NGC 3864	O	6997		-73	753	
11 42 36	20 7.0		O	7646	26	-71	564	
11 42 48	19 40.0	NGC 3867	O	7457		-73	753	
11 42 48	19 43.0	NGC 3868	O	6653	43	-72	559	
11 42 48	59 15.0	7 ZW421	O	2979	190	111	429	
11 42 52	31 50.5	3CR265	O	243300		-15	718	
			O	243000			766	
			O	243100	60		849	
11 42 53	31 50.5		O	117500		-15	849	
11 42 54	20 37.0		O	7243	170	-68	559	
11 43 6	20 3.0	NGC 3873	O	5552	34	-71	564	
11 43 6	20 18.0		O	5255		-70	753	
11 43 12	14 3.0	NGC 3872	O	3109	75	-98	13	

11ʰ43ᵐ

R.A. (1950)	DEC. (1950)	NAME	OBS	HEL VEL (C*Z)	EPR	GAL CORR	REF	COMMENTS
11 43 12	20 2.0	NGC 3875	O	6936		-71	753	
11 43 17	50 28.7	NGC 3870	R	755		72	368	
		MRK 186	O	500			233	
			O	740	80		300	
			O	630			341	
11 43 18	19 48.0		O	5486		-72	753	
11 43 24	19 43.0		O	5165		-72	753	
11 43 24	21 19.0		O	(6801)		-65	703	
11 43 24	56 1.0		O	11700	200	97	2	
11 43 29	47 56.2	NGC 3877	O	868	11	60	741	
11 43 30	19 55.0		O	6357		-71	753	
11 43 37	20 40.2	NGC 3884	O	6822	30	-68	644	
			O	7002	187		559	
11 43 46	71 54.2	MRK 187	O	9600		163	308	
11 43 49	35 7.8	MRK 429	O	1500		0	359	
			O	1290			438	
11 44 1	- 3 34.7	A 1145A	O	5108		-170	13	WLD TRI
		ARP 248A						
		VV 35						
11 44 5	69 39.6	NGC 3879	P	1431	10	154	829	
11 44 11	- 3 33.9	A 1145B	O	5008		-170	13	
		ARP 248B						
		VV 35						
11 44 11	21 33.0	MRK 640	O	6665		-63	557	
11 44 15	-27 38.7	NGC 3885	O	1948	32	-242	741	
11 44 16	- 3 32.4	A 1145C	O	5396		-170	13	
		ARP 248C						
		VV 35						
11 44 30	20 7.0	NGC 3886	O	5431	181	-70	559	
11 44 30	55 59.0		O	14982	50	97	13	
11 44 32	-16 34.6	NGC 3887	O	(1163)		-213	13	
11 44 42	- 3 34.0		O	5008	50	-170	13	

R.A. (1950)	DEC. (1950)	NAME	OBS	HEL VEL (C*Z)	ERR	GAL CORR	REF	COMMENTS
11 44 42	55 58.0		O	14688	60	97	13	
11 44 42	56 1.0		O	15459	50	97	13	
11 44 48	- 3 35.0		O	5396	75	-170	13	
11 44 55	56 14.9	NGC 3888	O	2049	75	98	741	
		MRK 188	O	2418			341	
			O	2400			308	
11 45 0	57 55.5	KDG 301A	O	9215	18	106	908	
11 45 0	57 55.5	KDG 301B	O	9263	10	106	908	
11 45 3	-47 0.7		O	7200		-269	675	
11 45 18	56 2.0		O	15260		98	255	
11 45 24	20 17.0		O	7280	83	-68	559	
11 45 28	-10 41.0	NGC 3892	O	1727	32	-194	741	
11 45 48	56 3.0		O	15572	60	98	13	
11 46 0	48 59.4	NGC 3893	R	980		66	283	
		KDG 302A	R	967			744	
		HOL 293A	R	970			693	
			O	868			13	
			O	932	25		283	
			O	999			748	
			O	955	22		908	
			O	1042	40		13	
11 46 2	32 54.8		R	6999		-9	901	
11 46 11	59 41.5	NGC 3894	O	3200		114	868	
		KDG 303A	O	3186			748	
		HOL 294A	O	3201	30		908	
			O	3237			910	
11 46 14	-28 1.0	DDO 239	R	1934	10	-242	492	
11 46 18	- 1 45.0		O	1751	34	-162	813	
11 46 19	48 57.2	NGC 3896	O	869	11	66	908	
		KDG 302B						
		HOL 293B						
11 46 24	35 17.0	NGC 3897	R	6412	10	2	582	
			O	6484	75		582	
11 46 24	59 42.5	NGC 3895	O	2986		114	748	
		KDG 303B	O	3266	63		908	
		HOL 294B	O	3226			910	

R.A. (1950)	DEC. (1950)	NAME	OBS	HEL VEL (C*Z)	ERR	GAL CORR	REF	COMMENTS
11 46 30	- 9 27.0	NGC 3905	R O	5794 5731	50 30	-190	582 582	
11 46 36	27 17.0	NGC 3900	O	1702	50	-35	13	
11 46 36	56 21.8	NGC 3898	R O	(1174) 1038	75	99	693 13	
11 46 42	-28 59.9	NGC 3904	R O	1496 1613	50 75	-244	617 13	
11 46 48	- 3 14.0		O	12300		-167	491	
11 46 54	- 3 11.0		O	8074	75	-167	843	
11 46 48	- 0 48.0	KDG 304A	O	6560	64	-158	908	
11 46 54	- 0 48.0	NGC 3907 KDG 304B	O	6206	59	-158	908	
11 47 24	21 38.0	NGC 3910	O	7829		-61	703	
11 47 30	25 13.0	NGC 3920 VV 367	O O O	3600 3600 3592		-45	491 535 703	
11 47 30	26 45.3	NGC 3912	O	1728	30	-37	741	
11 47 45	- 0 19.8		O	41670		-156	726	
11 47 47	- 0 18.9	NE	O	51440		-156	726	
11 47 47	- 0 18.9	SW	O	41490		-156	726	
11 48 0	20 17.0	NGC 3919	O	6119		-67	703	
11 48 1	55 37.9	NGC 3913 IC 740 HOL 296A	R O	956 830	10	97	829 449	
11 48 2	- 0 17.4		O	1757		-155	726	
11 48 6	52 6.0	NGC 3917	R	963		81	744	
11 48 7	56 44.0	DDO 98 VV 273	R	896	15	101	492	
11 48 14	36 38.7	4C 36.19	O	42360		9	400	
11 48 15	39 9.3	DDO 99	P	248	10	21	492	
11 48 29	55 21.5	NGC 3921 1 ZW 28 MRK 430 ARP 224 VV 31	R O O O O	5838 6000 5995 5930 5895	25	96	829 359 438 77 910	

R.A. (1950)	DEC. (1950)	NAME	OBS	HEL VEL (C*Z)	ERR	GAL CORR	REF	COMMENTS
11 48 30	-28 33.0	NGC 3923	O	1788	65	-242	13	
			O	2050			5	
11 48 30	43 23.0	3 ZW 63	O	6810		41	282	
		2 ZW 54						
11 48 36	52 16.7	NGC 3931	O	870		82	910	
11 48 54	22 19.0	NGC 3926N	O	8571		-57	703	
11 48 54	22 19.0	NGC 3926S	O	7521		-57	703	
11 49 0	36 52.0	KDG 306	O	(10774)	40	10	908	
11 49 6	21 17.0	NGC 3929	O	7103		-62	703	
11 49 10	38 17.6	NGC 3930	O	915		17	641	
11 49 10	48 57.7	NGC 3928	O	964		67	341	
		MRK 190	O	900			308	
11 49 12	36 52.0	KDG 306	O	10951	40	10	908	
11 49 18	46 2.0	1 ZW 29	O	5910		53	282	
11 49 24	21 24.0		O	6555	39	-61	559	
11 49 30	52 23.1	DDO 100	R	1019	10	83	492	
11 49 50	23 54.3	MRK 642	O	3450		-49	557	
11 49 52	35 10.4	MRK 641	O	2165		3	557	
11 50 0	23 53.0		O	6842		-49	703	
11 50 6	20 55.0	NGC 3937	O	6617		-63	703	
11 50 6	21 17.0	NGC 3940	O	6402		-61	703	
11 50 12	2 1.0	KDG 307A	O	6093	27	-145	908	
11 50 12	2 1.0	KDG 307B	O	6067	27	-145	908	
11 50 13	44 24.0	NGC 3938	R	809		46	693	
			R	812			744	
			R	812	25		183	
			O	742	25		283	
			O	874			13	
11 50 18	20 46.0	NGC 3943	O	6623		-64	703	
11 50 19	37 15.9	NGC 3941	R	917		13	610	
			O	927			13	
			O	972	50		13	

11ʰ50ᵐ

R.A. (1950)	DEC. (1950)	NAME	OBS	HEL VEL (C*Z)	ERR	GAL CORR	REF	COMMENTS
11 50 24	- 4 8.0	VV 457	O	1500		-169	535	
			O	1461	50		873	
11 50 29	70 42.7	MRK 191	O	9400		160	233	
11 50 35	22 45.9	4C 22.32	O	35520		-54	400	
11 50 42	60 57.0	NGC 3945	O	1220	75	120	13	
11 50 48	21 2.0	NGC 3947	R	6196		-62	749	
			O	6288	8		559	
11 50 55	-72 36.2		O	5160		-258	896	
11 51 5	46 29.0	MRK 42	O	7200		56	224	SEYF
11 51 6	21 10.0	NGC 3954	O	6861		-61	703	
11 51 6	23 40.0	NGC 3951	O	6480		-50	703	
11 51 6	48 8.0	NGC 3949 HOL 301A	O	681	150	64	13	
11 51 7	- 3 43.1	NGC 3952	O	1625	27	-167	741	
11 51 13	8 4.8		O	2550		-119	895	
11 51 13	52 36.5	NGC 3953	R	1065		84	934	
			O	1008			13	
			O	938	50		13	
11 51 24	-22 53.2	NGC 3955	O	1552	45	-227	432	
			O	1537			227	
11 51 24	20 52.0		O	6903		-63	703	
11 51 28	-20 17.3	NGC 3956	O	726		-220	932	
11 51 33	-19 17.3	NGC 3957	O	1838	39	-217	741	
11 51 38	29 32.8	4C 29.44	O	98770		-22	400	
11 51 42	0 25.0	IC 745 MRK 1308	O	1000		-150	491	
			O	1200			920	
11 51 44	45 40.2	4C 45.23	O	57600		53	864	
11 51 54	-16 13.0	1 ZW 37-S	O	6593	120	-208	735	
11 51 58	58 38.7	NGC 3958	R	3380	50	111	829	
			R	3380	50		885	
11 52 6	13 41.8	NGC 3962	R	1815	50	-95	781	
			O	1794	65		13	
			O	1843			106	

R.A. (1950)	DEC. (1950)	NAME	OBS	HEL VEL (C*Z)	ERR	GAL CORR	REF	COMMENTS
11 52 9	20 20.0		R	618	10	-65	860	
11 52 22	58 46.3	NGC 3963	R	3185	20	112	829	
			O	3234	67		741	
11 52 36	6 27.0		O	6973	25	-126	582	HII REG
11 52 42	33 25.0	KDG 309	O	3257	30	-3	908	
11 52 44	51 24.3	MRK 192	O	3600		79	308	
11 52 48	-18 34.0	1 ZW 18-S	O	7050	210	-215	735	
11 52 48	33 24.0	KDG 309	O	3149	30	-3	908	
11 52 52	57 56.4	MRK 193	O	5180	80	108	300	
			O	5300			233	
11 52 54	12 15.0	NGC 3968	R	6406	15	-101	752	
11 53 0	-19 15.0	1 ZW 15A-S	O	18000	250	-217	735	
11 53 0	30 16.0	NGC 3971	O	6880		-18	548	
11 53 0	80 30.0	7 ZW429A	O	13070		195	303	
11 53 0	80 30.0	7 ZW429B	O	12560		195	303	
11 53 6	-16 9.0	1 ZW 38-S	O	6671	89	-207	735	
11 53 6	1 32.0		O	1957	25	-145	813	
11 53 12	55 36.0	NGC 3972 HOL 304A	O	609	100	98	789	
11 53 18	23 41.0	A	O	42844	100	-49	13	
			O	42862			444	
11 53 18	23 41.0	B	O	42819	100	-49	13	
11 53 18	60 48.0	NGC 3975 HOL 306B	O	(10000)	100	120	789	
11 53 23	7 1.7	NGC 3976 HOL 305A	R	2496	8	-123	914	
			O	2521	18		741	
11 53 35	-19 36.9	NGC 3981 VV 8	O	1809		-217	741	
11 53 35	60 48.0	NGC 3978 HOL 306A	O	9990	53	121	741	
			O	10006	67		789	
11 53 36	-18 42.0	1 ZW 23-S	O	9900	400	-215	735	

R.A. (1950)	DEC. (1950)	NAME	ORS	HEL VEL (C*Z)	ERR	GAL CORR	REF	COMMENTS
11 53 50	55 40.0	NGC 3977 HOL 304B	O	5710		99	449	
11 53 53	55 24.0	NGC 3982	O	975	25	97	741	
11 54 0	48 36.0	KDG 310A	O	886	40	67	908	
11 54 7	48 36.8	NGC 3985 KDG 310B	O O O O	906 937 958 533	43 26	67	789 908 741 558	*
11 54 24	-20 20.0	1 ZW 54-S	O	9314	162	-219	735	
11 54 30	30 40.0		O	3209		-15	548	
11 54 42	25 29.0	NGC 3987	O	4548		-40	703	
11 54 48	28 9.0	NGC 3988	O	6562		-27	548	
11 54 54	-19 20.0	1 ZW 59-S	O	1781	41	-216	735	
11 54 56	32 36.8	NGC 3991A/B HOL 309C ARP 313	R O O O	3185 3269 3269 3302		-6	893 57 253 253	
11 54 56	32 36.8	NGC 3991A VV 523 KDG 311A	O O O O	3305 3164 3238 3304	22 20 34	-6	72 873 908 163	
11 54 56	32 36.8	NGC 3991B VV 523 KDG 311B	O O O	3040 3048 3092	41 30 34	-6	72 873 908	
11 55 0	-73 24.4	IC 2980	O	8387	145	-257	680	
11 55 0	25 31.0	NGC 3993 VV 229A HOL 308A	O O	4824 4800		-39	253 57	
11 55 0	53 39.0	NGC 3992	R R R R O	1047 1051 1053 3265 1059	100	90	693 934 744 893 13	M 109 * DIS
11 55 1	55 44.3	NGC 3990 HOL 310B	O	720		99	13	
11 55 2	32 33.5	NGC 3994 ARP 313 VV 249 HOL 309B	O O O O O	3118 3006 3016 3142 3024	60	-6	253 558 631 253 57	

R.A. (1950)	DEC. (1950)	NAME	OBS	HEL VEL (C*Z)	ERR	GAL CORR	REF	COMMENTS
11 55 10	32 34.3	NGC 3995	O	3336		-6	253	
		ARP 313	O	3347			13	
		VV 249	O	3360			57	
		HOL 309A	O	3360			253	
			O	3242			57	
			O	3393			253	
11 55 13	25 33.0	NGC 3997	O	4762		-39	253	
		VV 229B	O	4738			57	
		HOL 308B						
11 55 20	55 44.1	NGC 3998	O	1059		99	13	
		HOL 310A	O	1109	50		13	
			O	1100			244	
			O	1185			106	
			O	1173			281	
11 55 24	23 24.0	NGC 4003	O	6487		-49	703	
		KDG 312	O	6555	20		908	
11 55 24	23 29.0	NGC 4002	O	(6711)	80	-49	908	
		KDG 312						
11 55 24	36 40.0	VV 126	O	10415	50	12	873	
		ARP 194	O	10488			922	
11 55 31	28 9.0	NGC 4004	O	3300		-27	359	
		MRK 432	O	3440			438	
		VV 230	O	3447			548	
11 55 36	-17 37.0	1 ZW 74-S	O	2374	76	-210	735	
11 55 36	25 24.0	NGC 4005	O	4373		-40	703	
11 55 43	28 28.2	NGC 4008	O	3550	49	-25	741	
			O	3603			548	
11 55 48	-19 15.0	1 ZW 70-S	O	1491	86	-215	735	
11 55 48	27 48.0		O	3494		-28	548	
11 55 51	51 11.6	DDO 102	R	912	10	79	492	
11 55 52	-22 9.9	DDO 106	R	1784	10	-223	492	
11 55 54	38 21.0	DDO 105	R	917	10	20	492	ZWO N.2
11 55 58	-18 4.2	NGC 4024	O	1694	56	-212	741	
11 55 58	44 13.4	NGC 4013	O	661	42	48	741	
			O	750	71		789	
11 56 0	43 0.8	IC 749	O	658		42	748	
		KDG 313A	O	813	24		908	
		HOL 313A						

211

11ʰ56ᵐ

R.A. (1950)			DEC. (1950)		NAME		OBS	HEL VEL (C*Z)	ERR	GAL CORR	REF	COMMENTS
11 56	6	27	43.0		NGC	4016	O	3351		-28	548	
11 56	9	25	19.3		NGC ARP VV	4015 138 216	O	4800		-40	535	
11 56 12		- 1	10.0		KDG	315A	O	1481	11	-154	908	
11 56 12		- 1	10.0		KDG	315B	O	6384	25	-154	908	
11 56 17		43	0.1		IC KDG HOL	750 313B 313B	O O	713 765	30 24	42	741 908	
11 56 18		30	41.0		NGC	4020	O	814		-14	548	
11 56 24		-18	45.0		1 ZW	63-S	O	1016	47	-213	735	
11 56 54		51	14.0		NGC	4026	O	878	75	80	13	
11 56 55		35	10.3		MRK	434	O	9830		6	557	
11 56 57		-19	3.1		NGC	4027A	O	1700	55	-214	555	
11 56 57		-18	59.4		NGC ARP VV 1 ZW109-S	4027 22 66	O O O O O	1844 1637 1563 2000 1671	59 58 10	-214	72 150 735 5 148	
11 57 12		-21	6.0		1 ZW	91-S	O	2800	143	-219	735	
11 57 15		28	8.0				O	5663		-26	714	
11 57 18		-19	1.0		1 ZW108A-S		O	6260	230	-214	735	
11 57 50		- 0	49.4		NGC	4030	O	1509		-152	13	
11 57 54		-15	42.0		1 ZW122-S		O	1604	130	-203	735	
11 58 0		20	21.0		NGC	4032	R O	1268 1186	6 53	-62	914 789	
11 58 1		-17	34.0		NGC	4033	O	1521	30	-209	741	
11 58 6		-20	34.0		1 ZW	96-S	O	18639	142	-217	735	
11 58 10		-47	45.9				O	21900		-265	675	
11 58 14		0	15.3				R	1937	10	-147	860	
11 58 25		31	50.0		3CR268.2		O O	108230 108200	300	-8	585 899	

R.A. (1950)	DEC. (1950)	NAME	OBS	HEL VEL (C∗Z)	ERR	GAL CORR	REF	COMMENTS
11 58 30	-17 41.0	1 ZW114-S	O	17380	146	-209	735	
11 58 42	- 3 23.0	MRK 1310	O	5861		-161	920	
11 58 48	-33 35.0		O	3120	30	-247	655	
11 58 48	45 56.0	3 ZW 64	O	20525		57	262	
11 58 50	13 40.7	NGC 4037	O	926	87	-91	741	
11 58 54	62 10.0	NGC 4036	O	1382	50	128	13	
11 59 14	6 6.3		O	1320		-123	895	
11 59 16	-18 35.1	NGC 4038	O	1521	54	-211	72	KNOT
11 59 16	-18 35.1	NGC 4038	O	1588	69	-211	72	KNOT
11 59 18	-18 35.1	NGC 4038	R	1640		-211	934	
			O	1681	30		8	
			O	1586	60		72	
			O	1691			253	
			O	1673	75		13	
11 59 19	-18 35.7	NGC 4038/9 ARP 244 VV 245	R	1622		-211	498	
			R	1630	30		455	
			R	1630			861	
			O	1636			260	
			O	1700			5	
			O	1671			128	KNOT
			O	1677			128	KNOT
11 59 20	-18 36.4	NGC 4039	O	1647		-211	253	
			O	1631	30		8	
			O	1660	50		13	
11 59 30	30 7.0	KDG 316	O	3046	18	-16	908	
11 59 36	66 40.7	MRK 194	O	15100		146	233	
11 59 39	-10 24.2	PKS	O	79700		-185	600	
11 59 39	62 25.0	NGC 4041	R	1234		129	744	
			O	1186	34		148	
11 59 48	30 8.0	KDG 316	O	3224	43	-15	908	
12 0 0	41 20.0		R	6132	10	36	582	
			O	6137	20		582	
12 0 3	64 39.3	MRK 195	O	1100		138	233	
			O	1500			287	
			O	1362			341	
			O	1425	80		300	
			O	800			231	

R.A. (1950)			DEC. (1950)		NAME		OBS	HEL VEL (C*Z)	ERR	GAL CORR	REF	COMMENTS
12	0	6	2	14.0			0	4891		-138	694	
12	0	8	-43	55.9		A	0	6821	40	-261	680	
12	0	8	-43	55.9		B	0	6931	80	-261	680	
12	0	8	-43	55.9		C	0	6796	45	-261	680	
12	0	9	2	15.5	NGC 4045		0	2001	41	-138	741	
					HOL 320A		0	(1810)	190		555	
							0	1842			694	
12	0	18	18	19.0	NGC 4048		0	4639		-69	774	
					VV 384							
12	0	18	26	31.9			0	9633		-32	714	
12	0	18	48	55.2	NGC 4047		0	3355	17	71	741	
12	0	20	-16	5.4	NGC 4050		0	1904	44	-203	741	
12	0	27	39	42.1	MRK 43		0	6105	60	29	261	
12	0	34	51	57.3			0	18920		85	595	
12	0	36	2	14.0			0	5975		-138	694	
12	0	36	44	48.0	NGC 4051		R	706		52	693	SFYF
							R	766	28		390	
							R	690	20		267	
							R	701			707	
							0	646	50		112	
							0	627	20		13	
							0	658	32		148	
							0	709			641	
							0	650	40		2	
							0	677	23		50	
12	0	48	16	46.0		A	0	3510	150	-76	594	
12	0	51	39	5.4	MRK 44		0	6901	60	26	261	
					IC 2987							
12	0	54	2	19.0			0	5611		-138	694	
12	0	54	8	39.0	KDG 317		0	(6391)	90	-112	908	
12	0	54	16	46.0		B	0	3910	19	-76	594	
							0	4200			533	
12	0	54	22	30.0			0	4726		-50	703	
12	1	0	8	40.0	KDG 317		0	(6130)	90	-111	908	

R.A. (1950)	DEC. (1950)	NAME	OBS	HEL VEL (C★Z)	ERR	GAL CORR	REF	COMMENTS
12 1 2	60 48.5	MRK 45	O	4180	60	123	261	
12 1 11	23 59.0	MRK 645	O	7675		-43	557	
12 1 12	2 7.0		O	4990		-138	694	
12 1 12	2 18.0		O	6006		-138	694	
12 1 18	2 11.0		O	5381		-138	694	
12 1 20	25 42.7		O	3152		-35	714	
12 1 30	20 30.0	NGC 4061	O	7131		-59	703	
		VV 179	O	1604	83		594	DIS
12 1 30	32 10.6	NGC 4062	R	774		-5	693	
			O	747	10		936	
			O	748			322	
			O	670	48		741	
12 1 31	28 26.6		O	9083		-22	714	
12 1 33	2 8.0		O	4998		-138	750	
			O	5875			694	DIS
12 1 36	18 43.0	NGC 4064	O	1033		-67	13	
12 1 36	20 30.0	NGC 4065	O	6275		-59	703	
		VV 179	O	1181	152		594	DIS
12 1 36	20 38.0	NGC 4066	O	7352		-58	703	
12 1 36	20 42.0	NGC 4070	O	7201		-58	703	
12 1 42	2 8.0		O	6641		-138	694	
			O	7248			750	
12 1 42	2 9.0		O	6661		-138	694	
12 1 47	- 1 15.1	DDO 108	R	1463	15	-151	492	
12 1 48	66 52.6	MRK 196	O	17400		147	308	
12 1 54	-18 15.0	1 ZW182-S	O	14508	72	-209	735	
12 1 54	2 10.6	NGC 4073	O	5961	59	-138	741	
			O	5756			694	
			O	6008			750	
			O	5869	52		843	
12 2 0	2 5.0	IC 2989	O	5863	104	-138	594	
			O	5668			750	
			O	5329			694	

215

12ʰ2ᵐ

R.A. (1950)	DEC. (1950)	NAME	OBS	HEL VEL (C*Z)	ERR	GAL CORR	REF	COMMENTS
12 2 0	20 36.0	NGC 4074	0	6600		-58	533	SEYF
12 2 0	64 43.0	NGC 4081	0	1312	37	139	644	
12 2 6	2 4.0	NGC 4077	0	6921	140	-138	594	
			0	7296			694	
			0	6956			750	
12 2 6	31 26.0		0	7996		-8	322	
12 2 6	31 27.0		0	7469		-8	322	
12 2 7	-27 50.4	DDO 240	R	(1740)		-233	492	
12 2 18	27 16.3	NGC 4080	0	561		-27	714	
			0	749			548	
12 2 28	-43 27.1		0	4500	200	-259	794	
12 2 36	42 35.0	2 ZW 56	0	15018		43	262	
			0	16040			282	
12 2 42	21 30.0	NGC 4084	0	6675		-54	703	
12 2 48	18 9.0	KDG 318	0	4492	56	-69	594	
12 2 48	18 12.0	KDG 318	0	4699	68	-69	594	
12 2 51	50 38.0	NGC 4085	0	714	14	80	741	
		HOL 326B	0	756	47		789	
12 3 0	20 50.0	NGC 4089	0	6962		-57	703	
12 3 0	50 49.0	NGC 4088	0	739		80	13	
		ARP 18	0	722	62		72	
		HOL 326A	0	678	22		741	
			0	741			689	
			0	537	50		72	KNOT
			0	917	86		72	KNOT
12 3 1	-26 17.0	NGC 4087	0	3260		-229	574	
12 3 2	25 22.5		0	6853		-36	714	
12 3 6	9 16.0	KDG 319A	0	6330	47	-108	594	
12 3 6	9 16.0	KDG 319B	0	6166	54	-108	594	
12 3 13	31 20.3	VV 13B ARP 97	0	6894	34	-8	38	
12 3 13	31 21.5	VV 13A ARP 97	0	7010	33	-8	38	

R.A. (1950)			DEC. (1950)		NAME	OBS	HEL VEL (C*Z)	ERR	GAL CORR	REF	COMMENTS
12	3	17	35	27.5	MRK 646	O	16110		10	557	
12	3	18	-14	15.9	NGC 4094	O	1363		-196	741	
12	3	18	20	46.0	NGC 4092	O	6794		-57	703	
12	3	24	20	51.0	NGC 4095	O	7113		-56	703	
12	3	30	20	53.0	NGC 4098	O	7336		-56	703	
12	3	30	47	45.4	NGC 4096	R	474	40	67	238	
						R	570			934	
						R	575			693	
						R	500			363	
						O	513			426	
						O	518			540	
						O	483	30		502	
						O	479	37		741	
12	3	33	28	31.0		O	8127		-21	714	
12	3	37	25	50.1	NGC 4101	O	6744		-33	714	
						O	6122			703	
12	3	37	49	51.4	NGC 4100	O	1138	59	76	741	
						O	1145			910	
						O	861	58		789	
12	3	39	21	30.3	MRK 647	O	7670		-53	557	
12	3	48	52	59.0	NGC 4102	O	908	50	90	13	
						O	878			13	
12	3	54	64	30.3	3CR268.3	O	111220	300	138	585	
						O	111200			717	
12	3	58	28	25.0		O	9024		-21	714	
12	4	1	22	32.5	4C 22.33	O	19650		-48	400	
12	4	1	28	25.4		O	9202		-21	714	
12	4	6	-29	28.9	NGC 4105	R	1820	60	-236	782	
						O	2030			574	
						R	1873	50		642	
						O	1895	50		13	
12	4	6	28	27.2	NGC 4104	O	8400		-21	714	
						O	8123	78		843	
						O	8493			322	
12	4	11	-29	29.4	NGC 4106	R	2150	60	-236	782	
						O	2178	50		13	
						O	2190			574	

12ʰ4ᵐ

R.A. (1950)	DEC. (1950)	NAME	OBS	HEL VEL (C*Z)	ERR	GAL CORR	REF	COMMENTS
12 4 16	67 26.6	NGC 4108	O	2485	40	150	644	
			O	2488	45		789	
12 4 30	43 21.0	NGC 4111	O	870		47	13	
		HOL 333A	O	784	15		13	
			O	800	50		1	
12 4 34	24 11.1	B2	O	23100	250	-40	696	
12 4 36	17 16.0		O	6740		-72	54	
12 4 47	24 28.6	MRK 648	O	(15320)		-39	557	
12 4 51	40 5.4	DDO 109	R	877		32	492	
12 5 0	34 9.4	B2	O	23620	250	5	696	
			O	23720	50		741	
12 5 3	2 58.2	NGC 4116	R	1325		-133	117	
		KDG 322A	R	1309	10		829	
			R	1310	5		914	
			O	1302			13	
12 5 8	- 0 49.7		O	91800		-147	672	
12 5 18	67 39.8	MRK 197	O	2200		151	233	
			O	2194			341	
12 5 19	39 12.7	4C 39.35	O	73060		29	400	
12 5 27	25 49.8		O	10183		-32	714	
12 5 27	65 23.7	NGC 4121	O	1457	20	142	594	
		HOL 335A						
12 5 30	25 31.0		O	(6734)		-34	703	
12 5 32	25 52.1		O	6861		-32	714	
12 5 36	10 39.5	NGC 4124	O	1651		-100	841	
			O	1652	29		741	
			O	1582			694	
12 5 36	25 21.5		O	6961		-34	714	
12 5 37	25 30.9		O	7057		-34	714	
12 5 38	3 9.4	NGC 4123	R	1330	13	-131	889	
		KDG 322B	R	1328	10		829	
			O	1283			694	
			O	1198	47		789	
			O	1283	16		741	

R.A. (1950)			DEC. (1950)		NAME		ORS	HEL VEL (C*Z)	ERR	GAL CORR	REF	COMMENTS
12	5	38	65	27.0	NGC	4125	O	1485		143	13	
					HOL	335A	O	1380	36		594	
							O	1361	40		644	
							O	1305	50		13	
							O	1356			106	
12	5	39	26	2.2	IC	762	O	7452		-31	714	
12	5	40	25	13.3			O	7059		-35	714	
12	5	42	26	5.4	IC	763	O	7078		-31	714	
12	6	0	77	5.0	NGC	4127	R	1813	10	185	752	
12	6	6	69	3.0	NGC	4128	O	2395		157	13	
					HOL	337A						
12	6	12	29	35.0	NGC	4131	O	3710		-15	322	
12	6	14	25	13.5			O	6743		-35	714	
12	6	19	- 8	45.7	NGC	4129	O	1210	69	-176	741	
12	6	23	25	28.2			O	6885		-33	714	
12	6	30	-75	11.8			O	1962		-252	680	
12	6	30	29	31.0	NGC	4132	O	4069		-15	322	
12	6	36	29	27.0	NGC	4134	O	3860		-15	548	
12	6	36	31	51.0			O	6758		-4	322	
12	6	43	47	20.1	MRK	198	O	7195	80	66	300	SEYF
							O	7400			233	
							O	7165			341	
							O	7800			232	
12	6	45	25	14.7			O	6754		-34	714	
12	6	45	30	12.3	NGC	4136	R	596	13	-11	772	
							O	445	50		13	
12	6	47	44	22.2	NGC	4137	O	9300		53	535	
					VV	454						
12	6	56	26	30.3			O	11078		-28	714	
12	7	0	17	17.0	2 ZW	57	R	6676	19	-70	622	
							O	6660			282	
12	7	0	43	57.0	NGC	4138	O	1039	100	51	13	

R.A. (1950)			DEC. (1950)		NAME		OBS	HEL VEL (C*Z)	ERR	GAL CORR	REF	COMMENTS
12	7	6	42	36.0	1 ZW 31		O	7016		45	262	
12	7	6	42	49.0	NGC 4143		O	784	100	46	13	
12	7	14	25	18.3			O	2579		-34	714	
12	7	22	25	35.4			O	6921		-32	714	
12	7	28	46	44.1	NGC 4144		R	260	20	64	772	
							R	263			744	
							R	535	15		216	
12	7	30	40	10.0	NGC 4145		R	1019		34	693	
					KDG	324A	R	1026			614	
					HOL	342A	O	1001			748	
							O	829	21		741	
							O	1001			321	
							O	890	120		555	
12	7	36	36	9.0	NGC 4148		O	4750		16	220	
							O	5200			605	
12	7	42	25	35.2			O	5808		-32	714	
12	7	46	26	42.6	NGC 4146		O	6607		-27	714	
							O	6930			777	
							O	6488			548	
12	8	0	25	43.0			O	7861	250	-31	556	
12	8	0	30	41.0	NGC 4150		O	244	50	-9	13	
12	8	0	39	41.0	NGC 4151		R	991	10	32	390	SEYF
					KDG	324B	R	1013	20		544	
					HOL	345A	R	994			707	
							R	1005	25		267	
							R	1000			614	
							R	967	25		238	
							R	996	4		371	
							R	999			693	*
							O	962			910	
							O	980	50		1	
							O	940	50		1	
							O	875	61		594	
							O	930	26		514	
							O	953			419	
							O	972			514	
							O	927			327	
							O	980			179	
							O	862	85		140	
							O	952	10		50	
							O	907	22		112	
							O	990			264	
							O	980	20		194	
							O	934			13	
							O	960	8		13	

R.A. (1950)			DEC. (1950)		NAME		OBS	HEL VEL (C*Z)	ERR	GAL CORR	REF	COMMENTS
12	8	0	39	41.0			O	970		32	411	*
							O	957			748	
12	8	2	25	42.3			O	7043		-31	714	
12	8	2	26	12.4			O	6555		-29	714	
12	8	4	70	38.8	MRK	199	O	2154		163	341	*
12	8	8	16	18.7	NGC	4152	R	2161	10	-74	566	
							O	2018	58		789	
							O	2150	34		741	
12	8	11	39	57.2			O	1853	87	33	428	
12	8	18	39	45.0	NGC	4156	O	6765	25	33	582	
					KDG	325	O	6878	47		741	
					HOL	345B	O	6700			605	
							O	7600	100		452	DIS
							O	725	63		594	DIS
12	8	24	40	2.0			R	1139		34	614	
12	8	25	50	33.9	DDO	111	R	859	20	81	492	
12	8	30	2	17.0	DDO	110	R	1339	10	-133	492	
12	8	30	64	12.0			O	2642	30	138	644	
12	8	33	50	45.7	NGC	4157	R	782	20	82	238	
							R	805			363	
							O	681			741	
							O	920			449	
							O	878			426	
12	8	37	20	27.2	NGC	4158	O	2536	17	-55	741	
							O	2200			694	
12	8	52	48	48.6	MRK	200	O	4500		74	308	
12	8	54	3	12.0			R	1297	10	-129	860	
12	9	6	16	46.0			O	2400		-71	533	
12	9	24	24	24.0	NGC	4162	R	2571	12	-37	914	
							O	2546			13	
12	9	24	27	55.4			O	4267		-20	714	
12	9	25	40	55.7	MRK	435	O	6740		38	374	
							O	6600			359	
							O	6745			438	
							O	6660			557	

R.A. (1950)	DEC. (1950)	NAME	OBS	HEL VEL (C*Z)	ERR	GAL CORR	REF	COMMENTS
12 9 42	13 31.0	NGC 4165	O	1505		-85	694	
12 9 42	18 24.0	VV 147	O	7448	50	-64	873	
12 9 42	29 27.0	NGC 4169	O	3840		-13	322	
12 9 44	13 29.0	NGC 4168	O	2428	27	-85	741	
			O	2248			694	
12 9 48	29 5.0		O	4050		-15	322	
12 9 48	29 29.0	NGC 4173	O	1106		-13	548	
12 9 48	37 17.4		R	1048	10	22	860	
12 9 54	29 25.0	NGC 4174	O	4161		-13	322	
12 10 0	-17 15.0		O	2911		-201	694	
12 10 0	29 26.0	NGC 4175	O	4061		-13	322	
12 10 13	11 8.3	NGC 4178	R	329	25	-95	372	
		IC 3042	R	380	18		914	
			O	233			13	
			O	333			694	
12 10 18	1 34.7	NGC 4179	O	1279	50	-135	13	
12 10 30	52 32.0	VV 497	O	797	50	91	873	
12 10 42	29 7.0		O	3862		-14	548	
12 10 48	15 3.0	NGC 4186	O	2090		-78	694	
12 10 48	21 55.0		O	7267		-47	703	
12 10 48	28 47.0	NGC 4185	O	4074		-16	548	
12 10 48	43 59.0	NGC 4183	R	931		53	744	
12 11 0	-34 13.0		O	2850	60	-242	655	
12 11 6	16 24.0	KDG 326 ARP 260 VV 128	O	7130	10	-72	908	
12 11 12	15 10.0	NGC 4192 HOL 348A	R	-143	48	-77	372	M 98
			R	-171			805	
			O	-198			694	
			O	-124	40		13	
			O	1150	100		2	DIS
12 11 12	16 24.0	KDG 326 ARP 260 VV 128	O	7074	10	-72	908	

R.A. (1950)	DEC. (1950)	NAME	OBS	HFL VEL (C*Z)	ERR	GAL CORR	REF	COMMENTS
12 11 13	36 54.5	NGC 4190	R	234	4	21	772	
		VV 104	O	138	47		789	
12 11 14	13 42.2	NGC 4189	R	2101		-84	744	
		IC 3050	O	1951	53		789	
			O	2214	116		741	
			O	2040			449	
12 11 18	24 32.4		R	942	10	-35	860	
12 11 24	13 27.0	NGC 4193	O	2501	58	-85	789	
12 11 36	8 3.0		R	1221	10	-108	860	
12 11 40	54 48.3	NGC 4194	R	2515	35	101	455	
		MRK 201	R	2545			368	
		1 ZW 33	O	2400			268	
		ARP 160	O	2585			13	
		VV 261	O	2482	50		873	
			O	2495			208	
			O	2530			604	
			O	2520			910	
			O	2522	50		141	
			O	2488	50		873	KNOT
12 11 54	28 42.0	NGC 4196	O	3982		-16	322	
12 12 0	6 5.0	NGC 4197	O	2030	50	-116	873	
		VV 520						
12 12 5	13 5.4	DDO 114	R	(654)		-86	492	
12 12 18	64 44.0	7 ZW433	O	2520	100	142	820	
12 12 27	36 29.8	DDO 113	R	283	10	20	492	
12 12 30	14 19.0	IC 3061	O	2263	100	-80	789	
12 12 30	36 14.0		O	3980		18	220	
12 12 30	64 4.0	NGC 4205	O	1436	25	139	644	
12 12 34	33 28.7	NGC 4203	R	1074		6	610	
			O	1093			645	
			O	1001	150		13	
12 12 42	6 3.0		O	2042		-115	163	
12 12 42	13 18.0	NGC 4206	R	712	13	-84	889	
		HOL 353B	O	(1240)	260		555	
		IC 3064	O	373			694	
12 12 42	20 56.0	NGC 4204	O	741		-51	703	

R.A. (1950)	DEC. (1950)	NAME	OBS	HEL VEL (C*Z)	ERR	GAL CORR	REF	COMMENTS
12 12 48	66 15.0	NGC 4210	O	2501	25	148	644	
12 12 51	27 9.9		O	7382		-22	714	
12 12 54	6 5.0		O	2039	21	-115	72	
12 13 4	28 26.8	NGC 4211A KDG 327A ARP 106 VV 199	O O	6412 6695		-16	714 322	
12 13 5	27 17.5		O	7874		-21	714	
12 13 5	28 27.3	NGC 4211B KDG 327B ARP 106 VV 199	O	6838		-16	322	
12 13 6	-34 37.0		O	7741	20	-241	582	
12 13 6	14 11.0	NGC 4212	O	2125		-80	13	
12 13 6	24 15.0	NGC 4213	O O	7021 6565	200	-35	703 843	
12 13 8	36 36.5	NGC 4214	R R R R O O O O O O	286 289 290 388 302 300 315 295 295 318	5 3 50 30	20	121 93 773 156 170 1 170 483 13 13	
12 13 12	10 58.0	IC 3074	O	2215	47	-94	789	
12 13 18	48 24.5	NGC 4218	R O	725 1388	20	74	829 178	
12 13 20	13 25.4	NGC 4216 HOL 353A	R R O O O O	101 -103 75 32 59 -137	60 40	-84	805 372 910 13 13 694	
12 13 22	6 40.8	NGC 4215	O O	2092 2093	17	-112	841 741	
12 13 22	47 22.2	NGC 4217 HOL 354A	R R R	1032 962 985	10 100	69	734 238 363	

224

R.A. (1950)		DEC. (1950)		NAME		OBS	HEL VEL (C*Z)	ERR	GAL CORR	REF	COMMENTS
12 13 23		26 56.5				O	7722		-23	714	
						O	7712			548	
12 13 26		41 12.2		MRK	46	O	3718	60	42	261	
12 13 27		27 43.3				O	7852		-19	714	
12 13 29		40 51.0		MRK	47	O	5843	60	40	261	
12 13 36		66 30.0		NGC 4221		O	2816	45	149	644	
12 13 38		28 24.5				O	7963		-16	714	
12 13 42		48 10.0		NGC 4220		O	979	50	73	13	
12 13 46		28 25.4			A	O	6571	34	-16	38	
12 13 46		28 25.4			B	O	6618	79	-16	38	
12 13 50		-43 2.8		NGC 4219		O	1978	27	-254	741	
						O	1700	100		844	
12 13 50		13 35.2		NGC 4222		R	222	30	-83	889	
				HOL 353C		O	10			694	
						O	309	71		789	
12 13 54		-11 15.1		DDO 116		R	1163	15	-180	492	
12 14 0		33 48.0		NGC 4227		O	6516	43	8	594	
						O	6393			548	
						O	4820			220	DIS
12 14 0		52 30.3				R	172	10	92	860	
12 14 1		7 44.4		NGC 4224		O	2651	36	-107	741	
						O	2458			694	
12 14 1		59 9.4		MRK	48	O	4660	60	120	261	
12 14 6		33 50.0		NGC 4229		O	6663	62	8	594	
						O	6765			548	
12 14 11		29 0.4		DDO 117		R	1215	10	-13	492	
						R	1230			858	
12 14 12		-19 53.0				O	11600	300	-206	735	
12 14 12		4 24.0				O	22921		-121	683	
12 14 20		69 45.0		NGC 4236I		O	86		161	650	HII REG
12 14 20		69 45.0		NGC 4236III		O	92		161	650	HII REG

225

R.A. (1950)	DEC. (1950)	NAME	OBS	HEL VEL (C*Z)	ERR	GAL CORR	REF	COMMENTS
12 14 24	69 45.7	NGC 4236	R	5		161	934	
		HOL 357A	R	-3			693	
			R	-10	8		404	
			R	33			27	
			R	5			156	
			O	27			13	
12 14 35	7 54.1	NGC 4233	O	2224	75	-106	741	
12 14 36	3 57.7	NGC 4234	O	2147		-123	163	
		HOL 358A	O	2143	75		72	
12 14 36	12 34.0	IC 3100	O	610	80	-86	953	
12 14 36	12 44.0	IC 3099	O	2246	60	-86	953	
12 14 37	-11 24.1	DDO 118	R	(1306)		-180	492	
12 14 37	7 28.1	NGC 4235	O	2596	18	-108	741	SEYF
		IC 3098	O	2408	100		674	
		HOL 359A	O	2208	71		789	
			O	2410			712	
			O	2300			694	
12 14 39	15 36.1	NGC 4237	R	855	30	-73	889	
			O	945	18		741	
			O	800	63		789	
			O	891			694	
12 14 54	45 54.0	NGC 4242	O	661	87	63	72	
12 14 55	16 10.8		O	25000		-70	868	
12 14 58	3 53.8	PKS	O	23220		-123	143	
12 15 0	22 50.0		O	469	25	-41	813	
12 15 0	38 5.0	NGC 4244	R	240	10	28	74	
			R	240	10		93	
			R	245			156	
			R	242	2		385	
			R	230			121	
			O	265			13	
			O	241			641	
12 15 6	12 40.0	IC 3105	O	813	70	-86	953	*
		VV 432	O	325	50		953	KNOT
			O	-159	50		873	DIS
			O	-80	50		873	KNOT
12 15 6	29 53.0	NGC 4245	O	890	65	-8	13	
12 15 6	71 4.8	NGC 4250 7 ZW447	O	1940		166	910	

R.A. (1950)	DEC. (1950)	NAME	OBS	HEL VEL (C*Z)	ERR	GAL CORR	REF	COMMENTS
12 15 7	3 56.0	PKS	0	22770		-122	143	
12 15 24	7 28.0	NGC 4246	R	3728	10	-108	582	
		IC 3113	0	3926	100		789	
		HOL 359B	0	3702	50		582	
12 15 24	7 33.0	NGC 4247	0	3944	82	-107	789	
		HOL 359C						
12 15 24	20 57.0		0	60100	300	-49	583	
12 15 24	47 41.0	NGC 4248	R	484	10	71	664	
		HOL 363B	0	598	100		789	
12 15 29	58 56.1	MRK 202	0	6300		119	308	
12 15 36	28 27.0	NGC 4251	0	1014	75	-15	13	
12 15 44	44 27.0	MRK 203	0	7500		57	268	
12 15 54	1 19.3	PKS	0	35000		-132	672	
12 15 54	30 5.0	NGC 4253	0	3876		-7	548	SEYF
		MRK 766	0	3862	15		630	
12 16 0	12 8.0	IC 3127	0	8688	100	-87	953	
12 16 3	-10 3.9	PKS	0	26230		-174	494	
12 16 6	12 1.0	IC 3128A	0	11674	70	-88	953	
12 16 6	12 1.0	IC 3128B	0	11525	60	-88	953	
12 16 18	14 42.0	NGC 4254	R	2387	14	-76	372	M 99
			R	2407	13		889	
			0	2485	50		13	
			0	2451			13	
12 16 21	66 10.6	NGC 4256	0	2560	80	148	644	
			0	2583			13	
12 16 24	-19 43.0	1 ZW369-S	0	6000	210	-205	735	
12 16 30	9 8.0	VV 614	0	2364	50	-100	873	
12 16 30	47 34.9	NGC 4258	R	450		72	960	M 106
		HOL 363A	R	455	8		523	
			R	(450)			156	
			R	451			934	
			0	508			147	
			0	420	40		13	
			0	490	9		112	
			0	500	50		1	
			0	472	9		148	
			0	467	10		469	
			0	425	50		112	

227

12ʰ16ᵐ

R.A. (1950)	DEC. (1950)	NAME	OBS	HEL VEL (C∗Z)	ERR	GAL CORR	REF	COMMENTS
12 16 30	47 34.9	NGC 4258	O	510		72	79	
12 16 36	4 8.1	MRK 49	O	1564	52	−121	246	
			O	1260			224	
			O	1590	100		820	
			O	1717			163	
			O	1716	34		72	
12 16 36	14 15.0	IC 3142A	O	421	50	−78	953	
12 16 39	47 22.0		R	255	10	71	860	
12 16 42	− 1 33.0	MRK 1320	O	33033		−143	920	★
12 16 42	14 9.6		R	−235	15	−78	643	★
			O	−250	25		643	
			O	−225			158	
12 16 48	5 39.0	NGC 4259 HOL 368F	O	2426		−114	694	
12 16 48	6 23.0	NGC 4260	O	1935	43	−111	72	
			O	1795			694	
12 16 48	12 35.0	IC 3149	O	8127	100	−85	953	
12 16 49	6 6.3	NGC 4261 3CR270	O	2202	75	−112	13	
			O	2233	19		938	
			O	2202			227	
			O	2070			694	
12 16 54	28 35.0	IC 777	O	2620		−13	548	
12 16 58	15 9.3	NGC 4262	R	1354		−74	610	
			R	1354			572	
			R	1369	9		818	
			O	1351	36		148	
12 17 0	6 7.0	NGC 4264	O	2632		−112	694	
12 17 12	5 34.0	NGC 4268 HOL 368D	O	2317		−114	694	
12 17 12	6 17.0	IC 3155 HOL 365B	O	2044		−111	694	
12 17 12	6 18.0	NGC 4269 HOL 365A	O	2534		−111	694	
12 17 12	13 3.0	NGC 4267	O	1260	75	−83	13	
12 17 18	5 45.0	NGC 4270 HOL 368C	O	2347	50	−114	13	
			O	2235			694	

R.A. (1950)	DEC. (1950)	NAME	OBS	HEL VEL (C*Z)	ERR	GAL CORR	REF	COMMENTS
12 17 18	30 27.0	NGC 4272	O	8462		-4	322	
12 17 20	29 53.3	NGC 4274	R	933		-7	572	
			R	718	16		390	
			R	930	5		818	
			O	767	150		13	
12 17 24	-17 7.0	1 ZW399-S	O	900	120	-196	735	
12 17 24	5 37.0	NGC 4273	R	2375		-114	744	
		HOL 368A	O	2302	40		13	*
			O	2307			694	
12 17 24	26 3.0	IC 780	O	6804		-25	703	
12 17 24	27 54.0	NGC 4275	O	2200		-16	491	
			O	2400			533	
			O	2307			322	
12 17 30	5 37.2	NGC 4277 HOL 368F	O	2498		-114	694	
12 17 30	28 15.0	IC 3165	O	8443		-15	548	
12 17 36	29 33.6	NGC 4278	R	680	40	-8	616	
		HOL 369A	R	620			682	
		B?	R	610			637	
			R	700	50		499	
			R	670	17		637	
			O	702	80		8	
			O	590			281	
			O	703			253	
			O	624	40		13	
			O	601			842	
12 17 48	5 40.0	NGC 4281 HOL 368B	O	2602	50	-114	13	
12 17 48	5 51.0	NGC 4282	O	6542		-113	694	
12 17 48	58 22.2	NGC 4284 KDG 329 HOL 373B	O	3126	40	118	908	
12 17 50	29 35.3	NGC 4283	O	1071	65	-8	13	
		HOL 369B	O	1153			253	
			O	1152	80		8	
12 17 54	25 51.0	IC 3171	O	6960		-25	703	
12 17 54	31 27.0		O	6697		0	548	
12 18 6	31 4.0		O	1010		-1	548	

R.A. (1950)	DEC. (1950)	NAME	OBS	HEL VEL (C*Z)	ERR	GAL CORR	REF	COMMENTS
12 18 6	75 40.0	NGC 4291	O	1903		182	13	
			O	1785	50		13	
12 18 11	46 34.2	NGC 4288	R	534	15	68	492	
		DDO 119						
		HOL 371A						
12 18 23	58 22.3	NGC 4290	O	2766	41	118	741	
		HOL 373A	O	3032	20		908	
		KDG 329						
12 18 24	11 18.0	IC 3188	O	5402	100	-90	953	
12 18 42	18 40.0	NGC 4293	R	948	40	-57	372	
			O	750			13	
12 18 42	28 26.0	NGC 4295	O	8568		-13	322	
12 18 45	11 47.3	NGC 4294/9	R	312	18	-87	372	
		HOL 376						
12 18 45	11 47.3	NGC 4294A/B	R	351	15	-87	829	
		KDG 330	R	359			744	*
		HOL 376A	R	365			893	
			O	390	47		72	
12 18 45	11 47.3	NGC 4294A	O	466	16	-87	908	
12 18 45	11 47.3	NGC 4294B	O	431	16	-87	908	
12 18 48	46 5.3	DDO 120	R	464	10	66	492	
12 19 0	14 53.0	NGC 4298	R	1035	10	-74	566	
		KDG 332A	O	1173	27		741	
		HOL 377A	O	1106	25		283	
			O	1133	11		908	
12 19 8	11 46.8	NGC 4299	R	228	15	-87	829	
		HOL 376B	R	241	13		889	
			R	237			893	
			R	221	11		914	
			O	187	36		72	
			O	223			694	
12 19 12	5 39.0		O	2368		-113	694	
12 19 12	14 53.0	NGC 4302	R	1035	10	-74	566	
		KDG 332B	O	1036	60		789	
		HOL 377B	O	1339	50		741	
			O	1134	16		908	
			O	1049			694	
12 19 12	41 7.0		R	6901	15	44	582	
			O	6927	25		582	

R.A. (1950)	DEC. (1950)	NAME	OBS	HEL VEL (C*Z)	ERR	GAL CORR	REF	COMMENTS
12 19 18	-43 5.0		O	7102	70	-251	603	
12 19 22	4 45.1	NGC 4303	R	1560	5	-116	372	M 61
		HOL 369A	R	1670			117	
			R	1566			693	
			R	1568			744	
			R	1558	5		566	
			R	1561	7		251	
			O	1671	150		13	
			O	1560			264	
12 19 24	30 20.0	NGC 4308	O	606		-4	322	
12 19 31	13 1.1	NGC 4305	O	1861	15	-81	908	
		KDG 333						
12 19 32	-43 3.8		O	7200	200	-251	794	
12 19 32	13 3.9	NGC 4306	O	1780	76	-81	908	
		KDG 333						
12 19 32	75 35.2	MRK 205	O	20559	150	182	854	* SEYF
			O	20800			231	
			O	21000			265	
			O	21377	63		789	
			O	21448	80		300	
			O	20671	150		854	KNOT
12 19 33	9 19.1	NGC 4307	O	1305	81	-97	741	
		HOL 380A						
12 19 34	75 35.9	NGC 4319	O	1700	100	182	270	
12 19 35	-33 12.4	NGC 4304	O	2595	25	-235	741	
12 19 36	6 16.0	IC 3211	O	5971	100	-110	789	
12 19 42	7 25.0	NGC 4309	O	1103		-105	694	
		HOL 382A						
12 19 54	4 51.0	NGC 4303A	O	1252	32	-116	789	
		HOL 379B						
12 19 54	29 29.0	NGC 4310	O	901		-8	322	
12 20 0	15 49.0	NGC 4312	O	116		-69	694	
		HOL 387B						
12 20 0	30 10.0	NGC 4314	O	883	85	-4	13	
			O	977			910	
12 20 6	12 5.0	NGC 4313	O	1579		-85	694	
		HOL 385A						

12ʰ20ᵐ

R.A. (1950)	DEC. (1950)	NAME	OBS	HEL VEL (C∗Z)	ERR	GAL CORR	REF	COMMENTS
12 20 12	8 29.0	NGC 4318	O	-200		-100	476	
12 20 22	22 43.3	MRK 438	O	6600		-38	359	
			O	6685			438	
12 20 23	16 6.0	NGC 4321	R	1585		-67	363	M 100
		HOL 387A	R	1630			117	
			R	1576			744	
			R	1537	10		238	
			R	1663	35		372	
			R	1567	5		566	
			R	1580	30		889	
			O	1545	25		936	
			O	1617	75		13	
			O	1578	15		412	
12 20 30	66 7.0	NGC 4332	O	2844	46	149	644	
12 20 34	5 31.7	NGC 4324	R	1663		-112	572	
		HOL 388A	R	1675	50		818	
			O	1714	50		13	
			O	1669			694	
12 20 36	6 21.0	NGC 4326	O	7286		-109	694	
			O	7078			910	
12 20 36	63 30.0		O	17580	100	139	820	
12 20 42	14 40.0	IC 3244	O	12880	25	-74	283	
12 20 48	7 45.0	NGC 4334	O	4377		-103	694	
12 20 49	6 19.1	NGC 4333	O	6916		-109	694	
			O	6986			910	
12 20 51	2 57.3	MRK 50	O	7013	60	-123	261	SEYF
12 20 54	1 54.0		O	8227		-127	683	
12 20 54	5 7.0		O	5514		-114	683	
12 21 0	6 22.0	NGC 4339	O	1278	100	-109	13	
12 21 1	47 16.2	NGC 4346	O	849	38	72	741	
12 21 4	17 0.0	NGC 4340	O	854	23	-63	741	
		HOL 391B	O	974			694	
			O	893			910	
12 21 7	7 14.0	NGC 4343	R	1012	10	-105	829	
			O	1053			910	
			O	1251			694	
			O	714	50		13	

R.A. (1950)	DEC. (1950)	NAME	OBS	HEL VEL (C*Z)	ERR	GAL CORR	REF	COMMENTS
12 21 12	7 28.0	IC 3259	O	1575		-104	694	
12 21 12	12 45.0	IC 3258	R	-473		-82	805	
			O	-419	35		200	
12 21 20	- 3 10.0	NGC 4348	O	1951		-146	694	
			O	1988	15		813	
			O	2202	25		741	
12 21 24	7 23.0	NGC 4341 IC 3260	O	1102		-104	694	
12 21 24	16 58.0	NGC 4350 HOL 391A	O	1184	60	-63	13	
12 21 28	8 3.7	NGC 4353 IC 3265/6	O	1059		-101	694	
12 21 30	7 19.0	IC 3267	O	2130		-104	694	
12 21 30	12 29.0	NGC 4351	R	2284		-83	805	
			R	2297	10		566	
			O	2388			321	
12 21 36	6 53.0	IC 3268	O	764		-106	321	
12 21 36	49 3.0	NGC 4357	O	4252	32	80	813	
12 21 42	8 49.0	NGC 4356	O	1081		-98	694	
12 21 42	31 47.0	NGC 4359	O	1171		3	548	
12 21 48	9 34.0	NGC 4360A	O	7014		-95	694	
12 21 48	9 34.0	NGC 4360B	O	6676		-95	694	
12 21 55	7 35.7	NGC 4365	O	1171	30	-103	13	
			O	1233	11		938	
			O	1290			13	
			O	1270			694	
12 21 59	67 43.0	MRK 206	O	900		155	268	
12 22 8	39 39.7	NGC 4369 MRK 439	R	1052		39	693	
			O	900			359	
			O	1075			321	
			O	981			641	
			O	1020			438	
			O	1081	20		741	
12 22 10	3 35.0		O	960		-119	895	
12 22 10	70 36.5	DDO 122	R	470	10	166	492	

233

12ʰ22ᵐ

R.A. (1950)	DEC. (1950)	NAME	OBS	HEL VEL (C∗Z)	ERR	GAL CORR	REF	COMMENTS
12 22 20	75 48.3	NGC 4386	O	1811		183	13	
12 22 24	7 43.0	NGC 4370	O	468		-102	694	
12 22 24	11 59.0	NGC 4371	O	977	38	-84	148	
12 22 30	28 50.0	NGC 4375	O	9042		-9	322	
12 22 31	-39 29.7	IC 3290	O	3380	60	-245	741	
12 22 32	13 9.9	NGC 4374	R	910	17	-79	372	M 84
		3CR272.1	O	954	50		13	
		HOL 403B	O	957			227	
			O	1050	100		2	
			O	997			106	
			O	1052	15		938	
			O	1119			281	
12 22 40	-39 28.7	NGC 4373	O	3457	21	-245	741	
			O	3400			574	
12 22 40	15 2.4	NGC 4377	O	1325	72	-71	741	
		3 ZW 65	O	1329			262	
12 22 42	5 12.0	NGC 4378	R	2561	14	-112	818	
			R	2535			572	
			O	2510	14		741	
			O	2540	12		738	
12 22 43	15 53.0	NGC 4379	O	1039	21	-67	741	
			O	979			694	
12 22 48	4 37.9		O	22904	60	-115	952	SEYF
12 22 48	6 1.0		O	1172		-109	694	
12 22 48	54 46.9	MRK 207	O	2400		105	268	
12 22 50	10 17.6	NGC 4380	R	963	7	-91	818	
			R	(1108)	30		889	
			O	843			694	
			O	1000	26		741	
12 22 53	18 28.0	NGC 4382	R	770		-55	117	M 85
		KDG 334A	O	773	30		13	
		HOL 397A	O	713	13		938	
			O	500	50		1	
			O	754	49		908	
12 22 54	16 44.8	NGC 4383	O	1609	36	-63	741	
12 23 0	65 13.0	NGC 4391	O	1338	74	146	644	
		7 ZW454						

R.A. (1950)	DEC. (1950)	NAME	OBS	HEL VEL (C*Z)	ERR	GAL CORR	REF	COMMENTS
12 23 9	0 50.9	NGC 4385	R	2140	7	-130	818	
		MRK 52	R	2135			572	
			O	1860			224	
			O	2160			264	
			O	2300			604	
			O	2203	14		741	
12 23 9	2 26.1		R	1504	10	-123	860	
12 23 10	7 29.5	IC 3322A	O	1186		-103	694	
12 23 10	45 57.7	NGC 4389	O	752		67	741	
12 23 12	13 5.0	NGC 4387	O	511	65	-79	13	
12 23 15	12 56.3	NGC 4388	R	2642	15	-79	566	
		HOL 403C	R	2354	15		566	
			R	2472			805	
			O	2450	34		741	
			O	2458	26		789	
			O	2604	25		283	
12 23 18	27 50.0	NGC 4393	O	842		-13	548	
12 23 22	7 49.9	IC 3322	O	1177		-101	694	
12 23 24	15 57.0	NGC 4396	O	-84	44	-66	789	
		HOL 400A						
12 23 24	18 29.0	NGC 4394	R	945	10	-55	566	
		KDG 334B	O	772	150		13	
		HOL 397B	O	974	47		908	
12 23 24	33 49.0	NGC 4395	R	315		13	693	
			R	320			934	
			O	311			548	
12 23 30	-38 51.0		O	3720	60	-243	655	
12 23 30	4 45.0		O	5700		-114	533	
12 23 30	33 47.0	NGC 4401	O	294		13	13	
12 23 36	7 56.0	2 ZW 62	O	3930	200	-101	820	
12 23 36	13 23.0	NGC 4402	O	161	47	-77	789	
		HOL 403D	O	-11			694	
			O	-37	100		953	
12 23 36	16 28.0	NGC 4405	O	1764		-64	694	
12 23 42	- 1 1.0		O	1800		-137	535	

R.A. (1950)	DEC. (1950)	NAME	OBS	HEL VEL (C*Z)	ERR	GAL CORR	REF	COMMENTS
12 23 42	13 13.5	NGC 4406	O	-289		-78	227	M 86
		HOL 403A	O	-110	41		72	
			O	-374	50		13	
			O	-309			13	
			O	-228	10		938	
			O	-377			694	
12 23 46	58 35.8	DDO 123	R	724	10	121	492	
12 23 48	48 46.1	MRK 209	O	253	45	80	246	
12 23 56	9 8.9	NGC 4411A	R	1273	30	-95	889	
		KDG 336A	O	1254	41		789	★
			O	-109			694	DIS
12 23 56	9 17.8	NGC 4410A/B	O	7225		-95	426	
		IC 790	O	7468			694	
12 23 56	9 17.8	NGC 4410A	O	7456		-95	694	
		KDG 335A	O	7416	50		908	
			O	7875	100		789	
			O	7595			748	
12 23 57	9 17.8	NGC 4410B	O	7567		-95	694	
		KDG 335B	O	7284	50		908	
			O	7595			748	
12 24 0	12 53.0	NGC 4413	R	96	9	-79	818	
			R (	-79)			805	
			O	-60	53		741	
			O	-88			694	
12 24 0	31 30.0	NGC 4414	O	715	100	3	13	
			O	730	35		673	
12 24 3	4 14.4	NGC 4412	R	1735	10	-115	752	
			O	2301	29		741	
12 24 6	-37 18.0		O	3270	30	-241	655	
12 24 6	8 42.0	NGC 4415	O	496		-97	694	
12 24 15	9 9.7	NGC 4411B	R	1273	13	-95	889	
		KDG 336B	O	1196	100		789	
			O	842			694	
12 24 18	- 0 35.0	NGC 4418 KDG 337A	O	2045	69	-134	908	
12 24 18	9 19.0		O	7076		-94	694	
12 24 18	9 51.7	NGC 4417	O	826	25	-92	741	
			O	920			694	

R.A. (1950)	DEC. (1950)	NAME	OBS	HEL VEL (C*Z)	ERR	GAL CORR	REF	COMMENTS
12 24 18	13 27.2	IC 3355	0	162	53	-77	789	
		DDO 124	0	-10	100		953	
12 24 19	11 49.9	IC 3356	R	1098	10	-84	860	
12 24 24	15 19.0	NGC 4419	R	(0)		-68	805	
			R	(-273)	48		818	
			0	-274	17		741	
			0	-232			694	
12 24 25	2 46.3	NGC 4420	R	1678	10	-121	566	
			0	1724	23		741	
			0	1728			694	
12 24 29	-39 3.7	IC 3370	0	2973	33	-243	741	
12 24 30	- 0 37.0	KDG 337B	0	2233	21	-134	908	
12 24 30	9 14.0		0	7280		-94	694	
12 24 30	15 44.0	NGC 4421	0	1692	250	-67	13	
12 24 30	22 55.0	IC 791	0	6770		-35	703	
12 24 31	48 33.1	MRK 210	0	7800		79	268	
12 24 36	37 25.1	DDO 126	R	222	10	30	492	
12 24 38	7 32.3		R	928	10	-102	860	
12 24 39	9 41.8	NGC 4424	0	452		-92	426	
			0	507			694	
			0	462	51		741	
12 24 42	13 1.0	NGC 4425 HOL 403E	0	1883	50	-78	13	
12 24 48	11 9.0	IC 3371	0	929	60	-86	953	
12 24 53	- 7 53.9	NGC 4428 HOL 407B	0	3036	16	-161	741	
12 24 54	6 32.0	NGC 4430 KDG 338A	0	1436		-106	694	
			0	1494	35		908	
12 24 54	11 23.0	NGC 4429	0	1114	65	-85	13	
12 24 54	12 34.0	NGC 4431 HOL 408C	0	456		-80	694	
12 25 0	6 30.0	NGC 4432 KDG 338B	0	(6466)	30	-106	908	

237

R.A. (1950)	DEC. (1950)	NAME	OBS	HEL VEL (C*Z)	ERR	GAL CORR	REF	COMMENTS
12 25 0	8 26.0	NGC 4434	0	1007		-98	694	
12 25 0	65 5.0	NGC 4441	0	1440	44	146	644	
12 25 2	36 58.0	MRK 440	0	14700		28	359	
			0	14990			438	
12 25 4	- 8 0.2	NGC 4433	0	2880	42	-161	329	
		HOL 407A	0	2983			432	
			0	3056	36		741	
12 25 4	26 30.4	B2	0	19090	250	-18	696	*
12 25 6	13 21.0	NGC 4435 ARP 120 VV 188	0	869	100	-77	13	
12 25 6	57 12.0		0	6000		116	476	
12 25 14	13 17.2	NGC 4438 VV 188 ARP 120	R R 0 0 0	3 0 -32 321 -46	20 75	-77	497 805 13 106 694	
12 25 14	43 46.3	DDO 125	R R	197 198	2 10	59	754 492	
12 25 18	27 16.0	IC 3376	0	7093		-15	548	
12 25 24	12 34.0	NGC 4440 HOL 408B	0	707		-80	694	
12 25 30	10 5.0	NGC 4442	0	580	100	-90	13	
12 25 36	14 11.0	NGC 4446 KDG 339	0 0	7887 7342	70 28	-73	953 908	
12 25 38	44 46.1	MRK 211	0	12300		63	308	
12 25 42	9 43.0	NGC 4445	0	254		-92	694	
12 25 42	14 10.0	NGC 4447 KDG 339	0 0	7114 7243	28 80	-73	908 953	
12 25 47	44 22.3	NGC 4449	R R R R 0 0 0 0 0 0	190 203 204 204 200 216 309 206 204 200	3 6 10 50 28 30 50	62	156 944 216 93 1 149 72 13 204 142	

R.A. (1950)	DEC. (1950)	NAME	OBS	HEL VEL (C*Z)	ERR	GAL CORR	REF	COMMENTS
12 25 48	28 54.0	NGC 4448	O	693	65	-7	13	
12 25 48	44 43.0	MRK 212A/B	O	6900		63	308	
12 25 48	44 43.0	KDG 340A	O	7084	150	63	567	
		MRK 212A	O	7000	20		908	
12 25 48	44 43.0	KDG 340B	O	6834	80	63	567	
		MRK 212B	O	6929	19		908	
12 25 58	17 21.7	NGC 4450	R	2050		-58	117	
			R	1957	12		818	
			O	2048	150		13	
12 26 1	37 30.6	DDO 127	R	280	10	31	492	
12 26 6	9 32.0	NGC 4451	O	600		-92	476	
			O	977			694	
12 26 10	35 59.4		R	306	10	24	860	
12 26 12	12 2.0	NGC 4452	O	212	42	-81	741	
			O	62			694	
12 26 12	45 8.6	NGC 4460	O	558	11	65	741	
12 26 13	23 5.5	NGC 4455	O	631	33	-33	741	
			O	621			703	
12 26 16	- 1 39.7	NGC 4454	O	2358	68	-137	741	
			O	2450			694	
12 26 18	43 30.0	DDO 129	R	543	10	58	492	
12 26 24	3 51.0	NGC 4457	R	894	23	-116	818	
			O	738	26		148	
12 26 24	13 31.0	NGC 4458	O	383	250	-75	13	
		HOL 411B						
12 26 29	2 59.9	DDO 128	R	1571	10	-119	492	
12 26 30	13 28.0	NGC 4461	O	1887	40	-75	13	
		HOL 411A						
12 26 30	14 15.0	NGC 4459	O	1111	75	-72	13	
12 26 30	28 4.0	IC 3407	O	7072		-10	548	
12 26 42	11 58.0		O	150		-81	158	
12 26 44	-22 53.6	NGC 4462	O	1866	51	-207	741	

R.A. (1950)	DEC. (1950)	NAME	OBS	HEL VEL (C*Z)	ERR	GAL CORR	REF	COMMENTS
12 26 44	47 58.0		O	76100	300	78	583	
12 26 48	8 26.0	NGC 4464	O	1199	50	-96	13	
12 26 48	27 23.0		O	3819	174	-13	246	
12 26 56	9 1.5	NGC 4469	O	498	10	-94	741	
			O	784			694	
12 27 0	7 58.0	NGC 4466 HOL 412A	O	1011		-98	694	
12 27 0	8 6.0	NGC 4470	O	2441		-98	694	
			O	2337	54		741	
12 27 0	8 16.0	NGC 4467 HOL 413C	O	1474	300	-97	13	
12 27 6	28 17.0		O	6818		-9	593	
12 27 14	8 16.7	NGC 4472 ARP 134 HOL 413A	R	868	50	-97	372	M 49
			R	1010			117	
			O	850	50		1	
			O	876	41		72	
			O	1001	12		938	
			O	1034	45		554	
			O	1013	50		13	
			O	926			281	
12 27 18	13 42.0	NGC 4473	O	2241	75	-74	13	
			O	2281	34		392	
12 27 18	27 31.0	NGC 4475	R	7388	25	-12	582	
			O	7349	20		582	
12 27 24	14 21.0	NGC 4474	O	1526	50	-71	13	
12 27 27	12 37.5	NGC 4476	O	1978	28	-78	741	
12 27 30	13 55.0	NGC 4477	O	1263	75	-73	13	
12 27 46	12 36.3	NGC 4478	O	1488		-78	227	
			O	1482	75		13	
12 27 48	13 51.0	NGC 4479	O	822	100	-73	13	
12 27 54	4 31.0	NGC 4480	O	2285	58	-112	789	
12 27 54	12 47.0		O	1486	50	-77	13	
			O	1540	60		98	
12 27 54	14 27.0	IC 3432	O	(361)	100	-70	953	

240

R.A. (1950)	DEC. (1950)	NAME	OBS	HEL VEL (C*Z)	ERR	GAL CORR	REF	COMMENTS
12 28 6	41 56.0	NGC 4490/85	R	578	5	52	944	
		VV 30	R	570	5		905	
12 28 6	41 58.6	NGC 4485	R	480	15	52	961	
		ARP 269	R	480	15		962	
		KDG 341A	O	795	90		8	
		HOL 414B	O	786			253	
12 28 8	9 17.1	NGC 4483	O	875		-92	741	
12 28 10	41 54.9	NGC 4490	R	560		52	171	
		ARP 269	R	572			693	★
		VV 30	R	568	7		489	
		KDG 341B	R	590			156	
		HOL 414A	R	585	5		962	
			R	565	3		390	
			R	585	5		961	
			R	585	5		905	
			R	565	5		216	
			O	640	90		8	
			O	655			125	
			O	625	50		13	
			O	590			239	
			O	631			253	
			O	580	30		72	
			O	476	29		72	
12 28 18	-37 13.0		O	3238		-238	683	
12 28 18	11 37.0	IC 3437	O	1779	70	-82	953	
12 28 18	12 40.0	NGC 4486	O	1196		-77	13	M 87
		ARP 152	O	1183	41		72	VIR A
		3CR274	O	1260			165	★
			O	1040	50		165	
			O	(800)			1	
			O	1302	31		938	
			O	1287	21		742	
			O	1280	30		855	
			O	1267			227	
			O	1258			444	★
			O	1225			281	
			O	1290	20		13	
			O	1332	105		555	
12 28 18	17 2.0	NGC 4489	O	863		-58	694	
12 28 18	57 35.0	KDG 342	O	4582	20	118	908	
12 28 24	3 46.0		O	5400		-115	533	
12 28 24	8 21.0	NGC 4492	O	1735	200	-96	13	
		IC 3438						

12ʰ28ᵐ

R.A. (1950)	DEC. (1950)	NAME	OBS	HEL VEL (C*Z)	ERR	GAL CORR	REF	COMMENTS
12 28 24	10 50.0	N	O	50402	200	-85	13	
12 28 24	10 50.0	S	O	48788	200	-85	13	
12 28 24	11 46.0	NGC 4491	O	160		-81	694	
12 28 24	57 34.0	KDG 342	O	4567	14	118	908	
12 28 30	- 7 47.0	NGC 4487	R	1037		-158	744	
			O	1203	55		741	
			O	901			694	
12 28 30	12 33.0		O	81		-78	694	
12 28 48	11 46.0	IC 3446	O	1297	60	-81	953	
12 28 48	12 36.0	IC 3443	O	2254		-77	694	
12 28 54	26 3.0	NGC 4494	O	1303		-18	13	
			O	1308	40		392	
			O	1333	65		13	
12 28 54	29 25.0	NGC 4495	O	4364		-3	548	
12 29 0	11 54.0	NGC 4497	O	1342		-80	694	
12 29 0	15 8.0	IC 3453	O	2546	25	-66	283	
12 29 1	58 14.3	MRK 213	O	3000		121	268	
12 29 6	17 8.0	NGC 4498	R	1507	6	-57	914	
			O	1638			694	
12 29 7	4 12.9	NGC 4496A/B VV 76	R	1725	5	-112	566	
12 29 7	4 12.9	NGC 4496A KDG 343	O	1883	64	-112	148	
			O	1629	100		555	
12 29 8	4 12.1	NGC 4496B KDG 343	O	1772	50	-112	555	
			O	1773	69		72	
			O	(4706)	90		148	DIS
12 29 12	- 2 42.0		O	2100		-139	533	
12 29 16	66 2.3	MRK 214	O	9300		151	308	
12 29 27	14 41.7	NGC 4501	R	2281		-68	817	M 88
			R	2270	5		566	
			R	2016	60		372	
			R	2120			117	
			O	2120	100		13	

R.A. (1950)	DEC. (1950)	NAME	OBS	HEL VEL (C*Z)	ERR	GAL CORR	REF	COMMENTS
12 29 29	29 59.1	DDO 131	R	639	10	0	492	
12 29 30	20 25.0	MRK 771	O	19200		-43	533	SEYF
			O	18943			630	
12 29 34	11 27.2	NGC 4503	O	1417	37	-82	741	
			O	1348			694	
12 29 42	- 7 17.0	NGC 4504	R	998		-156	744	
			R	689			694	
12 29 42	13 42.0	NGC 4506	O	680		-72	694	
12 29 48	66 40.0	7 ZW466	O	14494	39	153	789	
			O	14490			591	
12 29 54	0 39.9	NGC 4517A	R	1529	10	-126	829	REI 80
		KDG 344A	O	1501	47		789	
12 29 54	11 32.0	IC 3470	O	2025		-81	694	
12 29 54	12 4.0		O	7810	50	-79	953	
12 30 0	9 26.0		O	120		-90	533	
12 30 6	9 27.0		O	1317	300	-90	13	
12 30 6	30 0.0	NGC 4514	O	8011		0	548	
12 30 11	14 19.5	IC 3476	O	-240		-69	535	
		VV 563	O	-28	44		789	
12 30 11	46 2.5	MRK 215	O	5910		71	287	
			O	5700			268	
12 30 12	0 23.3	NGC 4517	R	1125		-127	817	
		KDG 344B	R	1129	10		829	
			P	1078			805	
			R	1129			744	
			O	1218			13	
12 30 18	11 40.0	IC 3481 ARP 175 VV 43	O	7086	80	-80	13	ZW TRI
12 30 24	37 54.0		O	579	24	35	813	
12 30 24	64 10.0	NGC 4512 KDG 345	O	2929	30	144	908	
12 30 25	31 48.8	DDO 133	R	334	4	8	772	
			R	335	10		492	

12ʰ30ᵐ

R.A. (1950)	DEC. (1950)	NAME	OBS	HEL VEL (C*Z)	ERR	GAL CORR	REF	COMMENTS
12 30 28	11 12.0	IC 3490	O	80	50	-82	953	
12 30 30	11 40.0	IC 3481A	O	7304	65	-80	13	
12 30 34	64 12.9	NGC 4521	R	2971	20	145	829	
		KDG 345	O	2426	40		644	
			O	2446	63		908	
12 30 36	11 37.0	IC 3483	O	108	40	-80	13	
		ARP 175						
		VV 43						
12 30 42	12 31.0	IC 3489	O	7861	90	-76	953	
12 30 42	13 8.0	IC 3492	O	300		-74	476	
12 30 51	52 3.1	MRK 216	O	9883		97	437	
			O	10200			308	
12 30 54	13 32.0		O	7500		-72	476	
12 31 0	8 55.0	NGC 4519	R	1229	5	-92	566	
12 31 0	8 55.0	HOL 418A	O	1213		-92	13	
12 31 8	9 27.0	NGC 4522	P	2316	10	-89	566	
			O	2250	50		789	
			O	2404	39		741	
			O	2419			694	
12 31 17	15 26.3	NGC 4523	R	262	10	-63	492	
		DDO 135	P	253	13		889	
12 31 18	14 15.0	IC 3500	O	-555	100	-69	953	
12 31 18	30 34.0	NGC 4525	O	1131		3	322	
12 31 30	7 58.0	NGC 4526	R	448	24	-95	372	
			O	580	50		1	
			O	447	50		13	
12 31 35	2 55.7	NGC 4527	R	1700		-116	117	
			R	1737	10		829	
			R	1740			817	
			O	1727	75		13	
12 31 36	11 36.0	NGC 4528	O	1337		-80	694	
12 31 36	35 48.0	NGC 4534	O	796	15	26	813	
12 31 48	6 45.0	NGC 4532	R	1997	5	-100	566	
			R	2000			744	
			O	2154	21		148	

244

R.A. (1950)	DEC. (1950)	NAME	OBS	HEL VEL (C*Z)	ERR	GAL CORR	REF	COMMENTS
12 31 48	8 28.6	NGC 4535	R	1955	10	-93	566	
		HOL 420A	R	1950			117	
			R	1934			886	
			R	1942	13		372	
			O	1920			264	
			O	2097			13	
			O	1930	20		13	
12 31 48	13 21.0	NGC 4531	O	338		-72	694	
12 31 54	2 27.7	NGC 4536	R	1930		-117	117	
			R	1807	10		829	
			R	1804			693	
			R	1800			817	
			O	1927			13	
12 32 0	13 47.0	IC 3520	O	1093	70	-70	953	
12 32 6	3 36.0	NGC 4538	O	5100		-113	533	
12 32 9	83 40.6		O	62400		207	717	
12 32 12	6 34.3	DDO 137	R	(2026)	30	-101	492	HOL VII
12 32 15	15 29.8	IC 3522 DDO 136	R	661	15	-63	492	
12 32 18	15 50.0	NGC 4540	R	1286	10	-61	566	
		HOL 421A	O	1262	58		789	
			O	1177			694	
12 32 20	63 48.2	NGC 4545	R	2740	20	144	829	
			O	2703	60		644	
12 32 24	48 1.0	1 ZW 39	O	9360		80	303	
12 32 42	41 20.0		R	605	5	52	905	
12 32 54	-39 37.7	NGC 4507	O	3540		-240	574	SEYF
			O	3480	30		655	
			O	3523	12		741	
12 32 54	- 2 3.0		O	6000		-134	626	
12 32 54	14 46.0	NGC 4548	R	495	36	-65	372	M 91
			R	472	10		566	
			O	433	50		13	
12 32 54	21 37.0	3CR274.1	O	126600		-35	718	
			O	127000			766	
12 32 54	29 47.0		O	4653		0	548	
			O	4752			593	

R.A. (1950)	DEC. (1950)	NAME	OBS	HEL VEL (C✶Z)	ERR	GAL CORR	REF	COMMENTS
12 32 59	12 29.8	NGC 4550 HOL 422A	R O	381 350	15 50	-75	829 13	
12 33 0	- 3 31.1	NGC 4546	R O	1050 1014	40	-140	682 13	
12 33 0	26 8.0	NGC 4562	O O	1316 1394		-15	548 593	
12 33 6	12 31.0	NGC 4551 HOL 422B	O	978	300	-75	13	
12 33 6	12 50.0	NGC 4552	O O O O O	247 352 276 172 127	16 65	-74	13 938 13 281 694	M 89
12 33 8	4 29.8		O	21710		-108	895	
12 33 12	26 48.0	NGC 4555	O	6694		-12	322	
12 33 18	- 2 7.0		O	3000		-134	626	
12 33 18	27 12.0	NGC 4556	O O	7987 7402		-10	322 322	
12 33 29	28 14.1	NGC 4559 HOL 423A	R R R P R R O	813 816 807 791 810 790 856	17	-5	817 744 886 216 647 171 13	
12 33 37	16 55.0	PKS	O	23500		-56	511	
12 33 38	19 36.1	NGC 4561 IC 3569 KDG 346 VV 561	R O O O	1387 1413 1474 1494	12 50 56	-44	889 789 741 774	
12 33 38	19 36.1	NGC 4561A	O	1470	38	-44	908	
12 33 42	19 36.0	NGC 4561B	O	1443	17	-44	908	
12 33 42	54 30.0	NGC 4566	O	5290		108	322	
12 33 42	81 53.0	7 ZW475	O	9310		203	303	
12 33 52	26 15.6	NGC 4565 HOL 426A	R R R R O O	1227 1219 1230 1233 1100 1223	10 50 100	-14	817 647 906 934 1 13	

R.A. (1950)	DEC. (1950)	NAME	OBS	HEL VEL (C★Z)	ERR	GAL CORR	REF	COMMENTS
12 33 52	26 15.6	NGC 4565	O	1174		-14	13	
12 34 0	11 32.0	NGC 4567/8	R	2297	55	-79	372	
		VV 219A/B	R	2242	12		889	
		KDG 347	R	2253			744	
12 34 0	11 32.0	NGC 4567	O	2284		-79	13	
		HOL 427B	O	2175			253	
			O	2223	50		8	
12 34 0	11 43.0	NGC 4564	O	1015	34	-78	148	
12 34 2	26 28.6	IC 3582	O	7115		-13	557	
		MRK 649	O	7122			322	
12 34 3	11 30.8	NGC 4568	O	2223		-79	253	
		HOL 427A	O	2413			13	
			O	2167	50		8	
12 34 5	-72 19.0	PKS	O	7026	150	-251	410	
12 34 5	6 53.8	IC 3576	R	1080		-98	334	
		DDO 138	R	1076	10		492	
			R	1074	35		424	
12 34 6	24 42.0	IC 3581	O	6993		-21	703	
			O	6900			533	
12 34 6	27 7.0	IC 3585	O	7412		-10	322	
12 34 12	13 32.0	IC 3583	O	962	100	-70	789	
		ARP 76	O	1896	50		953	DIS
12 34 18	13 26.4	NGC 4569	R	-236	50	-70	372	★ M 90
		ARP 76	R	-286			805	
			R	985			117	DIS
			O	-316			910	
			O	-300	25		258	
			O	-310			273	
			O	-320			264	
			O	960	50		13	DIS
			O	-319			694	
12 34 24	7 31.0	NGC 4570	O	1730	75	-95	13	
12 34 24	14 29.0	NGC 4571	R	349	10	-66	566	
		IC 3588	O	266			694	
12 34 30	-38 16.0		O	2957	30	-237	680	
			O	3270	30		655	
			O	2920	30		719	
12 34 42	- 4 17.0		O	35000		-142	626	

12^h34^m

R.A. (1950)	DEC. (1950)	NAME	OBS	HEL VEL (C*Z)	ERR	GAL CORR	REF	COMMENTS
12 34 42	7 12.0		O	1896		-96	683	
12 34 48	28 30.0	IC 3598	O	7626		-3	548	
12 34 54	- 4 29.0		O	22000		-142	626	
12 35 0	9 50.0	NGC 4578	O	2282	50	-85	13	
12 35 6	- 4 27.0		O	49000		-142	626	
12 35 6	26 59.0	IC 3599	O	6442	100	-10	408	
12 35 6	27 24.0	IC 3600	O	4599	100	-8	408	
12 35 9	-40 15.8	NGC 4575	O O	3037 5250	 200	-240	576 629	 DIS
12 35 13	12 5.7	NGC 4579	R R O	1808 1750 1752	37 150	-75	372 117 13	M 58
12 35 13	27 24.2	MRK 650	O	4580		-8	557	
12 35 16	5 38.6	NGC 4580	O O	1290 1139	29	-102	741 694	
12 35 30	74 28.0	NGC 4589	O O	1825 2005	75 18	181	13 938	
12 35 36	-35 20.0		O	2931	20	-231	582	
12 35 42	- 4 30.0		O	22000		-142	626	
12 35 42	29 13.0	NGC 4585	O	7411		0	593	
12 35 48	- 4 52.0		O	45000		-143	626	
12 35 55	4 35.6	NGC 4586	O O	829 801	25	-106	741 694	
12 36 0	- 4 18.0		O	35000		-141	626	
12 36 18	- 4 21.0		O	28000		-141	626	
12 36 19	56 12.1	MRK 219	O	3000		116	308	
12 36 24	32 16.0		O	6959		13	548	
12 36 24	32 23.0		O	4361		14	548	
12 36 32	28 35.4	MRK 651	O	7480		-2	557	
12 36 42	- 0 15.0	NGC 4592	O O	1000 1281		-125	694 741	

R.A. (1950)	DEC. (1950)	NAME	OBS	HEL VEL (C*Z)	FRR	GAL CORR	REF	COMMENTS
12 36 42	26 57.0	IC 3618	O	6518	100	-9	408	
12 36 42	28 1.0		O	8040	100	-4	408	
12 36 43	28 16.8	MRK 652	O	7205		-3	557	
12 36 48	- 5 23.0		O	12000		-144	626	
12 36 48	- 4 36.0		O	9000		-141	626	
12 36 48	28 3.0		O	6698	100	-4	408	
12 36 48	28 11.0	IC 3620	O	6551	100	-4	408	
12 36 53	8 14.2	IC 3617 DDO 140	R	(2115)		-90	492	
12 36 54	- 4 25.0		O	32000		-141	626	
12 36 54	27 23.0	IC 3623	O	6967	100	-7	408	
12 37 0	- 5 4.0	NGC 4593	O O	2698 2610	19	-143	148 822	SEYF
12 37 0	- 4 40.0		O	22000		-141	626	
12 37 6	55 42.0	A	O	4971	57	114	246	
12 37 6	55 42.0	B	O	4912	52	114	246	
12 37 10	-40 50.1		O	4310	200	-239	629	
12 37 12	16 52.0	IC 803A	O	7985	71	-53	789	
12 37 12	16 52.0	IC 803B	O	7862	38	-53	789	
12 37 18	-11 21.0	NGC 4594	R R O O O O O O O	1090 1265 1076 1100 1180 1180 1207 1055 1089	10 50 50 50 41 15	-165	632 117 740 1 1 13 13 72 632	M 104 SOM GAL
12 37 21	15 34.3	NGC 4595	R O	630 660	6 34	-59	914 741	
12 37 24	10 27.0	NGC 4596	O O	1950 1975	46	-81	741 694	
12 37 30	- 4 42.0		O	35000		-141	626	

R.A. (1950)	DEC. (1950)	NAME	OBS	HEL VEL (C*Z)	ERR	GAL CORR	REF	COMMENTS
12 37 35	-40 29.4	NGC 4603B	O	3490	390	-239	629	
12 37 42	- 9 1.0		R	6360	25	-157	582	
			O	6351	25		582	
12 37 42	- 5 3.0		O	42000		-142	626	
12 37 47	61 53.1	NGC 4605	O	195		138	125	
			O	140			13	
			O	150	25		936	
12 38 0	- 4 55.0		O	32000		-142	626	
12 38 0	26 48.0	IC 3645	O	6541	100	-9	408	
12 38 2	- 4 51.5	NGC 4602	O	2564		-141	694	
			O	2575	63		741	
12 38 6	- 4 49.0		O	35000		-141	626	
12 38 12	-40 42.2	NGC 4603	O	2360	200	-239	629	
12 38 12	-36 29.0		O	3360	30	-231	655	
12 38 18	- 5 32.0		O	12000		-144	626	
12 38 18	- 5 31.0		O	35000		-144	626	
12 38 18	27 0.0	IC 3651	O	4785	100	-8	408	
12 38 24	- 5 10.0		O	22000		-142	626	
12 38 24	28 15.0	A	O	9390		-2	246	
			O	9491	100		408	*
12 38 24	28 15.0	B	O	7885	86	-2	246	
12 38 24	29 45.0		O	9374		3	593	
12 38 30	12 11.0	NGC 4606	O	1660		-73	694	
12 38 36	- 5 27.0		O	45000		-143	626	
12 38 36	- 4 55.0		O	42000		-141	626	
12 38 42	- 4 44.0		O	3000		-141	626	
12 38 42	1 40.0		O	9300		-116	533	
12 38 42	10 25.7	NGC 4608	O	1870	18	-80	741	
			O	1805			694	
12 38 42	11 40.0	IC 3653	O	449		-75	694	

R.A. (1950)	DEC. (1950)	NAME	ORS	HEL VEL (C*Z)	ERR	GAL CORR	REF	COMMENTS
12 38 42	12 9.0	NGC 4607	0	2440		-73	694	
12 38 42	18 51.0		0	21094	50	-44	13	
12 38 42	27 1.0		0	4485	100	-8	408	
12 38 48	18 52.0		0	22056	75	-44	13	
12 38 48	28 8.0		0	7634	100	-3	408	
12 38 48	50 42.0	NGC 4617	0	4590	31	94	813	
12 39 0	- 4 35.0		0	45000		-140	626	
12 39 0	- 4 32.0		0	19000		-140	626	
12 39 0	7 35.3	NGC 4612	R	1212		-92	610	
			0	1843			694	
			0	1832	23		741	
12 39 0	26 18.0	NGC 4614	0	4789		-11	548	
			0	4928			593	
12 39 6	26 20.0	NGC 4615	0	4687		-10	548	
			0	4805			593	
12 39 8	41 25.6	NGC 4618	R	545	5	55	905	
		ARP 23	0	573	120		567	
		VV 73	0	484			13	
		KDG 349	0	613	20		908	
		HOL 438A	0	462	40		555	
			0	562	22		148	
12 39 24	- 4 36.0		0	25000		-140	626	
12 39 30	-40 33.0	NGC 4603D	0	2690		-238	343	
			0	2880	190		629	
12 39 30	-40 22.0	NGC 4616	0	4365		-237	343	
			0	4530	230		629	
12 39 30	- 4 52.0		0	25000		-140	626	
12 39 30	32 51.0	NGC 4627	0	727		18	548	
		HOL 422B	0	660	120		758	
12 39 30	41 33.0	NGC 4625	R	610	5	56	905	
		IC 3675	0	847	100		567	
		ARP 23	0	553	60		789	
		HOL 438B	0 (	440)	320		555	
		KDG 349	0	636	19		908	
			0	362	70		555	

R.A. (1950)	DEC. (1950)	NAME	OBS	HEL VEL (C*Z)	ERR	GAL CORR	REF	COMMENTS
12 39 31	11 55.2	NGC 4621	O	439	15	-73	938	M 59
			O	414	125		13	
12 39 36	- 5 25.0		O	15000		-142	626	
12 39 36	12 1.0	IC 3672	O	229		-73	694	
12 39 36	12 52.0	IC 810	O	-100		-69	694	
12 39 38	7 57.0	NGC 4623	O	1963	42	-90	741	
12 39 41	32 48.8	NGC 4631	R	612		18	817	
		ARP 281	R	598	5		531	
		KDG 350A	R	600	10		226	
		HOL 422A	R	630	10		184	
			R	610	10		762	
			R	632			886	
			R	570	80		93	
			R	606			647	
			R	613			934	
			R	617			693	
			R	600			156	
			O	646			50	
			O	692	12		50	
			O	630			483	
			O	620	10		555	
			O	625			205	
			O	665	38		72	
			O	630	15		83	
		NGC 4631A	O	637	28		14	KNOT
			O	591	65		13	KNOT
			O	661			150	KNOT
12 39 42	- 5 30.0		O	6000		-143	626	
12 39 45	- 4 29.9	3CR275	O	143900		-139	850	
12 39 50	38 46.5	IC 3687	R	352	10	44	492	
		DDO 141	R	367	20		424	
12 39 54	-40 28.2	NGC 4622	O	4340	160	-237	629	
			O	4278			343	
12 39 54	74 3.0	NGC 4648	O	1429		181	910	
12 39 58	4 13.9	NGC 4630	O	692	24	-105	741	
			O	675	47		789	
12 40 0	0 11.0	NGC 4632	O	1688	30	-121	148	
			O	1816			777	
12 40 6	14 38.0	NGC 4633/4	R	303	30	-61	889	

R.A. (1950)	DEC. (1950)	NAME	OBS	HEL VEL (C*Z)	ERR	GAL CORR	REF	COMMENTS
12 40 6	14 38.0	NGC 4633	O	270	38	-61	789	
		KDG 351	O	382	30		908	
		IC 3688						
		HOL 445B						
12 40 9	20 13.2	NGC 4635	O	1022		-37	741	
			O	919	25		813	
12 40 12	14 34.0	NGC 4634	O	169	58	-61	789	
		KDG 351	O	253	30		908	
		HOL 445A						
12 40 12	27 24.0		O	6527	100	-5	408	
12 40 16	34 22.0	MRK 654	O	4235		25	557	
12 40 17	2 57.7	NGC 4636	R	970		-110	117	
			P	1100	40		617	
			R	1090	100		715	
			R	1100	40		685	
			R	(1305)			815	
			O	931	15		938	
			O	1000			106	
			O	973	75		13	
			O	954			13	
			O	736	41		72	
12 40 18	11 43.0	NGC 4638	O	1080	150	-73	13	
12 40 18	26 55.0		O	6804	100	-7	408	
12 40 22	-41 5.2	NGC 4645A	O	3240	160	-238	629	
12 40 22	13 31.9	NGC 4639	R	971	10	-66	566	
			O	755	32		741	
			O	953	25		283	
12 40 24	21 16.0	IC 3692	O	6444		-32	703	
12 40 24	55 25.0	NGC 4644A	O	4764	86	114	246	
		KDG 352						
12 40 24	55 25.0	NGC 4644B	O	4913	19	114	246	
		KDG 352						
12 40 30	12 34.0		O	2087		-70	694	
12 40 36	- 1 9.0		O	9000		-126	626	
12 40 36	12 19.0	NGC 4641	O	2305		-71	694	
12 40 36	28 0.0		O	7458	100	-2	408	

12ʰ40ᵐ

R.A. (1950)	DEC. (1950)	NAME	OBS	HEL VEL (C*Z)	ERR	GAL CORR	REF	COMMENTS
12 40 36	55 7.0	NGC 4646	O	4551		113	246	
12 40 47	-41 5.4	NGC 4645B	O	2610	200	-238	629	
12 40 48	2 15.0	NGC 4643	O	1432		-112	13	
12 40 48	31 22.0		O	7101		12	593	
12 41 1	11 51.3	NGC 4647	R	1396	30	-72	889	
		ARP 116	O	1448			13	
		VV 206	O	1073			253	
		KDG 353	O	1454	40		908	
		HOL 448B	O	1133	31		938	
			O	1100	100		8	
12 41 6	-40 26.3	NGC 4622A	O	4831		-237	343	
			O	5112	85		680	
			O	5010	160		629	
12 41 8	-40 26.4	NGC 4622B	O	4713		-237	343	
			O	5057	110		680	
12 41 9	11 49.4	NGC 4649	R	1320		-72	117	M 60
		ARP 116	O	1389	50		13	
		VV 206	O	961			253	
		KDG 353	O	1242	30		908	
		HOL 448A	O	1244			13	
			O	1193	41		72	
			O	840	100		8	
			O	1090	50		1	
12 41 12	16 40.1	NGC 4651	R	800	10	-52	497	
		ARP 189	R	827	5		566	
		VV 56	R	810			379	
			O	932			774	
			R	796	12		372	
			O	800			272	
			O	733	42		148	
12 41 17	- 0 17.2	NGC 4653	O	2577		-122	741	
12 41 18	27 8.0		O	7551	100	-5	408	
12 41 24	13 25.0	NGC 4654	R	1027	15	-66	372	
		IC 3708	R	1040			283	
			R	1037			744	
			O	1042	25		283	
			O	1022	45		148	
12 41 25	-41 28.7	NGC 4645	O	2651	39	-238	741	
			O	2460	140		629	
12 41 26	13 24.0	IC 3708	R	1032	30	-66	889	

R.A. (1950)	DEC. (1950)	NAME	OBS	HEL. VFL (C*Z)	ERR	GAL CORR	REF	COMMENTS
12 41 27	54 13.8		R	453	10	110	860	
12 41 29	- 5 24.3	DDO 142	R	1429	10	-141	492	
12 41 32	32 26.5	NGC 4656	R	625		17	171	
		KDG 350B	R	634			817	
			R	650	10		762	
			R	632			647	
			R	660	30		93	
			R	645			693	*
			R	644			934	
			R	600	30		184	
			R	627	7		216	
			R	638			531	
			O	758			748	
			O	827			50	
			O	630	10		226	
			O	635			483	
			O	721			13	
12 41 32	55 10.2	1 ZW 41	R	4922	31	114	430	
		MRK 220/1	O	4843	55		430	
			O	4912	52		246	
12 41 32	55 10.2	MRK 220	O	4920		114	287	
		KDG 354	O	4903			262	
12 41 34	55 10.8	MRK 221	O	4806		114	437	
		KDG 354	O	4903			262	
12 41 35	-40 27.5	NGC 4650	O	2689		-237	343	
			O	2966			728	
			O	2962	55		680	
			O	4080	180		629	DIS
12 41 48	45 17.0	VV 493	O	3766	50	73	873	
			O	3749			954	
12 41 48	45 17.0	VV 493A	O	3748		73	954	
12 41 48	45 17.0	VV 493B	O	3788		73	954	
12 41 48	45 17.0	VV 493C	O	3532		73	954	
12 41 48	45 17.0	VV 493F	O	3928		73	954	
12 41 48	45 17.0	VV 493H	O	(5202)		73	954	
12 41 48	45 17.0	VV 493I	O	19360		73	954	
12 41 55	0 44.6	DDO 144	R	1184	10	-117	492	
12 42 0	-40 26.2	NGC 4650A	O	2530		-236	343	
			O	2830	200		629	
			O	2871	60		680	
			O	2861	11		917	

255

R.A. (1950)	DEC. (1950)	NAME	OBS	HEL VEL (C*Z)	ERR	GAL CORR	REF	COMMENTS
12 42 0	- 2 2.9		R	1589	10	-128	860	
12 42 0	11 26.0	NGC 4660	O	1017	30	-74	13	
12 42 0	13 47.0	NGC 4659	O	380		-64	694	
12 42 0	28 45.0		O	1111	142	1	246	
12 42 0	34 39.8	DDO 143 VV 127	R	612	10	27	492	
12 42 2	- 9 48.7	NGC 4658	O	2407	17	-156	741	
12 42 6	28 10.0		O	6374	100	0	408	
12 42 6	28 45.0		O O	992 903	100	1	408 315	
12 42 6	37 23.0	NGC 4662	R O	6980 6869	50 50	39	582 582	*
12 42 6	41 1.0	IC 3723 MRK 441	O O	5402 5400		55	438 359	
12 42 12	-20 10.0		R O	6301 6323	25 50	-190	752 582	
12 42 18	12 37.0	IC 3718	O	953		-68	694	
12 42 18	56 25.0		O	4805		119	322	
12 42 24	23 19.0	IC 813	O	7071		-22	593	
12 42 30	-40 33.0	NGC 4650B	O O	2553 2546	140	-236	343 680	
12 42 30	55 9.0	NGC 4669	O	4999	300	114	246	
12 42 36	- 0 12.0	NGC 4666	O	1645		-121	13	
12 42 36	3 20.0	NGC 4665	O	785	50	-107	13	
12 42 36	21 27.0	IC 3730	O O	7002 6900	94	-30	246 533	
12 42 50	27 24.0	NGC 4670 ARP 163	R R R O O O O	1049 1084 1073 1180 1219 1113 1125	100	-3	538 707 693 288 55 321 398	

R.A. (1950)	DEC. (1950)	NAME	OBS	HEL VEL (C*Z)	ERR	GAL CORR	REF	COMMENTS
12 42 58	- 0 15.7	NGC 4668	O	1701		-121	741	
12 43 0	13 36.0	IC 3742	O	915		-64	694	
12 43 6	- 5 48.0	DDO 146	R	1470	10	-141	492	
12 43 8	27 20.0	NGC 4673 MRK 656	O	7000		-3	55	
12 43 18	- 1 24.0		O	9000		-125	626	
12 43 18	55 0.0	NGC 4675	O	4806		114	322	
12 43 19	-33 34.0		R	585	5	-223	865	
12 43 24	47 22.5	MRK 222	O	4800		83	308	
12 43 33	-41 16.1	NGC 4672	O	3430	130	-237	629	
12 43 41	71 35.5	MRK 223	O	1200		174	268	
12 43 42	31 0.0	NGC 4676A/B ARP 242 VV 224	O	6552	50	12	41	
12 43 42	31 0.0	NGC 4676A IC 819 KDG 355	O	6640		12	463	★
12 43 42	31 0.0	NGC 4676B IC 820 KDG 355	O	6560		12	463	
12 43 55	26 43.3	B2	O	26720	250	-6	696	
12 44 0	64 50.0		O	2345	30	151	813	
12 44 4	48 30.5	MRK 224	O	1200		88	308	
12 44 12	-41 13.0	NGC 4696A	O	2630	170	-236	629	
12 44 12	30 0.0	IC 818	O	4650	100	8	288	
12 44 14	-41 18.8	NGC 4677	O O O	3215 3120 3153	51 150	-236	741 629 667	
12 44 24	26 50.0		O	865		-5	321	
12 44 24	54 49.0	NGC 4686	O	5015	102	113	246	
12 44 29	26 50.2		O	890		-5	55	

12ʰ44ᵐ

R.A. (1950)	DEC. (1950)	NAME	OBS	HEL VEL (C*Z)	ERR	GAL CORR	REF	COMMENTS
12 44 30	-33 21.0		O	4590	60	-222	655	
12 44 36	36 45.0	DDO 147	R	333	10	38	492	
12 44 39	-40 58.0	NGC 4696B	O	3180	160	-236	629	
			O	3097			667	
12 44 39	- 9 47.4	NGC 4682	O	2307	10	-155	936	
12 44 42	51 55.0		O	322	20	102	246	
12 44 43	- 2 27.1	NGC 4684	O	1589	22	-128	741	
			O	1603			694	
12 44 54	3 18.0		O	5100		-105	533	
12 44 54	3 52.0		O	10200		-103	533	
12 44 54	27 44.0		O	6540		-1	496	
12 44 57	-41 15.5	NGC 4683	O	3627	24	-236	741	
			O	3580	140		629	
			O	3667			667	
12 44 59	35 37.5	NGC 4687	O	(570)		33	374	
		MRK 442	O	690			438	
12 44 59	47 26.0	MRK 225	O	6000		84	268	
12 45 0	3 52.3		O	9950		-103	895	
12 45 0	30 4.0	IC 821	R	6730	50	9	582	
			O	6582			593	
			O	6711	120		582	
			O	6850	150		288	
12 45 0	72 11.2	MRK 226	O	8100		176	268	
12 45 6	27 15.0		O	7127		-3	55	
12 45 14	4 36.6	NGC 4688	R	981		-100	744	
		HOL 461A	R	994	13		889	
			O	925	100		153	★
12 45 18	14 2.0	NGC 4689	R	1616	5	-60	914	
			R	1620	12		889	
			O	1491	71		789	
			O	1577			694	
			O	1776	59		741	
12 45 18	54 39.0	NGC 4695	O	4928	21	113	246	
12 45 24	-40 32.0	NGC 4696C	O	3760	220	-235	629	

R.A. (1950)	DEC. (1950)	NAME	OBS	HEL VEL (C*Z)	ERR	GAL CORR	REF	COMMENTS
12 45 24	27 30.0	NGC 4692	O	7918		-1	55	
12 45 39	- 3 3.6	NGC 4691	O	1125	24	-129	741	
			O	1118	20		148	
			O	1102	22		433	
12 45 42	-40 39.0	NGC 4696E	O	1860	150	-235	629	
12 45 42	-41 26.0	NGC 4696D	O	2650	170	-236	629	
12 45 48	-41 2.1		O	3160	210	-235	629	
12 45 48	11 15.0	NGC 4694	R	1176		-72	572	
			R	1185	31		818	
			R	(1172)			780	
			O	1146			694	
			O	1264	28		741	
12 45 48	27 8.0		O	6926		-3	55	
12 45 52	8 45.6	NGC 4698	R	872	35	-82	372	
			R	1006			645	
			R	966	10		566	
			R	1004			572	
			R	1008	4		818	
			O	1032	50		13	
12 46 0	-41 55.3		O	2180	120	-236	629	
12 46 0	- 5 31.7	NGC 4697	R	1320		-138	117	
			O	1240	7		938	
			O	1308	65		13	
			O	1206	13		479	
			O	1113			694	
12 46 0	54 18.0		O	3383		112	322	
12 46 4	-41 2.5	NGC 4696 PKS	O	3060	100	-235	629	
			O	2790			317	
			O	2794			136	
			O	2805			227	
			O	2807			667	
			O	2999	12		741	
12 46 6	-40 59.7		O	3500	330	-235	629	
12 46 7	51 26.2	NGC 4707 1 ZW 53 DDO 150	R	468	10	101	492	
12 46 12	26 42.0		O	7098		-4	496	
12 46 17	34 44.8	MRK 444	R	4248		30	366	
			O	4250			435	
			O	4500			359	
			O	4182	70		246	

R.A. (1950)	DEC. (1950)	NAME	OBS	HEL VEL (C*Z)	ERR	GAL CORR	REF	COMMENTS
12 46 22	35 36.4	NGC 4711 IC 3804	R	4062	25	34	829	
12 46 24	42 11.0	NGC 4704	R O	8174 8098	25 25	62	582 582	
12 46 26	- 8 23.6	NGC 4699	R O	1510 1511		-148	117 13	
12 46 28	47 59.2	MRK 229	O	7200		87	308	
12 46 29	-41 16.4		O	3020	200	-235	629	
12 46 32	-11 8.2	NGC 4700	O	1439	14	-158	741	
12 46 35	27 27.0	MRK 657	O	6015		-1	557	
12 46 37	40 52.1	IC 3808 MRK 445	O	4600		57	557	
12 46 42	-41 31.1		O	4299	24	-235	741	
12 46 42	- 9 1.0		O	3925		-150	694	
12 46 42	3 40.0	NGC 4701	O O	750 697	47	-102	694 789	
12 46 43	-41 13.1		O O	4200 4305	150	-235	629 667	
12 46 43	-41 11.5		O O	5050 4655	220	-235	629 667	
12 46 51	-40 47.1		O O	2040 2006	140	-234	629 667	
12 46 53	-41 7.0		O O O	3770 3536 3782	50	-235	574 667 741	
12 46 54	-40 6.9		O	3870	110	-233	629	
12 47 6	-40 57.5		O	2270		-234	667	
12 47 6	-36 21.0		O	9226		-226	683	
12 47 6	15 26.0	NGC 4710	O O	1086 1125	100 25	-53	288 148	
12 47 6	31 7.0		O	8159		15	593	
12 47 7	25 44.5	NGC 4712 HOL 468B	R R O O	4376 4384 4381 4542	20	-8	693 752 321 322	

260

R.A. (1950)	DEC. (1950)	NAME	OBS	HEL VEL (C*Z)	ERR	GAL CORR	REF	COMMENTS
12 47 9	-41 0.5	NGC 4706	O	3830	120	-234	629	
			O	3511			136	
			O	3624			667	
			O	3790	30		741	
			O	3715			574	
12 47 12	27 10.0		O	7437		-2	55	
12 47 14	4 7.0		R	694	10	-100	860	
12 47 19	-41 6.7	NGC 4709	O	4500		-235	136	
			O	4610	120		629	
			O	4759	31		741	
			O	4572			667	
12 47 24	28 5.0	NGC 4715	O	6897	100	1	288	
12 47 25	-41 7.5		O	5090	160	-234	629	
			O	4760			667	
12 47 25	5 35.0	NGC 4713	R	657		-94	744	
			O	664			13	
12 47 26	-40 57.0		O	2670		-234	667	
12 47 27	-41 14.8		O	2684	70	-235	741	
			O	2810	150		629	
			O	2974			667	
12 47 39	-40 24.5		O	4500	190	-233	629	
12 47 44	33 25.8	NGC 4719 MRK 446	O	7105		25	438	
			O	7080			374	
			O	7145			593	
12 47 48	27 36.0	NGC 4721	O	7858	100	0	495	
12 47 51	-41 12.0		O	5100	240	-234	629	
			O	5083			667	
12 47 53	-10 35.1	DDO 151	R	2407	10	-155	492	
12 47 54	- 9 12.0		O	4497		-150	694	
12 47 54	- 9 11.0		O	4472		-150	694	
12 47 59	-42 16.3		O	2730	90	-236	629	
12 47 59	-41 9.5	NGC 4729	O	2380	120	-234	629	
			O	2427			667	
12 48 0	25 46.5	NGC 4725	R	1212		-7	693	
			R	1210			803	
			R	1215	9		581	
			O	1114	65		13	
			O	1218			321	

R.A. (1950)	DEC. (1950)	NAME	OBS	HEL VEL (C*Z)	ERR	GAL CORR	REF	COMMENTS
12 48 0	27 42.0	NGC 4728	O	6528		0	55	
12 48 6	- 8 44.0	IC 3826	O	4017		-148	694	
12 48 12	- 8 30.0		O	11475		-148	694	
12 48 13	-41 7.4	NGC 4730	O	4320	150	-234	629	
12 48 18	-14 4.0	NGC 4727	O	7622		-167	520	
12 48 21	73 8.8	NGC 4750	O	1647		180	13	
12 48 24	- 6 7.0	NGC 4731	R	1495		-139	744	
			O	1438	49		72	
12 48 28	28 6.8		O	7410	100	2	408	
			O	7700			60	
			O	7437	150		288	
			O	7410			593	
			O	7560			321	
12 48 30	27 39.0		O	8355	100	0	495	
12 48 32	41 23.5	NGC 4736	R	307		60	356	M 94
			R	240			156	
			R	311			934	
			R	307	5		615	
			O	280	26		112	
			O	290	50		1	
			O	314	1		597	
			O	338			61	
			O	282	50		13	
			O	262	16		112	
			O	259	5		146	
			O	313			13	
			O	299			61	
			O	304	4		468	
12 48 36	11 11.0	NGC 4733	O	1033		-70	694	
12 48 36	29 12.0	NGC 4735	O	6677		7	322	
			O	6643	100		495	
12 48 42	5 8.0	NGC 4734	O	7548	58	-95	789	
12 48 42	29 4.0		R	4800		7	858	
			O	4575	100		288	
			O	4794	100		495	
			O	4784			593	
12 48 48	-40 57.8		O	2156		-233	667	
12 48 48	- 9 8.0		O	3963		-149	694	

R.A. (1950)	DEC. (1950)	NAME	OBS	HEL VEL (C*Z)	ERP	GAL CORR	REF	COMMENTS
12 48 48	27 23.0		O	5449		0	496	
			O	5450			321	
			O	5434	100		288	
12 48 54	-41 2.0		O	3440	110	-234	629	
12 48 54	27 42.0	NGC 4745	O	7596	100	1	495	
			O	7156	100		288	
12 48 54	31 19.0	IC 826	O	7136		17	593	
12 49 0	-40 51.8		O	3120		-233	667	
			O	3410	220		629	
			O	2268	34		741	DIS
12 49 0	- 8 8.3	NGC 4739	O	3741		-146	694	
12 49 0	18 20.0		O	6600		-39	533	
12 49 2	-38 33.6		O	3140	120	-229	629	
12 49 12	-10 12.0	NGC 4742	O	1321	50	-153	13	
12 49 15	-40 52.6		O	2050	180	-233	629	
			O	2204			667	
12 49 16	-41 30.6		O	4950	220	-234	629	
12 49 16	-41 11.6		O	3778	77	-234	741	
			O	4029			667	
			O	3990	190		629	
12 49 19	26 2.8	NGC 4747	R	1197		-5	693	
		ARP 159	R	1200	15		455	
			R	1179			803	
			O	1219			321	
12 49 24	12 21.0	NGC 4746	O	1447		-65	694	
12 49 30	-41 7.5	NGC 4743	O	2830	200	-233	629	
			O	2976			667	
			O	3054	47		741	
12 49 35	-40 47.4	NGC 4744	O	3398	21	-233	741	
			O	3060	260		629	
			O	3455			667	
12 49 36	- 9 14.0		O	4191		-149	694	
12 49 36	27 18.0		O	6546	100	0	495	
			O	6477			593	
12 49 36	27 51.0		O	8186	100	2	495	

12ʰ49ᵐ

R.A. (1950)	DEC. (1950)	NAME	OBS	HEL VEL (C*Z)	ERR	GAL CORR	REF	COMMENTS
12 49 38	-41 1.0		0	4604		-233	667	
12 49 40	-40 26.3		0	4230	190	-232	629	
12 49 48	- 0 55.0	NGC 4753	R	1360		-118	117	
			0	847	57		148	
			0	1364			13	
			0	1078	140		555	
			0	1202	80		370	
12 49 48	11 35.0	NGC 4754 KDG 356A	0	1461	75	-68	13	
12 49 54	-38 45.5		0	4060	260	-229	629	
12 50 1	- 6 1.2	DDO 152	R	1536	10	-137	492	
12 50 4	-42 23.4	NGC 4751	0	1920	140	-235	629	
12 50 5	-41 4.1		0	4920	210	-233	629	
			0	4675			667	
12 50 6	-40 11.0		R	3841	25	-231	752	
			0	3840	50		582	
12 50 6	27 1.0		0	7981	100	0	495	
12 50 12	26 45.0	IC 831	0	6366	100	-2	495	
12 50 12	29 8.0	5C 4.6	0	58805		8	663	
12 50 15	-15 8.6	NGC 4756	0	4164	36	-169	741	
12 50 18	27 40.0		0	7850		1	321	
			0	7836			496	
12 50 22	-16 44.1	NGC 4763	0	4190	23	-174	741	
12 50 24	11 31.0	NGC 4762 KDG 356B	R	939		-67	572	
			0	997			13	
			0	868	50		13	
			0	974			681	
			0	885	41		72	
12 50 28	28 38.2	5C 4.7	0	7103	100	6	288	
			0	7004			520	
			0	7006			663	
12 50 31	-10 13.4	NGC 4760 PKS	0	4640	53	-152	741	
			0	4290	150		355	
			0	4680			694	
12 50 36	32 22.0		0	6872		22	593	

R.A. (1950)	DEC. (1950)	NAME	OBS	HEL VEL (C*Z)	EPR	GAL CORR	REF	COMMENTS
12 50 36	47 1.0	1 ZW 44E	O	18110		85	262	
12 50 41	-40 55.9		O	4230	120	-232	629	
			O	4370			667	
12 50 42	4 44.1	NGC 4765	O	772		-95	741	
		VV 366	O	995			774	
			O	750			533	
12 50 42	27 21.0		O	6150		0	321	
12 50 48	1 32.0	NGC 4771	O	1218		-108	694	
12 50 48	37 5.0	NGC 4774A 1 ZW 45	O	8271	45	43	789	★
12 50 48	37 5.0	NGC 4774B 1 ZW 45	O	8373	35	43	789	★
			O	8345			591	
12 50 56	2 26.4	NGC 4772	O	1087	14	-104	741	
			O	992			694	
12 51 8	-39 27.1	NGC 4767	O	2930	120	-230	629	
			O	3090	15		741	
12 51 9	-28 45.0		O	16362	80	-207	525	
12 51 12	- 6 21.0	NGC 4775	P	1565		-138	744	
			O	1684			13	
12 51 18	10 0.0	NGC 4779	R	2826	10	-73	752	
12 51 20	-11 50.2	DDO 153	R	824	10	-157	492	
12 51 24	26 43.0	IC 832	O	7001		-1	321	
12 51 24	29 15.0		O	7881	100	9	495	
12 51 24	29 51.0		O	13849	100	12	495	
12 51 36	29 52.0		O	6517	100	12	495	
			O	6297			593	
12 51 39	27 25.1	NGC 4789A DDO 154	R	374	2	1	769	
			R	378	10		492	
			R	360	20		424	
12 51 42	27 20.0	NGC 4787	O	7635	100	1	495	
12 51 46	27 53.8	3CR277.3	O	25690	250	3	696	
			O	25710			120	
			O	25729			663	
			O	25729			520	

R.A. (1950)	DEC. (1950)	NAME	OBS	HEL VEL (C*Z)	ERR	GAL CORR	REF	COMMENTS
12 51 48	-10 16.0	NGC 4781	O	895		-151	13	
12 51 48	27 35.0	NGC 4788	O	6460	100	2	495	
12 51 57	- 6 35.3	NGC 4786	O	4647	28	-138	741	
12 51 59	27 21.0	NGC 4789	O	8374		1	55	
			O	8372			13	
			O	8223			663	
			O	8223			520	
12 52 0	-28 58.4	PKS	O	17194	80	-207	525	
			O	1680			612	DIS
12 52 0	-12 18.1	NGC 4782	O	3998		-158	253	
		VV 201	O	4010			52	
		HOL 485A	O	3974			444	
		3C 278	O	4018			227	
			O	3987	100		8	
12 52 0	-12 17.4	NGC 4783	O	4688		-158	227	
		VV 201	O	4626			253	
		HOL 485B	O	4611	46		444	
		3C 285	O	4680			52	
			O	4617	100		8	
12 52 6	28 39.0		O	7125	100	7	495	
12 52 12	27 12.0		O	6202	100	1	495	
12 52 14	29 12.5	NGC 4793	O	2555	79	9	741	SEYF
		5C 4.022	O	2379			520	
			O	2455			663	
			O	2535			55	
			O	2529			13	
12 52 15	- 9 58.6	NGC 4790	O	1266		-150	694	
			O	1491	44		741	
12 52 18	2 55.0	NGC 4810	R	878	10	-101	566	
		ARP 277	O	899			253	
		KDG 358	O	847	25		908	
		VV 313	O	894	100		8	
12 52 18	2 56.0	NGC 4809	R	878	10	-101	566	
		ARP 277	O	956			253	
		KDG 358	O	910	25		908	
		VV 313	O	954	100		8	
12 52 18	30 48.0		O	6332		16	496	
12 52 18	46 48.0	NGC 4800	O	747		85	641	
			O	746	50		13	

R.A. (1950)	DEC. (1950)	NAME	OBS	HEL VEL (C*Z)	ERR	GAL CORR	REF	COMMENTS
12 52 24	27 37.0		O	12048		3	663	
12 52 30	8 20.0	NGC 4795	O	2714		-79	694	
		KDG 359A	O	2689	52		908	
			O	3199	71		741	
12 52 30	27 41.0	NGC 4798	O	7673	50	3	13	
			O	7677			593	
12 52 30	28 41.0		O	7496	100	7	495	
12 52 36	8 20.0	NGC 4795A	O	2574		-79	694	★
12 52 36	8 20.0	NGC 4796	O	2542	67	-79	908	
		KDG 359B						
12 52 36	35 39.0		O	4500		38	476	
12 52 42	0 23.0		O	1340	82	-111	789	
12 53 0	28 5.0		O	7438	100	5	495	
			O	7493			663	
12 53 0	28 44.0		O	7177	100	8	495	
12 53 6	0 0.2		R	1116	10	-112	860	
12 53 6	27 47.0	NGC 4807	O	6846		4	383	
			O	6976			593	
12 53 6	27 48.0	NGC 4807A	O	7119	150	4	288	
12 53 6	27 56.0		O	7088	100	4	495	
12 53 9	64 32.0	MRK 230	O	21300		153	308	
12 53 12	27 32.0	IC 3900	O	7171	50	3	13	
12 53 13	-11 47.2	NGC 4802	O	1014		-156	694	
12 53 18	4 34.0	NGC 4808	R	778		-94	744	
			O	738	37		148	
12 53 18	58 37.0	NGC 4814	O	2531	65	132	13	
12 53 19	4 17.3		O	1020		-95	895	
12 53 42	-50 4.0	IC 3896	O	2274	44	-243	741	
			O	2098			925	
12 53 42	27 57.0	MRK 53	O	4802	60	5	261	
			O	4977			663	
			O	4882			496	
			O	(7419)	100		495	DIS

12ʰ53ᵐ

R.A. (1950)	DEC. (1950)	NAME	OBS	HEL VEL (C*Z)	ERR	GAL CORR	REF	COMMENTS
12 53 48	28 1.0	NGC 4816	O	6853	100	5	288	
12 53 54	44 20.0		O	59304	40	75	13	
12 53 55	4 4.9		R	617	10	-96	860	
12 54 0	26 38.0		O	6292	100	0	408	
12 54 0	27 15.0	NGC 4819 HOL 490A	O	6700		2	55	
12 54 0	28 1.0		O	7095	100	5	495	
12 54 4	-42 51.4		O	3840	360	-234	629	
12 54 5	-41 32.7	NGC 4812	O	3400	260	-232	629	
12 54 5	-41 31.7	NGC 4811	O	3180	200	-232	629	
12 54 5	57 8.6	MRK 231	O	12300		126	268	SEYF
			O	12300			304	*
			O	8100			304	DIS
12 54 6	27 14.0	NGC 4821 HOL 490B	O	6978		2	55	
12 54 6	28 6.0		O	6665	100	6	495	
12 54 10	28 32.8		O	6945		8	595	
			O	7251			520	
12 54 12	- 8 15.2	NGC 4818	O	952		-142	694	
			O	1155	54		741	
12 54 17	21 57.1	NGC 4826	R	404	24	-20	390	M 64
			R	381			886	
			R	417			934	
			R	408			817	
			R	367			310	
			O	488			119	
			O	150	50		1	
			O	382	30		13	
			O	385	75		555	
12 54 18	27 27.0	NGC 4827 5C 4.43	O	7653		3	55	
			O	7495			663	
12 54 18	28 17.0	NGC 4828	O	6134	100	7	495	
12 54 18	30 59.0		O	7844		18	496	
12 54 24	27 22.0		O	6849	100	3	495	

R.A. (1950)	DEC. (1950)	NAME	OBS	HEL VEL (C∗Z)	ERR	GAL CORR	REF	COMMENTS
12 54 30	26 45.0	IC 835	O	7747	100	0	408	
12 54 30	27 10.0		O O	6248 6304	100	2	288 321	
12 54 30	26 49.0	5C 4.44	O	36624		0	663	
12 54 30	29 12.0		O	8019	100	11	495	
12 54 32	32 43.1	MRK 54	O	13418	50	26	13	
12 54 36	-13 23.9	NGC 4825	O	4452	18	-160	741	
12 54 36	29 19.0		O	7517	100	11	495	
12 54 36	30 59.0		O	7264	150	19	288	
12 54 36	48 34.0	NGC 4837 1 ZW 46	O	8748		93	262	
11 54 48	23 38.9	MRK 644	O	14620		-48	557	
12 54 48	27 44.0		O	7336	100	5	495	
12 54 48	29 18.0		O	7457	100	11	288	
12 54 59	27 46.0	NGC 4839 B2	O O O	7371 7445 7444		5	520 55 663	
12 55 0	27 40.6	MRK 55	O	4861	60	5	261	
12 55 0	-39 29.6		O	3750	210	-228	629	
12 55 0	27 49.0		O	5518	100	5	495	
12 55 6	26 46.0	IC 837	O	7040	100	1	495	
12 55 6	27 53.0	NGC 4840	O O O	5783 6190 5943	100	6	321 288 496	
12 55 6	28 28.0		R	7070	100	8	495	
12 55 6	28 45.0	NGC 4841A/B HOL 492	O	6695	100	9	288	
12 55 6	28 45.0	NGC 4841A KDG 361	O (O	6486) 6883	70	9	908 593	
12 55 11	59 20.2	MRK 232	O	6465		135	437	
12 55 12	2 57.8	DDO 156	R (	926)		-99	492	

12ʰ55ᵐ

R.A. (1950)	DEC. (1950)	NAME	OBS	HEL VEL (C*Z)	ERR	GAL CORR	REF	COMMENTS
12 55 12	27 7.0		0	6512	100	2	495	
12 55 12	27 45.0	NGC 4842N	0	7297		5	593	
12 55 12	27 45.0	NGC 4842S	0	7491		5	593	
			0	7513			55	
12 55 12	28 46.0	NGC 4841B	0	6238	72	9	908	
		KDG 361	0	6215			593	
12 55 18	-45 59.0	NGC 4835	0	2243	56	-237	741	
			0	2270	100		844	
			0	2037			154	
12 55 18	28 6.0		0	6156	100	7	495	
12 55 18	28 9.0	A	0	6037	100	7	495	
12 55 18	28 9.0	B	0	6941	100	7	495	
12 55 24	28 27.0		0	7351		8	496	
			0	7385	100		495	
12 55 24	29 55.0		0	(5304)	100	15	495	
12 55 28	1 50.8	NGC 4845	0	1228	80	-103	741	
			0	1060	71		789	
			0	867			694	
12 55 30	27 53.2	W	0	20220		6	595	
			0	20250			520	
			0	20298			663	
12 55 33	28 19.9		0	8175		8	595	
			0	8172			663	
			0	8143			520	
12 55 36	27 8.0	3 ZW 68	0	5700		3	282	
			0	7368	100		495	DIS
12 55 36	27 11.0		0	7976	100	3	495	
12 55 36	28 5.0		0	7198	100	7	495	
12 55 38	27 8.8		0	24850	150	3	807	
12 55 41	28 30.9	NGC 4848	0	7262		9	520	
		5C 4.58	0	7263			663	
			0	7215			55	
			0	7209			13	
12 55 44	24 36.9	MRK 447	0	6600		-7	359	
			0	6685			438	

270

R.A. (1950)	DEC. (1950)	NAME	OBS	HEL VEL (C*Z)	ERR	GAL CORR	REF	COMMENTS
12 55 48	26 40.0	NGC 4849	0	6028	100	1	408	
		IC 3935	0	5828			55	
12 55 48	28 14.0	NGC 4850	0	5984	100	7	13	
12 55 48	28 24.0		0	7406	100	8	495	
12 55 48	28 59.0	W	0	7663		11	663	
			0	(7681)	100		495	
			0	7612			496	
12 55 48	29 13.0		0	7560	100	12	495	
12 55 49	3 3.7	DDO 158	R	(2725)		-98	492	
12 55 54	27 24.0		0	12050		4	229	
12 55 54	27 35.0	W	0	7473		5	663	
12 55 54	28 2.0		0	5372		7	349	
12 55 54	28 25.0	IC 839	0	6674	100	8	495	
12 55 54	28 25.0	NGC 4851	0	7786	100	8	495	
12 55 54	29 24.0		0	7837	100	13	495	
12 56 6	27 57.0		0	6900		6	349	
12 56 6	28 17.0		0	7242	100	8	288	
			0	7177			593	
12 56 6	28 19.0		0	5897		8	349	
12 56 10	14 29.2	DDO 155	R	216	5	-51	492	GR 8
		VV 558	R	222	5		443	
			0	257	30		443	
			0	239	50		873	
12 56 10	27 32.0	MRK 56	0	7288	60	5	261	
		W	0	7418			663	
			0	7417			321	
12 56 10	27 52.0	NGC 4853	0	7550	50	6	13	
		2 ZW 67	0	7600	100		2	
12 56 12	27 22.0		0	(7868)	100	4	495	
12 56 12	27 26.7	MRK 57	0	7599	60	4	261	
			0	7704			321	
			0	7658	100		288	
12 56 12	28 23.0	IC 3943	0	6819	100	8	495	

12ʰ56ᵐ

R.A. (1950)	DEC. (1950)	NAME	OBS	HEL VEL (C*Z)	ERR	GAL CORR	REF	COMMENTS
12 56 23	59 24.2	MRK 233	O	8294		136	437	
12 56 24	27 57.0	NGC 4854	O	8052	100	7	288	
			O	8269			593	
12 56 24	28 5.0	IC 3946	O	6101	75	7	13	
			O	6002			593	
12 56 24	28 6.0	IC 3949	O	7531		7	55	
			O	7412			593	
			O	7526			13	
12 56 24	28 21.0		O	5932		8	349	
12 56 30	28 3.0	IC 3947	O	5696		7	349	
12 56 30	28 4.0		O	6865		7	595	
			O	7001			407	
			O	6968			520	
12 56 30	28 5.0		O	5514		7	407	
12 56 30	28 14.0		O	5668		8	349	
12 56 30	28 24.0	NGC 4860	O	7858	40	9	13	
			O	7900	75		2	
			O	7878			55	
			O	7896			593	
			O	7896			520	
12 56 36	27 5.0	NGC 4859	O	7030	100	3	288	
12 56 36	28 14.0		O	8312		8	349	
12 56 37	28 23.1	NGC 4858	O	9482		9	520	
		W	O	9491			663	
			O	9405			595	
12 56 38	28 29.8		O	8019		9	593	
12 56 39	35 6.9	NGC 4861	R	837	9	38	430	
		1 ZW 49	O	789	30		567	
		MRK 59	O	810			224	
		ARP 266	O	828	9		430	
		VV 797E	O	851	50		873	
		KDG 362	O	776	13		908	
		IC 3961	O	790			77	★
			O	793			13	★
12 56 40	35 7.9	IC 3961	O	(624)	150	38	567	
		VV 797W	O	776	50		873	
		KDG 362	O	851	20		908	

R.A. (1950)	DEC. (1950)	NAME	OBS	HEL VEL (C*Z)	ERR	GAL CORR	REF	COMMENTS
12 56 42	-14 46.0	NGC 4856	O	1251	75	-163	13	
12 56 42	27 55.0	MRK 58	O	5356	100	7	288	
		W	O	5347			663	
			O	5347			520	
12 56 42	28 2.0	IC 3957	O	6310		7	407	
			O	(6283)			349	
12 56 42	28 3.0	IC 3959	O	7087		7	407	
			O	6999			593	
			O	7220			349	
12 56 42	28 8.0	IC 3960	O	6674		8	349	
			O	6868	65		13	
12 56 42	28 11.0		O	6443		8	349	
12 56 42	28 14.0		O	6897		8	349	
12 56 42	28 16.0	IC 3955	O	7860		8	349	
			O	7679			593	
12 56 48	27 40.0		O	5627		6	495	
12 56 48	27 54.0		O	5577		7	349	
12 56 48	28 3.0	IC 3963	O	6641		7	349	
12 56 48	28 14.0	NGC 4867	O	4815	100	8	13	
12 56 48	28 15.0	NGC 4864	O	6819	125	8	13	
			O	6761			593	
12 56 48	28 21.0		O	7802		9	349	
12 56 49	37 34.7	NGC 4868	O	4731	17	48	741	
12 56 54	28 9.0		O	6518		8	349	
12 56 54	28 15.0		O	6412		8	349	
12 56 54	28 21.0		O	7126		9	407	
12 56 54	28 28.0		O	9675		9	407	
12 56 55	28 21.2	NGC 4865	O	4643	50	9	13	
		W	O	4579			520	
			O	5000	75		2	
12 57 0	14 27.0	NGC 4866	R	1986		-50	645	
			O	1910	30		13	

12ʰ57ᵐ

R.A. (1950)	DEC. (1950)	NAME	OBS	HEL VEL (C*Z)	ERR	GAL CORR	REF	COMMENTS
12 57 0	28 1.0		0	6008		7	349	
12 57 0	28 7.0	IC 3976	0 0	6925 6769	150	8	288 593	
12 57 0	28 10.0		0	5092		8	349	
12 57 0	28 11.0	NGC 4869 B2	0 0	6703 6793	100	8	13 520	
12 57 0	28 12.0		0	7832		8	407	
12 57 0	28 14.0		0	7663		8	349	
12 57 0	28 15.0		0	6083		8	349	
12 57 0	28 16.0		0	6657		8	349	
12 57 0	28 27.0		0	7421		9	407	
12 57 0	28 33.0		0	7911		10	407	
12 57 6	27 59.0		0	7834		7	349	
12 57 6	28 9.0	IC 3973	0	4720	100	8	288	
12 57 6	28 14.0	NGC 4871	0 0	7088 6749	150	8	288 593	
12 57 6	28 15.0	NGC 4873	0	5637	100	8	288	
12 57 6	28 19.0		0 0	6907 (6871)		9	407 349	★
12 57 6	28 19.0		0	5579		9	349	
12 57 6	28 21.0		0	3651		9	407	
12 57 9	28 13.0	NGC 4872	0 0 0 0	6910 7009 7250 6900	300 100 200	8	13 288 593 2	
12 57 11	28 13.8	NGC 4874 B2	0 0 0 0 0 0	7171 7172 7000 7119 7170 6959	30 200 100	8	13 520 2 288 449 321	
12 57 12	28 8.0		0	6418		8	407	
12 57 12	28 11.0	NGC 4875	0	7862		8	349	

R.A. (1950)	DEC. (1950)	NAME	OBS	HEL VEL (C*Z)	ERR	GAL CORR	REF	COMMENTS
12 57 12	28 15.0		O	6848		8	407	
12 57 12	28 16.0	W	O	5428		9	663	
12 57 12	28 26.0		O	5428		9	407	
12 57 12	28 54.0		R	5822		11	749	
			O	5327			322	
12 57 12	29 10.0	IC 3990	O	6192	100	12	288	
12 57 12	29 12.0	IC 3991	O	5901	100	13	495	
12 57 18	27 58.0		O	8208	100	7	288	
12 57 18	28 7.0		O	3682		8	349	
12 57 18	28 8.0	W	O	17462		8	663	
12 57 18	28 11.0	NGC 4876	O	6967	100	8	288	
			O	6731			593	
12 57 18	28 12.0		O	6804		8	349	
12 57 18	28 13.0		O	7912		8	349	
12 57 18	28 14.0		O	7655		8	349	*
12 57 18	28 14.0		O	6763		8	349	
12 57 18	28 16.0		O	6992		9	349	
12 57 24	28 8.0		O	7908		8	349	
			O	8029			593	
12 57 24	28 12.0		O	9834		8	407	
12 57 24	28 15.0	IC 3998	O	9346	100	9	288	
			O	9391			593	
12 57 24	32 18.0		O	6818		26	593	
12 57 28	2 19.5		O	1020		-100	895	
12 57 30	28 6.0	W	O	7887		8	663	
12 57 30	28 12.0		O	7840		8	407	
12 57 30	28 18.0	NGC 4883	O	7936	100	9	288	
			O	8042			663	
12 57 30	28 24.0		O	7444		9	349	

12ʰ57ᵐ

R.A. (1950)	DEC. (1950)	NAME	OBS	HEL VEL (C*Z)	ERR	GAL CORR	REF	COMMENTS
12 57 30	28 31.0	NGC 4881	0	6691	50	10	13	
			0	6900	200		2	
			0	6715			593	
12 57 36	27 10.0	NGC 4892	0	5873	100	4	288	
12 57 36	28 13.0		0	5985		9	663	★
12 57 36	28 13.0		0	6940		9	663	★
12 57 36	28 15.0	NGC 4886	0	6214	150	9	13	
12 57 36	28 31.0		0	7422		10	407	
12 57 39	33 42.3	MRK 235	0	7500		32	268	SEYF
12 57 40	12 45.1	NGC 4880	0	1557	25	-57	741	
			0	1252			694	
12 57 40	28 4.6		0	6628		8	520	
			0	6640			595	
12 57 42	27 7.0		0	23746		4	663	
12 57 42	28 8.0	A	0	5128		8	349	
			0	5113			407	
12 57 42	28 8.0	B	0	9867		8	593	
12 57 42	28 15.0		0	7807		9	349	
12 57 42	28 16.0	IC 4011	0	7106		9	349	
12 57 42	28 18.0		0	5922		9	349	
12 57 42	28 21.0	IC 4012	0	7340		9	349	
			0	7239			593	
12 57 43	28 14.7	NGC 4889 NGC 4884	0	6481		9	407	
			0	6512			593	
			0	6700	75		2	
			0	6416	40		13	
			0	6585			13	
			0	6588			55	
12 57 44	28 26.0		0	(6764)		10	55	
12 57 45	28 8.1	MRK 60W	0	5149	60	8	261	
12 57 48	27 39.0	W	0	10865	100	6	495	
			0	11177			663	
12 57 48	28 8.0		0	9870		8	349	

R.A. (1950)	DEC. (1950)	NAME	OBS	HEL VEL (C*Z)	ERR	GAL CORR	REF	COMMENTS
12 57 48	28 18.0	IC 4021	0	5789	75	9	13	
12 57 48	28 21.0		0	7381	100	9	288	
12 57 48	29 6.0		0	7386	100	13	495	
12 57 52	28 14.1	NGC 4894	0	4551		9	349	
12 57 53	28 13.5	NGC 4898	0	6935	50	9	13	
12 57 53	28 13.5	NGC 4898W	0	6759		9	349	
			0	6803			593	
12 57 53	28 13.5	NGC 4898E	0	6504		9	349	
			0	6500			593	
12 57 53	28 28.1	NGC 4895	0	8406	75	10	13	
			0	8391			593	
			0	8500	200		2	
12 57 54	28 13.0		0	5137		9	349	
12 57 54	28 20.0		0	6137		9	407	
			0	6166	150		288	
12 57 58	28 40.2		0	27250	60	11	826	SEYF
			0	27600	600		365	X COM
12 58 0	26 56.0		0	7209	100	3	495	
12 58 0	27 47.0	W	0	7842		7	663	
12 58 0	27 47.0	E	0	7460		7	663	
12 58 0	29 8.0	IC 4032	0	6774	100	13	495	
12 58 0	29 19.0	IC 4026	0	8148		14	349	
12 58 5	-15 27.0	DDO 159	R	(1407)		-164	492	
12 58 6	2 46.1	NGC 4900	R	962	6	-98	914	
			R	973			744	
			0	1054			13	
12 58 6	27 55.0	W	0	7497		8	321	
			0	7473			663	
12 58 6	28 14.0		0	6927		9	349	
12 58 6	28 15.0		0	7744		9	349	
12 58 6	28 36.8	NGC 4896	0	5820	150	11	13	

R.A. (1950)	DEC. (1950)	NAME	OBS	HEL VEL (C*Z)	ERR	GAL CORR	REF	COMMENTS
12 58 11	28 24.9		0	5526		10	593	
12 58 12	28 17.0		0	7578		9	349	
12 58 12	28 20.0	IC 4040	0	7744		9	322	
		5C 4.108	0	7604			663	
			0	7635			520	
			0	7515			13	
			0	7518			55	
12 58 12	29 17.0	IC 842	0	7179	100	14	495	
12 58 14	28 17.0		0	7020		9	595	
12 58 15	-13 10.9	NGC 4891	0	2632		-157	741	
12 58 15	28 47.5	5C 4.109	0	8959		11	520	
			0	8896			663	
12 58 16	28 16.0		0	7560		9	595	
12 58 18	-36 19.0		0	4930		-220	683	
			0	4890	60		655	
12 58 18	28 11.0	NGC 4906	0	7469		9	349	
12 58 18	28 14.0		0	8043	300	9	288	
12 58 18	28 14.3	IC 4042	0	6230	100	9	288	
			0	6594			593	
12 58 18	28 16.0	IC 4041	0	7078		9	407	
			0	7043			349	
12 58 18	28 22.0		0	6611		10	349	
12 58 18	28 36.0		0	7997	100	11	495	
12 58 18	61 55.5	MRK 236	0	15600		145	306	SEYF
			0	15000			308	
12 58 19	-13 40.6	NGC 4899	0	2653		-158	741	
12 58 19	28 41.2	5C 4.113	0	6303		11	520	
			0	6381			663	
12 58 22	-14 14.7	NGC 4902	R	2638	15	-160	752	
			0	2758			13	
12 58 23	37 35.0	NGC 4914	0	4778	23	49	741	
12 58 24	0 14.0	NGC 4904	0	1111	45	-108	789	
			0	1470			741	

R.A. (1950)	DEC. (1950)	NAME	OBS	HEL VEL (C*Z)	ERR	GAL CORR	REF	COMMENTS
12 58 24	27 40.0		O	6925	100	7	495	
12 58 24	28 11.0		O	8683		9	349	
12 58 24	28 18.0	NGC 4908	O	8838	150	9	13	
			O	8841			593	
12 58 24	28 22.0	IC 4045	O	6527	200	10	13	
			O	6855			593	
			O	6600	200		2	
12 58 24	28 25.5	NGC 4907	O	5868		10	13	
			O	5873			55	
12 58 30	28 6.0		O	6206		9	349	
12 58 30	28 17.0	IC 4051	O	4932	150	9	13	
			O	4917			593	
12 58 30	28 38.0		O	7680	100	11	288	
12 58 32	28 3.7	NGC 4911	O	8006		8	13	
		HOL 449A	O	8013			55	
		5C 4.117	O	7823			663	
			O	7890			520	
12 58 36	28 10.0		O	6908		9	407	
			O	6885			349	
12 58 36	40 7.0	IC 4062	O	10589	81	60	843	
12 58 37	64 42.8	MRK 234	O	2100		155	308	
12 58 42	28 5.0		O	5977	100	9	495	
12 58 42	28 18.0		O	6915		10	407	
12 58 48	- 4 16.0	NGC 4915	O	3152	40	-125	13	
12 58 48	28 4.0		O	6513	100	9	495	
12 58 48	28 11.0		O	6031		9	349	
			O	6006			663	
			O	6006			407	
12 58 51	- 1 41.2		R	1419	10	-115	860	
12 58 54	28 4.0	NGC 4919	O	7085	100	9	288	
			O	7269			593	
12 58 58	- 4 30.4	DDO 160	R	2971	20	-125	492	
12 59 0	28 16.0		O	7472		10	407	

279

R.A. (1950)	DEC. (1950)	NAME	OBS	HEL VEL (C*Z)	ERR	GAL CORR	REF	COMMENTS
12 59 0	29 35.0	NGC 4922N 5C 4.130 KDG 363A	0 0	7012 7014		15	520 663	
12 59 0	29 35.0	NGC 4922S KDG 363B	0 0	7217 7236		15	520 593	
12 59 2	-41 36.4		0	3400	120	-229	629	
12 59 2	28 9.3	NGC 4921	0 0 0	5459 5322 5465		9	13 663 55	
12 59 2	48 19.8	MRK 237	0	9300		94	308	
12 59 6	28 6.0	NGC 4923	0	5433	100	9	288	
12 59 6	28 57.0		0 0	(8909) 8759	20	13	495 663	
12 59 12	28 25.0	NE	0	6022		10	663	
12 59 12	28 25.0	SW	0	5377		10	663	
12 59 12	29 24.0	IC 843	0 0	7500 7401	100	15	288 593	
12 59 13	-42 30.1	NGC 4909	0	3240	130	-230	629	
12 59 21	65 16.1	MRK 238	0	15000		157	308	
12 59 24	28 9.0		0	7617	100	9	495	
12 59 24	29 19.0	IC 4088	0	7013	100	14	495	
12 59 24	28 22.0		0	5918	100	10	495	
12 59 24	32 14.70		0	28480	330	27	196	
12 59 26	-50 3.9		0	1358		-240	925	
12 59 29	-49 12.2		0	1093		-239	925	
12 59 30	27 53.6	NGC 4926	0 0	(7668) 7888	100	8	55 288	
12 59 32	32 15.00		0	28720	330	27	196	
12 59 34	32 16.50		0	27970	330	27	196	
12 59 36	28 3.0		0	7202		9	322	
12 59 36	28 16.0	NGC 4927 W	0 0	7558 7563	100	10	288 663	

R.A. (1950)	DEC. (1950)	NAME	OBS	HEL VEL (C*Z)	ERR	GAL CORR	REF	COMMENTS
12 59 42	29 31.0		O	7266	100	16	288	
			O	7230			593	
			O	7412			321	
12 59 42	32 20.60		O	28600	330	28	196	
12 59 44	27 55.0	NGC 4926A	O	(7177)		9	55	
12 59 48	28 29.0		O	8308		11	322	
12 59 48	28 40.0		O	7613	100	12	495	
13 0 0	28 30.0		O	5722	100	11	495	
13 0 0	28 32.0		O	7138	100	11	495	
13 0 0	32 14.00		O	18920	330	27	196	
13 0 6	16 39.0	MRK 783	O	19984		-39	874	SEYF
			O	20439			630	
13 0 12	28 22.0	IC 4106	O	7443	100	11	495	
13 0 18	28 18.0	NGC 4929	O	6338		11	322	
13 0 18	28 24.0	W	O	49239	999	11	663	
13 0 18	28 39.0		O	7442	100	12	495	
13 0 24	- 7 49.0	NGC 4928	O	1602	45	-136	789	
			O	1759			741	
13 0 24	28 7.0		O	8170	100	10	495	
13 0 30	28 20.0	IC 4111	O	7893	100	11	495	
13 0 36	26 47.0		O	5966	100	4	495	
13 0 36	28 17.0	NGC 4931	O	5824	100	11	288	
13 0 38	-17 9.2	DDO 161	R	746	10	-168	492	
13 0 43	36 7.8		O	17990		45	172	*
13 0 48	28 50.0		O	6914		13	322	
13 0 53	32 6.4	B2	O	49100	250	27	696	
			O	48990	330		196	
13 0 54	28 17.0	NGC 4934	O	6093	100	11	495	
13 1 0	26 49.0		O	6724	100	5	495	

R.A. (1950)		DEC. (1950)		NAME	OBS	HEL VEL (C★7)	ERR	GAL CORR	REF	COMMENTS
13 1 0		32	1.60		O	49800	330	27	196	
13 1 17		-40	58.6	NGC 4930	O	2560	140	-227	629	
13 1 18		26	21.0		O	11203	100	3	495	
13 1 20		-11	13.8	NGC 4933NE ARP 176 HOL 502B	O	3188	27	-148	741	
13 1 20		-11	13.8	NGC 4933SW ARP 176 HOL 502A	O	3306	16	-148	741	
13 1 24		28	21.0	NGC 4943	O	5739	100	12	495	
13 1 24		29	29.0	5C 4.157	O	49997		16	663	
13 1 25		38	12.3	4C 38.35	O	35500		54	400	
13 1 30		28	15.0	IC 4133	O	6331	100	11	495	
13 1 34		-30	15.4	NGC 4936	O	3309	33	-204	741	
13 1 37		-10	4.4	NGC 4939	R	3109		-144	707	
					R	3110			693	
					R	3106			744	
					O	3182			201	
					O	2998	22		148	
					O	2483			201	KNOT
13 1 37		- 5	17.0	NGC 4941	R	648		-126	310	
					O	846	40		13	
13 1 48		28	31.0		O	5971	100	13	495	
13 1 54		28	44.4	5C 4.161	O	7773		14	520	
					O	7818			663	
13 2 0		- 3	18.0		O	1350		-119	13	
13 2 0		26	56.0		O	10814	100	6	495	
13 2 0		27	34.0		O	5618	100	9	495	
13 2 0		29	5.0		O	8047		15	321	
13 2 18		- 7	41.0	NGC 4948	O	1279	63	-135	789	
13 2 29		- 7	53.6	DDO 162	R	1553	10	-135	492	
13 2 30		- 6	14.0	NGC 4951	R	1180		-129	744	
					O	1195			741	
					O	1045			694	

R.A. (1950)	DEC. (1950)	NAME	OBS	HEL VEL (C*Z)	ERR	GAL CORR	REF	COMMENTS
13 2 32	-49 12.0	NGC 4945	R	570		-238	416	
		PKS	R	563			671	
			O	571			728	
			O	550			317	
			O	108			154	
13 2 34	-35 4.2	NGC 4947	O	2367	52	-215	741	
			O	2760	100		844	
			O	2460			574	
13 2 35	29 23.5	NGC 4952	O	5865		17	13	
			O	5869			55	
13 2 38	- 7 37.1	DDO 163	R	1123	15	-134	492	
13 2 48	26 13.0		O	6534	100	3	288	
13 2 48	27 50.0	NGC 4957	O	6981	100	10	288	
13 2 56	30 32.6	MRK 62	O	9998		22	437	
13 2 58	-49 3.5		O	8830		-237	925	
13 3 0	-49 16.0		O	15330		-238	925	
13 3 0	-41 17.0	NGC 4930	O	2599		-226	728	
13 3 0	29 34.0		O	7158	100	18	495	
13 3 12	- 7 45.0	NGC 4958	R	1112		-134	645	
			O	1515	75		13	
13 3 24	28 0.0	NGC 4961	R	2535		11	744	
		5C 4.175	O	2580			663	
			O	2577			520	
13 3 28	31 10.3	82	O	54410	250	25	696	
			O	54887	100		13	
13 3 30	37 52.5	IC 4182	R	326	6	54	772	
			R	319	10		860	
			R	321			693	
			O	(280)			472	
			O	206			641	
			O	324	38		789	
13 3 34	-49 33.7		O	1218		-238	925	
13 3 36	28 51.0		O	7995	100	15	495	
13 3 36	29 33.0		O	7982	100	18	495	
13 3 38	-49 25.4	NGC 4945A	O	1367		-238	925	

13ʰ3ᵐ

R.A. (1950)	DEC. (1950)	NAME	OBS	HEL VEL (C*Z)	ERR	GAL CORR	REF	COMMENTS
13 3 38	-17 14.9	DDO 169	R	1463	10	-166	492	
13 3 48	25 43.0		O	7200		2	533	
13 3 50	11 29.6	MC 2	O	25800		-58	465	
13 3 54	29 20.0	NGC 4966	O O	7077 (7164)	100	17	288 496	
13 3 56	53 45.5	MRK 242	O	7500		118	308	
13 3 58	33 14.3	MRK 241	O	7800		34	308	
13 4 6	27 26.0		O	7852	100	9	495	
13 4 12	29 6.0		O	4837	100	17	495	
13 4 30	28 48.0	NGC 4971	O	6374	100	16	288	
13 4 37	67 58.2	DDO 165 7 ZW499	R	(34)	10	167	492	
13 4 48	28 18.0		O	6640	100	14	495	
13 5 0	26 59.0	ARP 139	O O O	11326 11509 11293	50 100	8	141 408 496	
13 5 12	27 45.0		O	6655	100	12	495	
13 5 30	-12 42.0	2 ZW 4-S	O	12335	300	-150	735	
13 5 34	29 52.4		O	284000		21	646	*
13 5 36	27 1.0		O	10343	100	9	408	
13 5 36	28 58.0		O	7713	100	17	495	
13 5 42	-49 14.5	NGC 4976	O O O	1370 1356 1425		-236	136 84 925	
13 5 42	-11 33.0	2 ZW 3-S	O	11246	130	-146	735	
13 5 42	-11 8.0	2 ZW 2-S	O	11900	300	-144	735	
13 5 48	27 46.0		O	7373	100	12	495	
13 5 51	47 5.4		R	214	10	93	860	
13 6 0	-10 45.0	2 ZW 1-S	O	21129	150	-143	735	
13 6 0	28 35.0	NGC 4983	O	6631	100	16	495	

284

R.A. (1950)			DEC. (1950)		NAME	OBS	HEL VEL (C*Z)	ERR	GAL CORR	REF	COMMENTS
13	6	12	- 0	32.0		0	4500		-105	533	
13	6	13	- 6	30.8	NGC 4981	0	1797	24	-127	741	
						0	1692			777	
13	6	18	-15	15.0	NGC 4984	0	1353		-158	694	
					2 ZW 6-S	0	1259	12		741	
						0	1233	160		735	
13	6	30	28	27.0		0	5971	100	15	495	
13	6	36	28	18.0		0	7261	100	15	495	*
13	6	36	29	18.0		0	9571	100	19	495	
13	6	36	53	12.0	IC 853	R	7152	20	117	582	
13	6	48	- 5	0.0	NGC 4490 MRK 1344	0	3181		-121	920	
13	6	54	29	38.0		0	6366		21	321	
						0	6509	300		288	
13	7	0	- 7	34.0	NGC 4995	0	1835		-131	13	
13	7	0	1	56.4	NGC 4999	0	3105		-94	741	
13	7	24	29	10.0	NGC 5000	R	5621		19	779	
						R	5615	10		752	
						0	5706			321	
						0	5620	100		288	
13	7	36	19	59.0		0	3900		-19	533	
13	8	0	25	11.3		R	2587	10	2	860	
13	8	18	-33	44.0		0	12508		-208	683	
13	8	24	29	59.0		0	6427		23	496	
13	8	38	37	19.4	NGC 5005	R	1003	35	54	390	
						0	900	50		1	
						0	1041			13	
						0	1013	65		13	
						0	1025			46	
13	8	41	27	44.0	3CR284	0	71760		14	318	
13	8	42	29	50.0	NGC 5004A	0	7163	150	23	288	
13	8	42	29	54.0	NGC 5004	0	6957	100	23	288	
13	8	46	60	51.3	MRK 243	0	8400		145	308	

13ʰ8ᵐ

R.A. (1950)	DEC. (1950)	NAME	OBS	HEL VEL (C*Z)	ERR	GAL CORR	REF	COMMENTS
13 8 48	3 41.0	A 1309	R	3042		-86	744	
			O	2990	40		71	
13 9 0	84 53.0	7 ZW501	O	4641	165	213	429	
			O	4705	100		820	
13 9 12	23 10.9	NGC 5012	O	2815	26	-4	741	
			O	2681	58		789	
			O	2570			593	
13 9 12	36 33.0	NGC 5014	R	1043	47	51	885	
		MRK 449	O	1140			438	
			O	900			359	
13 9 18	- 1 4.0		O	52458	300	-105	13	
13 9 18	31 46.0		O	7003		31	593	
13 9 32	21 3.8	4C 21.39	O	9150		-13	244	
13 9 34	5 44.2		R	914	10	-77	860	
13 9 42	24 21.7	NGC 5016	O	2583	14	0	741	
			O	2769			593	
13 9 42	27 35.0		O	6127	100	14	408	
			O	8400			476	DIS
13 9 48	-15 31.0	2 ZW 8-S	O	6400	500	-156	735	
13 9 54	26 57.0		O	861	100	11	288	
13 10 0	-42 50.1	NGC 5011	O	3101	50	-225	741	
			O	3025			574	
13 10 0	28 47.0		O	6919		19	496	
13 10 11	12 51.9	NGC 5020	R	3354	15	-47	752	
13 10 12	23 6.0	NGC 5012A	O	2527	55	-4	789	
			O	2596			593	
13 10 12	50 39.5	MRK 244	O	8700		109	306	
13 10 15	-16 30.1	NGC 5017	O	2543	27	-159	741	
13 10 18	-19 15.0	NGC 5018	O	2897	75	-168	13	
13 10 23	67 46.0	MRK 245	O	6000		168	268	
13 10 36	-15 5.0	2 ZW 10-S	O	11965	150	-154	735	
13 10 24	32 4.0	NGC 5025	O	6361		33	593	

R.A. (1950)	DEC. (1950)	NAME	OBS	HEL VEL (C*Z)	ERR	GAL CORR	REF	COMMENTS
13 10 30	-10 52.0	2 ZW 11-S	O	10210	350	-140	735	
13 10 36	27 25.0		O	6863	100	14	408	
13 10 48	-49 11.4		O	1261		-234	925	
13 10 59	36 28.6	DDO 166	R	956	10	52	492	HOL IIX
13 11 0	-49 12.9		O	3518		-234	925	
13 11 0	28 1.0		O	6253	100	17	408	
13 11 0	28 4.0	NGC 5032	O	6519		17	593	
13 11 10	46 35.0	DDO 167	R	165	10	93	492	
13 11 12	-11 52.0	2 ZW 12-S	O	14147	650	-143	735	
13 11 12	36 51.0	NGC 5033	R	913	25	54	183	
			R	878			934	
			R	872			693	
			O	881	5		759	
			O	794	54		945	
			O	908			13	
			O	924	40		13	
			O	813	37		72	
			O	894	25		283	
13 11 13	56 21.7	MRK 246	O	12300		130	308	
13 11 16	-42 41.8	NGC 5026	O	3664	36	-224	741	
			O	3490			574	
			O	4340	400		629	
13 11 26	-45 51.3		O	3009	80	-229	680	
13 11 36	35 35.0	1 ZW 53	O	5022		49	262	
13 11 48	-12 15.0	2 ZW 13-S	O	23085	210	-144	735	
13 12 0	4 10.0		O	21382		-82	683	
13 12 3	2 50.3		O	6210		-87	895	
13 12 12	30 58.0	NGC 5041	O	7441		30	593	
13 12 15	46 11.0	DDO 168	R	198	10	92	492	
			R	193	35		424	
13 12 18	-13 15.0	2 ZW 15-S	O	13867	500	-147	735	
13 12 22	-16 19.6	NGC 5037	O	1887	18	-157	741	
			O	1985			694	

13ʰ12ᵐ

R.A. (1950)	DEC. (1950)	NAME	OBS	HEL VEL (C*Z)	ERR	GAL CORR	REF	COMMENTS
13 12 30	35 8.0	MRK 450	O	840		48	438	
13 12 33	55 3.8	MRK 247	O	9680		126	287	
			O	9600			268	
13 12 36	24 53.0	IC 860	O	3867		4	593	
13 12 42	-10 19.0	2 ZW 22-S	O	14163	300	-136	735	
13 12 43	-16 7.3	NGC 5044	O	2712		-156	694	
			O	2704	33		741	
13 13 0	44 40.0	KDG 368A	O	10231	130	87	567	
			O	10676	20		908	
13 13 6	44 40.0	KDG 368B	O	10579	70	87	567	
		MRK 248	O	10800			308	
			O	10626	20		908	
13 13 6	59 22.0	MRK 65	O	13010		142	224	
13 13 12	29 55.0	NGC 5052	O	6747		26	593	
13 13 18	-16 8.0	NGC 5049	O	2744	65	-156	13	
13 13 19	47 45.7	DDO 169	R	258	10	99	492	
13 13 30	25 42.0	DDO 170	R	936	10	8	492	
13 13 35	42 17.9	NGC 5055	R	510		78	156	M 63
			R	503			934	
			R	511			693	
			O	450	50		1	
			O	504	5		759	
			O	500	30		13	
			O	560			31	
			O	538			13	
13 13 36	-14 53.0	2 ZW 17-S	O	9580	400	-151	735	
13 13 38	62 23.6	ARP 238	O	9471		152	748	
		VV 250	O	9471	65		500	
		KDG 369	O	9453	18		908	
13 13 42	62 23.3	ARP 238	O	9241	71	152	500	
		VV 250	O	9241			748	
		KDG 369	O	9313	14		908	
13 13 48	41 45.0		O	6300		76	476	
13 13 51	31 12.8	NGC 5056	R	5607	15	32	752	
			O	5449			593	

R.A. (1950)	DEC. (1950)	NAME	OBS	HEL VEL (C*Z)	ERR	GAL CORR	REF	COMMENTS
13 13 54	-15 1.0	2 ZW 18-S	0	10282	500	-151	735	
13 13 54	-13 24.0	2 ZW 25-S	0	6405	110	-146	735	
13 13 54	30 31.0		0	14810		29	593	
13 14 6	31 17.0	NGC 5057	0	5824		32	593	
13 14 18	-16 22.3	NGC 5054	R	1743		-156	744	
			0	1892			694	
			0	1764	25		741	
13 14 24	12 49.0	NGC 5058A MRK 786A KDG 370A	0 0	945 895	10	-45	908 929	★
13 14 24	12 49.0	NGC 5058B MRK 786B KDG 370B	0 0	986 835	22	-45	908 929	★
13 15 12	31 20.0	NGC 5065	0	5699		33	593	
13 15 18	-26 36.0	NGC 5061	0	2065		-186	13	
13 15 25	44 4.4	MRK 250	0	8400		86	308	
13 15 36	-31 22.0	IC 4219	0	3647	25	-198	582	
13 15 42	-12 42.0	2 ZW 32-S	0	10638	85	-142	735	
13 16 4	-47 38.8	NGC 5064	0	2982	10	-229	741	
13 16 4	- 8 11.0	DDO 171	R	1310	10	-126	492	
13 16 6	31 44.0	NGC 5074	0	5684		36	593	
13 16 13	-20 46.6	NGC 5068	R	669		-168	671	
			0	570			13	
13 16 18	-31 55.0		0	4800	60	-199	655	
13 16 31	42 12.7	DDO 172	R	1215	20	79	492	
13 16 42	-14 36.0	NGC 5073	0	2715		-148	694	
13 16 44	29 54.5	4C 29.47	0	21790	250	28	696	
			0	21573	85		741	
			0	21840			400	
13 16 48	-14 52.0	2 ZW 34-S	0	2832	120	-149	735	
13 16 48	28 46.0	NGC 5081	0	6707		24	593	
			0	6669	60		813	

13ʰ16ᵐ

R.A. (1950)	DEC. (1950)	NAME	OBS	HEL VEL (C*Z)	ERR	GAL CORR	REF	COMMENTS
13 16 53	-12 23.7	NGC 5077	O	2787		-140	281	
			O	2647	100		13	
			O	2858			106	
13 16 54	-14 30.0	2 ZW 33-S	O	10452	295	-147	735	
13 17 18	30 31.0	NGC 5089	O	2159		31	593	
13 17 28	36 16.2	4C 36.23	O	21840		55	400	
13 17 30	16 7.0	IC 881	O	6843	34	-28	594	
13 17 34	-24 10.7	NGC 5085	R	1949		-177	671	
			O	2073	20		741	
13 17 34	-21 33.9	NGC 5084	O	1739	30	-170	741	
13 17 36	16 9.0	IC 882	O	6782	47	-28	594	
13 17 42	-12 49.0	2 ZW 36-S	O	7000	200	-141	735	
13 17 42	-12 18.8	NGC 5088	O	1464	16	-139	741	
			O	1478			910	
13 17 45	33 20.0	VV 633	O	8400	150	43	776	
13 17 48	-20 21.0	NGC 5087	O	1832	150	-166	13	
13 17 56	33 24.3	NGC 5098 B2	O	11320		44	485	
13 17 56	52 18.8	MRK 251	O	4500		119	306	
			O	4602			279	
13 18 0	31 47.0		O	5146		37	593	
13 18 9	10 3.0	DDO 173	R	1133	15	-53	492	
13 18 17	-43 26.5	NGC 5090	O	3326	56	-221	741	
			O	2981	20		680	
13 18 18	34 24.0	IC 883 ARP 193 B2 1 ZW 56	O O	6946 6878		48	262 90	
13 18 23	-43 27.5	NGC 5091	O	3701	160	-221	680	
13 18 24	-12 12.0		O	2268	80	-139	735	
13 18 24	1 45.0		O	6000		-87	533	
13 18 27	56 41.9	MRK 253	O	6600		135	308	

R.A. (1950)	DEC. (1950)	NAME	OBS	HEL VEL (C*Z)	ERR	GAL CORR	REF	COMMENTS
13 18 30	-13 23.0	2 ZW 39-S	O	6538	500	-143	735	
13 18 30	-10 38.0	2 ZW 38-S	O	8990	500	-133	735	
13 18 40	-35 32.2		O	15156	78	-206	741	
13 18 55	57 54.3	NGC 5109	R	2131	15	139	829	
13 19 0	31 39.0		O	7084		37	593	
13 19 1	-27 10.1	NGC 5101	O	1850	16	-185	741	
13 19 5	42 51.0	3CR285	O	23811	50	83	741	
			O	23800			159	
13 19 7	-36 22.1	NGC 5102	R	453	20	-207	544	
			R	411	10		238	
			O	500	73		148	
			O	546			84	
13 19 10	38 48.0	NGC 5107	R	940	15	67	734	
			R	943			744	
13 19 18	31 37.0		O	5051		37	593	
13 19 21	-40 16.2		O	3770	200	-215	794	
13 19 24	-43 3.6		O	7340	200	-220	794	
13 19 24	-38 2.0		O	6060	90	-211	655	
13 19 24	31 30.0		O	7235		37	593	
13 19 42	39 0.0	NGC 5112	R	965		68	744	
			R	965	10		829	
			O	937	63		789	
13 20 2	21 41.2	MRK 659	O	6850		-3	557	
13 20 12	-11 6.0	2 ZW 40-S	O	7174	154	-133	735	
13 20 14	-41 18.5		O	11630	200	-216	793	
13 20 24	8 25.0	MRK 1347	O	15288		-58	920	SEYF
13 20 36	27 15.0	NGC 5116	O	2839		20	593	
13 20 36	28 35.0	NGC 5117	O	2440		25	593	
13 20 42	-11 5.0	2 ZW 41-S	O	7180	200	-133	735	
13 20 42	27 23.0	IC 4234	O	10357		20	593	

R.A. (1950)	DEC. (1950)	NAME	OBS	HEL VEL (C*Z)	ERR	GAL CORR	REF	COMMENTS
13 20 46	39 0.0		O	26287		69	741	
13 20 47	51 59.9	MRK 254	O	9000		119	308	
13 20 57	53 13.7	MRK 255	O	9000		123	306	
			O	9123			279	
13 21 18	-24 24.3	DDO 241	R	2056	20	-175	492	
13 21 18	-10 24.0	2 ZW 43-S	O	2939	250	-130	735	
13 21 26	31 49.5	NGC 5127	O	4790		39	485	
		B2	O	4684			703	
			O	4876	62		741	
13 21 27	70 46.4	NGC 5144	O	3000		180	268	
		MRK 256						
13 21 53	-37 25.3	NGC 5121	O	1532	50	-208	741	
13 22 4	36 51.0	MRK 451	O	4800		61	359	
			O	4800			438	
13 22 11	58 8.7	MRK 264	O	18900		141	306	
13 22 18	16 23.0	KDG 372A	O	7247	20	-24	500	
13 22 18	16 23.0	KDG 372B	O	7175	28	-24	500	
13 22 32	-42 45.5	NGC 5128	R	530		-218	302	CEN A
		ARP 153	R	563			256	
		PKS	O	677			145	
			O	630			62	*
			O	559	24		876	
			O	605			18	
			O	547	5		701	
			O	492			290	
			O	540			799	
			O	468	40		13	
			O	434	75		51	
			O	465	50		222	
			O	470	50		62	
			O	462	25		221	
13 22 35	36 38.3	NGC 5141	O	5220	120	60	350	
		KDG 373A	O	5250			400	
		4C 36.24	O	5020			748	
13 22 36	-20 52.6	NGC 5134	O	1696	50	-164	741	
13 22 45	36 39.6	NGC 5142	O	5220		60	748	
		KDG 373B	O	5150	50		908	

R.A. (1950)	DEC. (1950)	NAME	OBS	HEL VEL (C*Z)	ERR	GAL CORR	REF	COMMENTS
13 22 48	-12 53.0	2 ZW 45-S	0	6882	400	-138	735	
13 22 52	-29 34.3	NGC 5135	0	4112		-188	925	
			0	4157	14		741	
13 23 12	-14 3.0	2 ZW 44-S	0	10384	380	-141	735	
13 23 18	-11 55.0	2 ZW 47-S	0	6477	240	-134	735	
13 23 22	-33 36.4	NGC 5140	0	3771	40	-198	741	
13 23 32	58 5.0	DDO 175	R	1519	15	142	492	
13 23 47	2 21.7	NGC 5147	R	1088	4	-80	914	
			R	1107			744	
			R	1093			693	
			0	(1115)	64		148	
13 23 48	33 16.0	MRK 453	0	13940		47	438	
13 23 51	-21 58.8	DDO 174	R	1440	20	-166	492	
13 23 54	4 44.0	KDG 374	0	6524	61	-70	908	
13 23 54	36 12.0	NGC 5149 KDG 375	0	5577	50	59	908	
13 23 58	-26 52.5		0	13130		-180	255	
13 23 58	57 31.0	MRK 66	0	6690		140	224	
			0	6360			264	
13 24 0	-29 37.0		0	4230	60	-188	655	
13 24 0	4 43.0	KDG 374	0	6620	50	-70	908	
13 24 12	36 16.0	NGC 5154 KDG 375	0	(5533)	100	60	908	
13 24 18	-27 41.0		0	1860	60	-182	655	
13 24 18	20 12.0		R	7149	20	-6	582	
			0	7144	50		582	
13 24 24	-11 19.0	2 ZW 49-S	0	4300	200	-131	735	
13 24 30	26 50.7	MRK 454	R	7059		21	779	
			0	6900			359	
			0	6985			438	
13 24 32	-27 47.7		R	590	5	-183	865	
13 24 36	32 27.0		R	5275	10	44	582	
			0	5293	50		582	

13ʰ24ᵐ

R.A. (1950)	DEC. (1950)	NAME	OBS	HEL VEL (C∗Z)	ERR	GAL CORR	REF	COMMENTS
13 24 42	−41 13.2		R	512	5	−214	865	
13 24 50	−24 18.2	NGC 5150	O	4376	32	−172	741	
13 24 54	−12 14.0	2 ZW 50-S	O	5838	500	−134	735	
13 24 52	−40 54.4		O	11310	200	−213	794	
13 25 0	32 17.0	NGC 5157	O	7190		40	703	
13 25 1	−40 47.4		O	(16580)	200	−213	793	
13 25 11	−40 52.8		O	15220		−213	793	
13 25 11	−40 52.7		O	15260	200	−213	793	
13 25 11	−40 52.6		O	14330	200	−213	793	
13 25 11	−40 52.4		O	15230	200	−213	793	
13 25 14	55 44.7	MRK 257	O	4800		134	268	
13 25 19	−40 47.3		O	(15050)	200	−212	793	
13 25 41	−48 39.3	NGC 5156	O	3000		−226	574	
			O	2832	64		741	
13 26 4	31 2.1	4C 31.42	O	7180		39	166	
13 26 18	46 50.0	NGC 5173	O	2404	50	102	13	
13 26 24	−37 58.0		O	8750	60	−206	655	
13 26 24	38 0.0	1 ZW 58	O	18629		68	262	
13 26 25	−32 54.9	NGC 5161	O	2252	68	−194	741	
13 26 35	44 11.4	MRK 259	O	8400		92	306	
13 26 49	53 42.1	MRK 258	O	7500		128	306	
			O	7597			279	
13 26 53	17 18.6	NGC 5172	R	4028	10	−16	752	
			R	4033	6		914	
			R	4037	10		829	
			O	4367			741	
13 26 54	12 0.0	NGC 5171	O	6723	63	−38	843	
13 27 6	−38 1.0		O	8450	150	−206	655	
13 27 7	−17 42.4	NGC 5170	O	1585		−150	694	
			O	1583	20		741	

R.A. (1950)	DEC. (1950)	NAME	OBS	HEL VEL (C*Z)	ERR	GAL CORR	REF	COMMENTS
13 27 15	45 15.7	MRK 260	O	10500		97	268	
13 27 18	-32 33.0		O	15830	90	-193	655	
13 27 30	- 1 28.0	NGC 5183 KDG 378	O	4213	15	-92	908	
13 27 32	45 38.8	DDO 176	R R	1297 1304	10 20	99	492 424	
13 27 36	- 1 25.0	NGC 5184 KDG 378	O	4147	17	-92	908	
13 27 44	58 40.7	NGC 5204	R R R R R O O	210 200 208 210 205 272 200	 2 32 25 	145	693 676 216 171 183 13 483	
13 27 47	47 27.3	NGC 5194/5	R R	474 468	30	106	93 934	
13 27 47	47 27.3	NGC 5194 VV 1 ARP 85 KDG 379 HOL 526A	R R R R R R R R R O O O O O O O O O O O O O	449 461 463 458 455 460 454 460 465 485 499 413 438 270 460 482 562 474 464 510 468 472 381 (616)	 11 5 110 35 50 2 4 70 3 31 175	106	27 257 693 257 156 216 489 171 415 53 910 149 13 1 797 797 8 87 467 264 202 253 72 555	M 51 WHI NEB *
13 27 52	47 31.8	NGC 5195 VV 1 ARP 85 KDG 379 HOL 526B	R O O O O O O	(465) 508 562 606 542 606 240	9 29 35 50	106	581 72 253 87 13 106 1	

295

13ʰ27ᵐ

R.A. (1950)	DEC. (1950)	NAME	OBS	HEL VEL (C*Z)	ERR	GAL CORR	REF	COMMENTS
13 27 52	47 31.8		O	472	70	106	8	
13 28 5	46 55.5	NGC 5198	O O	2482 2562	50	104	13 13	
13 28 18	31 32.0	MRK 455 VV 326	O	10195		43	438	
13 28 30	-13 2.0	2 ZW 66-S	O	11599	999	-134	735	
13 28 34	19 41.7	VV 88	R O	1000 2005	15	-5	829 774	DIS
13 28 37	-34 32.0	NGC 5188	O	2366	17	-197	741	
13 29 0	-32 58.9		O	3558	50	-193	741	
13 29 0	25 52.5		O	7557		20	714	
13 29 4	-32 58.6	NGC 5193	O	3684	53	-193	741	
13 29 7	75 49.7	MRK 261	O O	9000 9300		195	568 268	
13 29 7	75 52.0	A	O	9000		195	420	KNOT
13 29 7	75 52.0	B	O	9000		195	420	KNOT
13 29 25	75 49.8	MRK 262	O O	9000 9000		195	568 308	
13 29 28	20 15.8		R	1011	15	-2	829	
13 29 30	20 15.0		O	992	23	-2	813	
13 29 51	28 38.8		O	8860		32	714	
13 29 54	11 22.0	MRK 789A/B	O	9639		-39	630	
13 29 54	11 22.0	MRK 789A	O	9357	30	-39	831	SEYF
13 29 54	11 22.0	MRK 789B	O	9355	39	-39	831	SEYF
13 30 0	13 5.0	MRK 1352	O	7532		-32	920	
13 30 17	25 41.8		O	10216		20	714	
13 30 21	2 16.1	3CR287.1	O O	64780 64758	60 30	-75	134 741	SEYF
13 30 30	63 1.0	NGC 5218 ARP 104 VV 33	O	2724	100	160	789	

R.A. (1950)	DEC. (1950)	NAME	OBS	HEL VEL (C*Z)	ERR	GAL CORR	REF	COMMENTS
13 30 36	9 46.0	MRK 1354	O	7005		-45	920	
13 30 36	42 7.0	NGC 5214A KDG 381A	O	8056	20	87	908	
13 30 36	42 8.0	NGC 5214B KDG 381B	O	8217	20	87	908	
13 30 39	50 49.0		O	83800	300	119	583	
13 30 48	9 47.0	MRK 1355	O	7185		-45	920	
13 30 48	37 26.9	MRK 456	O O	10200 10250		68	359 438	
13 31 40	- 9 54.1	PKS	O	24300		-120	511	
13 31 40	20 30.9		O	50400		0	717	
13 31 41	69 7.0	MRK 263	O	1500		178	268	
13 31 45	-48 17.6		R O O	825 850 1047	5 14	-222	865 701 378	* FFG
13 32 4	35 8.6	B2	O	7154	47	60	741	
13 32 6	34 57.0	NGC 5223	O	7180	50	59	141	
13 32 11	52 8.7	MRK 264	O	18734		125	279	
13 32 12	9 36.0	IC 900	P R O	7062 7067 7080	20 25	-44	582 779 582	
13 32 24	4 29.0		O	6965		-65	683	
13 32 24	33 13.0		O	17400		52	476	
13 32 27	-33 38.5		O	3873	30	-192	741	
13 32 30	14 0.0	NGC 5222A KDG 383A	O	6860	30	-26	908	
13 32 30	14 0.0	NGC 5222B KDG 383B	O	6903	25	-26	908	
13 32 33	26 27.7		O	7957		25	714	
13 32 36	34 18.8		O	7400		57	422	
13 32 42	10 56.0	KDG 385A VV 211	O O	11939 11400	30	-38	908 535	

R.A. (1950)	DEC. (1950)	NAME	OBS	HEL VEL (C★Z)	ERR	GAL CORR	REF	COMMENTS
13 32 42	10 57.0	KDG 385B VV 211	O	11802	11	-38	908	
13 32 42	34 18.1		O	7130		57	422	
13 32 42	51 53.0	NGC 5238A/B VV 828	O	272	30	124	776	
13 32 42	51 53.0	NGC 5238A KDG 384A	O	254	12	124	908	
13 32 42	51 53.0	NGC 5238B KDG 384B	O	194	25	124	908	
13 32 48	35 44.0	1 ZW 63	O	19079		63	262	
13 32 54	-75 59.0		O	2914	19	-236	599	
13 32 54	1 40.0	NGC 5227	R O	5254 5198	10 20	-76	582 582	
13 32 54	34 17.2	MRK 459	O O	7150 7130		57	435 422	
13 32 59	26 40.8	IC 4279	O	7389		26	714	
13 33 2	-34 2.5		O	1800	200	-193	731	SEYF
13 33 6	13 56.0	NGC 5230	R O	6838 6893		-26	744 741	
13 33 6	34 2.0		O O	2323 2460	15 300	56	808 931	
13 33 15	9 13.4		R	1161	10	-45	860	
13 33 17	26 7.8		O	8022		24	714	
13 33 18	29 29.0		R O	838 855	45	38	366 246	
13 33 37	46 11.0	DDO 178	R	1447	10	104	492	
13 33 47	-33 42.4	IC 4296 PKS	O O O	3729 3881 3541	55 42	-192	741 84 148	
13 33 56	-33 48.7	IC 4299	O	4063	59	-192	741	
13 34 10	-29 36.8	NGC 5236	R R R R R R R	530 535 509 600 340 505 580	15 8 20	-181	93 456 489 171 219 182 216	M 83

R.A. (1950)	DEC. (1950)	NAME	OBS	HEL VEL (C*Z)	ERR	GAL CORR	REF	COMMENTS
13 34 10	-29 36.8	NGC 5236	R	515	15	-181	367	
			O	445	20		732	
			O	480			175	
			O	500	50		1	
			O	504			145	
			O	516			155	
			O	418			84	
			O	497	18		148	
			O	516			227	
			O	491	30		13	
13 34 10	46 27.1	DDO 177	R	2427	10	105	492	
13 34 16	27 29.8	IC 4307	O	10724		30	714	
13 34 24	17 43.0	A	O	6737	87	-9	594	
13 34 24	17 43.0	B	O	6704	88	-9	594	
13 34 32	-27 47.7		R	590	5	-176	865	
13 34 42	-32 45.0		O	3990	90	-189	655	
			O	4029			683	
13 34 46	-42 35.7		O	300	200	-210	794	
13 34 48	-33 33.4		O	3920	20	-191	741	
13 34 52	-42 5.2		O	14750	200	-209	793	
13 34 56	7 53.9	DDO 179	R	1053	10	-49	492	
13 34 57	-42 7.8		O	15530	200	-209	793	
13 34 57	-42 7.7		O	15560	200	-209	793	
13 35 5	27 40.4	NGC 5251	O	11129		32	714	
13 35 6	9 8.0	NGC 5248	R	1125		-44	363	
			R	1155			744	
			R	1152	5		390	
			O	1190			65	
			O	1232			13	
			O	1176	50		13	
13 35 12	-42 6.7		O	13100	200	-209	793	
13 35 12	-33 37.3		O	3861	29	-190	741	
13 35 21	-17 38.1	NGC 5247	R	1655		-144	363	
			R	1354			744	
			O	1386	31		741	

13ʰ35ᵐ

R.A. (1950)	DEC. (1950)	NAME	OBS	HEL VEL (C*Z)	ERR	GAL CORR	REF	COMMENTS
13 35 32	- 9 33.0	DDO 180	R	1300	10	-116	492	
13 35 47	4 48.3	NGC 5252	0	6570		-61	895	
13 35 58	28 1.4	MRK 265	0	10200		34	306	
13 36 5	26 59.8	IC 4314	0	8719		30	714	
13 36 11	26 21.5		0	8452		27	714	
13 36 15	48 31.8	NGC 5256A/B KDG 388	0	8400		114	268	
13 36 15	48 31.8	NGC 5256A	0	8486		114	748	
		1 ZW 67A	0	8340	40		197	
		MRK 266A	0	8437			927	
			0	8437	49		830	
13 36 15	48 31.9	NGC 5256B	0	8486		114	748	
		1 ZW 67B	0	8130	103		197	
		MRK 266B	0	8310			927	
			0	8310			830	
13 36 24	26 34.9		0	3994		28	714	
13 36 36	25 14.0		0	7987		23	714	
13 36 39	39 6.4	3CR288	0	73700		79	946	
13 37 1	- 0 1.2	PKS	0	43200		-79	828	
13 37 6	-39 36.2		0	15290	200	-203	793	
13 37 6	-31 23.4	NGC 5253	R	407	20	-183	544	
			R	382	8		390	
			R	380	10		314	
			R	409			671	
			R	229			219	
			0	432	30		13	
			0	412	5		344	
			0	396			84	
			0	396			313	
			0	405	6		7	
			0	383	5		225	
			0	551			145	*
13 37 9	-39 36.5		0	14870	200	-203	793	
13 37 9	-39 34.3		0	15020	200	-203	793	
13 37 15	-39 36.0		0	14960	200	-203	793	
13 37 20	1 5.6	NGC 5257	0	6820		-74	253	
		ARP 240	0	6697	62		594	
		VV 55	0	6810	70		8	
		KDG 389						

300

R.A. (1950)	DEC. (1950)	NAME	OBS	HEL VEL (C*Z)	ERR	GAL CORR	REF	COMMENTS
13 37 25	1 5.1	NGC 5258	O	6645		-74	253	
		ARP 240	O	6638	26		594	
		VV 55	O	6635	70		8	
		KDG 389						
13 37 25	28 1.8		O	8540		35	714	
13 37 28	43 18.3	MRK 267	O	3600		95	268	
13 37 30	-28 38.0		O	720	90	-176	655	
13 37 37	28 39.2	NGC 5263	O	4932		37	714	
13 37 44	40 59.4	DDO 181	R	200	7	87	492	
13 37 58	26 36.1		O	8331		29	714	
13 38 21	6 32.5		O	8810		-52	895	
13 38 24	26 28.9		O	19630	100	29	806	
13 38 27	26 8.9		O	8875		28	714	
13 38 37	26 44.4		O	22460	100	30	806	
13 38 41	2 2.3		O	8540		-69	895	
13 38 42	54 35.0		R	2024	10	136	582	HOL V
			O	1977	50		582	
13 38 47	-29 39.5	NGC 5264	R	478	10	-178	865	
		DDO 242	R	484	20		492	
13 38 54	30 37.8	MRK 268	O	12300		46	268	SEYF
			O	11700			232	
13 39 19	66 4.5	MRK 269	O	14700		172	308	
13 39 26	27 15.4		O	8589		33	714	
13 39 30	26 37.6	NW	O	22559	100	31	806	
			O	20590	250		696	
			O	22657	29		444	
			O	22486			714	
			O	22288			321	
13 39 31	26 37.3	SE	O	20621		31	714	
		4C 26.41	O	22660	250		696	
			O	20782	100		806	
			O	20786			444	
			O	20719			321	
13 39 39	30 46.3	MRK 67	R	953		47	779	
			O	1080			224	

R.A. (1950)	DEC. (1950)	NAME	OBS	HEL VEL (C*Z)	ERR	GAL CORR	REF	COMMENTS
13 39 41	67 55.5	NGC 5283 MRK 270	O O	2700 2700		178	232 268	SEYF
13 39 48	55 55.3	NGC 5278 MRK 271A ARP 239 KDG 390 VV 19	O O O O	7570 7800 7518 7545	69	141	594 268 253 57	
13 39 51	-47 56.4	NGC 5266	O O	3030 3201	43	-217	574 741	
13 39 52	55 55.5	NGC 5279 MRK 271B KDG 390 ARP 239 VV 19	O O O O	7561 7500 7588 7585	42	141	253 268 57 594	
13 39 53	43 0.8	MRK 272	O	10500		96	306	
13 39 55	35 54.5	NGC 5273 KDG 391A	R O O	1032 1022 1021	20 22	68	610 13 594	
13 40 6	26 27.9		O	19630	100	30	806	
13 40 6	30 4.0	NGC 5275 VV 543	O	1395		45	774	
13 40 6	35 53.0	NGC 5276 KDG 391B	O	5278	44	68	594	
13 40 12	5 19.4	4C 05.57	O	40020		-55	143	
13 40 12	26 44.2		O O	19757 19529	100	31	714 806	
13 40 23	39 54.4	DDO 182	R	663	10	84	492	
13 40 39	26 30.9		O	20008	100	31	806	
13 40 43	-37 57.5		O	(9740)	200	-197	794	
13 40 43	-37 55.5		O	(10790)	200	-197	794	
13 40 58	26 34.8		O	23309	100	31	806	
13 41 1	26 34.7		O	23441	100	31	806	
13 41 24	25 38.5		O	8183		28	714	
13 41 44	25 54.8		O	5802		29	714	

R.A. (1950)	DEC. (1950)	NAME	OBS	HFL VEL (C*Z)	ERR	GAL CORR	REF	COMMENTS
13 42 1	-41 36.8		R	540	5	-204	865	
13 42 24	37 25.0	KDG 393A	O	8002	78	76	908	
13 42 36	37 25.0	KDG 393B	O	7991	26	76	908	
13 42 51	56 8.3	MRK 273	O	11400		143	268	SEYF
		1 ZW 71	O	11150	100		820	
		VV 851	O	12200			535	
13 42 59	27 22.2	MRK 68	R	5194		36	779	
			O	5172	60		261	
13 43 2	41 45.2	NGC 5289	R	2516	15	93	829	
			O	2387	65		644	
13 43 3	4 57.5		O	9263		-54	895	
13 43 12	41 57.9	NGC 5290	R	2583	15	94	829	
			O	2518	57		644	
13 43 34	27 1.5		O	8630		35	714	
13 43 51	29 53.0	MRK 69	O	22830	60	47	261	SEYF
13 44 13	44 6.1	NGC 5296	O	2221	107	102	594	
		KDG 394	O	2212	50		908	
13 44 14	44 5.6		O	25870		102	536	
13 44 18	44 7.0	NGC 5297	O	2581	44	102	594	
		KDG 394	O	2653	70		908	
13 44 21	46 21.4	NGC 5301	R	1523	100	111	544	
			R	1487	100		238	
			O	1705			641	
			O	1702			13	
13 44 24	11 53.0	KDG 395	O	10656	15	-25	908	
13 44 30	-30 6.3	B	O	4110	10	-175	821	KNOT
13 44 30	-30 5.8	A	O	4100	15	-175	821	KNOT
13 44 30	11 53.0	KDG 395	O	10670	10	-25	908	
13 44 30	16 31.0	NGC 5293	R	5790	10	-6	582	
			R	5787			779	
			O	5789	35		582	
13 44 31	-30 13.4	H	O	4680	10	-175	821	KNOT
13 44 34	-30 13.1	G	O	4750	50	-175	821	KNOT

303

13ʰ44ᵐ

R.A. (1950)	DEC. (1950)	NAME	OBS	HEL VEL (C*Z)	ERR	GAL CORR	REF	COMMENTS
13 44 34	-30 12.5	F	O	4645	10	-175	821	KNOT
13 44 35	-30 12.1	E	O	4490	15	-175	821	KNOT
13 44 35	-30 11.4	D	O	4630	50	-175	821	KNOT
13 44 35	-30 9.6	NGC 5291	R	4386	10	-175	821	
			O	4335	25		821	
			O	4366	25		741	
13 44 35	-30 9.0	NGC 5291B	O	3730	50	-175	821	
13 44 37	-30 11.0	C	O	4515	20	-175	821	KNOT
13 44 40	3 54.0		O	6870		-57	895	
13 44 44	25 14.6		O	3188		29	714	
13 44 48	34 8.5	A 1345B	O	4914		64	253	
		KDG 396	O	4968	30		908	
		HOL 541B	O	5144			57	
		VV 317						
13 44 48	60 37.2		R	2055	10	158	860	
13 44 49	-30 41.6	NGC 5292	O	4482	33	-176	741	
13 44 53	34 7.8	A 1345A	O	4450		64	253	
		KDG 396	O	4480	10		908	
		HOL 541A	O	4680			57	
		VV 317						
13 44 55	-30 5.0		O	4420	20	-175	821	
13 45 4	26 47.8		O	18140	65	35	902	
13 45 4	34 24.0	MRK 461	O	4850		66	435	
			O	4800			359	
13 45 6	12 32.3	4C 12.50	O	36520		-22	240	
			O	36420			638	
13 45 21	61 13.3	NGC 5308	O	2039	46	160	644	
			O	2046	75		13	
			O	2035			13	
13 45 36	7 38.0	KDG 398A	O	6956	30	-42	908	
13 45 36	38 30.0	KDG 397B	O	1428	21	82	908	
13 45 36	38 33.0	NGC 5303	O	1285	60	82	644	
		KDG 397A	O	1401	10		908	

304

R.A. (1950)	DEC. (1950)	NAME	OBS	HEL VEL (C*Z)	ERR	GAL CORR	REF	COMMENTS
13 45 39	26 52.5		O	19361	65	36	902	
13 45 42	7 38.0	KDG 398B	O	7072	15	-41	908	
13 45 44	4 11.9	NGC 5300	R	1174	10	-55	752	
13 45 48	-30 34.0		O	5275	32	-175	741	
13 45 58	-30 15.8	NGC 5302	O	3329	35	-174	741	
13 46 0	-35 49.0		O	272		-189	925	
13 46 6	25 26.0	ARP 221 VV 190	O	7300		30	535	
13 46 7	26 29.5		O	18938	65	35	902	
13 46 14	-30 2.8	IC 4329	O	4456	27	-174	741	
13 46 21	-35 48.7		R	330	4	-188	865	
			O	272			728	
13 46 24	27 7.8		O	17397	65	38	902	
13 46 25	31 42.5	MRK 275	R	7884		56	779	
			O	8100			308	
13 46 28	-30 3.7	IC 4329A	O	4751	154	-173	827	SEYF
			O	4870	30		869	
			O	4855	24		741	
			O	4310			376	
			O	4720			574	
13 46 34	26 50.5	4C 26.42	O	18940		37	485	
			O	19019	65		902	
			O	18900			400	
			O	18946			444	
			O	19014			444	
13 46 37	26 9.3		O	17706	65	34	902	
13 46 37	26 51.9		O	18779	65	37	902	
13 46 50	27 2.9		O	20272	65	38	902	
13 46 57	26 53.5		O	17790	65	37	902	
13 46 59	26 60.0		O	19481	65	37	902	
13 47 4	26 42.2		O	17925	65	36	902	
13 47 6	26 44.2		O	18272	65	36	902	

R.A. (1950)	DEC. (1950)	NAME	OBS	HEL VEL (C*Z)	ERR	GAL CORR	REF	COMMENTS
13 47 10	-30 20.0	NGC 5304	O	3723	74	-174	741	
13 47 22	2 20.3		O	9830		-61	895	
13 47 27	39 10.0		R	2264	10	86	860	
13 47 36	60 26.4	NGC 5322	O	1902	75	159	13	
			O	1629	65		644	
13 47 37	27 7.6		O	18398	65	38	902	
13 47 37	40 14.0	NGC 5313	R	2537	25	90	829	
			O	2606			741	
13 47 48	-48 49.0		O	3142	17	-214	724	
13 47 48	-48 48.5	P	O	3099	39	-214	765	*
13 47 48	-48 48.5	C	O	2939	23	-214	765	
13 47 48	-48 48.5	D	O	2882	12	-214	765	
13 47 56	28 31.6	B2	O	21660	250	44	696	
13 48 40	38 15.8	DDO 183	R	189	7	83	492	
			R	206	20		424	
13 48 42	39 49.2	NGC 5326	O	2554		89	741	
13 48 54	71 25.0	7 ZW527 IC 954	O	9000		16	429	
13 49 18	40 27.0	MRK 462	O	2370		92	438	
13 49 30	- 5 48.7	NGC 5324	O	2937	40	-91	741	
13 49 48	2 20.0	NGC 5331A KDG 401A VV 253	O	9995	32	-59	908	
13 49 48	2 21.0	NGC 5331B KDG 401B VV 253	O	9943	20	-59	908	
13 50 3	-28 14.6	NGC 5328	O	4816	50	-166	741	
13 50 3	31 41.5	3CR293	O	13493	40	58	741	
			O	13560			134	
			O	13500			485	
			O	13500			143	
13 50 6	-38 20.0		O	17920	120	-192	655	

R.A. (1950)	DEC. (1950)	NAME	OBS	HEL VEL (C*Z)	ERR	GAL CORR	REF	COMMENTS
13 50 8	-28 14.7	NGC 5330	O	4911	51	-166	741	
13 50 22	38 3.7	NGC 5341	R	3648	10	83	829	
13 50 25	64 37.1	MRK 277	O	1800		172	306	
13 50 42	-31 13.0		O	11390	150	-174	655	
13 50 42	17 35.0		O	7500		2	533	
13 51 6	33 44.3	NGC 5347	R	2386	20	67	829	
			O	2296			741	
13 51 14	40 36.7	NGC 5350	O	2251	28	93	741	
		HOL 555C	O	2357	66		644	
13 51 18	-37 32.0		O	15500	30	-189	655	
13 51 19	-48 15.6	NGC 5333	O	2750		-211	574	
13 51 19	38 9.6	NGC 5351	R	3611	15	84	752	
		HOL 554A	R	3630	15		829	
			O	3875			741	
13 51 20	40 31.8	NGC 5353	O	1990		93	253	
		HOL 555B	O	2298	51		644	
			O	2188	65		13	
13 51 20	40 33.0	NGC 5354	O	2980		93	253	
		HOL 555A	O	2414	50		644	
13 51 21	51 31.5		O	66000		132	717	
13 51 22	72 58.7	MRK 278	O	10800		193	268	
13 51 36	40 35.0	NGC 5355	O	2368	90	94	644	
13 51 46	23 40.5	MRK 662	O	16533		28	874	
			O	16560			557	
13 51 52	69 33.2	MRK 279	O	9050		185	287	SEYF
			O	9600			268	
13 51 54	33 50.0	1 ZW 75	O	13640		68	303	
13 52 6	15 17.0	MRK 1365	O	5495		-5	920	
			O	5700			533	
13 52 48	41 33.7	NGC 5362	O	2262		98	741	
13 52 56	54 8.8	DDO 185	R	139	2	142	676	HOL IV
			R	142	4		891	
			R	141	10		492	
			O	149			13	
			O	126	31		741	

13ʰ53ᵐ

R.A. (1950)	DEC. (1950)	NAME	OBS	HEL VEL (C*Z)	ERR	GAL CORR	REF	COMMENTS
13 53 1	18 2.2	DDO 184	R	965	10	6	492	
13 53 6	-30 5.8	NGC 5357	O	5019	46	-169	741	
13 53 6	58 55.0	NGC 5372	O	1711	25	157	644	
13 53 8	5 13.7	NGC 5360	O	1200		-45	446	
		HOL 557B	O	1174	28		433	
13 53 33	40 42.4	NGC 5371	R	2557		95	744	
			O	2633			13	
			O	2551	40		13	
			O	(2584)	31		644	
13 53 36	5 30.0	NGC 5363	O	1138		-44	13	
			O	1138	50		13	
13 53 36	17 45.0	IC 960A KDG 402A VV 335	O	6523	44	5	908	
13 53 36	17 45.0	IC 960B KDG 402B VV 335	O	6197	64	5	908	
13 53 36	59 45.0	NGC 5376	O	2078	64	159	644	
			O	2064	33		741	
13 53 40	18 36.7	MRK 463A/B	R	14702		9	707	
			O	15140			438	
			O	15250			374	
			O	15200			868	
13 53 40	18 36.7	MRK 463A	O	14931	85	9	831	SFYF
13 53 40	18 36.7	MRK 463B	O	14881	54	9	831	
13 53 42	5 15.6	NGC 5364	R	1545		-44	886	
			O	1393	150		13	
			O	1107	71		789	
13 53 45	38 48.9	MRK 464	O	15140		88	374	SEYF
			O	15300			438	
13 53 56	59 59.2	NGC 5379	O	1728		160	910	
13 54 18	47 28.0	NGC 5377	R	1784	30.	120	885	
			O	1830	100		13	
13 54 24	64 37.0	7 ZW528	O	1568	100	173	820	
13 54 30	29 25.0	NGC 5375	O	2083	50	53	555	
			O	2424	50		813	

308

R.A. (1950)	DEC. (1950)	NAME	OBS	HEL VEL (C*Z)	ERR	GAL CORR	REF	COMMENTS
13 54 30	59 59.0	NGC 5389	O	1836	56	160	644	
13 54 33	32 34.2	4C 32.46	O	14000		65	899	
13 54 42	38 2.0	NGC 5378	O	2968	100	86	644	
13 54 46	-43 41.7	NGC 5365	O	2370		-201	574	
			O	2497	20		741	
13 54 47	-38 47.9		O	10730	90	-190	680	
13 54 47	-38 47.7		O	10890	80	-190	680	
13 54 49	37 51.1	NGC 5380	O	3156	38	85	741	
			O	2849	105		644	
13 55 0	42 5.6	NGC 5383	R	2268	10	101	752	
		MRK 281	R	2264	5		840	
			R	2264	6		730	
			O	2264	20		730	
			R	2250	15		418	
			O	2100			268	
			O	2220			483	
			O	2264			63	
			O	(2265)	50		644	
13 55 0	42 5.6	NGC 5383A	O	2379	20	101	730	
13 55 2	-29 4.3	IC 4351	O	2891	29	-164	152	
			O	2587			741	
13 55 4	29 2.5	MRK 280A	O	11148		52	437	
13 55 5	29 2.1	MRK 280B	O	10929		51	437	
13 55 18	17 45.0	IC 964	O	6700		6	533	
13 55 22	10 7.0		R	6833		-24	786	
13 55 48	-30 49.0		O	7280	30	-169	655	
13 56 0	57 14.0	VV 792	O	1729	50	153	873	
13 56 12	14 24.0		O	24520		-6	282	
13 56 25	37 41.8	NGC 5395/4	R	(3490)		86	914	
13 56 25	37 41.8	NGC 5394	O	3355	15	86	741	
		1 ZW 77	O	3360	30		195	
		ARP 84	O	3510	90		195	
		VV 48W	O	3467	50		873	
		KDG 404	O	3282	30		500	
		HOL 563A	O	3390	30		195	
			O	3420	30		195	
			O	3558	100		13	
			O	3714			748	

13ʰ56ᵐ

R.A. (1950)	DEC. (1950)	NAME	OBS	HEL VEL (C∗Z)	ERR	GAL CORR	REF	COMMENTS
13 56 25	37 41.8		O	3451	12	86	908	
			O	(3558)	100		644	
13 56 30	8 58.0		O	12140		-27	282	
13 56 30	37 40.0	NGC 5395	R	3459	24	86	622	
		ARP 84	O	3542	12		741	
		VV 48E	O	3501	50		873	
		KDG 404	O	3493	10		908	
		HOL 563B	O	3514			748	
13 57 42	5 16.0	NGC 5364	R	1251		-41	780	
13 57 43	4 20.0		O	11690		-45	895	
13 57 45	9 12.5		R	4106		-25	786	
			R	4101	20		955	
13 57 45	28 44.5	B2	O	18800	250	52	696	
13 57 48	13 12.0	VV 339W	O	7142	50	-9	873	
		KDG 405	O	7142	14		908	
13 57 52	27 33.6	4C 27.27	O	46410		48	851	
			O	46500			899	
13 57 54	13 12.0	VV 339E	O	7382	50	-9	873	
		KDG 405	O	7382	15		908	
13 58 0	- 2 37.0		O	7404	59	-72	843	
13 58 0	13 12.0		O	(480)		-9	282	
13 58 3	8 53.5		P	4363		-26	786	
			R	4609	20		955	
13 58 14	39 9.4	NGC 5406	O	5151	40	93	644	
			O	5278	59		741	
13 58 27	-32 49.3	NGC 5398	O	1272	9	-172	741	
13 58 33	30 26.4	B2	O	33030	250	60	696	
13 58 42	41 14.0	NGC 5410 KDG 406A	O	3685	15	100	908	
13 58 48	10 22.0		R	6844	20	-20	939	
13 58 48	41 15.0	KDG 406B	O	3777	15	101	908	
13 58 51	28 54.9	B2	O	24610	250	54	696	
13 58 52	41 26.0		O	3600		101	591	

R.A. (1950)	DEC. (1950)	NAME	OBS	HEL VEL (C*Z)	ERR	GAL CORR	REF	COMMENTS
13 58 54	-40 48.0		O	1920	60	-192	655	
13 58 57	55 24.3	NGC 5422	O	1837	17	149	741	
			O	1765			910	
13 58 59	-11 22.1	PKS	O	7500		-104	189	
13 59 8	-37 45.8		O	(3300)	200	-185	794	
13 59 9	59 34.1	NGC 5430	O	3115	83	161	741	
		MRK 799A/B	O	4398	170		732	DIS
			O	2820	74		644	
13 59 9	59 34.1	MRK 799A	O	3179		161	929	★
13 59 9	59 34.1	MRK 799B	O	3151		161	929	★
13 59 15	37 2.4	MRK 465	O	2700		85	359	
			O	2700			438	
13 59 18	9 43.8	NGC 5409	R	6258	20	-22	955	
			R	6256			786	
13 59 24	9 1.6		R	6071		-25	786	
			O	6276	100		751	
13 59 30	9 11.0	NGC 5411	O	5850	100	-24	751	
13 59 30	34 4.0	NGC 5421A	O	(8049)	84	74	500	
		MRK 665A	O	7915			557	
		KDG 407A	O	8021	40		908	
		ARP 111						
		VV 120						
13 59 30	34 4.0	NGC 5421B	O	7839	40	74	908	
		MRK 665B						
		KDG 407B						
		ARP 111						
		VV 120						
13 59 36	10 10.1	NGC 5414	R	4279		-20	786	
			O	4224	100		751	
13 59 43	9 41.0	NGC 5416	R	6240		-22	786	
			R	6235	20		955	
			O	6187	100		751	
13 59 59	2 30.2	4C 02.39	O	54000		-50	600	
14 0 4	9 19.1		R	6018		-23	786	
			R	6005	20		955	
14 0 6	-30 0.0		O	6920	90	-163	655	

14ʰ0ᵐ

R.A. (1950)	DEC. (1950)	NAME	OBS	HEL VEL (C∗Z)	ERR	GAL CORR	REF	COMMENTS
14 0 14	9 24.1		R	5864		-22	786	
			R	5859	20		955	
14 0 18	-41 10.0	NGC 5408A/B	0	600	30	-192	655	
14 0 18	-41 10.0	NGC 5408A	0	505		-192	313	
14 0 18	-41 8.0	NGC 5408B	0	546		-192	313	
14 0 18	9 35.0	NGC 5423	0	6176	100	-22	751	
14 0 22	9 1.0		R	5872		-24	786	
			R	5865	20		955	
14 0 30	9 40.0	NGC 5424	0	5772	100	-21	751	
14 0 36	9 37.0	NGC 5431	0	5744	100	-21	751	
14 0 36	51 57.0	1 ZW 79	0	12360		138	262	
14 0 42	-33 44.3	NGC 5419	0	4290		-173	317	
		PKS	0	4040	120		355	
			0	4179	57		741	
14 0 43	69 6.9		R	448	10	187	860	
14 0 48	- 5 49.8	NGC 5426/7 ARP 271 VV 21	R	2625	20	-82	455	
14 0 48	- 5 49.8	NGC 5426 HOL 573B	R	2618		-82	744	*
			0	2364			253	
			0	2500	60		8	
14 0 48	- 0 18.0		0	7500		-60	533	
14 0 49	- 5 47.5	NGC 5427 HOL 573A	R	2460		-82	253	
			0	2596	60		8	
14 0 51	16 14.4		0	73200		5	607	
14 0 55	9 41.2	NGC 5434 KDG 410	R	4633		-21	786	
			R	4634	20		955	
14 0 59	9 42.3	KDG 410	R	5642		-21	786	
			R	5637	20		955	
			0	5550	100		751	
14 1 0	49 25.0	NGC 5448	R	2023	20	130	885	
			0	1973			641	
			0	1970	50		13	
14 1 3	69 43.3	MRK 282	0	6300		188	268	

R.A. (1950)			DEC. (1950)		NAME		OBS	HFL VEL (C∗Z)	ERR	GAL CORR	REF	COMMENTS
14	1	6	9	55.0	IC	971	R	3316	10	-20	752	
14	1	12	9	49.0	NGC	5436	O	6613	100	-20	751	
14	1	12	41	50.6	MRK	283	O	10500		104	308	∗
14	1	15	35	22.3	NGC	5444	O	3878		80	281	
					4C	35.32	O	3910			485	
							O	3994	18		741	
14	1	18	9	51.0	NGC	5438	R	6068	20	-20	955	
							O	7066	100		751	DIS
14	1	28	54	35.6	NGC	5457	R	240		147	340	M 101
					ARP	26	R	266			198	M 102
					VV	344	R	239	4		489	PIN NEB
					4C	54.30	R	240	10		295	
							R	238	3		891	
							R	233	4		944	
							R	238			934	
							R	268	5		93	
							R	235			156	
							R	253	3		26	
							R	(247)			69	
							R	249			693	
							R	225			248	
							R	260			296	
							R	247			934	∗
							O	247	30		13	
							O	160	36		72	
							O	300	25		2	
							O	238	4		788	
							O	79	100		555	
14	1	37	33	48.5	MRK	666	O	5700		75	557	
14	1	54	54	33.0	NGC	5461	O	298	30	147	13	HII REG
14	1	56	4	58.0			O	8780		-39	895	
14	2	7	54	36.1	NGC	5462	O	293	2	148	741	HII REG
							O	270			150	
14	2	12	-33	32.0	IC	4366	R	4613	15	-172	752	
							O	4609	50		582	
14	2	18	- 0	24.0	KDG	413	O	(7332)	40	-60	908	
14	2	18	- 0	22.0	KDG	413	O	7273	60	-59	908	
14	2	18	0	23.0			O	7424	20	-57	582	
14	2	24	12	57.0	KDG	414	O	4164	34	-6	908	
					VV	328						

313

R.A. (1950)	DEC. (1950)	NAME	OBS	HEL VEL (C*Z)	ERR	GAL CORR	REF	COMMENTS
14 2 24	12 58.0	KDG 414	O	4124	34	-6	908	
		VV 328	O	4200			533	
14 2 33	21 52.3	MRK 667	O	4915		29	557	
14 2 38	9 34.7		R	4600		-20	786	
14 2 42	54 38.0	NGC 5471	O	281	3	148	741	HII REG
		VV 394W	O	297	50		873	
			O	283	50		873	
			O	289			650	
14 2 56	3 17.0		O	5100		-45	895	
14 3 0	55 8.0	NGC 5473	O	2141		150	13	
			O	1976	50		13	
14 3 8	9 9.2		R	7044		-21	786	
14 3 16	53 54.1	NGC 5474	R	280		146	171	
		VV 344B	R	288			693	
			R	279			934	
			R	275	2		891	
			R	275	5		862	
			R	280	19		216	
			R	273			744	
			P	247			198	
			R	275	2		676	
			O	234	4		741	
			O	247			13	
14 3 28	9 15.9		R	7001		-20	786	
14 3 42	2 23.0		O	7548		-48	683	
14 3 47	54 41.9	NGC 5477	O	343	15	149	741	
		DDO 186	R	312	2		676	
14 3 58	- 5 12.8	NGC 5468	R	2845		-77	744	
			O	2856			13	
14 4 9	69 22.5	MRK 284	O	9305	45	188	382	
14 4 11	-29 46.8	NGC 5464	O	2686	15	-160	741	
14 4 17	- 5 13.3	NGC 5472	R	2910	50	-76	885	
14 4 23	9 33.5		R	7208		-18	786	
14 4 24	8 28.0		O	13790		-23	282	
14 4 30	50 57.9	NGC 5480	R	1860	20	137	829	
		KDG 416	R	1850			744	
		HOL 588B	O	1787			253	
			O	1870			57	
			O	2028	40		594	

R.A. (1950)	DEC. (1950)	NAME	OBS	HEL. VEL (C*Z)	ERR	GAL CORR	REF	COMMENTS
14 4 31	6 28.5		O	7410		-31	895	
14 4 46	28 41.6	MRK 668	O	22970		57	318	SEYF
		OQ 208	O	23835			557	
14 4 51	50 57.8	NGC 5481	O	2092		137	253	
		KDG 416	O	2338	102		594	
		HOL 588A	O	2175			57	
14 4 54	15 22.0	A	O	7415	56	4	594	
14 4 54	15 22.0	B	O	12017	41	4	594	
14 5 8	-33 3.9	IC 4375	O	4360		-168	574	
14 5 30	10 4.0		O	26080		-15	282	
14 5 30	55 14.0	NGC 5485	O	1985	50	151	13	
14 5 42	55 20.3	NGC 5486	R	1383	10	152	829	
			R	1813	50		782	
			O	1347	21		741	
14 5 54	- 4 48.0	NGC 5493	O	2627	75	-74	13	
14 6 24	49 5.0	1 ZW 81	O	15480		132	262	
14 6 30	-29 44.0		O	5670	180	-158	655	
14 7 19	-43 5.2	NGC 5483	R	1773		-192	671	
			R	1773	10		752	
			O	1660	100		844	
			O	1830	25		741	
14 7 24	- 1 1.2	4C 01.32	O	7520		-58	166	
14 7 36	17 56.0	IC 982/3	O	5053	50	17	141	
		ARP 117						
14 7 40	71 54.2	MRK 285	O	10200		195	306	
14 8 5	-39 52.2		O	8619	85	-184	680	
14 8 48	-45 51.0	NGC 5489	O	2970		-197	574	
14 9 0	- 0 56.0	NGC 5496	R	1527		-56	363	
14 9 18	-65 6.3		R	439	2	-224	636	CIR GAL
			O	400	18		636	
14 9 29	-30 24.8	NGC 5494	O	2638		-158	925	
14 9 30	52 26.0		O	73100		145	37	F G

14ʰ9ᵐ

R.A. (1950)	DEC. (1950)	NAME	OBS	HEL VEL (C*Z)	ERR	GAL CORR	REF	COMMENTS
14 9 33	52 26.2	3CR295	O	138300	60	145	37	
			O	138000	999		292	
14 9 48	52 35.0		O	8733	75	145	13	
14 9 52	16 4.6	NGC 5504	R	5253	10	11	752	
14 9 54	38 25.0		O	6229	40	97	813	
14 10 22	34 46.9	MRK 467	O	9480		84	435	
14 10 30	45 55.0	KDG 418A	O	8051	15	124	594	
			O	8053	45		908	
14 10 36	45 55.0	KDG 418B	O	8273	24	124	594	
			O	8333	45		908	
14 10 39	- 2 58.5	NGC 5506	O	1750		-63	253	SEYF
		KDG 419	O	1845	29		908	
		HOL 604A	O	2205			748	
			O	1854	20		737	
			O	1814	50		601	
			O	1825	174		945	
			O	2040			57	
14 10 40	29 28.2		O	76100	300	65	583	
14 10 40	29 28.5		O	66500	300	65	583	
14 10 42	- 2 54.9	NGC 5507	O	2220		-62	253	
		KDG 419	O	2016	30		908	
		HOL 604B	O	1893			748	
			O	1729			57	
14 11 10	7 53.6	NGC 5514A	O	7219		-19	748	
		KDG 420	O	7375	40		908	
		VV 70						
14 11 11	7 53.5	NGC 5514B	O	7419		-19	748	
		KDG 420	O	7511	28		908	
		VV 70						
14 12 0	3 25.0	IC 988	O	7088	73	-36	594	
14 12 18	3 21.0	IC 989	O	7490	34	-36	594	
14 12 30	-35 17.0		O	4050	60	-169	655	
14 12 35	-47 53.0	NGC 5516	O	4460		-199	574	
14 12 36	25 33.0	NGC 5523	R	1047		51	744	
			R	1040	5		914	
			R	760			216	

R.A. (1950)	DEC. (1950)	NAME	OBS	HEL VEL (C★Z)	ERR	GAL CORR	REF	COMMENTS
14 12 54	4 38.0	NGC 5521	O	12300		-31	533	
14 13 13	34 45.3	MRK 671	O	24560		86	557	
14 13 18	16 47.0	DDO 188	R	2277	10	17	492	
14 13 30	36 27.0	NGC 5529	R	2882		93	744	
14 13 38	23 17.1	DDO 187	R	153	5	43	492	
14 13 46	41 13.2	MRK 468	O	12300		110	359	
			O	(12440)			435	
14 14 0	35 34.5	NGC 5533	R	3864	10	90	752	
			O	3781	60		13	
14 14 18	39 44.0	NGC 5536	O	5142	50	105	594	
14 14 24	9 54.0		O	25180		-9	282	
14 14 24	39 49.0	NGC 5541	O	6000		105	476	
			O	7477	41		594	DIS
			O	1764	64		644	DIS
14 14 26	11 2.3	NGC 5532	O	7117	39	-4	741	
		3CR296	O	7114			227	
			O	7110			134	
14 14 57	36 48.2	NGC 5544	R	3072	17	95	745	
		KDG 422	O	3006	70		8	
		ARP 199	O	3172			253	
		VV 210	O	3127	53		594	
14 15 0	36 48.4	NGC 5545	O	3182		95	253	
		KDG 422	O	3016	70		8	
		ARP 199	O	3261	68		594	
		VV 210						
14 15 1	- 7 11.2	NGC 5534	O	2600		-75	741	
		MRK 1379	O	2805			920	
		VV 615	O	2590	50		873	
14 15 6	2 16.0		O	16340	150	-38	843	
14 15 6	27 5.3	MRK 673A	O	10767	32	59	831	
14 15 6	27 5.3	MRK 673B	O	10933	54	59	831	
14 15 16	56 5.7		R	1880	10	158	860	
14 15 18	-43 9.7	NGC 5530	R	1196		-187	671	
			O	491	49		152	
			O	0			574	★

R.A. (1950)	DEC. (1950)	NAME	OBS	HFL VEL (C*Z)	ERR	GAL CORR	REF	COMMENTS
14 15 24	7 47.0	NGC 5542	O	0		-16	533	
14 15 42	25 22.0	NGC 5548	R	5147		53	779	SEYF
			R	5204			707	
			O	5120	60		112	
			O	4980			910	
			O	4980			287	
			O	4930	50		13	
14 15 48	11 27.0	IC 993	O	18600		-1	476	
14 16 6	34 35.0	MRK 469	O	20660		88	438	
14 16 24	36 43.0	NGC 5557	O	3195	60	96	13	
14 16 25	22 2.9	MRK 674	O	2510		40	557	
		VV 557	O	2500			535	
14 16 30	-26 25.0		O	6683	20	-141	582	
14 16 54	26 32.0	IC 4405	O	11025	14	58	908	
		KDG 423A						
14 17 0	26 32.0	KDG 423B	O	11093	14	58	908	
14 17 3	-19 14.6	PKS	O	35850		-117	143	
			O	35970	150		709	
14 17 34	4 13.3	NGC 5560	R	1711	15	-29	829	
		ARP 286						
		HOL 630B						
14 17 39	-29 1.0	NGC 5556	R	1384	10	-148	492	
		DDO 243						
14 17 48	4 11.0	NGC 5566	O	1455	150	-29	13	
		ARP 286	O	1643	45		72	
		HOL 630A						
14 18 12	56 57.5	NGC 5585	R	317		162	693	
			R	305			171	
			R	303	2		676	
			R	298	25		183	
			R	304			744	
			R	307	20		216	
			O	305			641	
			O	304			13	
14 18 24	3 28.0	NGC 5574	O	1716	50	-31	13	
		HOL 632B						
14 18 24	22 10.0		O	4679	17	42	813	

R.A. (1950)	DEC. (1950)	NAME	OBS	HEL VEL (C*Z)	ERR	GAL CORR	REF	COMMENTS
14 18 30	3 30.0	NGC 5576 HOL 632A	O	1528	100	-31	13	
14 18 42	3 40.0	NGC 5577	R	1485		-30	744	
14 18 46	71 48.8	NGC 5607 MRK 286	O	7800		198	268	
14 18 54	12 44.0		O	28390		5	282	
14 19 7	41 58.6	3CR299	O	110100	300	116	405	
14 19 21	2 15.8	8 ZW	O	7410		-35	895	
14 19 50	- 0 9.6	NGC 5584	R	1635		-44	744	
			O	1498	22		813	
			O	1651	36		554	
14 20 0	46 57.0	1 ZW 84	O	8101		133	262	
14 20 12	13 57.0	NGC 5591A KDG 424A	O	7625	12	11	908	
14 20 12	13 57.0	NGC 5591B KDG 424B	O	7631	10	11	908	
14 20 24	48 43.0	1 ZW 85NE	O	21663		139	262	
			O	21527			321	
			O	21694			444	
14 20 24	48 43.0	1 ZW 85SW	O	21298		139	262	
			O	21563			321	
			O	21742			444	
14 20 24	48 48.0		O	21277		139	321	
			O	21497	93		444	
14 20 25	37 20.9	NGC 5596 MRK 470	O	4500		101	359	
14 20 37	45 36.7	DDO 189	R	690	10	129	492	
14 20 40	19 49.2	3CR300	O	80940		35	783	
14 20 47	33 4.6	MRK 471	R	10256		85	707	SEYF
			O	10200			359	
14 21 2	-28 27.9	NGC 5592	O	13511		-144	741	
14 21 18	42 0.0	NGC 5608	O	736	30	117	813	
14 21 26	14 51.8	NGC 5600	O	2363	10	16	741	
			O	2700			533	
			O	2243	37		789	

14ʰ21ᵐ

R.A. (1950)	DEC. (1950)	NAME	OBS	HEL VEL (C*Z)	ERR	GAL CORR	REF	COMMENTS
14 21 28	-16 29.9	NGC 5595	O	2500		-104	535	
		VV 530	O	2672	18		741	
		HOL 638A	O	2716	50		789	
14 21 42	-16 32.0	NGC 5597	O	2579	50	-103	789	
		HOL 638B	O	2677			741	
			O	2930	130		732	
14 21 57	33 16.5	NGC 5611	O	1924		87	910	
14 22 0	27 55.0	VV 739	O	10109	50	67	873	
14 22 0	35 5.1	NGC 5614	R	3899	64	94	885	
		VV 77	O	3872	75		13	
		ARP 178						
14 22 6	- 2 59.0	NGC 5604	R	2751	10	-53	752	
14 22 25	-12 56.3	NGC 5605	R	3363	15	-90	752	
14 22 26	26 51.3	PKS	O	10310		63	244	
			O	11030			485	
14 22 48	44 45.0	DDO 190	R	153	7	128	492	
14 23 12	14 3.0		O	5520		14	282	
14 23 12	32 42.0		O	3900		86	476	
14 24 24	- 2 3.0	NGC 5618	R	7148	10	-47	582	
14 24 48	-46 4.8		R	393	6	-188	865	
14 24 48	63 31.0		O	41797		182	444	★
14 24 54	20 3.0	MRK 813	O	39261		39	630	★
14 25 0	33 28.6	NGC 5623	O	3456		90	910	
14 25 0	56 48.0	NGC 5631	R	(1950)		165	645	
			O	1979	60		13	
14 25 18	12 59.0		O	(0)		12	282	
14 25 24	13 9.0		O	7800		12	282	
14 25 24	13 40.0		O	5760		14	282	
14 25 36	- 1 27.0	MRK 1382	O	9184		-44	920	
14 25 36	1 13.0	IC 1011	O	7500		-34	533	
14 25 36	21 32.0	VV 371	O	1043	50	45	873	

R.A. (1950)	DEC. (1950)	NAME	OBS	HEL VEL (C*Z)	ERR	GAL CORR	REF	COMMENTS
14 25 36	46 22.0	NGC 5633	R	2332	45	134	622	
		1 ZW 89	O	2350			641	
			O	2390			13	
			O	2316	50		13	
			O	2350	38		644	
14 25 58	- 1 10.8	3CR300.1	O	92340	300	-43	585	
14 26 18	27 29.0		O	3819	174	69	246	
14 26 30	1 31.0	MRK 1383	O	26502		-32	920	SEYF
14 26 34	27 28.4	MRK 682	O	4370		69	557	
14 25 54	26 3.0		O	4540	200	63	843	
14 26 56	36 9.8	MRK 472	O	4500		101	359	
			O	4500			435	
14 27 5	28 2.6	NGC 5641	O	4467	58	71	741	
14 27 6	-34 1.0		O	3014	50	-156	582	
14 27 6	3 27.0	NGC 5638	O	1677	50	-24	13	
			O	1733	61		594	
14 27 28	-43 57.2	NGC 5643	O	1180	100	-181	844	
14 27 52	4 53.3	8 ZW	O	7740		-17	895	
14 27 54	31 26.0	NGC 5653	R	3589	43	85	914	
			O	3557			13	
14 27 58	44 40.0	DDO 192	R	2745	25	130	492	
14 28 0	49 52.0	NGC 5660A	O	3730	82	147	789	
14 28 3	49 50.7	NGC 5660	R	2323		147	693	
			O	2314	42		644	
			O	2342	78		741	
			O	2286	53		789	
14 28 11	7 29.9	NGC 5645	R	1363	8	-7	914	
			O	1455	89		741	
14 28 12	-78 9.8	NGC 5612	O	2764	59	-226	741	
14 28 30	-43 12.0	IC 4444	R	1960	15	-179	752	
			O	1934	24		724	
			O	1979	14		741	
			O	2010	100		844	
14 28 53	20 30.5	MRK 684	O	(13500)		44	557	

14ʰ28ᵐ

R.A. (1950)	DEC. (1950)	NAME	OBS	HEL VEL (C∗Z)	ERR	GAL CORR	REF	COMMENTS
14 28 57	27 27.5	MRK 685	R	4460	20	71	366	
			O	4512	36		246	
			O	4495			557	
14 29 28	-43 57.2	NGC 5643	R	1196		-180	671	
			O	1178	8		741	
			O	1185			145	
			O	1145			574	
14 29 36	6 28.0		O	2409		-9	683	
14 29 46	50 10.8	NGC 5673	R	2082	15	149	829	
			O	2140	20		644	
14 29 48	5 24.0		O	2414		-14	683	∗
14 29 58	8 18.0	NGC 5665	O	2220	60	-2	195	
		ARP 49	O	2249	21		148	
			O	2310	30		195	
14 30 0	8 17.9	NGC 5665A	O	2130	30	-2	195	KNOT
			O	2280	30		195	
14 30 17	10 6.6	NGC 5669	R	1371	10	4	829	
			R	1372	5		914	
			R	1371	10		752	
			R	1371			744	
			O	1338	45		789	
14 30 24	3 8.0		O	1522		-22	683	
14 30 27	25 8.4	4C 25.46	O	24300	250	63	696	
			O	24390			400	
14 30 30	31 53.0	NGC 5672	O	3701	65	88	13	
14 30 36	31 47.0		O	39046	50	88	13	
14 30 36	31 49.0		O	39496	65	88	13	
14 30 38	58 8.4	NGC 5678	O	2300		171	13	
14 30 43	10 43.8	NGC 5666	R	2221	10	7	829	
14 30 43	50 7.8	IC 1029	R	2381	10	149	829	
			O	2377	46		644	
14 30 52	57 4.0	MRK 473	O	12900		168	359	
14 30 54	4 40.0	NGC 5668	R	1583		-15	283	
			R	1560			216	
			R	1585	4		914	
			R	1581	7		251	
			R	1583			744	
			R	1577	30		116	
			O	1780	50		13	

322

R.A. (1950)	DEC. (1950)	NAME	OBS	HEL VEL (C*Z)	FPR	GAL CORR	REF	COMMENTS
14 30 54	4 40.0	NGC 5668	O	1665		-15	13	
			O	1550	25		283	
14 31 0	49 41.0	NGC 5676	R	2196	50	148	238	
			R	2225			363	
			R	2117			744	
			R	2122			693	
			O	2157	20		644	
			O	2244			13	
14 31 24	5 40.0	NGC 5674	R	1508	5	-11	752	
14 31 30	4 0.0		O	8672	150	-18	843	
14 31 41	52 59.5	MRK 816	O	26457		158	874	
14 31 48	40 17.0	KDG 426A	O	7647	17	118	908	
14 31 48	40 18.0	KDG 426B	O	7803	18	118	908	
14 32 9	4 28.9		R	1690	10	-15	860	
14 32 20	55 21.50		O	(22634)	100	165	633	
14 32 22	-45 44.5	NGC 5670	O	2920		-183	574	
14 32 31	34 30.1		R	3035	10	106	860	
14 32 36	55 17.35		O	(28240)	100	164	633	
14 32 39	5 34.7	NGC 5679A/B	R	1641	25	-10	829	
		VV 458	O	3110			774	DIS
		ARP 274						
		KDG 427A/B						
14 32 39	5 34.7	KDG 427A	O	7457	20	-10	908	
14 32 39	5 34.7	KDG 427B	O	8641	30	-10	908	
14 32 54	0 32.0		O	10200		-30	533	
14 32 54	5 29.4		R	1634	10	-11	860	
14 32 59	48 53.4	NGC 5682	O	2288	38	146	148	
		HOL 663A	O	2230			361	
14 33 6	13 7.0	KDG 428	O	1932	15	19	908	
14 33 6	48 52.8	NGC 5683	O	12320		147	361	SEYF
		MRK 474	O	10650			435	
		HOL 663B	O	(2208)	90		148	DIS
14 33 13	59 33.3	DDO 193	R	1920	15	176	492	

R.A. (1950)	DEC. (1950)	NAME	OBS	HEL. VEL (C*Z)	ERR	GAL CORR	REF	COMMENTS
14 33 18	13 23.0	KDG 428	O	1736	30	20	908	
14 33 18	54 42.0	NGC 5687	O	2119	75	163	13	
14 33 20	55 18.91		O	40487	100	165	633	
14 33 29	55 24.85		O	41668	100	165	633	
14 33 30	55 24.85		O	42136	100	165	633	
14 33 35	55 18.11	A	O	42117	100	165	633	
14 33 35	55 18.11	B	O	50425	100	165	633	
14 33 36	55 20.25		O	41058	100	165	633	
14 33 37	55 16.82		O	40893	100	165	633	
14 33 40	55 19.80		O	51277	100	165	633	
14 33 44	48 57.6	NGC 5689	O O	2205 2029	50 33	147	13 644	
14 33 51	55 17.65		O	23162	100	165	633	
14 33 55	55 20.90	4C 55.29	O	41962	100	165	633	*
14 33 56	57 28.4	DDO 194	R R	222 215	10 5	171	492 676	
14 34 5	55 21.18		O	41552	100	165	633	
14 34 6	55 23.70		O	40735	100	165	633	
14 34 12	-78 36.0	IC 4448	O	4580	100	-225	844	
14 34 12	55 23.22		O	42052	100	166	633	
14 34 13	2 36.3	8 ZW	O	6840		-21	895	
14 34 26	48 48.2	NGC 5693	R	2276	15	147	829	
14 34 26	55 27.99		O	42040	100	166	633	
14 34 45	-36 39.5	IC 4464	O	4490		-158	574	
14 34 48	3 16.0		O	1518		-18	683	
14 34 54	59 1.0	MRK 817	O	9275		175	630	SEYF
14 34 56	25 1.6		O	25840		66	726	
14 34 56	55 27.27		O	43500	100	166	633	

R.A. (1950)	DEC. (1950)	NAME	OBS	HEL VEL (C*Z)	ERR	GAL CORR	REF	COMMENTS
14 34 56	55 31.28		O	22732	100	166	633	
14 35 1	24 58.5		O	26140		66	726	
14 35 15	2 30.4	NGC 5690	O	(2190)	220	-20	555	
			O	1778	20		813	
14 35 19	25 3.5		O	27160		67	726	
14 35 20	- 0 10.9	NGC 5691	O	1876		-31	741	
14 35 21	36 47.2	NGC 5695 MRK 686	O	4209		109	874	
14 35 24	25 0.4		O	26920		66	726	
14 35 34	25 4.4		O	25960		67	726	
14 35 42	48 50.3		R	2306	10	148	860	
14 35 48	3 37.0	NGC 5692	O	1800		-15	533	
14 35 48	3 51.4		O	67200		-14	717	
14 35 51	3 53.3	PKS	O	67200		-14	717	
14 36 14	- 8 24.9	DDO 195	R	1823	15	-62	492	
14 36 32	3 9.7		R	1569	10	-17	860	
14 36 41	5 34.8	NGC 5701	R	1505		-7	693	
			R	1496	15		885	
			O	1586	22		741	
14 36 54	46 42.0	NGC 5724	O	5793	150	142	843	
14 37 3	37 1.1	MRK 475	O	(550)		111	435	
14 37 38	- 0 4.6	NGC 5713	R	1862	10	-28	238	
			R	1910	20		544	
			R	1878			693	
			R	1801	25		183	
			R	1908			744	
			O	1870	100		13	
			O	1965			13	
			O	1865	25		283	
14 37 48	3 0.0		O	8716		-16	683	
14 38 12	3 4.0	NGC 5718	O	7958	150	-15	843	
14 38 24	34 12.0	NGC 5727	O	1523	181	102	813	

R.A. (1950)	DEC. (1950)	NAME	OBS	HEL VEL (C*Z)	ERR	GAL CORR	REF	COMMENTS
14 38 53	28 50.0		O	74300	300	83	583	
14 38 54	28 50.0		O	74000	300	83	583	
14 38 56	28 50.2		O	42200		83	583	
14 39 0	53 42.4	1 ZW 92	O	11470	224	163	123	* SEYF
			O	11280	300		819	
			O	11450	240		348	
14 39 3	53 42.9	MRK 477 1 ZW 92	O	11400		163	359	
14 39 9	- 1 35.9		R	1833	10	-33	860	
14 39 24	44 42.0	KDG 432	O	3298	30	137	908	
14 39 37	-17 2.4	NGC 5728	R	2813	100	-91	885	
			O	2970	16		741	
			O	2976	20		732	
			O	2800			935	
14 39 41	44 43.6	KDG 432	R	3250	10	138	752	
			O	3261	38		908	
14 39 55	59 30.7		R	2319	10	179	860	
14 40 6	35 39.0	MRK 478	O	23700		108	388	SEYF
			O	23426			447	
			O	23700			358	
14 40 24	28 56.3	NGC 5735	R	3744	15	85	752	
14 40 34	42 3.3	NGC 5739	O	5608	47	130	741	
14 41 11	31 23.4		O	72200	300	94	583	
14 41 25	52 14.3	3CR303	O	42570	300	161	585	
			O	42319	30		741	
14 41 30	16 41.0	MRK 1387	O	16270		40	920	
14 41 52	1 53.4	NGC 5740 KDG 434A	R	1567	10	-17	829	
			O	1802			741	
14 41 54	26 13.9	B2	O	18540	250	76	696	
14 42 0	12 21.0	IC 4493A KDG 435A	O	8562	70	23	908	
14 42 0	12 21.0	IC 4493B KDG 435B	O	8923	10	23	908	

R.A. (1950)	DEC. (1950)	NAME	OBS	HEL VEL (C*Z)	ERR	GAL CORR	REF	COMMENTS
14 42 8	31 28.2		O	70100	300	95	583	
14 42 23	2 9.9	NGC 5746	R	1710		-15	934	
		KDG 434B	R	1724	10		829	
			O	1964	54		72	
			O	1789	40		13	
			O	1882			13	
14 42 30	50 48.0		O	21350		157	303	
14 42 54	8 4.3	DDO 196	R	1691	5	7	492	
14 43 18	31 38.6	VV 842S	R	1521	10	97	860	
14 43 18	31 39.0	VV 842N	O	1751	50	97	873	
14 43 18	51 35.0	1 ZW 96 VV 517	O	26997		160	262	
14 43 37	- 0 0.8	NGC 5750	O	2023		-23	741	
14 43 38	17 51.0	4C 17.60	O	19580		46	228	
14 43 48	8 42.0	VV 109	O	10470		10	77	
14 43 53	77 20.1	3CR303.1	O	80000		212	946	
14 44 0	51 47.0	VV 713	O	9249	100	161	556	
			O	(9400)			776	
14 44 6	50 38.0	IC 1056 IC 1057	O	4014	15	158	813	
14 44 48	-14 38.4	NGC 5756	O	2183		-77	741	
14 44 54	13 40.0	NGC 5759	O	8400		31	533	
14 44 57	-18 52.1	NGC 5757	O	2771		-93	741	
14 45 12	-14 4.0		R	2045	10	-75	752	
14 46 33	20 38.2	4C 20.34	O	76210		59	400	
14 46 48	- 9 57.7	DDO 197	R	1849	20	-58	492	
		ARP 261	R	1890	30		455	
		VV 140	O	1880			77	
14 47 18	27 59.2	B2	O	9080	250	87	696	*
14 47 49	77 8.8	3CR305.1	O	136710		213	848	
14 48 0	26 23.0		O	35084	60	81	13	

14ʰ48ᵐ

R.A. (1950)	DEC. (1950)	NAME	OBS	HEL VEL (C⋆Z)	ERR	GAL CORR	REF	COMMENTS
14 48 0	26 23.1		O	35506	60	81	13	
14 48 12	11 37.0		O	2074		26	683	
14 48 18	63 28.6	IC 1065	O	12590		191	177	
		3CR305	O	12373	53		741	
			O	12300	42		134	
14 48 24	5 19.0	NGC 5765A	O	8340	35	1	908	
		KDG 437A						
14 48 24	5 19.0	NGC 5765B	O	8204	19	1	908	
		KDG 437B						
14 48 48	17 24.0		O	13581	20	48	582	
14 48 54	35 46.7	2 ZW 70	R	1210		115	679	
		VV 324B	R	1209	50		389	
		KDG 438A	R	1219			893	
			O	1121			262	
			O	1110			303	
			O	1167	10		336	
14 49 0	9 31.0	KDG 439A	O	8456	110	18	567	
		ARP 173	O	8725	38		908	
14 49 0	9 32.0	KDG 439B	O	8871	25	18	908	
		ARP 173	O	8700			535	*
			O	8776	110		567	
14 49 13	35 44.8	2 ZW 71	R	1255		115	679	
		VV 324	R	1238			389	
		KDG 438B	R	1257			893	
			O	1205			262	
			O	1200			303	
14 49 45	40 48.2	NGC 5772	O	4877		132	910	
			O	4926	47		813	
14 50 2	17 20.7	MC 3	O	12000		49	465	
14 50 24	28 10.3	B2	O	37820	250	90	696	
14 50 50	74 1.7	MRK 288	O	7500		210	308	
14 51 12	3 47.1	NGC 5774	R	1680	50	-1	216	
		KDG 440	R	1575			744	
		HOL 685B	R	1562			893	
			O	1535			748	
			O	1585			57	
			O	1542			253	
14 51 27	3 44.9	NGC 5775	R	1707	21	-1	914	
		KDG 440	R	1677			893	
		HOL 685A	O	1582			253	
			O	1545			57	

R.A. (1950)	DEC. (1950)	NAME	OBS	HEL VEL (C*Z)	ERR	GAL CORR	REF	COMMENTS
14 51 27	3 44.9	KDG 440	O	1575		-1	748	
14 52 3	16 33.7	3C 306	O	13670	200	48	350	
14 52 25	- 4 8.8	3CR306.1	O	133540		-31	946	
14 52 30	18 18.0	IC 1075	O	6259	150	55	567	
		KDG 444	O	6131	20		908	
14 52 36	18 14.0	IC 1076	R	6045		55	779	
		MRK 479	O	6339	100		567	
		KDG 444	O	6091	20		908	
			O	6135			437	
14 52 54	31 1.6		R	1727	10	102	860	
15 53 15	61 14.7		R	622	10	213	860	
14 53 24	42 42.0	NGC 5787 1 ZW 98	O	5453		140	262	
14 53 32	12 3.7	KDG 445A	O	9768		32	748	
		MC 2	O	9600			465	
14 53 33	12 4.6	KDG 445B	O	8768		32	748	
14 54 29	30 26.0	NGC 5789	R	1800	15	101	829	
14 54 30	-42 56.1		O	4555	70	-162	603	
14 54 45	49 53.8	NGC 5797	O	3922		162	910	
14 54 48	24 48.4	KDG 446	O	10019		81	748	
		ARP 302	O	9758	13		908	
		VV 340						
14 54 48	24 49.1	KDG 446	O	10219		81	748	
		ARP 302	O	10086	20		908	
		VV 340						
14 55 27	49 52.1	NGC 5804	O	4097		162	910	
14 55 32	30 10.1	NGC 5798	R	1787	10	101	829	
14 55 46	28 44.3	4C 28.38	O	42190	250	96	696	
14 55 48	- 6 35.0		R	7598	15	-38	582	
14 55 48	- 0 53.4	NGC 5792	R	1929		-16	934	
			O	1985	34		148	
14 55 48	- 6 35.0		O	7606	50	-38	582	

R.A. (1950)	DEC. (1950)	NAME	OBS	HEL VEL (C*Z)	ERR	GAL CORR	REF	COMMENTS
14 55 50	-19 3.9	NGC 5791	O	3339	29	-84	741	
14 56 36	-16 25.4	NGC 5796	O	2946	12	-74	741	
14 56 37	-16 29.9	NGC 5793	O	3521	72	-74	741	
14 56 54	53 26.0	1 ZW 99	O	9275		172	262	
14 56 55	59 4.4		R	2267	10	185	860	
14 57 12	54 5.0	NGC 5820 ARP 136	O	3269	60	174	13	
14 57 28	2 5.3	NGC 5806	R	1353	10	-3	829	
			R	1369	8		914	
			O	1301	65		13	
14 57 30	33 2.4		R	2584	10	112	860	
14 57 43	-48 5.6		R	586	4	-173	865	
14 57 48	42 13.0	1 ZW 97	O	2506		142	262	
14 57 54	1 48.0	NGC 5811 KDG 450A	O	1527	30	-3	908	
14 57 54	71 53.0	NGC 5832	O	390		208	272	
14 58 0	83 44.0	KDG 451	O	3901	30	220	908	
14 58 18	- 7 16.0	NGC 5812	O	2066	50	-38	13	
14 58 30	17 9.0	MRK 1391	O	9084		56	920	
14 58 42	1 54.0	NGC 5813	O	1882	65	-2	13	
14 59 41	52 47.5	DDO 198	R	2422	10	172	492	
15 0 0	83 44.0	KDG 451	O	3917	25	221	908	
15 0 20	-72 14.0		O	3150	200	-215	794	
15 0 30	23 32.0	NGC 5829	R	5684	10	81	582	
		VV 7	R	5600			858	
		ARP 42	O	5689	20		582	
15 1 36	1 24.0	NGC 5831	O	1684	50	-2	13	
15 1 36	10 38.0	MRK 841	O	10897		33	630	SEYF
15 1 54	42 53.0	IC 1090	R	4936		146	389	
		1 ZW101	R	4907	28		622	
			O	4863			262	
			O	4800			533	
			O	4840			389	

R.A. (1950)			DEC. (1950)		NAME	OBS	HEL VEL (C*Z)	ERR	GAL CORR	REF	COMMENTS
15	2	47	26	12.6	A	O	16051	41	92	444	
					3CR310	O	16200	600		824	*
						O	16190			120	
15	2	47	26	12.6	B	O	16717		92	444	
15	2	54	2	18.0	NGC 5838	O	1427	50	2	13	
15	3	8	16	33.8		O	13070		57	255	
15	3	28	1	46.9	NGC 5845	O	1785	53	0	554	
15	3	36	12	55.0	KDG 452A/B	O	16800		44	476	
15	3	36	12	55.0	KDG 452B	O	6315	24	44	908	
15	3	36	12	56.0	KDG 452A	O	6851	37	44	908	
15	3	56	1	47.8	NGC 5846A/B	O	1774		1	13	
						O	1723	12		938	
15	3	56	1	47.8	NGC 5846A	R	1826	50	1	642	
						O	2250	20		938	
						O	1795	55		594	
						O	1768	50		13	
15	3	56	1	47.1	NGC 5846B	O	1936	61	1	594	
						O	2278	40		13	
						O	2321			13	
15	4	30	-75	49.0		O	2506	40	-217	599	
15	4	30	13	4.0	NGC 5851	O	16389	78	45	594	
15	5	0	1	44.2	NGC 5850	R	2541		2	693	
						O	2476			13	
						O	2319	50		13	
15	4	36	13	3.0	NGC 5852	O	16351	72	45	594	
15	4	42	42	50.0	NGC 5860A/B KDG 454	O	5400		148	358	
15	4	42	42	50.0	NGC 5860A MRK 480A	O	5460	90	148	567	
						O	5417	71		830	
15	4	42	42	50.0	NGC 5860B MRK 480B	O	5530	90	148	567	
						O	5368	80		830	
15	4	49	-36	8.3	NGC 5843	O	3970		-135	574	
15	5	8	55	57.2	NGC 5866	O	740	40	182	13	
						O	650	50		1	
						O	850			13	
						O	970			444	
						O	950			106	

15ʰ5ᵐ

R.A. (1950)	DEC. (1950)	NAME	OBS	HEL VEL (C*Z)	ERR	GAL CORR	REF	COMMENTS
15 5 8	55 57.2		O	692	42	182	149	
			O	725	30		346	
15 5 12	19 47.0	NGC 5857	O	4602	43	71	594	
		KDG 455	O	4714			748	
			O	4616	150		13	
			O	4721			13	
15 5 18	2 45.0	NGC 5854	O	1626	65	6	13	
15 5 19	19 46.4	NGC 5859	O	4664	150	71	13	
		KDG 455	O	4674			748	
			O	4605	30		594	
15 5 36	1 25.0	A	O	13582	49	1	594	
15 5 36	1 25.0	B	O	13349	36	1	594	
15 6 5	34 34.7	B2	R	13500		123	678	★
			O	13430			485	
15 6 13	51 38.2	MRK 845	O	13862		172	874	
			O	12428			630	
15 6 24	54 56.0	NGC 5874	R	3134	10	180	582	
15 6 28	-64 29.3	NGC 5844	O	-60		-203	574	
15 6 33	-11 8.0	NGC 5861	R	1855	10	-46	752	
			R	1851	50		544	
			O	1814	50		741	
			O	1864	46		789	
15 7 2	3 14.7	NGC 5864	R	1623		9	780	
			O	1618	33		148	
15 7 18	0 39.0	NGC 5869 KDG 456B	O	2198	30	0	908	
15 7 18	0 43.0	NGC 5868 KDG 456A	O	11858	57	0	908	
15 7 43	52 43.1	NGC 5875	R	3527	20	176	829	
			O	3470	53		789	
15 7 57	6 1.82		O	22409	100	21	633	
15 8 4	6 4.85		O	22243	100	21	633	
15 8 6	5 50.78		O	24519	100	20	633	
15 8 12	-77 5.0		O	2648	11	-218	599	

R.A. (1950)			DEC. (1950)		NAME	OBS	HEL VEL (C*Z)	ERR	GAL CORR	REF	COMMENTS
15	8	14	5	54.46		0	22005	100	21	633	
15	8	22	5	59.36		0	22297	100	21	633	
15	8	23	5	50.80		0	22517	100	21	633	
15	8	25	67	27.9	DDO 199	0	-188	70	205	704	UMI DWF
15	8	27	5	55.99	PKS	0	23383	100	21	633	
						0	23300			401	
						0	23384			444	
15	8	29	5	57.30		0	23305	100	21	633	
15	8	29	57	11.4	NGC 5879	R	790		186	363	
						R	769			693	
						R	773			744	
						0	876	65		13	
15	8	30	5	55.96		0	26780	100	21	633	
15	8	30	5	57.56		0	22314	100	21	633	
15	8	31	5	58.29		0	23684	100	21	633	
15	8	32	5	51.01		0	24290	100	21	633	
15	8	33	8	2.8	3CR313	0	138200		29	946	
15	8	34	6	4.01		0	26634	100	22	633	
15	8	41	6	0.70		0	23064	100	22	633	
15	8	42	6	0.18		0	23998	100	21	633	
15	8	51	6	3.98		0	21992	100	22	633	
15	8	54	6	0.40		0	22369	100	22	633	
15	8	59	5	56.97		0	19997	100	22	633	
15	9	38	32	49.9		R	2267	10	120	860	
15	9	44	2	14.5	PKS	0	(65900)		8	600	
15	10	11	70	57.2	3CR314.1	0	35890		211	946	
15	11	0	-15	17.0		0	2375	71	-58	789	FAT 703
15	11	0	-14	5.0	NGC 5878	0	2111	65	-53	13	
15	11	9	-38	37.7		0	20240	200	-138	793	

15h11m

R.A. (1950)	DEC. (1950)	NAME	OBS	HEL VEL (C*Z)	ERR	GAL CORR	REF	COMMENTS
15 11 12	-38 37.8		O	18800	200	-138	793	
15 11 12	-39 37.3		O	19070	200	-141	793	
15 11 31	26 18.6	3CR315	O	32490		99	120	
			O	32210	250		696	
15 11 46	42 8.7	NGC 5893	R	5381	15	150	829	
15 11 52	57 9.2		R	832	10	188	860	
15 11 59	30 19.9	B2	O	27790	250	114	696	
15 12 12	44 47.0		O	5157	25	158	813	
15 12 24	- 9 54.0	NGC 5885	R	2002		-36	744	
			O	1984	71		789	
15 12 48	37 1.9		O	112000		136	406	
15 13 6	4 33.0		O	28300	60	20	13	
15 13 15	42 14.0	NGC 5899	R	2554	15	152	829	
			O	2549	50		13	
15 13 56	6 23.1		O	8100		28	255	
15 14 3	55 42.1	NGC 5905	R	3393	10	186	829	
			O	3386	20		741	
			O	3244			426	
			O	3357	32		789	
			O	3451			910	
15 14 6	7 10.0		O	9963	74	31	72	*
15 14 6	7 10.0		O	10159	92	31	72	
15 14 7	0 29.5	4C 00.56	O	15900		5	189	
15 14 17	7 12.26	3CR317	O	10143	10	31	651	
			O	10480			120	
			O	10504			444	
15 14 24	19 16.6	MRK 688	O	11505		77	557	SEYF
15 14 26	7 7.35		O	9488	60	31	651	
15 14 30	7 12.0	A	O	10354	40	31	594	*
15 14 30	7 12.0	B	O	10324	38	31	594	
15 14 36	7 10.21		O	10560	45	31	651	

R.A. (1950)	DEC. (1950)	NAME	OBS	HEL VEL (C*Z)	ERR	GAL CORR	REF	COMMENTS
15 14 36	56 30.4	NGC 5907'	R	650		188	171	
		NGC 5906	R	660	60		216	
			R	670			934	
			R	669	5		944	
			R	666			693	
			O	553	75		13	
			O	522			13	
15 14 37	7 4.24		O	10324	80	31	651	
15 14 43	7 7.45		O	10870	15	31	651	
15 15 17	-23 55.0	NGC 5898	O	2251	53	-86	741	
			O	2304	200		13	
15 15 23	55 35.6	NGC 5908	R	3309	10	186	829	
			O	3318			910	
			O	3433	40		741	
			O	3379	71		789	
15 15 38	-23 56.4		O	2380	66	-86	741	
			O	2350	66		510	
15 15 40	-23 53.1	NGC 5903	O	2499	51	-86	741	
			O	2612	150		13	
15 16 19	42 55.6	1 ZW107A/B	O	11978		155	262	★
			O	12110			77	
15 16 24	14 51.0		O	10738		62	683	
15 16 36	28 30.0	3 ZW 71	O	22870		111	282	
15 17 30	-36 45.0		O	3069	50	-128	582	
15 17 51	20 26.9	3CR318	O	225700	300	84	589	
15 17 51	28 45.4	MRK 849	O	24721		113	874	★
			O	23587			630	
15 18 18	23 29.0	IC 4538	R	2323	10	95	752	
15 18 48	-12 54.9	NGC 5915	R	2338	60	-42	885	
			O	2267	18		148	
15 18 53	4 31.2	PKS	O	15380		25	511	
15 19 0	2 58.0	3 ZW 72	O	15020		19	282	
15 19 6	7 55.0		O	12984	60	38	741	
15 19 11	7 53.7		O	13643	45	38	741	
			O	13230	210		912	

R.A. (1950)	DEC. (1950)	NAME		OBS	HEL VEL (C*Z)	ERR	GAL COPR	REF	COMMENTS
15 19 24	7 54.8			O	14235	49	38	741	
				O	14090	470		912	
15 19 24	41 54.0	NGC 5923		R	5570	10	155	582	
				O	5571	20		582	
15 19 25	7 53.2	NGC 5920		O	13176	61	38	741	
		3CR318.1		O	13790	300		585	
				O	13022	200		843	
				O	13510	200		912	
15 19 30	5 15.0	NGC 5921		R	1480		28	744	
				R	1478	4		914	
				R	1480	40		216	
				O	1389	150		13	
15 19 42	42 6.0	1 ZW110		O	20490		155	262	
15 19 53	27 52.3			O	19522	65	111	13	
15 20 6	27 52.33			O	20984	100	112	13	
15 20 9	27 52.61			O	21388		112	662	
15 20 12	- 6 19.7	4C-06.41		O	38390		-16	318	
15 20 12	27 53.48			O	23380		112	662	
15 20 16	27 54.90			O	23812	65	112	13	
15 20 17	27 51.49			O	19860		112	662	
15 20 17	27 53.45			O	20775	50	112	13	
15 20 19	27 52.29			O	21155		112	662	
15 20 21	27 52.54			O	21392		112	662	
15 20 21	27 53.19			O	21179		112	662	
15 20 22	27 51.84			O	20840	75	112	13	
15 20 22	27 53.09			O	22447		112	662	
15 20 22	27 53.35			O	21841	150	112	13	
15 20 24	8 47.0		A	O	10540		42	13	
15 20 24	8 47.0		B	O	10546		42	13	
15 20 32	27 50.96			O	22380	75	112	13	
15 20 32	27 51.27			O	22088	75	112	13	

R.A. (1950)	DEC. (1950)	NAME	OBS	HEL VEL (C*Z)	ERR	GAL CORR	REF	COMMENTS
15 20 37	27 50.63		O	20075		112	662	
15 20 42	29 57.0		O	6907		119	777	
15 21 21	28 48.1	B2	O	24610	250	116	696	
15 21 30	66 49.0	7 ZW599	O	(3630)		209	282	
15 22 42	58 14.0		R	2660	10	195	582	
			O	2674	25		582	
15 22 44	54 38.6	3CR319	O	57600		188	946	
15 23 42	16 30.0		R	7030	10	74	582	
			O	7007	25		582	
15 24 0	67 20.0	MRK 1096	O	6870		210	921	
15 24 6	71 5.0	MRK 1097	O	17535		215	921	
15 24 19	41 50.7	NGC 5929	O	2525		158	748	
		KDG 466	O	2502	14		908	
		ARP 90	O	2436	29		594	
		HOL 710B	O	2522			253	
			O	2527	50		8	
15 24 21	41 51.0	NGC 5930	O	2710		158	748	
		KDG 466	O	2619	14		908	
		ARP 90	O	2698			253	
		HOL 710A	O	2714	26		594	
			O	2712	50		8	
15 24 37	43 34.3	MRK 854	O	46558		163	874	
15 24 52	40 44.3		R	2622	10	155	752	
15 25 40	29 6.3	B2	O	19460		120	485	
			O	19985			651	
15 26 24	43 5.0	NGC 5934	O	(5403)	100	162	789	
15 26 47	55 42.8	MRK 482	O	3370		192	435	
			O	3000			359	
15 27 18	64 56.0	NGC 5949	R	435		208	744	
			O	380			13	
			O	601	50		741	
15 27 40	13 10.0	NGC 5936	R	3986	22	65	914	
			O	3840	50		789	
			O	4138			741	
15 27 42	30 39.0	MRK 1098	O	10454		126	921	SEYF

337

15ʰ27ᵐ

R.A. (1950)	DEC. (1950)	NAME	OBS	HFL VEL (C*Z)	ERR	GAL CORR	REF	COMMENTS
15 27 43	30 52.8	B2	O	34120	250	127	696	
15 28 6	29 10.7	B2	O	25140	250	122	696	
15 28 36	23 13.9		R	1980	10	102	860	
15 29 7	47 29.2		R	2565	10	175	860	
15 29 30	35 43.8	3CR320	O	102530	300	144	585	
15 29 33	24 14.4	3CR321	O	28780	300	107	585	
15 29 38	54 51.5	MRK 484	O	11705		192	435	
			O	11700			358	
15 29 42	14 30.0		O	9228		72	683	
15 30 0	4 51.0		O	11467	91	36	843	
15 30 18	51 56.0	MRK 485	O	5960		186	438	
			O	5700			358	
15 30 30	- 1 28.0	IC 1125 KDG 467B	O	2768	43	11	908	
15 31 18	46 38.0	1 ZW115	R	655	50	174	241	
			R	665	10		622	
			O	644			262	
15 31 23	58 3.0	MRK 289	O	12000		199	308	
15 31 24	15 11.0	NGC 5951	O	1670	71	76	789	
15 31 36	-66 44.0	NGC 5938	O	3580		-197	574	
15 31 39	68 24.9	IC 1129	R	6540	5	214	752	
15 31 45	35 54.3	4C35.37	O	46950		146	400	SEYF
15 32 13	15 21.6	NGC 5953	O	2048	43	77	594	
		VV 244	O	2146	30		8	
		ARP 91	O	2164			253	
15 32 16	56 43.5	NGC 5963	R	655	10	197	829	
		KDG 469A	O	870	63		789	
15 32 18	15 22.0	NGC 5954	R	2000		77	744	
		ARP 91	O	1956	39		594	
		VV 244	O	2123			253	
			O	2106	30		8	
15 32 36	11 55.0	NGC 5956	R	1899	10	65	829	

R.A. (1950)	DEC. (1950)	NAME	OBS	HEL VEL (C*Z)	ERR	GAL CORR	REF	COMMENTS
15 32 47	23 40.0	IC 4554 KDG 470 ARP 220	O O	5493 5348	53	107	748 908	
15 32 47	23 40.3	IC 4553 KDG 470 ARP 220	O O	5593 5451	28	107	748 908	
15 32 50	56 51.1	NGC 5965 KDG 469B	R O	3416 3421	10	197	829 910	
15 33 0	57 28.0	7 ZW611	O	3280	100	198	820	
15 33 1	12 12.8	NGC 5957	R	1828	15	66	734	
15 33 12	31 1.0	NGC 5961	O	1800		132	476	
15 33 24	46 59.0	1 ZW116	O	5660	100	177	820	
15 33 48	82 35.7		O	6177	18	224	381	
15 34 0	38 50.0	1 ZW117	R O	5589 5492	23	156	622 262	
15 34 6	82 35.8		O	6148	140	224	381	
15 34 12	16 46.0	NGC 5962 HOL 716A	R R R R O	1932 1955 1976 1963 1993	100 5 100 75	84	238 914 544 744 13	
15 34 12	43 40.0	1 ZW118	O	5508		169	262	
15 34 18	30 50.8	MRK 689	O	1655		132	557	
15 34 24	37 42.0		O	46114	75	153	13	
15 34 24	37 48.0		O	45706	100	153	13	
15 34 36	54 28.0	1 ZW119	O	13559		194	262	
15 34 45	58 4.0	MRK 290	O O	8700 9020		200	268 287	SEYF
15 34 48	37 51.0		O	45557	200	153	13	
15 35 0	43 42.0	IC 4566	O	5588	25	169	582	
15 35 0	82 38.0	IC 1143	O	6395	143	224	381	
15 35 5	44 24.1	DDO 200	R	2628	10	171	492	

15ʰ35ᵐ

R.A. (1950)	DEC. (1950)	NAME	OBS	HFL VEL (C*Z)	ERR	GAL CORR	REF	COMMENTS
15 35 6	6 8.0	NGC 5964	R	1450		45	744	
			O	(1367)	30		813	
			O	1382	43		789	
15 35 9	30 14.4		R	1872	10	131	860	
15 35 22	54 43.1	MRK 486	O	11700		194	435	SEYF
		1 ZW121	O	11690			262	
			O	11648			447	
			O	11700			358	
			O	11700			388	
15 35 31	43 27.7	IC 4567	R	5722	15	169	752	
15 35 48	55 25.6	1 ZW123	R	827		196	389	
		MRK 487	O	(705)			435	
			O	619			389	
			O	620			262	
			O	11580			206	DIS
15 36 0	13 7.1		R	1861	10	72	860	
15 36 9	12 21.0	NGC 5970	R	1973	50	69	544	
			R	1960	6		914	
			R	1964	200		693	
			O	2034	50		13	
			O	2127			13	
15 36 51	-30 24.2	NGC 5968	O	5120		-93	574	
15 36 52	59 33.2	NGC 5981 HOL 719C	O	1717	145	204	555	
15 37 0	31 55.0	NGC 5974	O	1900		137	491	
			O	1800			533	
15 37 39	59 31.0	NGC 5982 HOL 719A	O	2981		204	13	
			O	2864	10		13	
15 37 43	34 35.2		O	68200	300	146	583	
15 38 36	59 29.6	NGC 5985 HOL 719B	R	2521	10	204	734	
			R	2516			693	
			O	2467	40		13	
15 39 31	34 20.5	4C 34.42	O	120560		147	400	
15 39 47	0 37.9	DDO 201	R	1978	20	28	492	
15 40 33	59 54.8	NGC 5989	R	2878	15	206	752	
15 40 36	14 23.0	NGC 5984	R	1118		81	744	
			O	1013	71		789	

R.A. (1950)	DEC. (1950)	NAME	OBS	HEL VEL (C*Z)	ERR	GAL CORR	REF	COMMENTS
15 40 48	66 25.7		O	73700	300	215	583	
15 41 24	67 25.0		R	423	10	216	860	
15 42 10	-75 31.2	NGC 5967	O	2904	36	-209	741	
15 42 12	43 20.0	2 ZW 73	O	5515		173	303	
15 42 36	41 15.0	NGC 5992	O	9414	90	168	567	
		KDG 471	O	9578	16		908	
		MRK 489	O	9300			359	
			O	10022			447	
15 42 42	41 17.0	NGC 5993	O	9396	58	168	594	
		KDG 471	O	9567	18		908	
			O	9591	150		567	
15 43 23	12 39.9		R	1116	10	77	860	
15 44 36	18 1.0	NGC 5994	O	3309	70	97	567	
		KDG 472	O	3247	20		908	
		ARP 72						
		VV 16						
15 44 42	18 2.0	NGC 5996	O	3064	30	97	567	
		MRK 691	O	3310			557	
		KDG 472	O	3303	10		908	
		ARP 72						
		VV 16						
15 44 54	49 9.1	MRK 490	O	2700		188	358	
			O	2660			447	
15 44 55	61 42.6		R	925	10	210	860	
15 45 9	21 4.9		O	80900	999	108	337	
15 45 48	54 30.0	1 ZW125	O	20159		199	262	
15 45 54	37 21.0	1 ZW126	O	11935		160	262	
15 46 37	-25 29.6		O	11960		-67	255	
15 46 45	-29 15.5	NGC 6000	O	2110		-81	574	
15 47 12	30 56.6	4C 30.29	O	33330		142	400	
15 47 52	81 57.9	DDO 203	R	1499	15	225	492	
15 47 54	69 38.0	MRK 1099	O	11200		220	921	
		8 ZW623	O	(2180)			774	DIS
		VV 291						

15ʰ48ᵐ

R.A. (1950)	DEC. (1950)	NAME	OBS	HEL VEL (C*Z)	EPR	GAL CORR	REF	COMMENTS
15 48 17	26 4.2		R	2145	10	127	860	
15 48 23	-74 36.0		O	5286	125	-206	680	
15 48 58	16 28.8	DDO 202	R	2087	15	95	492	
15 49 42	43 34.0	MRK 491	O	11881		178	714	
		IC 1144	O	(18900)			358	DIS
15 49 54	20 14.3	3CR326	O	29530	300	109	585	
15 50 18	41 53.0	MRK 1100	O	9475		175	921	
15 50 42	62 28.0	NGC 6015	R	831		214	744	
			R	850	40		216	
			R	852	25		183	
			R	830			693	
			O	822	8		546	
			O	732			13	
			O	646	50		13	
			O	675			641	
15 51 0	12 6.0	NGC 6007	R	10560	15	81	582	
			O	10487	40		582	
15 51 8	40 47.6	NGC 6013	O	4492		172	714	
15 51 46	41 43.6		O	9897		175	714	
15 51 48	18 47.0		O	14286	100	106	857	
15 51 54	14 44.0	NGC 6012	R	1846	30	91	885	
			O	1938	100		857	
			O	1990			641	
15 51 54	19 15.0		O	10949	100	107	857	
15 51 54	23 16.7	MRK 693	O	11500		121	557	
15 52 6	18 47.0		O	13925	100	106	857	
15 52 12	18 40.0		O	9684	100	105	857	
15 52 18	16 45.0		O	2206	100	99	857	
			O	2400			533	
15 52 21	41 45.8		O	6983		176	714	
15 52 54	19 20.3	MRK 291	O	10830	80	108	300	SEYF
			O	10562	45		382	
			O	10500			232	
15 53 19	41 43.4		O	9960		176	714	

R.A. (1950)	DEC. (1950)	NAME	OBS	HEL VEL (C∗Z)	ERR	GAL CORR	REF	COMMENTS
15 53 24	18 25.0		O	5242	100	106	857	
15 53 36	17 18.0		O	4845	100	102	857	
15 53 44	45 34.2		R	1852	10	185	860	
15 53 45	19 2.3	MRK 292	O	10678	100	108	857	
			O	9900			308	
15 53 57	24 35.5	4C 24.35	O	12873		127	748	
			O	12650			485	
15 54 6	16 40.0		O	4630	100	100	857	
15 54 18	20 11.0		O	9640	100	112	857	
15 54 42	18 47.0	E	O	18138	100	108	857	
15 54 42	18 47.0	W	O	18138	100	108	857	
15 54 48	18 19.0		O	9545	100	106	857	
15 54 51	42 2.5		O	10850		178	714	
15 54 54	42 1.0	MRK 1101	O	10162		178	921	
15 54 54	42 1.5	1 ZW129	O	10400		178	714	
			O	10414	10		336	
			O	10334			262	
			O	10200			533	
15 55 12	16 0.0	NGC 6018	O	5121		99	312	
15 55 12	16 5.0	NGC 6021	O	(4486)		99	312	
15 55 18	18 10.0		O	9410	100	106	857	
15 55 24	16 21.0		O	10120	100	100	857	
15 55 24	30 12.0		O	(9703)	40	146	813	
15 55 30	16 25.0	NGC 6022	O	11225		100	312	
15 55 30	16 27.0	NGC 6023	O	11140		100	312	
15 55 30	30 12.0		R	9848	10	146	582	
			O	9923	75		582	
15 55 34	-14 1.4	PKS	O	29100		-15	828	
15 55 36	41 41.0	MRK 1102	O	10433		177	920	
15 55 42	16 29.0		O	13502	100	101	857	
			O	13495	100		857	

343

R.A. (1950)	DEC. (1950)	NAME	OBS	HEL VEL (C*Z)	ERR	GAL CORR	REF	COMMENTS
15 55 48	41 40.0	MRK 1103	O	10312		178	920	
15 55 50	30 50.8	B2	O	22260	250	148	696	
15 56 9	27 23.4		O	28819	150	138	807	
15 56 10	27 24.8		O	28711	150	138	807	
15 56 12	17 35.0	IC 1151	O	2083	100	105	857	
15 56 12	40 10.0		O	9300		174	476	
15 56 12	42 4.8		O	10387		179	714	
15 56 14	27 24.7		O	25590	150	138	807	
15 56 15	27 41.2		O	9611	150	139	807	
15 56 16	27 22.5		O	27329	150	138	807	
15 56 17	27 28.6		O	26879	150	138	807	
15 56 28	27 36.9		O	9219	150	139	807	
15 56 30	15 5.0		O	10690	100	96	857	
15 56 33	27 39.2		O	28669	150	139	807	
15 56 36	15 6.0		O	10527	100	96	857	
15 56 39	26 57.3	MRK 492	O	4200		137	358	
15 56 39	27 47.0		O	26712	150	139	807	
15 56 43	27 46.9		O	25509	150	139	807	
15 56 46	27 35.1		O	25419	150	139	807	
15 56 48	15 3.0		O	12527	100	96	857	
15 56 48	27 32.3		O	28261	150	139	807	
15 56 48	27 32.5		O	26631	150	139	807	
15 56 52	27 39.2		O	26861	150	139	807	
15 56 55	58 18.2	MRK 865	O	10349		211	874	
15 56 59	20 53.7	NGC 6027GRP 7 ZW631	R	4563		117	779	SEY SEX
15 56 59	20 53.7	NGC 6027A VV 115	O O O O	4024 4285 4036 4031	 50	117	274 322 13 13	

R.A. (1950)	DEC. (1950)	NAME	OBS	HEL VEL (C★Z)	ERR	GAL CORR	REF	COMMENTS
15 56 59	20 54.3	NGC 6027B VV 115	0 0	4313 4187		117	274 322	
15 57 0	15 18.0		0	12774	100	98	857	
15 57 0	20 53.2	NGC 6027C VV 115	0 0	4464 4438		117	274 322	
15 57 1	20 53.9	NGC 6027D VV 115	0 0	19813 20009		117	274 322	
15 57 1	20 54.2	NGC 6027 VV 115	0 0 0 0	4351 4392 4468 4415	50	117	274 322 13 13	★
15 57 6	42 4.8		0	11586		179	714	
15 57 10	79 7.7	NGC 6068A HOL 727B KDG 476A	0 0	3996 4035		227	253 57	
15 57 17	35 10.0	MRK 493	0 0	9729 9386		161	447 874	
15 57 30	18 56.0		0	8961	100	111	857	
15 57 46	26 4.8	B2	0	13110	250	135	696	
15 57 49	79 8.4	NGC 6068 KDG 476B HOL 727A	0 0 0	3925 4200 3964		227	253 476 57	
15 58 0	15 54.0		0	4782	100	101	857	
15 58 0	16 17.0	NE	0	9914	100	102	857	
15 58 0	16 17.0	SW	0	9990	100	102	857	
15 58 12	16 46.0		0	10346	100	104	857	
15 58 12	21 0.0		0 0	4759 4753	100	118	857 484	
15 58 18	15 50.0	IC 1155	0 0	10629 11013	100	101	312 857	
15 58 24	19 52.0	IC 1156	0	9475	100	115	857	
15 58 30	15 17.0		0	10182	100	99	857	
15 58 31	30 30.8	MRK 494	0	9510		149	435	

R.A. (1950)	DEC. (1950)	NAME	OBS	HEL VEL (C★Z)	ERR	GAL CORR	REF	COMMENTS
15 58 36	16 28.0		0	12346		103	312	
			0	12382	100		857	
15 58 36	17 41.0		0	13349	100	107	857	
15 58 36	19 4.0		0	9437	100	112	857	
15 58 42	15 38.0	IC 1160	0	10970	100	100	857	
15 58 42	16 31.0		0	43000	100	103	857	
15 58 48	19 35.0		0	4347	100	114	857	
15 59 0	15 47.0	IC 1161	0	10852	100	101	857	
15 59 0	16 27.0		0	11282	100	103	857	
15 59 0	17 49.0	IC 1162	0	13273	100	108	857	
15 59 0	17 54.0		0	10378	100	108	857	
15 59 6	16 21.0		0	8566	100	103	857	
15 59 6	16 49.0		0	9473	100	105	857	
15 59 7	15 45.4	MC 3	0	10490		101	661	
15 59 12	15 38.0		0	10503	100	101	857	
15 59 12	16 53.0		0	9598	100	105	857	
15 59 12	17 23.0		0	10765	100	107	857	
15 59 12	19 29.0	NGC 6028 1 ZW133	0	4480		114	303	
15 59 18	16 34.0		0	13052	100	104	857	
15 59 36	18 6.0	NGC 6030	0	4356	100	110	857	
15 59 37	27 26.6		0	49640	150	140	807	
15 59 41	27 26.5		0	50401	150	140	807	
15 59 42	16 35.0		0	10589	100	104	857	
15 59 42	16 35.3	MRK 694	0	9120		104	557	
			0	9112	100		857	
15 59 48	15 50.0	IC 1165A/B VV 90	0	14898		102	774	
15 59 48	15 50.0	IC 1165A VV 90	0	10136		102	312	

R.A. (1950)	DEC. (1950)	NAME	OBS	HEL VEL (C*Z)	ERR	GAL CORR	REF	COMMENTS
15 59 48	15 50.0	IC 1165B VV 90	0	10109		102	312	
15 59 48	17 13.0		0	11036	100	107	857	
15 59 49	18 57.2	MRK 294	R	2525		113	779	
			R	2515			368	
			0	2509	100		857	
			0	2522	45		382	
			0	2400			268	
15 59 54	16 2.0	VV 159 ARP 324 3 ZW 75	0	10372		103	274	
15 59 54	16 2.0	VV 159C	0	10497		103	312	
15 59 54	16 2.0	VV 159DN	0	9973	100	103	857	
15 59 54	16 2.0	VV 159DS	0	9878	100	103	857	
15 59 54	16 4.0	VV 159B	0	13170		103	312	
			0	6000			533	DIS
15 59 54	16 34.0		0	9276		104	312	
15 59 56	2 6.2	3CR327	0	31170		51	120	
16 0 0	16 6.0	VV 159A	0	10337	100	103	857	
			0	10483			312	
			0	10430			255	
16 0 0	16 9.0	NW	0	10256	100	103	857	
16 0 0	16 9.0	S	0	10121	100	103	857	
			0	10135			595	*
16 0 0	16 17.0		0	12894		104	312	
16 0 0	16 29.0		0	11449		104	312	
16 0 0	16 30.0		0	12054		104	312	
16 0 24	16 15.0		0	9933	100	104	857	
16 0 30	16 49.0		0	33338	100	106	857	
16 0 34	16 5.9	MRK 695	0	10465		103	557	
16 0 36	16 42.0		0	10470	100	105	857	
16 0 37	43 3.3		0	7301		184	714	

R.A. (1950)	DEC. (1950)	NAME	OBS	HEL VEL (C*Z)	ERR	GAL CORR	REF	COMMENTS
16 0 42	15 59.0		O	9704	100	103	857	
16 0 48	21 6.0	NGC 6032	O	4284		121	484	
			O	4285	100		857	
16 1 0	16 32.0		O	10966	100	105	857	
			O	11145			595	
16 1 0	41 20.1	KDG 479A/B	O	9955		180	714	
16 1 0	41 20.1	KDG 479A	O	9862	24	180	908	
16 1 0	41 20.1	KDG 479B	O	9884	22	180	908	
16 1 12	16 28.0		O	11497	100	105	857	
			O	11497			312	
16 1 12	21 2.0	NGC 6035	O	2241		121	484	
			O	2219	100		857	
16 1 13	19 17.9	MRK 296	R	4685		115	779	
			R	4614	22		785	
			O	4746	100		785	
			O	4800			268	
			O	4738	100		857	
16 1 13	19 19.0	MRK 295	O	11434	100	115	857	
			O	11465	100		785	
16 1 15	17 20.0	NGC 6034	O	10112	100	108	857	
		PKS	O	10200			465	
			O	10150			595	
16 1 18	16 2.0		O	30000	100	104	857	
			O	32820			595	
16 1 24	16 28.0	W	O	10645	100	105	857	
16 1 24	39 46.0	VV 611	O	9024		176	774	
16 1 25	15 55.9		O	10185		103	595	
16 1 30	17 23.0		O	10953	100	109	857	
16 1 36	16 8.0		O	9577	100	104	857	
16 1 36	16 30.0		O	11592	100	106	857	
16 1 38	29 5.6		O	83900		147	717	
16 1 48	17 25.0		O	9908	100	109	857	
16 2 6	16 50.0		O	9366	100	107	857	

R.A. (1950)	DEC. (1950)	NAME	OBS	HEL VEL (C*Z)	ERR	GAL CORR	REF	COMMENTS
16 2 7	40 7.2		0	9319		178	714	
16 2 8	41 39.5		0	10210		182	714	
16 2 10	-63 21.4	PKS	0	17720		-179	627	
16 2 11	17 52.8	NGC 6040A/B ARP 122	0	5300		111	535	
16 2 11	17 52.8	NGC 6040A VV 212	0	12404	100	111	857	
			0	12386			322	
			0	12219	100		857	
			0	12283			595	
16 2 12	14 57.0		0	4708	100	101	857	
16 2 12	17 36.0		0	11987	100	110	857	
16 2 12	17 53.3	NGC 6040B VV 212	0	12612	100	111	857	
			0	12618			322	
16 2 18	16 37.0	NW	0	13527	100	107	857	
16 2 18	16 37.0	SE	0	11795	100	107	857	
16 2 18	17 1.0		0	9222	100	108	857	
16 2 18	17 51.0	IC 1170	0	9587	100	111	857	
16 2 18	17 51.0	NGC 6041C	0	10115	100	111	857	★
16 2 21	17 51.4	NGC 6041A/B VV 213	0	10438	100	111	22	
			0	10469	50		13	
16 2 21	17 51.4	NGC 6041B	0	11248	100	111	857	
			0	11233	100		857	
16 2 21	17 51.5	NGC 6041A	0	10272	100	111	857	
			0	10571	100		857	
16 2 24	16 34.0		0	9989	100	106	857	
16 2 24	17 50.0	NGC 6042	0	10542	100	111	857	
			0	10318	100		857	
16 2 30	15 52.0		0	10670	100	104	857	
16 2 30	16 43.0		0	9347	100	107	857	
16 2 30	17 1.0		0	12553	100	108	857	
16 2 30	17 35.0		0	10413	100	110	857	

R.A. (1950)			DEC. (1950)		NAME	OBS	HEL VEL (C*Z)	ERR	GAL CORR	REF	COMMENTS
16	2	36	18	1.0	NGC 6044 IC 1172	O	9936	50	112	13	
16	2	42	17	55.0	NGC 6043E	O	9798	100	111	857	
16	2	44	24	6.5		O	9570		132	747	
16	2	46	28	13.7	MRK 696	O	9155		145	557	
16	2	49	24	3.9	NGC 6051 4C 24.36	O O O O	9540 9690 9381 9557	 54 	132	400 747 843 910	
16	2	49	24	5.7		O	16020		132	747	
16	2	52	24	6.0		O	11580		132	747	
16	2	52	34	45.2		O	9310		164	485	
16	2	53	17	53.5	NGC 6045 ARP 71	O O	9850 9935	100 50	111	22 13	
16	2	54	17	33.0	IC 1173	O O	12839 10871	100 100	110	857 22	*
16	2	54	17	51.9	NGC 6047 4C 17.66	O O	9470 9435	50 100	111	13 857	
16	2	55	24	3.0		O	15950		132	747	
16	3	0	20	40.5	NGC 6052A/B MRK 297 ARP 209 VV 86	R R R O O O O O O O O O O	4700 4640 4738 4800 4770 4700 4744 4731 4593 4714 4760 4620 4671	 44 20 45 30 24 100 60 14 19	121	779 785 785 268 287 382 149 213 555 560 741 150 148	
16	3	0	20	40.5	NGC 6052A	O	4500	57	121	594	
16	3	0	20	40.5	NGC 6052B	O	4541	50	121	594	
16	3	6	14	46.0		O	12228	100	101	857	
16	3	6	15	10.0	IC 1174	O	4652	100	102	857	

R.A. (1950)			DEC. (1950)		NAME	OBS	HEL VEL (C*Z)	ERR	GAL CORR	REF	COMMENTS
16	3	6	16	20.0		0	10088	100	106	857	
16	3	8	17	53.3	NGC 6050A	R	11159		112	749	
					ARP 272	0	11086	100		22	
					KDG 481	0	11049	100		857	
					VV 220						
					IC 1179						
16	3	8	17	53.6	NGC 6050B	0	9544	100	112	857	
					ARP 272	0	9511	100		857	
					KDG 481						
					VV 220						
16	3	9	18	16.6		0	11408		113	23	*
						0	11220	100		22	
16	3	9	32	9.1	B2	0	112100		157	586	
16	3	12	16	35.0	NW	0	13598	100	107	857	
16	3	12	16	35.0	SE	0	13217	100	107	857	
16	3	12	17	55.0	NGC 6054	0	11017	100	112	857	
16	3	17	18	6.0	NGC 6056	0	11687	100	113	22	
16	3	18	17	44.0	IC 1178	0	10141	100	111	857	
					ARP 172	0	10200	100		857	
					VV 194						
16	3	18	17	54.0	IC 1183	0	10038	65	112	13	
					VV 220						
16	3	18	18	17.7	NGC 6055	0	11212	100	113	857	
						0	11315			910	
16	3	18	18	25.0		0	11285	100	114	857	
16	3	19	17	43.5	IC 1181A/B	0	10285	100	111	857	
					ARP 172	0	10739	100		857	
					VV 194						
16	3	19	17	43.5	IC 1181A	0	10354	100	111	22	
16	3	19	17	43.5	IC 1181B	0	10720	100	111	22	
16	3	22	17	56.1	IC 1182	0	10127	100	112	857	SEYF
					MRK 298	0	10200			232	
					VV 220	0	10220			287	
						0	10180			347	
16	3	24	18	12.0		0	11993	100	113	857	

R.A. (1950)	DEC. (1950)	NAME	OBS	HEL VEL (C∗Z)	ERR	GAL CORR	REF	COMMENTS
16 3 24	18 18.0	NGC 6057	0	10443	100	113	857	
16 3 25	17 56.1		0	9973		112	347	KNOT
16 3 26	17 56.1		0	9979		112	347	KNOT
16 3 29	17 29.8	IC 1186	0	11043	100	111	22	
16 3 30	15 55.0		0	12450	100	105	857	
16 3 30	16 20.0		0	12512	100	107	857	
16 3 30	17 51.0	IC 1185	0 0	10297 10452	100 200	112	857 13	
16 3 30	18 9.0		0	12212	100	113	857	
16 3 31	17 26.4	MRK 299	0	11100		110	308	
16 3 36	18 21.0		0	10877	100	114	857	
16 3 36	18 40.0		0	11139	100	115	857	
16 3 42	21 38.0	NGC 6060	0 0	4555 4554	 100	125	484 857	
16 3 48	18 15.0		0	11030	100	113	857	
16 3 48	18 20.0		0	11215	100	114	857	
16 3 49	42 45.7		0	11549		185	714	
16 3 54	18 49.0		0	11751	100	115	857	
16 3 54	20 55.0		0 0	4595 4602	 100	123	484 857	
16 4 0	16 34.0		0	10130	100	108	857	
16 4 0	16 40.0		0	10785	100	108	857	
16 4 0	18 19.0	IC 1189 MRK 300	0 0 0	11858 11924 12000	100 100	114	857 857 268	
16 4 0	18 23.0	NGC 6061	0 0	11305 11197	100 100	114	857 22	
16 4 0	18 33.0		0	11161	100	115	857	
16 4 0	18 45.0		0	11330	100	115	857	
16 4 0	41 27.0	KDG 482B	0 0	2100 1947	 43	182	476 594	

R.A. (1950)			DEC. (1950)		NAME		OBS	HFL VEL (C*Z)	ERR	GAL CORR	REF	COMMENTS
16	4	0	41	29.0	KDG	482A	O	1980	70	182	567	
					MRK	1104	O	1918			921	
							O	2030			714	
							O	2039	25		594	
16	4	5	41	27.1			O	14111		182	714	
16	4	6	19	55.0	NGC	6062N	O	11685	100	119	857	
16	4	9	15	49.0	VV	215A	O	12071	106	105	594	
							O	11993	100		857	
							O	15600			535	DIS
16	4	9	15	49.0	VV	215B	O	12344	55	105	594	
							O	13237	100		857	
							O	16400			535	DIS
16	4	12	17	50.0			O	11987	100	112	857	
16	4	18	17	51.0			O	10466	100	112	857	
16	4	18	18	1.0			O	11049	100	113	857	
16	4	24	16	27.0			O	11012	100	108	857	
16	4	24	17	19.0	IC	1195	O	12121	100	111	857	
16	4	24	17	55.0	IC	1194	O	11642	65	113	13	
16	4	30	17	38.0			O	13634	100	112	857	
16	4	42	15	44.0			O	11793	100	105	857	
16	6	18	12	28.0	IC	1198	O	9890		95	630	
					MRK	871						
16	7	10	0	0.9			R	1505	10	49	860	
16	7	12	36	45.0			O	9054	47	173	813	
16	7	25	0	50.4	NGC	6070	R	2013	50	52	544	
					HOL	729A	R	2002			744	
							O	2091	125		13	
							O	2120			13	
16	7	36	41	52.5	1	ZW134	O	7494		186	714	
							O	7676			262	
16	7	54	16	50.0	NGC	6073	O	4590	100	112	857	
16	8	36	17	11.0			O	10184	100	114	857	
16	8	40	41	59.0			O	12047		186	714	

R.A. (1950)	DEC. (1950)	NAME	OBS	HEL VEL (C*Z)	ERR	GAL CORR	REF	COMMENTS
16 9 2	41 16.6		O	9670		185	714	
16 9 12	42 0.8		O	10922		187	714	
16 9 14	66 4.4	S 3CR330	O	164590		224	588	
16 9 14	66 4.4	N	O O	159500 162000		224	717 746	
16 9 42	31 10.7	B2	O	28130	250	159	696	
16 10 24	52 35.0	NGC 6090A/B MRK 496 1 ZW135	O O O	8700 8780 8732		208	359 435 262	
16 10 24	52 35.0	NGC 6090A KDG 486A	O O O	8883 8690 8846	30 25 34	208	567 594 830	
16 10 24	52 35.0	NGC 6090B KDG 486B	O O O O	8612 8927 8962 8695	37 30 71 55	208	594 567 500 830	
16 10 31	26 36.2	NGC 6086 B2	O	9380		146	255	
16 10 43	-60 46.9	PKS	O O O	5010 5450 (8510)	30	-169	627 355 317	DIS
16 10 54	60 43.0	KDG 488A	O	4136	27	220	908	
16 10 54	60 43.0	KDG 488B	O	4144	20	220	908	
16 11 13	-60 32.4	PKS	O	4860		-168	627	
16 11 25	-60 47.6		O	3720		-169	627	*
16 11 42	4 6.0	KDG 490	O	9890	20	68	908	
16 11 48	4 7.0	KDG 490	O	10033	13	69	908	
16 12 0	0 56.7		R	1988	10	56	860	
16 12 0	26 13.0	TON 256	O	29213	60	146	741	
16 12 9	26 11.5		O	39510	300	146	286	
16 12 14	26 17.3		O	39540	300	146	286	
16 12 23	26 18.2		O	36150	999	147	286	

R.A. (1950)	DEC. (1950)	NAME	OBS	HEL VEL (C*Z)	ERR	GAL CORR	REF	COMMENTS
16 12 48	12 53.0	MRK 1105	O	9278		102	921	
16 13 24	19 35.0	NGC 6098 VV 192	O	12300		126	535	
16 13 24	19 35.0	NGC 6099 VV 192	O	12600		126	535	
16 13 29	27 34.3	B2	O	19230	250	151	696	
16 13 36	65 50.6	MRK 876	O	38742		226	874	SEYF
16 13 50	31 4.9		O	124300	300	162	583	
16 14 36	18 38.0	MRK 1106	O	10187		123	921	
16 14 49	47 10.0	DDO 204 ARP 2	R	715	10	201	492	
16 14 54	35 20.0		O	9770		174	756	
16 15 0	34 47.0		O	9380		173	756	
16 15 0	35 4.0		O	10606		174	756	
16 15 0	35 6.0		O	10408		174	756	
16 15 12	35 0.0	NGC 6105	O	8654		173	756	
16 15 18	34 51.0		O	10400		173	756	
16 15 18	34 55.0		O	21695		173	756	
16 15 24	35 1.0	NGC 6107	O	9182		174	756	
16 15 28	52 8.1	MRK 497	O O	9785 9900		210	435 359	
16 15 30	34 56.0		O	9880		173	756	
16 15 30	35 15.0	NGC 6108	O	9116		174	756	
16 15 36	35 15.0		O	9171		174	756	
16 15 42	35 10.0		O	9710		174	756	
16 15 47	32 29.8	3CR332	O O	45370 45410	250	167	696 318	
16 15 48	34 56.0		O	10037		174	756	
16 15 48	35 13.0	NGC 6110	O	9112		174	756	

16ʰ15ᵐ

R.A. (1950)	DEC. (1950)	NAME	OBS	HEL VEL (C*Z)	ERR	GAL CORR	REF	COMMENTS
16 15 49	35 7.5	NGC 6109	O	8700		174	485	
			O	8850			756	
16 16 0	46 13.0	KDG 494A	O	5885	100	200	567	
16 16 6	35 14.0	NGC 6112	O	9319		175	756	
16 16 6	35 23.0		O	8928		175	756	
16 16 6	46 12.0	KDG 494B	O	5865	50	200	567	
16 16 18	7 31.9	NGC 6106	R	1456		85	744	
			O	1470	52		148	
16 16 30	35 18.0	NGC 6114	O	8691		175	756	
16 16 30	35 25.0		O	8917		175	756	
16 17 0	35 17.0	NGC 6116	O	8800		175	756	
16 17 36	2 8.0		O	5100		66	533	
16 17 56	17 31.6	MRK 877	O	33708		122	874	SEYF
16 18 0	37 54.0	NGC 6120 1 ZW141	O O	9367 9180		183	321 262	
16 19 12	- 2 10.0	NGC 6118	R	1571		50	744	
			O	1660	63		789	
16 19 12	40 14.0		O	10143		189	777	
16 20 18	40 36.0		O	9308		191	321	
16 20 36	55 12.0	NGC 6143	O	5235	54	217	789	
16 20 36	65 30.0	NGC 6140	P	912		228	744	
			O	793	43		789	
16 20 48	20 39.0		O	7940		135	282	
16 21 17	38 2.3	NGC 6137	O	9110		185	485	
			O	9310			756	
16 21 54	39 20.0		O	8370		189	449	
16 22 0	54 16.0	1 ZW147	O	5408		217	262	
			O	5410			389	
16 22 6	41 12.0	3 ZW 77 MRK 699	O O O O	10250 10250 10110 10250	300 220	193	123 122 557 328	SEYF

356

R.A. (1950)	DEC. (1950)	NAME	OBS	HEL VEL (C*Z)	ERR	GAL CORR	REF	COMMENTS
16 22 30	65 33.0		R	822	10	229	860	
16 23 0	41 8.0	1 ZW148	O	8079		193	262	
16 24 12	40 28.0		O	8753		193	321	
16 24 36	40 5.0		O	9724		192	321	
16 25 0	48 25.0	NGC 6155	O	2424		209	728	
16 25 6	40 48.0	3 ZW 78	O	9380		194	303	
16 25 18	39 26.0		O	9028	150	191	288	
16 25 18	41 22.0		O	8580		195	449	
16 25 24	39 38.0		O	9969	100	192	288	
16 25 24	41 16.6		O	3715		195	753	
16 25 36	20 10.0		O	3720		137	282	
16 25 48	51 40.0	VV 472	O	6185		215	774	
16 25 54	39 23.0		O	10514		191	321	
16 25 54	39 28.6		O	8743		192	466	
16 25 54	39 41.6		O	7968		192	466	
16 25 55	41 1.7		O	9440		195	255	
16 26 0	38 31.0	1 ZW152	O	10017		189	262	
16 26 0	39 30.0	NGC 6158	O	8884	150	192	288	
16 26 0	39 39.0		O	9128		192	466	
16 26 0	41 2.0	NGC 6160	O	9344		195	321	
16 26 2	27 48.2	3CR341	O O	134300 134400		161	946 718	
16 26 12	39 39.7		O	9146		192	466	
16 26 12	39 42.8		O	7683		192	466	
16 26 18	39 38.1		O	9131		192	466	
16 26 36	39 40.5		O	8878	75	192	56	
16 26 36	39 42.8		O	9497	75	193	56	

16ʰ26ᵐ

R.A. (1950)	DEC. (1950)	NAME	OBS	HEL. VEL (C*Z)	ERR	GAL CORR	REF	COMMENTS
16 26 42	39 37.1		0	10208	75	192	56	
			0	10288			466	
16 26 42	39 40.5		0	9355		193	466	
16 26 42	39 51.8		0	9458		193	466	
16 26 48	39 30.6		0	9158		192	466	
16 26 48	39 38.3		0	8417	75	193	56	
			0	8481			466	
16 26 48	39 39.0		0	9897	75	193	56	
16 26 48	41 20.0		0	8700		196	476	
			0	8502			321	
16 26 56	39 39.6	NGC 6166 3CR338	0	8889	75	193	56	
16 26 56	39 39.6	NGC 6166A	0	9287	75	193	56	
16 26 56	39 39.6	NGC 6166B	0	7767	75	193	56	
16 26 56	39 39.6	NGC 6166C	0	9857	75	193	56	
16 27 0	39 33.8		0	7598	75	192	56	
			0	7659			466	
16 27 0	39 35.0		0	9167	75	193	56	
			0	8183			466	DIS
16 27 0	39 35.5		0	10077	75	193	56	
			0	10064			466	
16 27 0	39 37.5		0	8317	75	193	56	
			0	8125			466	
16 27 0	39 38.4		0	7657	75	193	56	
			0	7537			466	
16 27 0	39 43.5		0	9347		193	466	*
			0	8217W	75		56	
16 27 0	39 44.0		0	8107	75	193	56	
			0	8203			466	
16 27 0	39 46.0		0	9027	75	193	56	
			0	9104			466	
16 27 2	39 32.6		0	10548	75	192	56	
16 27 2	39 44.9		0	8587	75	193	56	

R.A. (1950)	DEC. (1950)	NAME	OBS	HEL VEL (C*Z)	ERR	GAL CORR	REF	COMMENTS
16 27 3	39 38.2		O	8267	75	193	56	
16 27 5	39 45.5		O	8017	75	193	56	
16 27 6	39 39.6		O	8746	75	193	56	
16 27 6	39 57.0		O	10798		193	321	
16 27 12	39 41.0		O	8794		193	322	
16 27 12	39 47.5		O	8490		193	466	
16 27 12	39 48.5		O	9389		193	466	
16 27 18	39 35.1		O	8647	75	193	56	
			O	8609			466	
16 27 18	39 48.0		O	8115		193	322	
16 27 18	40 38.0	3 ZW 80	O	20840		195	282	
16 27 30	23 26.8	3CR340	O	92900		149	946	
			O	93000			718	
16 27 30	39 29.9		O	8578		193	466	
16 27 30	39 31.8		O	8185		193	466	
16 27 30	41 6.0		O	10330		196	449	
16 27 36	17 12.0		O	4590		129	282	
16 27 42	40 59.0	3 ZW 82	O	8182		196	322	
16 27 47	26 33.1	MRK 883	O	11407		159	874	
			O	11241			630	
16 28 4	27 48.3		R	2617	10	163	860	
16 28 6	40 55.0	NGC 6173	O	8727		196	321	
16 28 42	41 2.0		O	9092		197	322	
16 28 48	40 50.0		O	8900		196	322	
16 28 54	40 40.0		O	9110		196	322	
16 28 54	41 36.0		O	9072		198	321	
16 29 43	67 29.1	MRK 885	O	7636		232	874	
16 30 9	19 55.9	NGC 6181	R	2379	20	140	544	
			R	2376			744	
			R	2371	7		914	
			O	2350			107	
			O	2158	250		13	

359

R.A. (1950)	DEC. (1950)	NAME	OBS	HEL VEL (C*Z)	ERR	GAL CORR	REF	COMMENTS
16 30 30	5 38.0		O	45716		90	444	
			O	45467			444	
16 30 48	59 45.0	NGC 6189	O	5523		227	728	
16 31 12	58 33.0	NGC 6190	R	3355	15	226	752	
			O	3416			728	
16 31 18	35 1.0	1 ZW156	O	10793		184	262	
16 31 48	29 4.0	VV 625	O	931		169	774	
16 32 24	41 27.0		O	8682		200	321	
16 33 12	46 30.0	IC 1221 KDG 500A	O	5479	22	210	908	
16 33 42	46 19.0	IC 1222 KDG 500B ARP 73	R O O	9222 9123 9298	10 19 40	210	582 908 582	
16 34 0	52 20.0	1 ZW159	O O	2642 2650		220	262 389	
16 34 12	1 47.0		O	7358	25	78	582	
16 34 33	81 38.6		O	11411	98	231	381	
16 34 48	39 7.0	NGC 6195	O	9000	30	196	582	
16 35 3	78 18.0	NGC 6217 ARP 185	R R R R R R O O	1359 1355 1325 1363 1370 1356 1386 1382	10 25 25 30	233	744 455 363 390 693 183 13 13	
16 35 39	66 18.6		O	51300		233	717	
16 36 0	85 36.0	7 ZW653	O	18780	100	228	820	
16 36 1	66 20.1		O	51000		233	717	
16 36 5	66 20.4		O	6900		233	717	
16 36 14	-69 17.1	NGC 6183	O	4880		-182	574	
16 36 42	42 2.0	ARP 125A	O	8572	50	204	141	
16 36 42	42 2.0	ARP 125B	O	8118		204	141	

R.A. (1950)	DEC. (1950)	NAME	OBS	HEL VEL (C*Z)	EPR	GAL CORR	REF	COMMENTS
16 37 3	-68 48.0		O	(4200)	200	-181	794	
16 37 6	-77 10.1	PKS	O	12890	150	-201	355	
			O	13130			317	
16 37 22	29 56.8	B2	O	26040	250	175	696	
16 37 37	-68 50.9		O	4350	200	-181	794	
16 37 55	62 40.6	3CR343.1	O	224800		232	717	
16 37 57	82 38.3	NGC 6251	O	6900		231	668	
		NB 82.22	O	6900	600		824	
16 38 5	-68 52.7		O	4800	200	-181	794	
16 38 6	37 16.0		O	2100		194	476	
16 38 21	72 28.2		R	4299	10	235	752	
16 38 24	53 52.2	4C 35.37	O	32920		224	595	
16 38 35	32 11.1	B2	O	41700	250	182	696	
16 38 38	11 49.8	MC 2	O	23400		119	652	
16 39 36	-77 24.0	IC 4608	O	2928	30	-201	599	
16 39 36	58 11.0		R	5418	10	229	582	
			O	5405	20		582	
16 41 18	36 55.7	NGC 6207	R	852	35	196	238	
			R	851	7		914	
			R	854			744	
			R	870	50		544	
			O	851	15		813	
			O	869	40		13	
			O	835			547	
16 41 35	17 21.3	3CR346	O	48570	300	140	585	
16 41 49	-72 32.8		O	13490	200	-189	794	
16 41 55	-72 33.8		O	12290	200	-189	794	
			O	12839	120		680	
16 43 27	27 25.5	B2	O	30300	250	173	696	
16 44 3	11 50.0	MC 2	O	25500		123	465	
16 45 0	70 53.0	NGC 6236	R	1279	10	237	752	
16 45 18	68 1.0	KDG 501	O	8089	45	237	908	

R.A. (1950)	DEC. (1950)	NAME	OBS	HEL VEL (C*Z)	ERR	GAL CORR	REF	COMMENTS
16 45 24	68 2.0	KDG 501	O	7500	45	237	908	
16 45 28	17 25.5	4C 17.71	O	94135		143	471	
16 46 6	58 32.0	IC 1231	R	2938	30	232	752	
			O	5107			777	DIS
16 46 24	-61 43.8		O	4528	45	-158	680	
16 46 26	-61 43.6		O	4593	80	-158	680	
16 46 44	-58 54.5	NGC 6215	R	1564		-149	671	
			O	1510			574	
			O	1547			84	
16 47 3	48 47.6	1 ZW166	O	7634		222	262	
		MRK 499	O	7680	100		521	
			O	7730			435	
			O	7800			359	
16 47 14	48 48.0	MRK 500	O	7710		222	435	
			O	7800			359	
16 47 54	53 30.0	ARP 330A	O	8865		228	274	MRK CH
16 47 58	45 32.5	ARP 103A	O	9418	20	217	59	*
16 48 0	45 32.6		O	9187		217	59	KNOT
16 48 0	45 32.6		O	9360		217	59	KNOT
16 48 0	45 32.7	ARP 103B	O	9405	30	217	59	
16 48 0	53 30.0	ARP 330B	O	8009		228	274	
16 48 0	53 31.0	ARP 330C	O	8720		228	274	
16 48 0	53 32.0	ARP 330D	O	8942		228	274	
16 48 5	45 34.8	ARP 103C	O	9449	68	217	59	
16 48 6	53 33.0	ARP 330E	O	8603		228	274	
16 48 6	53 33.0	ARP 330F	O	7706		228	274	
16 48 12	45 33.0		O	9386	50	217	13	
16 48 18	46 48.0	VV 197A/B ARP 312	O	11642		219	274	
16 48 25	-59 8.0	NGC 6221	R	1482		-149	671	
			O	1418			84	
			O	1320			574	

R.A. (1950)	DEC. (1950)	NAME	OBS	HEL VEL (C*Z)	ERR	GAL CORR	REF	COMMENTS
16 48 30	42 49.4	NGC 6239	R	946	20	212	544	
			R	931	16		238	
			O	964			13	
16 48 38	-50 7.5	NGC 6221A	O	1530		-118	574	
16 48 40	5 4.6	3CR348	O	46200		102	70	
16 48 48	28 56.0	MRK 1108	O	9229		181	921	
16 48 54	55 38.0	NGC 6246	O	5214	71	231	789	
16 49 36	53 45.0		O	9600		229	476	
16 50 30	2 29.6	NGC 6240	O	7298	33	94	789	
		VV 617	O	7198	30		871	
		4C 02.44	O	7400	30		355	
16 50 42	-76 55.0	IC 4618	O	3022	27	-198	599	
16 51 30	81 31.0		O	11497	107	233	381	
16 51 36	63 12.0	IC 1235 MRK 1109	O	2642		238	921	
16 51 48	69 1.0	MRK 1110	O	3451		239	921	
16 52 0	81 30.0		O	10838	188	233	381	
16 52 12	39 50.4	MRK 501 4C 39.49	O O	9890 10040		209	485 471	
16 53 0	36 35.0	NGC 6255	O	824	41	202	789	
			O	(841)	40		813	
16 53 0	81 56.0		O	20208	184	233	381	
16 53 12	53 11.6	DDO 206	R	1094	10	230	492	
16 53 38	66 50.1		O	41100		239	717	
16 54 0	81 42.0		O	11527	55	233	381	
16 54 30	28 4.0	NGC 6261	O	10270		182	950	
16 54 42	27 54.0	NGC 6263	O	9890		182	950	
16 55 8	63 19.2	NGC 6275 MRK 503	O O	6600 6790		239	358 435	
16 55 12	-77 37.5	PKS	O	(20090)	210	-199	355	
16 55 18	27 56.0	NGC 6264	O	10140		182	950	

R.A. (1950)	DEC. (1950)	NAME	OBS	HEL VEL (C*Z)	EPR	GAL CORR	REF	COMMENTS
16 55 18	28 16.0	MRK 1112	O	10367		183	920	
16 55 24	34 6.0	1 ZW169A	O	25440		198	303	
16 55 24	34 6.0	1 ZW169P	O	25215		198	303	
16 55 30	27 55.0	NGC 6265	O	9245		182	950	
16 56 0	27 56.0	NGC 6269	O	9970		183	950	
			O	10350	33		843	
16 56 0	81 39.6		O	11409	98	233	381	
16 56 0	81 40.0		O	11490	69	233	381	
16 56 0	81 40.4		O	11153	243	233	381	
16 56 18	20 7.0	IC 1236	R	6030	15	160	582	
		VV 442	O	6100			774	
			O	6073	25		582	
16 56 18	38 17.0	2 ZW 75	O	10075		208	303	
16 56 46	70 31.2		R	1328	10	240	860	
16 57 0	81 41.0		O	11486	119	234	381	
16 57 9	32 34.1	4C32.52A	O	18890		196	595	
			O	18890			587	
			O	18710	250		696	
16 57 10	32 33.9		O	18770		196	587	
16 57 30	81 36.0		O	11022	117	234	381	
16 57 30	81 43.3		O	12016	148	234	381	
16 57 51	32 41.2	4C32.52C	O	29500		196	595	
			O	29620			587	
16 58 0	81 44.0		O	11062	56	234	381	
16 58 2	23 3.6	NGC 6267	R	2974	15	170	829	
16 58 4	47 7.3	3CR349	O	61460	300	225	585	
16 58 19	32 37.1		O	30940		197	587	
16 58 19	32 39.6	4C32.52E	O	30360	250	197	696	
			O	30700			587	
16 58 30	81 45.0		O	12094	156	234	381	

R.A. (1950)			DEC. (1950)		NAME	OBS	HEL VEL (C*Z)	ERR	GAL CORR	REF	COMMENTS
16	58	42	32	45.0	MRK 1114	O	8163		197	921	
16	58	44	23	5.0	NGC 6278	R	2776	15	171	829	
16	58	49	30	12.5	4C 30.31	O	10320	250	191	696	
16	59	0	81	44.4		O	11848	45	234	381	
16	59	0	81	44.7		O	12074	34	234	381	
16	59	10	29	28.8	MRK 504	O	11000		189	435	SEYF
						O	10800			359	
17	0	6	59	4.0	NGC 6290	O	12300		239	476	
17	0	52	70	21.6		R	443	10	241	860	
17	1	6	33	8.0	MRK 1115 1 ZW172	O	18960		200	921	
17	1	21	31	31.4	MRK 700	O	10100		196	557	SEYF
17	4	54	34	8.0	1 ZW173	O	9096		204	262	
17	4	1	78	47.80		O	16038	100	237	633	
17	5	3	78	39.90		O	18362	100	237	633	
17	5	12	42	30.0		O	11700		221	476	
17	5	16	-74	4.7		O	11778	65	-188	680	
17	5	19	-74	4.7		O	12763	75	-188	680	
17	5	29	78	48.55		O	20071	100	237	633	
17	5	52	78	39.99		O	16303	100	238	633	
						O	16820	150		545	
17	6	19	78	41.96	NB 78.26	O	17584	100	238	633	
						O	17570			595	
						O	17000			868	
						O	17090	150		545	
17	6	21	78	43.95		O	17602	100	238	633	
						O	17420	120		545	
17	6	24	78	41.83	NGC 6331A	O	16912	100	238	633	
						O	16850	150		545	
17	6	26	78	41.78	NGC 6331B	O	15830	100	238	633	
						O	16850	150		545	

17ʰ6ᵐ

R.A. (1950)	DEC. (1950)	NAME	OBS	HEL VEL (C*Z)	ERR	GAL CORR	REF	COMMENTS
17 6 30	78 51.70		0	19800	100	237	633	
17 6 50	78 48.7		0	16580	180	238	545	
17 6 59	60 47.5	NGC 6306	0	3064	35	242	148	
		HOL 769B	0	2820			445	
		KDG 504						
17 7 0	75 29.0	NGC 6324	0	4800		240	491	
17 7 0	81 48.0		0	11598	126	235	381	
17 7 3	60 48.8	NGC 6307	0	2820		242	445	
		HOL 769A	0	3283	44		148	
		KDG 504						
17 7 5	78 41.71		0	16903	100	238	633	
17 7 18	78 42.40		0	17720	100	238	633	
17 7 24	61 4.0	NGC 6310	0	3386	71	242	789	
17 7 40	78 42.25		0	19276	100	238	633	
17 7 42	63 43.0	KDG 505A	0	8060	30	243	908	
17 7 50	34 29.6	4C 34.45	0	24030		207	400	
17 8 25	78 39.09		0	18746	100	238	633	
17 9 2	78 36.68		0	18494	100	238	633	
17 9 17	39 45.2	B2	0	18620		219	595	
17 9 18	46 5.1	3CR352	0	241600	300	229	849	
17 9 54	23 26.4	NGC 6308	0	8693	43	180	789	
			0	8866			910	
17 10 30	-73 6.0		0	5152	43	-184	599	
17 10 33	23 19.8	NGC 6314	0	6748		180	13	
17 10 40	23 16.9	NGC 6315	0	6672	58	180	789	
17 10 42	64 5.0		0	25591	100	244	853	
			0	24730			303	
17 10 54	63 4.0		0	21440		244	303	
17 10 54	63 56.0		0	23820		244	303	
			0	23615			401	
			0	23820			215	

R.A. (1950)	DEC. (1950)	NAME	OBS	HEL VEL (C∗Z)	ERR	GAL CORR	REF	COMMENTS
17 10 54	64 3.0		O	21440		244	215	
			O	21306			401	
17 11 0	64 8.0		O	24683		244	303	
17 11 12	63 54.0		O	24858		244	401	
17 11 12	63 56.0		O	23452	100	244	853	
17 11 16	72 21.9	NGC 6340	R	1902	60	243	238	
			R	1903	100		544	
			R	1193			538	
			O	2109	300		13	
17 11 18	64 7.0		O	26250		245	853	
17 11 30	38 8.0	1 ZW176	O	10818		217	262	
17 11 36	64 7.0		O	25460	100	245	853	
17 11 42	64 10.0		O	24493	150	245	590	
17 11 57	64 5.6		O	21761	150	245	590	
			O	21315	100		853	
			O	21440	100		853	
			O	21505			903	
17 12 0	64 23.0		O	23110	100	245	853	
17 12 5	64 5.6		O	23922	150	245	590	
			O	23820	100		853	
			O	23920	100		853	
			O	23982			903	
17 12 6	64 6.0		O	25930	100	245	853	
17 12 6	64 8.0		O	21818		245	303	
			O	21786			401	
			O	22215	100		853	
			O	21820	100		853	
17 12 12	64 4.0		O	25390	100	245	853	
17 12 12	64 7.0		O	24680	100	245	853	
			O	24940	100		853	
17 12 12	64 8.0		O	24561	150	245	590	
			O	24850	100		853	
17 12 17	-62 45.9	NGC 6300	R	1110		-152	671	
			O	1053	23		741	
			O	1270	90		113	

17ʰ12ᵐ

R.A. (1950)	DEC. (1950)	NAME	OBS	HEL VEL (C*Z)	ERR	GAL CORR	REF	COMMENTS
17 12 18	64 8.0		0	25200	100	245	853	
17 12 24	23 6.0		0	8700		181	533	
17 12 30	59 23.0	KDG 506A	0	1171	27	244	812	
		ARP 32	0	1126	41		812	
		VV 89						
17 12 30	64 7.0		0	22060	100	245	853	
17 12 30	64 8.0		0	21840	100	245	853	
17 12 36	59 24.0	KDG 506B	0	1028	74	244	812	
		ARP 32						
		VV 89						
17 12 36	64 4.0		0	24960	100	245	853	
17 12 43	64 6.6		0	24700		245	903	
			0	24800	100		853	
			0	24600	100		853	
17 12 46	64 10.4		0	24292	150	245	590	
			0	24034	100		853	
			0	24161			903	
			0	24160	100		853	
17 12 56	63 51.1		0	24875		245	903	
17 13 0	64 7.0		0	22460	150	245	590	
17 13 0	64 8.0		0	23994	150	245	590	
17 13 10	64 6.2		0	23682	150	245	590	
			0	23606			903	
			0	23530	100		853	
17 13 12	63 54.0		0	25650	100	245	853	
17 13 24	64 8.3		0	24938	150	245	590	
			0	24834			903	
			0	24730	100		853	
17 13 32	64 3.6		0	26512		245	903	
17 13 42	-59 6.4	NGC 6305	0	2770		-138	574	
17 14 24	64 4.0		0	24533	100	246	853	
17 14 42	21 41.0	KDG 507	0	8520	18	178	908	
17 14 48	21 40.0	KDG 507	0	(8771)	60	178	908	

R.A. (1950)	DEC. (1950)	NAME	OBS	HEL VEL (C*Z)	ERR	GAL CORR	REF	COMMENTS
17 14 50	64 6.1		O	23967		246	903	
17 14 54	64 6.0		O	23837	100	246	853	
17 15 16	64 43.0		O	23813		246	903	
17 15 30	40 54.0	NGC 6339	R	2112	10	224	752	
17 15 36	75 15.4		R	1228	10	242	860	
17 15 48	64 6.0		O	23229	100	246	853	
17 15 54	64 13.0		O	22698	100	246	853	
17 16 36	48 31.0	1 ZW178	O	8431		236	262	
17 17 3	22 48.1	4C 22.45	O	75760		183	400	
17 17 22	61 49.9	NGC 6359	O	2948	75	246	13	
			O	3000	75		2	
17 17 35	14 27.0	DDO 207	R	1557	7	157	492	
17 17 53	- 0 55.8	3CR353	O	9120		102	120	
17 17 57	49 1.9	KDG 508A	O	7250	24	238	59	
		ARP 102	O	7255			748	
		VV 10						
17 18 4	49 5.5	KDG 508B	O	7182	60	238	59	
		ARP 102	O	7187			748	
		VV 10						
17 18 46	-64 57.8	PKS	R	4260	20	-157	635	
			O	4360	90		635	
17 18 46	-64 54.8		O	4330	250	-157	635	
17 19 24	57 58.1	DDO 208	O	-277	35	246	704	DRA SYS
17 19 57	24 16.6	4C 24.41	O	26130		189	400	
17 20 27	24 48.0		O	19270	60	191	826	*
			O	19111			427	
17 20 45	30 55.0	KDG 510A	O	13487	60	207	908	
			O	13500			568	
17 20 46	30 55.5	MRK 506	O	13172	50	207	567	SEYF
		KDG 510B	O	12900			568	
			O	12833			447	
			O	12900			388	
			O	12300			358	
			O	13008	40		908	

R.A. (1950)	DEC. (1950)	NAME	OBS	HEL VEL (C*Z)	ERR	GAL CORR	REF	COMMENTS
17 21 12	-64 54.1	NGC 6328	O	4240		-156	574	*
17 22 18	62 13.0	NGC 6365A KDG 511A ARP 30 VV 232	O	7904	26	248	908	
17 22 18	62 13.0	NGC 6365B KDG 511B ARP 30 VV 232	O	8343	40	248	908	
17 23 33	75 53.5	IC 4660	R	1225	15	243	829	
17 23 36	46 4.0	1 ZW183	O	18650	100	237	820	
17 24 12	45 41.0	1 ZW184	O	11050		236	303	
17 24 12	51 31.0	1 ZW185	O	18460		243	303	
17 24 18	13 58.0	KDG 515	O	(6494)	94	161	812	
17 24 24	13 57.0	KDG 515	O	6914	56	161	812	
17 24 36	58 52.0	NGC 6376 KDG 516	O O	8721 8562	52 28	248	500 908	
17 24 42	58 52.0	NGC 6377 KDG 516	O	8434	31	248	908	
17 24 48	11 35.0	NGC 6368	R	2765	10	153	752	
17 25 18	37 53.0		O	7800		224	476	
17 25 30	26 30.0	NGC 6372	R R O	4746 4751 4739	15 25	199	779 582 582	
17 26 27	31 48.4	3CR357	O O O	50103 49850 49810	160 250	212	741 318 696	
17 26 30	58 35.0	IC 1258/59 ARP 310 VV 101	O	7916		249	274	
17 26 36	58 35.0	IC 1259A/B ARP 311 KDG 517	O	(13600)		249	535	
17 26 36	58 35.0	IC 1259A	O	7676	46	249	908	
17 26 36	58 35.0	IC 1259B	O	8092	123	249	908	

R.A. (1950)	DEC. (1950)	NAME	OBS	HEL VEL (C*Z)	ERR	GAL CORR	REF	COMMENTS
17 26 36	60 4.0	NGC 6381A	O	3430	58	249	789	
		KDG 518	O	3267	17		908	
17 26 42	60 3.0	NGC 6381	O	3275	50	249	789	
		KDG 518	O	3164	18		908	
17 27 6	50 15.0	1 ZW187	O	16610	90	243	722	
17 27 12	45 35.0	1 ZW188	O	11135		238	262	
17 27 36	57 36.0	1 ZW189	O	8540	100	249	820	
17 27 48	-60 42.0		O	5085	26	-139	765	
17 27 52	-60 42.0		O	5278	53	-139	765	
17 28 24	16 20.0	NGC 6379	R	5966	20	171	582	
			O	5964	90		582	
17 29 59	7 5.8	NGC 6384	R	1649	50	141	544	
			R	1664			886	
			R	1665			744	
			O	1784	50		13	
			O	1781	38		72	
			O	1717			13	
17 30 30	16 26.0	NGC 6389	R	3118	10	173	752	
			O	3131	30		813	
			O	3059			910	
17 31 22	75 44.3	NGC 6412	R	1320	10	244	829	
		ARP 38	R	1333			693	
			R	1320			744	
			O	1508			13	
17 31 36	59 59.0		O	9000		251	476	
17 31 41	23 52.9		R	4040	10	196	860	
17 35 30	68 7.0		O	5400		251	476	
17 36 0	86 47.0	MRK 1116	O	7509		231	921	
17 36 45	32 57.6	B2	O	21980	250	221	696	
17 38 0	59 25.0	IC 1267	R	9308	15	253	582	
			O	9311	50		582	
17 38 36	39 16.0	MRK 1117	O	12986		234	921	
17 39 0	47 46.0	1 ZW191	O	5800		246	262	
17 39 0	51 4.0	KDG 521	O	(6221)	150	249	908	

371

R.A. (1950)	DEC. (1950)	NAME	OBS	HEL VEL (C*Z)	ERR	GAL CORR	REF	COMMENTS
17 39 0	51 6.0	KDG 521	O	6451	37	249	908	
17 39 18	38 44.0		O	12300		234	533	
17 40 0	25 39.0		O	7800		206	476	
17 41 6	39 2.0	4C 39.50	O	12680		235	400	
17 42 12	-64 37.3	IC 4662	O	375		-149	84	
			O	406			254	
			O	377			145	
			O	269			227	
17 42 25	33 0.6	NRAO	O	22694		225	595	
			O	22800			868	
17 42 30	38 5.0	1 ZW193	O	15765		234	262	
17 42 30	40 53.0		O	1800		239	476	
17 44 0	55 43.0	NGC 6454	O	9110	200	254	350	
		4C55.331	O	9180			522	
17 44 18	35 35.0	NGC 6446	O	7500		231	476	
		KDG 523A	O	6850	39		908	
17 44 18	39 43.0	1 ZW194	O	12330		238	303	
17 44 30	35 35.0	NGC 6447 KDG 523B	O	6734	15	231	908	
17 44 36	30 43.0	VV 426A/B KDG 524	O	4579		221	774	
17 44 36	30 43.0	KDG 524A	O	4655	20	221	908	
17 44 42	30 43.0	KDG 524B	O	4459	20	221	908	
17 44 49	69 42.3		O	54600		252	717	
17 45 13	69 41.6		O	10800		252	717	
17 46 36	67 21.0		R	1460	10	254	860	
17 47 0	20 49.0	NGC 6458 KDG 525	O	3200	50	197	908	
17 47 12	18 35.0		O	3300		191	533	
17 47 24	20 46.0	NGC 6460 KDG 525	O	3307	33	197	908	
17 47 28	51 10.2	NGC 6478 NGC 6466	R O	6776 6857	15 50	253	752 13	

R.A. (1950)	DEC. (1950)	NAME	OBS	HEL VEL (C*Z)	ERR	GAL CORR	REF	COMMENTS
17 47 42	36 9.0		O	900		234	476	
17 47 56	30 18.9	B2	O	38630	250	222	696	
17 48 55	68 42.8	MRK 507	O	16750		254	435	*
			O	15900			358	
17 49 12	56 41.0	1 ZW199N/S	R	5198		256	389	
			O	5298			262	
17 49 12	56 41.0	1 ZW199N	O	5416		256	303	
17 49 12	56 41.0	1 ZW199S	O	5468		256	303	
17 49 43	24 29.6	NGC 6484	R	3112	10	209	829	
		MRK 1118	R	3112	15		752	
			O	2611			921	
17 49 44	23 5.0	NGC 6482	O	3922	60	205	13	
17 49 57	70 9.5	NGC 6503	R	60		253	216	
			R	35			934	
			R	30			744	
			R	80	25		183	
			O (	0)	210		555	
			O	33			13	
			O	13	32		72	
			O	60			89	
17 50 0	21 35.0	IC 1269	R	6109		201	779	
			R	6121	10		582	
			O	6072	20		582	
17 50 36	49 2.0	1 ZW200	O	21650		253	303	
17 50 36	54 0.0	1 ZW201	O	15921		256	262	
17 50 54	37 45.0	MRK 1119	O	2672		238	921	
17 51 36	47 45.0	1 ZW203	O	7290	100	252	820	
17 52 45	32 34.8	B2	O	13220	250	230	696	
17 53 0	34 47.0		O	4800		234	491	
			O	5100			476	
17 53 48	18 20.7	NGC 6500	O	3100		194	868	
		KDG 526	O	2937			910	
			O	2939	32		812	
			O	3006			748	
			O	2925	160		473	
			O	3003	50		473	
			O	2950	72		500	

17ʰ53ᵐ

R.A. (1950)	DEC. (1950)	NAME	OBS	HEL VEL (C*Z)	ERR	GAL CORR	REF	COMMENTS
17 53 52	18 22.8	NGC 6501	R	2856		194	780	
		KDG 526	O	2950			748	
			O	3068			910	
			O	2851	64		812	
17 54 24	62 39.0	NGC 6512	O	8420		258	595	
17 54 42	12 15.0	KDG 527	O	2941	20	176	908	
17 54 54	12 11.0	KDG 527	O	2852	20	175	908	
17 55 0	32 38.0		R	4752	10	231	582	
			O	4757	100		582	
17 55 0	40 15.0	MRK 1120	O	12316		244	921	
17 55 54	21 17.0		O	6000		204	476	
17 56 13	-81 58.3		O	17990	200	-202	794	
17 57 0	6 17.0	NGC 6509	R	1815	10	156	752	
17 58 30	34 38.0		O	7800		237	491	
17 59 30	6 58.0		O	1931	71	160	789	
18 2 42	-57 44.0		O	5180	130	-115	670	
18 2 50	46 43.7		R	1565	10	256	860	
18 3 11	23 5.9		R	2426	10	213	860	
18 3 25	50 2.2		O	8572		259	708	
18 3 36	18 32.0	NGC 6549 KDG 529	O	6562	25	201	908	
18 3 48	18 35.0	NGC 6550 KDG 529	O	2193	41	201	908	
18 3 54	46 52.0	NGC 6560	R	7040	20	257	582	
			O	7034	25		582	
18 5 24	65 54.0		O	6220		260	445	
18 5 37	17 36.0	NGC 6555	R	2226	10	199	752	
		HOL 774A	R	2224			744	
			O	2257	53		789	
18 5 40	35 33.4		R	1608	10	242	752	
18 6 48	28 2.0		O	6900		228	476	

R.A. (1950)	DEC. (1950)	NAME	OBS	HEL VEL (C*Z)	ERR	GAL CORR	REF	COMMENTS
18 7 4	50 1.2		O	15281		260	708	
18 7 18	69 49.0	3CR371	O	15300		257	236	
		7 ZW768	O	15000			134	
			O	15000	30		741	
			O	13700			177	
18 8 28	49 56.2		O	15778		261	708	
18 8 36	30 59.0	IC 1277 KDG 530	O	7003	18	235	908	
18 8 50	14 4.8	NGC 6570	R	2294		190	744	
			R	2283	10		829	
			O	2189	52		789	
			O	1970	70		555	
18 9 0	31 5.0	NGC 6575 KDG 530	O	6992	15	235	908	
18 9 15	49 54.6		O	15433		261	708	
18 9 18	-57 45.0	IC 4686	O	4950	220	-113	598	
18 9 18	-57 44.0	IC 4687	O	5110	260	-113	598	
18 9 18	-54 46.0	IC 4689	O	5260	290	-101	598	
18 9 24	-85 26.0	NGC 6438SYS	O	4401		-211	135	*
			O	2578			316	DIS
18 9 24	-85 26.0	A	O	6300		-211	135	
			O	2522			316	DIS
			O	2563			440	DIS
18 9 24	-85 26.0	B	O	4300		-211	135	
			O	2596			316	DIS
			O	2656			440	DIS
18 9 24	-85 26.0	C	O	2680		-211	135	
			O	2600			316	
18 9 30	31 51.0	MRK 1121	O	4533		237	920	
18 9 35	14 58.2	NGC 6574	R	2261	20	194	829	
			O	2355	50		13	
			O	2387			13	
			O	2270			212	
18 9 41	50 0.0		O	14821		261	708	
18 9 43	49 31.0		O	14240		261	708	

R.A. (1950)	DEC. (1950)	NAME	OBS	HEL VEL (C*Z)	ERR	GAL CORR	REF	COMMENTS
18 9 47	49 54.0	NGC 6582A KDG 531	O O	14120 14167	63	261	708 908	
18 9 51	49 53.8	NGC 6582B KDG 531	O O	14355 14525	30	261	708 908	
18 9 52	50 9.4		O	14325		262	708	
18 9 55	49 51.1		O	16215		261	708	
18 11 6	-58 14.0	IC 4694	O	2500	70	-115	598	
18 11 24	33 48.0	KDG 532	O	(6825)	60	242	908	
18 11 31	49 50.8		O	15254		262	708	
18 11 36	33 43.0	KDG 532	O	7022	45	242	908	
18 11 40	18 48.6	NGC 6587	O	3002		206	910	
18 12 36	-57 15.0		O	5320	80	-110	670	
18 13 6	29 45.0		O	5700		235	476	
18 13 12	68 20.0	NGC 6621/2 ARP 81 VV 247	O	6230		260	77	
18 13 12	68 20.0	NGC 6622 KDG 534	O O	5941 6456	15 30	260	500 908	
18 13 12	68 21.0	NGC 6621 KDG 534	O O	6194 6235	85 30	260	500 908	
18 13 47	49 24.5		O	9871		263	708	
18 15 7	-77 43.0		O	(1350)	200	-187	794	
18 15 7	-77 44.5		O	(1350)	200	-187	794	
18 15 39	70 59.0		R	1503	10	258	860	
18 15 47	48 55.4		O	9980		263	708	
18 16 0	30 38.0	KDG 535A/B VV 569	O	5362		238	774	
18 16 0	30 38.0	KDG 535A	O	5438	65	238	812	
18 16 0	30 38.0	KDG 535B	O	5584	15	238	812	
18 17 25	50 15.3		O	7809		265	708	

R.A. (1950)	DEC. (1950)	NAME	OBS	HEL VEL (C*Z)	ERR	GAL CORR	REF	COMMENTS
18 20 24	15 40.4	NGC 6627	O	5206	100	202	13	
18 20 36	12 24.0		R	2646	5	192	752	
18 21 14	74 32.7	NGC 6643	R	1482	10	254	752	
			O	1440			904	
			R	1506	50		544	
			R	1501			693	
			R	1507	35		238	
			R	1482			744	
			R	1477	25		183	
			O	1494	50		13	
			O	1682			13	
18 21 48	40 55.0		R	1451	10	258	860	
18 22 0	66 35.0	NGC 6636A/B 7 ZW790	O	4500		264	814	
18 22 0	66 35.0	NGC 6636A KDG 536A	O O	4193 3995	51 37	264	500 908	
18 22 0	66 36.0	NGC 6636B KDG 536B	O O	4366 4214	82 27	264	500 908	
18 22 6	66 35.0		O	4800		264	814	*
18 23 0	27 30.0	NGC 6632	O	4691		235	910	
18 24 54	34 18.0	1 ZW206	O	7912		250	262	
18 25 6	42 39.0	MRK 1122	O	11938		262	921	
18 25 6	71 34.3	NGC 6651	O	5769	58	259	789	
18 25 14	73 9.2	NGC 6654 7 ZW793	R O	(1821) 1924	9	257	581 13	
18 25 21	14 47.2	NGC 6635	O	5071		202	13	
18 25 56	74 19.1	3CR379.1	O	77050	300	255	585	
18 27 5	32 18.0	B2	O	19490	250	247	696	
18 27 24	48 13.0		O	4930		267	449	
18 29 21	-60 9.1	IC 4718	O	3770		-117	574	
18 29 24	-41 32.0		O	5703	70	-33	603	
18 29 46	33 54.0		R	6122	25	251	829	
18 30 6	-58 32.0	IC 4721	O	5954	37	-110	741	

R.A. (1950)	DEC. (1950)	NAME	OBS	HEL VEL (C*Z)	ERR	GAL CORR	REF	COMMENTS
18 30 12	55 14.0	1 ZW207	O	5535		270	389	
			O	5537			262	
18 31 29	73 10.9		O	36900		258	868	
18 31 36	54 29.0	1 ZW208	O	8710		271	262	
18 31 49	22 50.9	NGC 6658	O	4270	50	228	13	
			O	4100	75		2	
18 32 11	-65 18.2		O	4650	200	-138	794	
18 32 24	47 24.6	3CR381	O	48149	30	269	741	
			O	48120	60		159	
18 32 30	22 52.0	NGC 6661	O	4370	50	229	13	
			O	4193			13	
			O	3900	100		2	
18 32 32	-65 17.5		O	4710	300	-138	832	
18 33 12	32 39.3	3CR382	O	17340		251	575	SEYF
			O	17340			120	
18 33 22	-65 28.3		O	3900		-138	697	SEYF
			O	3990	90		832	
18 33 30	67 4.0	VV 672	O	5334		266	774	
18 33 42	67 4.0	IC 4763	O	6600		266	534	
18 34 12	-72 13.7		O	5975		-165	526	
18 34 29	19 41.0	PKS	O	4740	120	221	350	
18 35 18	59 50.0	7 ZW822	O	8420	100	271	820	
18 35 24	70 29.0	NGC 6689' NGC 6690	O	526	20	262	813	
18 35 48	40 2.0	NGC 6675	R	2491	10	263	752	
18 36 13	17 9.1	3CR386	O	5100	300	215	585	*
			O	5070	30		291	
18 36 30	-64 3.6		O	4950	200	-132	794	
18 36 31	25 19.8	NGC 6674	R	3430	10	237	752	
			O	3502	50		13	
18 36 53	-63 59.8	IC 4741	O	4950	200	-131	794	
18 37 2	-63 54.7	IC 4742	O	4800	200	-131	794	

R.A. (1950)	DEC. (1950)	NAME	OBS	HEL VEL (C*Z)	ERR	GAL CORR	REF	COMMENTS
18 37 57	-64 7.4	IC 4748	O	4650	200	-132	794	
18 38 18	55 35.0	NGC 6691	R	5886	10	273	582	
			O	5858	20		582	
18 38 37	-64 9.2		O	3300	200	-132	794	
18 38 39	-62 9.5	IC 4751	O	4800	200	-123	794	
18 38 42	23 38.0		R	3700	10	234	752	
18 38 48	-62 9.4	IC 4753	O	3750	200	-123	794	
18 39 18	-62 2.4	IC 4754	O	4950	200	-123	794	
18 40 15	-62 25.0		O	4295	30	-124	764	
18 40 18	-62 20.9	NGC 6673	O	1200	200	-124	794	
18 40 35	73 31.9	NGC 6654A	R	1420		259	744	
18 41 4	-63 19.8		O	11270	200	-128	793	
18 41 10	-63 17.3		O	4500	200	-127	793	
18 41 28	-63 26.1		O	(4110)	200	-128	793	
18 41 39	-63 22.6		O	3810	200	-128	793	
18 42 5	-63 24.6		O	4050	200	-128	793	
18 42 5	-63 22.0		O	10790	200	-128	793	
18 42 6	-63 15.5		O	3510	200	-127	793	
18 42 27	-63 24.5		O	4230	200	-128	793	
18 42 30	-63 18.9		O	4470	200	-127	793	
18 42 33	-63 21.7		O	4080	200	-127	793	
18 42 34	-63 12.5		O	5010	200	-127	793	
18 42 35	45 30.4	3CR388	O	27230		271	120	
			O	27284			444	
18 42 36	-63 23.0	IC 4765	O	4400	300	-127	505	
			O	4800	200		793	
18 42 36	60 37.0	NGC 6701	O	3983		273	910	
18 42 40	-63 23.1		O	4800	200	-127	794	

18ʰ42ᵐ

R.A. (1950)	DEC. (1950)	NAME	OBS	HEL VEL. (C*Z)	ERR	GAL CORR	REF	COMMENTS
18 42 45	−63 16.4		0	3420	200	−127	793	
18 42 48	−63 20.6	IC 4766	0	5010	200	−127	793	
18 42 57	−63 27.5	IC 4767	0	3600	200	−128	793	
18 43 1	−63 14.2		0	(3150)	200	−127	793	
18 43 9	−63 12.6	IC 4769	0	4500	100	−127	505	
			0	4500	200		794	
18 43 12	−63 8.8		0	(10910)	200	−126	793	
18 43 15	−63 14.7		0	(4320)	200	−127	793	
18 43 17	−63 8.4		0	11030	200	−126	793	
18 43 23	−63 15.7		0	4800	200	−127	793	
18 43 23	−63 9.7		0	3900	200	−126	793	
18 43 26	−63 26.2	IC 4770	0	5040	200	−128	793	
18 43 27	−63 16.3		0	10730	200	−127	793	
18 43 45	−63 18.2	IC 4771	0	4860	200	−127	793	
18 44 5	−65 13.9	NGC 6684	0	901	28	−135	741	
			0	870			574	
			0	838			84	
18 44 6	22 34.0		R	4696	15	234	752	
18 45 26	74 41.5		0	17400	999	258	394	
18 45 31	45 39.0	NGC 6702	0	4749	65	272	13	
			0	4706	2		13	
			0	2250	75		2	DIS
18 45 37	79 43.0	3CR390.3	0	16829	45	248	741	SEYF
		7 ZW838	0	16790			177	
			0	16830			134	
			0	16830			575	
18 45 44	74 44.0		0	9300	900	258	394	
18 45 52	45 29.7	NGC 6703	0	2316	40	272	13	
			0	2000	75		2	
			0	2382	10		938	
			0	2394			13	
18 46 6	−78 57.3		0	8540	200	−188	794	

R.A. (1950)	DEC. (1950)	NAME	OBS	HEL VEL (C★Z)	ERR	GAL CORR	REF	COMMENTS
18 47 36	47 36.0	NGC 6711	R	4678	20	274	582	
			O	4639	30		582	
18 47 48	-57 23.0	NGC 6699	O	2957		-100	145	
			O	3420			574	
			O	3411	20		741	
			O	3440	100		844	
18 48 34	26 46.7	NGC 6710	O	4556	50	246	13	
			O	5100	100		2	
18 51 20	-53 53.0	NGC 6707	O	2759	31	-83	741	
18 51 35	-53 47.0	NGC 6708	O	2622	17	-82	741	
			O	2571	24		724	
18 52 26	-54 16.8	IC 4796	O	3076	28	-84	741	
18 52 28	-54 22.3	IC 4797	O	2620	34	-85	741	
			O	2685			84	
18 52 53	-54 36.9	A 1853	O	2761	45	-86	741	
18 54 36	25 10.0		R	4410	15	245	752	
18 55 54	37 56.5	B2	O	16280		268	485	
18 56 30	-57 51.0	NGC 6721	O	4500		-100	574	
			O	4416	50		741	
19 0 0	40 41.0	NGC 6745	O	4545		273	811	★ KNOT
19 1 59	33 46.2		R	4524	15	265	752	
19 3 18	-61 29.5	NGC 6739	O	4272	48	-115	741	
19 5 2	-63 56.3	NGC 6744	R	835		-125	671	
			O	659			84	
			O	790			574	
19 6 48	42 59.0		R	4556	10	278	582	
			O	4562	20		582	
19 7 2	50 51.1	NGC 6764	R	2414	15	281	752	SEYF
			R	2426			744	
			R	2412			707	
			O	2431	8		508	
			O	2417			910	
19 7 12	-57 8.0	NGC 6753	O	3066		-94	84	
			O	3170			574	
			O	3147			85	
			O	3145	39		741	

R.A. (1950)			DEC. (1950)		NAME		OBS	HEL VEL (C*Z)	ERR	GAL CORR	REF	COMMENTS
19	7	34	-50	43.4	NGC 6754		0	3282	46	-63	741	
19	8	27	-60	28.2			0	4664	160	-109	680	
19	8	40	-53	56.8			0	7328	110	-78	680	
19	8	41	-53	57.3			0	7078	45	-78	680	
19	8	54	-60	57.2	IC 4827		0	4349	53	-111	741	
							0	4450	200		670	
19	9	0	65	54.0	7 ZW		0	9890		274	282	
19	9	26	52	8.0			0	8150	60	282	826	*
19	9	44	-56	23.7	NGC 6758		0	3367	43	-90	741	
19	10	14	-62	22.8	IC 4831		0	4312	84	-117	741	
19	11	12	-54	45.1	IC 4837		0	2708	24	-82	741	
19	11	23	-50	45.6	NGC 6761		0	5650		-62	574	
19	11	33	-54	43.0	IC 4839		0	2700	100	-81	844	
19	11	42	-60	17.0	IC 4836		0	4310	100	-107	844	
							0	6800	120		670	DIS
19	11	54	73	18.0	NGC 6786		0	7997	20	263	500	
					KDG	538A	0	7471	34		812	
19	12	0	73	19.0	KDG	538B	0	7500	13	263	812	
19	12	15	-61	42.2	IC 4838		0	4364	95	-114	680	
19	13	57	-60	36.2	NGC 6769		0	3820		-108	574	
					VV	304	0	3920	120		670	
							0	3530			17	
							0	3943	85		741	
19	14	14	-60	36.2	NGC 6770		0	3863	16	-108	741	
					VV	304	0	3950	120		670	
							0	3630			17	
							0	3910			574	
19	14	18	-60	38.0	NGC 6771		0	4273	28	-108	741	
							0	4120			17	
							0	4290			574	
							0	4070	60		670	
19	15	1	-60	44.4	IC 4842		0	4089	23	-109	741	
							0	4275	50		670	

R.A. (1950)	DEC. (1950)	NAME	OBS	HEL VEL (C*Z)	ERR	GAL CORR	REF	COMMENTS
19 16 1	-60 28.6	IC 4845	O	3863	38	-107	741	
			O	4050	140		670	
19 16 49	70 50.7		O	51000		268	717	
19 16 57	-58 45.9		O	11129	150	-99	761	
19 17 30	44 8.32		O	11584	100	282	633	
19 17 37	44 4.01		O	12064	100	282	633	
19 17 43	44 4.42		O	11666	100	282	633	
19 18 45	43 42.95		O	19242	100	282	633	
19 18 45	43 57.04		O	15817	100	282	633	
19 18 46	-55 52.4	NGC 6780	O	3516	32	-85	741	
19 18 47	43 52.93		O	18229	100	282	633	
19 19 0	43 59.81		O	19201	100	282	633	
19 19 3	43 56.56		O	15049	100	282	633	
19 19 3	44 1.07		O	18293	100	282	633	
19 19 9	43 50.92		O	14954	100	282	633	
19 19 11	43 53.12		O	15914	100	282	633	
19 19 11	43 54.87		O	19879	100	282	633	
19 19 26	43 53.09		O	18402	100	282	633	
19 19 29	43 2.0	NGC 6792	R	4643	15	282	752	
			R	4628	10		829	
19 19 34	43 13.9		R	5478	10	282	829	
19 19 34	43 47.33		O	15631	100	282	633	
19 19 36	43 51.02		O	16229	100	282	633	
			O	16160			255	
19 19 38	-60 1.0	NGC 6782	O	3876	50	-104	741	
			O	3920			574	
19 19 38	43 44.30		O	18065	100	282	633	
19 19 38	43 48.73		O	16755	100	282	633	
19 19 40	43 37.54		O	15158	100	282	633	

R.A. (1950)	DEC. (1950)	NAME	OBS	HEL VEL (C*Z)	ERR	GAL CORR	REF	COMMENTS
19 19 42	43 49.16		0	14687	100	282	633	
19 19 44	43 37.92		0	16321	100	282	633	
19 19 45	43 49.73		0	15059	100	282	633	
19 19 46	43 54.16		0	10240	100	282	633	
19 19 47	43 39.29		0	16855	100	282	633	
19 19 49	43 34.86		0	15590	100	282	633	
19 19 52	43 47.54		0	16516	100	282	633	
19 19 53	43 55.85		0	18209	100	282	633	
19 19 54	43 29.61		0	16651	100	282	633	
19 19 57	43 50.34		0	15168	100	282	633	
19 19 58	43 53.97		0 0	13732 13701	100	282	633 910	
19 20 8	43 52.03		0	14743	100	282	633	
19 20 40	44 9.46		0	16926	100	283	633	
19 20 42	43 42.56		0	15735	100	282	633	
19 20 43	-63 57.9	NGC 6776	0 0	5736 5330	50	-122	741 574	
19 20 48	43 46.72		0	18978	100	282	633	
19 20 51	61 2.9	NGC 6796	0	2090		281	910	
19 20 59	43 57.82		0	14782	100	283	633	
19 21 35	-53 5.9		0	18575	33	-71	763	
19 22 0	-60 28.0	IC 4852	0 0	4770 4498	100 20	-106	844 582	
19 22 0	63 4.0	7 ZW880	0	6120		279	282	
19 24 27	-59 20.4		0	4200		-100	675	
19 24 29	-41 40.6		0 0 0	2884 2874 2894	20 19	-14	680 765 515	
19 25 12	50 1.0	IC 4867 KDG 539A	0	4088	67	286	908	

R.A. (1950)	DEC. (1950)	NAME	OBS	HEL VEL (C*Z)	ERR	GAL CORR	REF	COMMENTS
19 25 18	49 39.0	IC 1301	R	3992	10	286	582	
			O	3969	20		582	
19 25 18	50 2.0	KDG 539B	O	4005	10	286	908	
19 26 48	-39 31.0		O	2728	28	-3	765	
19 26 50	-39 31.0		O	2924	81	-3	765	
19 26 51	-39 31.0		O	2853	37	-3	765	
19 26 54	65 12.0	KDG 540	O	5791	10	278	908	
19 27 0	65 13.0	KDG 540	O	5704	10	278	908	
19 27 5	-17 47.0		R	-79	5	103	720	
19 28 24	-34 1.2	PKS	O	29380	60	24	355	
19 29 0	35 39.0	IC 1302	O	4575	50	277	13	
19 30 0	54 0.0	KDG 542A	R	3850	10	287	582	
			O	3922	20		582	
19 32 48	-65 55.5	IC 4870	O	889	29	-129	763	SEYF
19 33 10	-38 28.1	NGC 6805	O	6027	105	3	680	
19 33 42	-42 26.0	NGC 6806	R	5728	20	-15	752	
			O	5717	25		582	
19 34 48	-63 49.6	PKS	O	54600		-119	338	
19 35 42	40 35.5		R	3123	10	284	752	
			R	3128	20		829	
19 38 29	-70 46.6	NGC 6808	O	3460		-150	574	
			O	3468	21		741	
19 39 23	-58 46.6	NGC 6810	O	2025		-95	574	
			O	1823			84	
19 39 39	60 34.5	3CR401	O	60000		285	717	
			O	60260	300		585	
19 39 55	-10 26.6	NGC 6814	R	1560		141	707	SEYF
			R	1588	50		238	
			R	1565			744	
			R	1560	50		544	
			O	1437	40		13	
			O	1540			574	
19 40 23	50 30.9	3CR402 A	O	7902	40	290	741	*

19ʰ40ᵐ

R.A. (1950)	DEC. (1950)	NAME	OBS	HEL VEL (C*Z)	ERR	GAL CORR	REF	COMMENTS
19 40 23	50 30.9	C	0	7310		290	318	
19 40 23	50 30.9	D	0	8260		290	318	
19 40 26	50 28.7	B	0	7565	60	290	741	
		3CR402	0	7642	37		741	
			0	7160			318	
19 41 3	-54 22.3		0	5625	33	-73	763	SEYF
19 41 19	-54 18.6	IC 4891	0	2520		-73	574	
19 41 19	-54 27.6	IC 4889	0	2531	27	-74	741	
19 41 42	- 6 57.0	NGC 6821	R	1523		157	744	
19 42 7	-14 55.7	NGC 6822	R	-60		121	82	
		IC 4895	R	-62	5		93	
		DDO 209	R	-57			156	
			R	-50			58	
			R	-23			27	
			0	-32	2		11	
			0	-150	25		2	
19 42 7	-14 55.7	NGC 6822I	0	61		121	650	HII REG
		NGC 6822III	0	53			650	HII REG
		NGC 6822V	0	-36			13	HII REG
			0	-34	20		13	HII REG
		NGC 6822VI	0	-21	38		72	HII REG
19 42 18	-14 51.0	IC 1308	0	-30	30	122	13	HII REG
19 42 36	55 59.0	NGC 6824	0	3386	30	289	13	
			0	3200	75		2	
19 43 54	43 1.0		R	4651	10	288	752	
19 49 12	53 10.0	1 ZW209A	0	11290		291	303	
19 49 12	53 10.0	1 ZW209B	0	11425		291	303	
19 49 45	2 22.6	3CR403	0	16700		197	333	
19 49 55	- 1 25.1	3CR403.1	0	16600		182	318	
19 51 36	57 20.0		R	3562	10	289	582	
			0	3602	30		582	
19 51 45	-12 41.8	NGC 6835	R	1546	20	134	544	
			R	1568	10		238	
			0	1720			449	
			0	1720	23		148	
			0	1790	34		741	

R.A. (1950)	DEC. (1950)	NAME	OBS	HEL VEL (C*Z)	ERR	GAL CORR	REF	COMMENTS
19 51 54	-12 49.0	NGC 6836	R	1628		134	744	
19 52 50	67 31.9		R	2105	10	278	860	
19 53 51	-32 24.5		0	5400		40	577	
19 54 0	62 14.0	7 ZW921	0	15300	100	285	820	
19 54 20	-55 17.7	PKS	0	(18000)	90	-75	355	
19 54 29	40 18.0	A 1955A	0	4794		288	13	
19 55 6	40 16.5	A 1955B	0	4708		289	13	
19 57 18	-47 14.5	D	0	7070	43	-34	836	
19 57 20	-47 13.4	C	0	6938	73	-34	613	
			0	6755	17		836	
19 57 20	49 53.9		0	7366	26	293	741	
19 57 22	-47 12.5	NGC 6845	0	6559	102	-34	613	KLE 30
		A	0	6410			836	
			0	6350			380	*
19 57 22	-47 12.5		0	6000		-34	380	KNOT
			0	6170			836	
19 57 30	-47 11.9	B	0	6320	28	-34	613	
			0	6750			380	
			0	6777	4		836	
19 57 44	40 35.8	3CR405	0	16830		289	9	
			0	16924			444	
			0	16804	30		13	
19 59 32	-55 7.2	IC 4933	0	4990		-74	574	
19 59 38	-54 59.2	NGC 6850	0	4950		-73	574	
19 59 53	-48 25.2	NGC 6851	0	3000		-40	574	
			0	3117	48		741	
20 1 25	49 10.6		0	7071	49	294	741	
20 1 45	-54 32.3	NGC 6854	0	5700		-70	574	
			0	5680	50		741	
20 2 41	-48 32.3	IC 4943	0	2930		-40	574	
20 3 42	-48 30.8	NGC 6861 IC 4949	0	2859	32	-40	741	

20ʰ4ᵐ

R.A. (1950)	DEC. (1950)	NAME	OBS	HEL VEL (C*Z)	ERR	GAL CORR	REF	COMMENTS
20 4 12	62 38.0		R	3253	5	286	752	
20 4 42	-29 17.0		0	6986	25	58	582	
20 4 43	-48 21.2	NGC 6861D	0	2534	44	-39	741	
20 6 19	-48 31.4	NGC 6868	0	2763		-39	136	
			0	2776			84	
20 6 25	-56 39.9	PKS	0	17630	120	-80	350	
			0	(12860)	270		355	DIS
20 6 29	-48 26.4	NGC 6870	0	2800		-39	574	
			0	2660	75		741	
20 8 4	-51 32.3		0	3114	50	-54	680	
20 9 36	5 36.0		0	5232	20	214	582	
20 9 40	-46 18.5	NGC 6875	0	3103		-27	136	
20 10 5	-70 41.6	IC 4960	0	3519	41	-146	741	
20 10 25	-44 40.4	NGC 6878	0	5855	30	-19	582	
			0	5870	48		741	
20 11 6	-62 1.0	IC 4974	0	(11000)		-106	670	
20 11 7	-70 43.0	IC 4967	0	4153	7	-146	741	
20 11 14	-45 44.6		0	5071	85	-24	741	
20 11 24	-62 2.0	IC 4976	0	(11000)		-106	670	
20 11 41	-70 55.5	NGC 6872	0	4737	58	-147	741	
		VV 297	0	4840			574	
			0	4857			526	
			0	4850	250		894	
			0	4897	60		680	
20 11 44	-70 54.2	IC 4970	0	4755	59	-147	741	
		VV 297	0	4782	60		680	
			0	4702			227	
20 12 50	-44 27.3		0	5412	30	-17	680	
20 12 50	-44 27.0		0	5247	60	-17	680	
20 13 6	-71 1.0	NGC 6876	0	3728		-148	227	
			0	3999	83		741	
			0	3943	105		680	
20 13 16	-46 41.2	A	0	5348	41	-29	765	

388

R.A. (1950)	DEC. (1950)	NAME	OBS	HEL VEL (C*Z)	ERR	GAL CORR	REF	COMMENTS
20 13 16	-46 41.2	B	O	5393	21	-29	765	
20 13 16	-46 41.2	C	O	5379	47	-29	765	
20 13 21	-71 0.6	NGC 6877	O	4172	39	-148	741	
			O	4528	45		680	
20 13 53	-71 1.9		O	3928		-148	526	
20 14 5	-71 1.7		O	4018		-148	526	
20 14 6	-55 49.2	PKS	O	18170	40	-75	350	
			O	18220	30		355	
20 14 16	0 26.4		R	3742	10	196	860	
20 14 17	-71 1.0	NGC 6880	O	3969	30	-148	741	
			O	4288	65		680	
20 14 26	-71 0.3	IC 4981	O	3812	20	-147	680	
20 14 46	-44 57.7	NGC 6890	O	2475		-20	574	
			O	2459	34		741	
20 15 12	-39 30.0		O	2719	25	7	582	
20 17 14	-48 23.7	NGC 6893	O	3175	38	-37	741	
20 19 36	-44 0.0		O	2975	20	-14	582	
20 19 44	9 51.5	3CR411	O	140600	300	230	518	
20 20 31	-37 4.8		O	8100		21	577	
20 20 33	-44 9.4	A 2021	O	2942	25	-15	741	
20 20 42	0 30.0	IC 1317 2 ZW 82	O	3975		197	13	
20 21 6	6 16.9	NGC 6906	O	4890		218	910	
20 21 7	-43 49.0	NGC 6902	R	2781		-13	538	
			O	2723			741	
			O	2749	50		724	
			O	2691	34		510	
20 22 7	-24 58.3	NGC 6907	O	3155	32	83	741	
20 22 42	5 6.0		R	4846	15	215	582	
			O	4760	30		582	
20 23 43	-55 15.3		O	(831)		-71	680	

389

R.A. (1950)	DEC. (1950)	NAME	OBS	HEL VEL (C∗Z)	ERR	GAL CORR	REF	COMMENTS
20 24 0	-18 47.0	NGC 6912	R	7087	30	113	582	
			O	7119	30		582	
20 24 9	-47 11.6	NGC 6909	O	2680	80	-30	741	
20 24 12	2 32.0		O	5289	20	205	582	
20 26 24	25 33.0	NGC 6921	O	4317	40	273	13	
20 27 18	- 2 21.0	NGC 6922	R	5665	10	187	752	
20 29 42	- 2 25.0		R	5954	10	187	582	
			O	5943	20		582	
20 30 12	9 42.0		O	4419	250	232	13	
20 30 12	9 43.0	NGC 6927	O	4277	50	232	13	
20 30 21	-23 3.5	PKS	O	39540	150	93	709	
20 30 25	9 45.4	NGC 6928	R	4727	30	232	885	
			R	4754			779	
			O	4754	75		13	
20 30 30	- 2 11.0	NGC 6926	R	5970	20	188	752	
		VV 621	O	6112			774	
			O	(5500)			776	
20 30 34	9 42.1	NGC 6930	R	4694	25	232	885	
			O	4182	75		13	
20 31 14	-32 9.2	NGC 6925	R	2767		48	671	
			R	2784	15		752	
			O	2619			84	
			O	2569			85	
20 31 42	-50 2.2	A	O	2424	55	-44	613	
20 31 42	-50 2.2	C	O	2729	55	-44	613	
20 31 47	-50 1.7	B	O	4689	65	-44	613	
20 33 48	59 59.0	NGC 6946	R	85	45	292	216	
		4C 59.31	R	45			156	
		ARP 29	R	100			171	
			R	55			934	
			R	50	3		181	
			R	95	30		93	
			R	40			340	
			R	(50)	9		489	
			R	40	5		397	
			O	38	50		13	
			O	2	7		11	
			O	-70			13	
			O	-70	25		13	KNOT

R.A. (1950)	DEC. (1950)	NAME	OBS	HEL VEL (C*Z)	ERR	GAL CORR	REF	COMMENTS
20 34 41	-52 17.2	NGC 6935	O	4794	80	-55	741	
20 35 4	-52 20.3	NGC 6937	O	4680		-55	574	
20 35 54	6 49.0	NGC 6944	O	4598		223	13	
			O	4375	40		13	
			O	3000			289	DIS
20 36 30	65 56.0	NGC 6951'	R	1396	90	284	238	
		NGC 6952	R	1380			363	
			R	1426			744	
			O	1364			13	
20 36 53	-54 28.8	NGC 6942	O	3964	100	-66	741	
20 37 12	1 52.0		O	3900		205	534	
20 37 32	- 2 58.2	4C 03.72	O	57600		186	600	
20 39 52	-68 55.6	NGC 6943	O	3114	99	-137	741	
20 40 42	-67 42.8	IC 5031A/B	O	9751	48	-131	724	
20 40 42	-67 42.8	IC 5031A	O	10334	15	-131	613	
20 40 44	-67 43.6	IC 5031B	O	10354	35	-131	613	
20 40 44	-26 43.8	PKS	O	11730	60	76	355	
			O	12080	120		350	
20 41 26	-10 54.3	MRK 509	O	10491		152	447	SEYF
			O	10200			388	
20 41 36	3 1.0	NGC 6954	O	4011	100	210	13	
20 43 18	0 0.0		O	3805	87	198	863	
20 44 6	0 9.0		O	3821	166	199	863	
20 44 8	-13 2.0	DDO 210	R	-131	10	143	492	
20 44 11	-69 16.7		O	11413	65	-138	680	
20 44 30	0 15.0	NGC 6959	O	3693	57	199	863	
20 44 36	0 11.0	NGC 6961	O	3803	74	199	863	
20 44 45	0 8.2	NGC 6962	R	4219	15	199	829	
			R	4204	60		885	
			O	4183	75		13	
			O	4288	40		863	
20 44 47	0 17.9	NGC 6963	O	4351	50	200	13	
			O	4350	77		863	

R.A. (1950)	DEC. (1950)	NAME	OBS	HEL VEL (C★Z)	ERR	GAL CORR	REF	COMMENTS
20 44 50	0 6.8	NGC 6964	O	3832	100	199	13	
			O	3823	40		863	
20 45 0	0 13.0	NGC 6965	O	3669	43	199	863	
20 45 24	−38 11.0	NGC 6958	O	2757		18	84	
20 45 43	6 49.9	W	O	37500		224	717	★ F G
20 45 43	6 49.9	E	O	36900		224	717	F G
20 45 44	6 50.2		O	66600		224	717	F G
20 45 44	6 50.2	3CR424	O	38070	300	224	585	
20 46 0	7 33.0	NGC 6969	O	4000		226	289	
20 46 0	79 58.0		O	4731	36	254	813	
20 47 22	−69 23.5	IC 5052	O	423	40	−138	741	
			O	190	100		844	
			O	966			145	
20 47 36	9 44.0	NGC 6972	O	4442	57	234	148	
20 48 6	−57 15.3	IC 5063	O	3402	6	−80	741	
		PKS	O	3400	60		355	
			O	3395			574	
20 48 33	−48 58.7	NGC 6970	O	5240		−37	574	
			O	5187	38		741	
			O	5543	110		113	
			O	5155			145	
			O	5287	30		680	
			O	5487	65		680	HII REG
20 49 6	18 47.0	ZW	O	(8444)	100	260	500	
20 49 12	18 46.0	ZW	O	8714	55	260	500	
20 49 30	−69 14.0	A/B	O	11185	46	−138	724	
20 49 54	− 5 54.0	NGC 6978	O	5869	50	175	148	
20 52 56	−22 16.0	W	O	10400		100	878	
20 52 59	−22 16.0	E	O	46860		100	878	
20 54 19	−52 3.8	NGC 6984	O	4522	45	−53	741	
20 55 0	−49 28.0	A/B	O	6912	28	−39	724	
20 55 0	16 55.0	2 ZW 96	O	10900	100	256	820	

R.A. (1950)			DEC. (1950)		NAME	OBS	HEL VEL (C*Z)	ERP	GAL CORR	REF	COMMENTS
20	55	9	-42	50.6		O	12788	30	-5	763	
20	56	9	-42	57.6		O	8855	36	-5	613	
20	56	12	-42	58.1		O	9105	47	-5	613	
20	57	12	- 2	4.0	2 ZW 97A	O	5898	30	191	500	
					KDG 551	O	5842	20		908	
20	57	12	- 2	3.0	2 ZW 97B	O	(5883)	225	191	500	
					KDG 551	O	5834	34		908	
20	58	30	16	7.0		O	9148	250	254	13	
20	58	39	-28	13.8	PKS	O	11740	90	70	355	
						O	11300	200		350	
20	58	48	15	56.0		O	11255	50	253	13	
20	58	57	-13	30.6	IC 1347	O	8730	90	142	355	
					PKS						
20	59	36	15	56.0		O	11965	50	253	13	
21	0	8	-48	25.0		O	5300	63	-34	741	
21	0	24	36	30.0	4 ZW 67A/B	O	3000		293	429	
21	1	20	-48	24.0		O	4848	49	-34	741	
21	1	26	29	42.0	NGC 7013	R	830	100	284	544	
						R	786			780	
						R	852	100		238	
						R	780	35		775	
						O	570	28		148	
21	1	30	-21	59.0		O	8724	50	101	582	
21	1	54	-52	45.0	NGC 7007	O	2954	24	-56	741	
21	2	6	-47	14.8		O	4985	37	-28	741	
21	3	18	11	13.0	NGC 7015	R	4876	15	239	752	
21	3	24	-47	23.6		O	5204	41	-28	741	
21	3	47	-47	45.4		O	4617	42	-30	613	
21	3	49	-47	45.4		O	4490	55	-30	613	
21	4	29	-47	22.8	NGC 7014	O	4790	25	-28	741	
21	4	45	76	21.1	3CR427.1	O	352300		264	948	*

21ʰ5ᵐ

R.A. (1950)	DEC. (1950)	NAME	OBS	HEL VEL (C*Z)	ERR	GAL CORR	REF	COMMENTS
21 5 6	3 41.0	2 ZW101	O	7680		214	262	
		KDG 552A	O	7854	23		500	
21 5 18	3 40.0	2 ZW102	O	7982	46	214	500	
		KDG 552B						
21 6 48	-37 42.0		O	2751	35	21	765	
21 6 48	-37 42.3		O	2479	48	21	765	KNOT
21 7 12	14 55.0	NGC 7033	O	9056	30	250	908	
		KDG 554						
21 7 16	-64 14.0	NGC 7020	O	3029	64	-114	741	
21 7 18	14 57.0	NGC 7034	O	8903	46	251	908	
		KDG 554						
21 8 27	-49 29.6	NGC 7029	O	2857	33	-39	741	
			O	2890			84	
21 8 36	- 2 15.0		O	24300		191	534	
21 9 14	- 1 35.0	MRK 512	O	10296		194	447	
21 9 36	- 1 41.0		R	9672	50	193	582	
			O	9684	50		582	
21 11 24	13 21.0	NGC 7042	O	5180	30	246	908	
		KDG 555						
21 11 42	13 24.0	NGC 7043	O	(5331)	60	246	908	
		KDG 555						
21 11 48	-47 25.4	NGC 7038	O	4844	67	-29	741	
21 12 0	21 20.0	2 ZW109	O	13570		267	262	
21 12 6	-64 41.0	IC 5092	O	3470	100	-116	844	
21 13 9	-48 34.2	NGC 7041	O	1915	42	-34	741	
			O	1935			84	
21 13 10	-59 32.6		O	18131		-91	680	
21 14 24	-41 28.0		O	8359	20	2	765	
21 15 38	-48 46.4	NGC 7049	O	2198	43	-36	741	
			O	2181			84	
21 16 18	2 3.0	MRK 513	O	5400		208	569	
21 16 21	26 14.1	B2	O	4640		278	485	

394

R.A. (1950)	DEC. (1950)	NAME	OBS	HEL VEL (C*Z)	ERR	GAL CORR	REF	COMMENTS
21 16 24	5 38.0	KDG 556	O	8560	33	221	908	
21 16 48	5 48.0	KDG 556	O	3902	50	221	908	
21 17 42	- 1 53.0		O	4874	30	192	813	
21 17 45	60 35.5	3CP430	O	16200		292	318	
21 17 54	-66 4.0	IC 5101	O	5166	20	-122	582	
21 17 57	2 33.7	2 ZW123	O	14100		209	553	
21 18 52	44 11.5		O	3695	90	299	640	
			O	4095			563	
21 19 13	29 14.3		R	4617	10	283	860	
21 19 48	18 27.0	NGC 7056	O	5288	20	260	813	
21 19 51	43 51.2		O	3518	60	299	640	
21 20 51	- 7 57.7	MRK 515	O	8990		167	569	
21 20 54	15 35.2	3CR434	O	96800		252	717	
			O	96530	300		585	
21 21 12	-40 45.1	IC 5105	O	5363	64	5	741	
21 21 31	24 51.5	3CP433	O	30480		275	120	
21 22 36	-40 29.0	IC 5105A	O	5012		7	728	
21 23 6	-22 59.0		O	10697	50	96	582	
21 23 35	-60 13.9	NGC 7059	O	1797	33	-94	741	
			O	(1700)			670	
21 23 42	-61 3.0		O	4200	50	-98	670	
21 24 10	-38 4.4		O	2626	25	19	680	
21 25 12	-38 5.0		O	2567	25	19	582	
21 25 34	-52 59.1	NGC 7064	O	970	100	-58	844	
			O	797	3		741	
21 26 37	7 19.8	3CR435	O	141200		226	946	
21 27 0	-60 13.0	IC 5110	O	9000	110	-94	670	
21 27 18	-43 19.0	NGC 7070	O	(2050)	210	-7	555	
21 27 18	27 6.0		R	4823	10	279	752	

21ʰ27ᵐ

R.A. (1950)	DEC. (1950)	NAME	OBS	HEL VEL (C*Z)	ERR	GAL CORR	REF	COMMENTS
21 27 24	-43 21.6	NGC 7072	O	4925		-8	574	
21 27 34	2 12.0	NGC 7077	O	1050		208	534	
21 27 36	26 30.0	NGC 7080	R	4838	15	278	582	
			O	4806	20		582	
21 28 48	2 16.0	NGC 7081	R	3280	8	208	914	
21 29 18	-44 18.0	NGC 7079	O	2630	99	-13	203	
			O	3007	70		152	
21 29 48	-83 33.0		O	17230	200	-199	794	
21 30 0	9 56.0	2 ZW136	O	18317		235	262	SEYF
			O	18740			282	
			O	18300			186	
21 30 45	-53 51.5	PKS	O	22940	180	-62	355	
21 30 45	-53 51.5		O	23480	90	-62	355	F G
21 30 51	7 46.7		R	3487	10	227	860	
21 31 24	-41 2.3		O	5084	70	3	603	
21 31 48	8 26.0	2 ZW140	O	8795		229	303	
21 31 52	-64 7.6	NGC 7083	O	3089	50	-113	741	
21 32 59	-54 46.9	NGC 7090	R	846		-67	671	
			O	790			84	
21 37 4	-42 46.1	NGC 7097	O	2404	32	-5	741	
21 37 28	-64 8.2	NGC 7096	O	2958	14	-114	741	
21 37 49	12 7.0		O	5833	23	241	813	
21 39 37	10 3.6	MC 2	O	22200		234	652	
21 39 40	-52 55.0		O	5296	58	-58	613	
21 39 42	25 4.0	4 ZW 74	O	17929		274	262	
21 41 57	27 56.5	3CR436	O	64360		279	120	
21 42 0	46 24.0	KDG 559	O	3100	18	298	908	
21 42 18	46 24.0	KDG 559	O	3273	18	298	908	
21 44 24	-35 7.0	IC 5131	O	2610		33	574	

R.A. (1950)	DEC. (1950)	NAME	OBS	HEL VEL (C*Z)	ERR	GAL CORR	REF	COMMENTS
21 44 24	1 29.0	IC 1401	R	4718	10	203	582	
			O	4648	35		582	
21 44 48	-50 48.0	NGC 7124	O	5027	61	-47	741	
21 45 21	-35 11.2	IC 5135	O	4812	40	33	741	
			O	4780			574	
			O	4877	23		148	★
21 45 38	-60 56.8	NGC 7125	O	3031	20	-99	741	
21 45 41	-60 50.4	NGC 7126	O	3009	26	-98	741	
21 45 48	- 1 54.0		O	8100		190	534	
21 45 54	21 55.6	NGC 7137	R	1694	5	266	752	
			R	1568	100		238	
			R	1654			610	
			R	1674			610	
			R	1538	100		544	
			R	1679			744	
			R	1734			780	
			O	1505			13	
21 45 54	26 13.0	4 ZW 77	O	19860		276	303	
21 46 48	-35 7.0	NGC 7135	O	2718		33	486	
21 47 4	13 59.8		R	1102	10	245	860	
21 48 48	25 38.0		R	5678	10	274	582	
			O	5696	20		582	
21 49 30	-48 29.4	NGC 7144	O	2113		-36	84	
			O	1910	140		555	
21 50 8	-48 7.1	NGC 7145	O	1918	24	-34	741	
			O	1950	40		113	
21 52 0	2 42.0	NGC 7156	R	3979	5	207	752	
			R	3983	17		914	
			O	3966	60		148	
21 52 56	-49 45.8	NGC 7155	O	1893	44	-43	741	
21 52 58	-69 55.7	PKS	O	8100		-142	627	
21 53 35	1 9.2		O	66600		201	717	
21 53 46	37 46.2	3CR438	O	87500		292	717	
			O	86900			946	
21 53 53	7 8.0	MRK 516	O	8922		222	447	

R.A. (1950)	DEC. (1950)	NAME	OBS	HEL VEL (C*Z)	ERR	GAL CORR	REF	COMMENTS
21 56 10	11 48.0	MRK 518	O	9506		237	447	
21 56 26	-32 7.3	NGC 7163	R	2875	100	47	885	
21 57 27	-43 37.7	NGC 7166	O	2407	29	-12	741	
21 57 42	-24 52.0	NGC 7167	R O	2575 2583	15	83	752 728	
21 58 14	10 18.7	MRK 520	O O	8162 8390		232	447 569	
21 58 18	-60 11.0		O	30250	100	-96	670	
21 58 18	-13 31.0	NGC 7171	O	2632	50	138	13	
21 58 18	17 29.9	NGC 7177	R R R O	1225 1150 1192 1105	20 10 20 75	253	238 752 544 13	
21 58 53	-51 59.0	NGC 7168	O O	2783 3005	42	-55	741 113	
21 59 6	-32 13.0	NGC 7173	O	2496	39	46	148	
21 59 7	-32 6.6	NGC 7172	O O	2662 2585	44	47	148 458	
21 59 12	-32 14.0	NGC 7176	O	2520	29	46	148	
21 59 12	-32 13.9	NGC 7174	O O	2777 2773	 31	46	458 148	
21 59 32	-20 47.4	NGC 7180	R O	1238 1347	100 55	103	885 555	
21 59 36	-51 32.0	IC 5152	R O O O	122 100 58 110	 50	-53	671 845 84 17	
21 59 54	-21 3.0	NGC 7184	O	2747	100	102	789	
22 0 0	18 5.0	2 ZW160A KDG 560A	O O	7900 7553	48 20	255	500 908	
22 0 0	18 5.0	2 ZW160B KDG 560B	O O	8017 7680	54 10	255	500 908	
22 0 12	19 30.0	IC 1420A	O	1413	30	258	567	
22 0 12	19 30.0	IC 1420B	O	1476	60	258	567	

R.A. (1950)	DEC. (1950)	NAME	OBS	HEL VEL (C*Z)	ERR	GAL CORR	REF	COMMENTS
22 0 20	-34 5.0	IC 5156	R O	2872 2584	30 77	36	885 741	
22 1 9	-64 18.3	NGC 7179	O	2890		-116	574	
22 1 24	-58 32.0		O	6449	16	-88	724	
22 1 30	43 30.6	5 ZW380	R	458	10	295	860	
22 1 46	4 25.5	4C 04.77	O	(8400)		211	600	
22 2 46	-50 22.0	NGC 7196	O O	3022 3016	75	-47	510 84	
22 3 8	-64 33.4	NGC 7192	O	2879	17	-117	741	
22 3 58	-50 14.6	NGC 7200	O	2937	18	-47	741	
22 4 0	-31 22.0	NGC 7204	O	2630		50	458	
22 4 36	-59 48.0	A	O	48400	90	-95	670	
22 4 36	-59 48.0	B	O	48400	90	-95	670	
22 5 6	-57 40.0	NGC 7205	O	1482		-84	84	
22 5 6	44 3.0	MRK 1123	O	(6005)		295	920	
22 5 36	31 7.0	NGC 7217	R R O R O O O	840 929 911 954 955 1050 950	 38 30 10 6 100 25	281	216 885 13 729 729 2 283	
22 5 42	4 27.0		O	4057		211	777	
22 5 54	-28 6.0	IC 5168	O	5098	141	66	789	
22 6 6	40 56.0		R O	5350 5313	15 25	293	582 582	
22 6 12	-47 25.0	NGC 7213	O O	1765 1810		-32	136 84	
22 6 18	-28 3.0	NGC 7214	O	6794	63	66	789	
22 6 24	-28 1.0	NGC 7214B	O	7275	71	66	789	
22 7 12	17 25.0	2 ZW168	O O	7800 8011		252	28 262	

R.A. (1950)			DEC. (1950)		NAME	OBS	HEL VEL (C★Z)	ERR	GAL CORR	REF	COMMENTS
22	7	28	-19	6.8	DDO 211	R	(1729)		110	492	
22	7	29	-16	54.6	NGC 7218	R	1670	20	120	544	
						R	1662			744	
						R	1710	20		238	
						O	1808			13	
22	7	50	-46	20.4		O	2723	46	-27	741	
22	8	0	40	46.0	NGC 7223	R	4634	15	292	752	
22	8	24	-30	52.0	NGC 7221	O	4313		51	728	
22	8	43	-62	58.3		O	8430	10	-110	680	
22	8	48	29	22.0	KDG 563	O	6881	20	278	908	
22	8	48	29	24.0	KDG 563	O	11352	30	278	908	
22	9	18	0	9.0		O	10022	30	194	582	
22	9	36	17	39.4		R	1737	10	252	860	
22	9	36	46	4.0		O	5472	30	295	582	
22	10	18	-46	16.4	IC 5181	O	2110	47	-27	741	
						O	1950			574	
22	11	12	-29	40.0	NGC 7229	R	4303	15	57	752	
						O	4236			728	
22	11	42	-17	16.8	3C 444	O	45900		118	699	
22	12	19	13	35.8	NGC 7236/7 3CR442	O	7855		240	227	
22	12	19	13	36.1	NGC 7236	O	7867	72	240	444	
					2 ZW172	O	7867			748	
					ARP 169	O	7617	30		70	
22	12	20	13	35.5	NGC 7237	O	7866	72	240	444	
					2 ZW172	O	7866			748	
					ARP 169	O	7612	30		70	
22	12	35	-46	6.0	NGC 7232	O	2061	82	-27	741	
						O	1887			227	
						O	1747	67		765	
						O	1612	55		680	
						O	1690			574	
22	12	46	-46	6.0	NGC 7233	O	1881	9	-27	741	
						O	1792	26		765	
						O	1822	20		680	
						O	1810			574	

R.A. (1950)	DEC. (1950)	NAME	OBS	HEL VEL (C*Z)	ERR	GAL CORR	REF	COMMENTS
22 13 12	37 2.0		O	5984	75	288	13	
22 13 14	-37 5.7	IC 5186	O	3447	10	19	741	
22 13 18	-37 5.0	IC 5179	O	3403		19	728	
22 13 24	18 59.0	NGC 7241 2 ZW174	R R	1447 1423	10 10	254	752 914	
22 13 24	39 43.0	5 ZW381NE	O	20976		290	321	
22 13 24	39 43.0	5 ZW381SW	O	20890		290	321	
22 13 30	37 3.0	NGC 7242 4 ZW 90	O O	5684 5000	100 200	288	13 2	
22 13 48	22 41.0	4 ZW 93	O	3885		263	321	
22 14 1	16 13.2	NGC 7244 MRK 303	O	7500		247	307	
22 14 6	-21 30.0	DDO 212	R	2599	25	97	492	
22 14 36	13 7.0		O	26400		238	534	
22 14 45	13 59.5	MRK 304 2 ZW175	O O O	19500 19840 19750	 200 45	240	307 820 382	SEYF
22 17 6	29 9.0	NGC 7253A KDG 566	O	4718	28	276	908	
22 17 6	44 59.0	NW	O	13138		293	321	
22 17 6	44 59.0	SE	O	13502		293	321	
22 17 12	29 8.0	NGC 7253B KDG 566	O	4492	33	276	908	
22 17 55	-46 17.0	IC 5201	O O	952 590	43 100	-29	741 844	
22 18 0	-24 56.0	NGC 7252 ARP 226	O	4733	65	79	13	
22 18 24	47 27.0		R O	5473 (5567)	10	294	582 582	
22 19 30	36 6.0	NGC 7263 4 ZW 97	O	6168		286	321	
22 20 0	- 0 23.0		O	1550		189	115	

22ʰ20ᵐ

R.A. (1950)	DEC. (1950)	NAME	OBS	HEL VEL (C*Z)	ERR	GAL CORR	REF	COMMENTS
22 20 48	30 40.0	KDG 567A	O	6646	35	278	500	
			O	6717			321	
22 20 48	30 40.0	KDG 567B	O	6552	51	278	500	
			O	6702			321	
22 21 15	− 2 21.5	3CR445	O	16860		181	120	
			O	16860			575	
22 21 35	−62 28.0		O	12890	200	−109	794	
22 21 52	35 52.1	NGC 7274	O	5820		285	255	
22 22 25	22 43.0		R	1243	10	261	860	
22 22 36	38 34.0	4 ZW 99N	O	8827		288	321	
22 22 36	38 34.0	4 ZW 99S	O	8847		288	321	
22 23 32	13 44.5		O	112100		237	717	
22 23 43	17 6.1	4C 17.89	O	32100		247	717	
22 24 2	15 53.6	NGC 7280	R	1857		243	610	
		KDG 568A	R	1817			780	
			R	1903	20		829	
			O	1812	80		148	
			O	1830	65		555	
22 25 32	−22 41.5	NGC 7287	O	6002		88	714	
22 26 1	16 53.6	NGC 7290	R	2896	20	246	829	
			R	2901	4		914	
22 26 6	30 3.0	NGC 7292	O	930		275	449	
22 26 17	−65 54.8	IC 5222	O	3150	200	−126	794	
22 26 50	−20 4.7		O	7589		101	714	
22 26 56	−20 0.3		O	7386		101	714	
22 28 12	−14 27.0	NGC 7298	R	5030		127	744	
		MRK 1124	R	5040	10		582	
			O	4640			921	
			O	5378	100		789	
			O	5053	25		582	
22 28 18	33 34.0		R	886	8	280	460	
22 28 21	−14 15.7	NGC 7300	O	3324	100	127	741	
			O	4895	50		789	DIS

R.A. (1950)	DEC. (1950)	NAME	OBS	HEL VEL (C*Z)	ERR	GAL CORR	REF	COMMENTS
22 28 43	-19 17.4		O	7263		104	714	
22 29 8	36 6.1	3CR449	O	5122	106	283	741	
			O	5130	150		159	
22 29 24	19 26.4	MRK 305	O	5700		252	568	
22 29 26	19 26.1	MRK 306A/B	O	5700		252	568	
			O	5609	45		382	
			O	5700			307	
22 29 26	19 26.1	MRK 306A	O	5605	81	252	830	
22 29 26	19 26.1	MRK 306B	O	5551	81	252	830	
22 29 42	-14 23.0	NGC 7302 IC 5228	O	2586	65	126	13	
22 30 17	-20 21.0		O	5706		98	714	
22 30 21	-22 29.8		O	15700		88	714	
22 30 36	7 50.0	2 ZW181	O	2100		216	534	
22 30 59	-41 14.7	NGC 7307	O	1880	24	-5	741	
22 31 18	5 19.0	NGC 7311	R	4536	18	207	914	
22 31 35	-62 8.6		O	7940	200	-109	794	
22 31 36	-10 37.0	NGC 7309	O	3938	33	143	148	
22 31 52	-22 44.6	NGC 7310	O	9686		86	714	
22 31 53	32 36.0	DDO 213	R	806	10	278	492	
			R	815	20		424	
22 33 0	-26 18.5	NGC 7314 ARP 14	R	1467	70	68	238	
			R	1422	100		544	
			O	1766	50		13	
			O	1374	45		789	
22 33 24	-61 40.0	A	O	12560	170	-107	670	
22 33 24	-61 40.0	B	O	12560	170	-107	670	
22 33 31	20 3.9	NGC 7316 MRK 307	R	5556	11	252	914	
			O	5511	45		382	
			O	5700			307	
22 33 36	33 41.0	NGC 7317 ARP 319 VV 288	O	6736	65	279	13	STP QUI

22ʰ33ᵐ

R.A. (1950)	DEC. (1950)	NAME	ORS	HEL VEL (C*Z)	ERR	GAL CORR	REF	COMMENTS
22 33 39	33 42.4	NGC 7318A	O	6724		279	13	
		ARP 319	O	6638	50		13	
		VV 288	O	6677			42	
22 33 41	33 42.4	NGC 7318B	R	5700		279	403	
		ARP 319	O	5638	40		13	
		VV 288	O	5867			13	
22 33 45	33 41.4	NGC 7320	R	774	12	279	460	
		ARP 319	R	755	25		362	
		VV 288	R	776			744	
			R	750	20		230	
			R	784	10		875	
			R	779	5		798	
			O	1002	50		555	
			O	795			42	
22 33 46	33 43.0	NGC 7319	R	6590	15	279	362	
		ARP 319	R	6620	23		460	
		VV 288	O	6657	50		13	
22 33 59	- 3 9.9	DDO 214	R	1693	10	174	492	
		ARP 3						
22 34 3	33 43.6	NGC 7320C	R	6030		279	461	
		VV 288	O	6000			332	
22 34 3	34 17.0		O	8200		280	332	
22 34 7	-12 48.3	MRK 915	O	7225		132	874	
22 34 9	-20 4.0		O	2351		99	714	
22 34 12	-22 28.8		O	10101		87	714	
22 34 18	34 15.0		O	8400		280	332	
22 34 47	34 9.5	NGC 7331	R	810		279	151	
		HOL 795A	R	815	20		362	
			R	822			934	
			R	821			817	
			R	852	25		183	
			R	812	12		944	
			R	838			886	
			R	815	20		216	
			R	817			744	
			R	815			171	
			O	830			904	
			O	780	20		13	
			O	919			13	
			O	500	50		1	
			O	849			118	
			O	673	110		555	
			O	863	38		72	

R.A. (1950)	DEC. (1950)	NAME	OBS	HEL VEL (C*Z)	ERR	GAL CORR	REF	COMMENTS
22 35 1	23 32.3	NGC 7332	R	1200		260	610	
		KDG 570	O	1307	28		908	
		HOL 796A	O	1190	25		335	
			O	1204	50		13	
			O	871	40		148	
22 35 1	34 11.3	NGC 7335 HOL 795C	O	6298	60	279	13	
22 35 4	34 14.0		O	8800		280	332	
22 35 6	34 7.0	NGC 7337	O	6900		279	332	
22 35 12	33 40.0		O	6200		279	332	
22 35 18	34 36.0		O	6800		280	332	
22 35 24	23 31.6	NGC 7339	R	1339	15	260	829	
		KDG 570	O	1237	26		908	
		HOL 796B	O	1271	27		148	
22 35 24	34 10.0	NGC 7430	O	6400		279	332	
22 36 6	35 7.0		O	7400		280	332	
22 36 12	33 59.0		O	6000		279	332	
22 36 12	35 4.2	B2	O	8020		280	485	
			O	8240			255	
			O	8000			332	
22 36 19	33 48.5	NGC 7343	O	6500		279	450	*
			O	5401			624	DIS
			O	(1216)	200		13	DIS
22 36 24	-19 55.4		O	9427		99	714	
22 36 24	35 17.0		O	7800		281	332	
22 36 30	34 3.0		O	5800		279	332	
22 36 34	- 5 1.5	DDO 215	R	829	10	165	492	
22 36 56	-66 44.6	NGC 7329	O	3189	25	-131	741	
22 37 6	11 31.0		R	7348	15	226	582	
			O	7318	20		582	
22 37 24	34 7.0		O	7390		279	449	
22 37 34	37 57.3		R	4705	25	283	582	
			R	4769	15		752	
			O	4694	20		582	

22ʰ38ᵐ

R.A. (1950)	DEC. (1950)	NAME	OBS	HEL VEL (C*Z)	ERR	GAL CORR	REF	COMMENTS
22 38 12	11 39.0	NGC 7348	O	7311	63	226	789	
22 38 25	-62 16.5		O	(27730)	200	-111	794	
22 38 31	-22 3.5		O	4639		88	714	
22 38 48	23 8.0	IC 5242	O	7173	18	258	500	
22 38 48	31 54.5	MRK 917	O	7261		275	874	
22 39 0	-45 4.0	IC 5240	O	1633	81	-27	203	
			O	1690			574	
			O	1834	55		741	
22 39 0	0 8.3		R	1757	10	185	860	
22 39 0	23 7.0	IC 5243	O	(7048)	100	258	500	
		2 ZW185	O	7105			262	
		KDG 571B	O	7200			534	
22 39 30	19 59.9	MRK 308	O	6999	45	250	382	
			O	7200			307	
22 39 31	-30 19.2	NGC 7361	R	1240	40	47	216	
			O	1174	37		148	
22 40 5	-21 25.7		O	3070		90	714	
22 40 18	29 27.8	B2	O	7340		270	485	
			O	7500			534	SEYF
22 40 23	24 43.7	4C 24.60	O	27150		261	400	
22 40 30	-46 12.9		O	9963	17	-33	763	
22 40 48	31 24.0	4 ZW111	O	9711		274	262	
22 41 0	33 44.0	NGC 7363	R	830		277	744	
22 43 18	10 52.0	NGC 7372	O	11614		222	659	
22 43 24	-11 16.0	NGC 7371	O	2384	45	136	148	
22 43 30	10 35.0		O	7564		221	659	
22 43 33	39 25.4	3CR452	O	24330		283	120	
22 43 36	10 35.0	NGC 7374	O	7099		221	659	
22 43 54	-65 19.3	IC 5250W	O	3150	200	-126	794	
			O	3616	20		680	
22 43 54	37 47.0		R	4758	25	282	582	
			O	4787	20		582	

R.A. (1950)	DEC. (1950)	NAME	OBS	HEL VEL (C*Z)	ERR	GAL CORR	REF	COMMENTS
22 44 0	-65 19.3	IC 5250E	O	3526	50	-126	680	
		IC 5250ENW	O	3450	200		794	
		IC 5250ESE	O	450	200		794	⋆ DIS
22 44 0	11 40.0		O	7078		225	659	
22 44 10	-84 22.5		O	(21880)	200	-204	794	
22 44 13	36 40.7	4C 36.47	O	24450		280	400	
22 44 38	- 2 21.5		O	98300		174	717	
22 45 5	-22 34.6	NGC 7377	O	3416	65	83	13	
22 46 18	27 21.0	KDG 573	O	9625	30	265	908	
22 46 24	27 19.0	KDG 573	O	9566	20	265	908	
22 47 0	11 3.0		O	7840		222	659	
22 47 6	11 17.0	NGC 7383	O	8112		222	659	
22 47 6	11 44.0		O	7184		224	659	
22 47 6	19 9.0	MRK 1125	O	9555		245	920	
22 47 25	11 20.6	NGC 7385	O	7829	65	222	13	
		4C 11.71	O	7764			659	
			O	7828			227	
			O	7770			595	
			O	7730	290		847	
			O	7290			350	⋆
22 47 26	11 20.8	NGC 7385	O	8080	120	222	847	KNOT
22 47 32	11 26.0	NGC 7386	O	7198	65	223	13	
			O	7197			227	
			O	7317			659	
22 47 36	11 47.0		O	6662		224	659	
22 47 48	11 19.0	NGC 7389	O	7916		222	659	
22 47 48	11 22.0	NGC 7387	O	6929		222	659	
22 47 54	11 16.0	NGC 7390	O	7763		222	659	
22 48 8	28 52.4		R	892	10	267	860	
22 48 30	11 48.0		O	11855		223	659	
22 48 41	-67 40.9	IC 5257	O	11840	200	-137	794	

22ʰ49ᵐ

R.A. (1950)	DEC. (1950)	NAME	OBS	HEL VEL (C*Z)	ERR	GAL CORR	REF	COMMENTS
22 49 0	- 5 49.0	NGC 7393 VV 68 ARP 15	O	3813		158	13	
22 49 12	-20 53.0	NGC 7392	O	2941		90	13	
23 49 22	- 1 25.9		O	47685		148	471	
22 50 0	11 23.0		O	8622		222	659	
22 50 10	24 27.9	MRK 309 4 ZW121	O O O	12600 12603 12630	45	257	307 874 382	
22 51 6	31 23.0	4 ZW122	O	6300		271	534	
22 51 17	-17 53.4		O	19601	208	103	930	
22 51 20	33 26.6		R	4141	20	274	752	
22 51 23	-17 52.1		O	19716	318	103	930	
22 51 26	-17 50.3		O	20074	138	103	930	
22 51 30	-17 51.9		O	19741	318	103	930	
22 51 39	11 20.5		O O	97210 96830		221	339 286	
22 51 39	11 21.0		O	98620		221	339	
22 51 48	32 14.0	4 ZW123A	O	6744		272	321	
22 52 0	32 13.0	4 ZW123B	O	6611		272	321	
22 52 11	-39 55.7	NGC 7410	O O	(1776) (1472)		-4	136 136	
22 52 12	11 27.0		O	8240		221	659	
22 52 34	12 57.2	NGC 7413	O O O	9767 9706 9710	127 180	226	309 741 159	*
22 52 55	-42 54.6	NGC 7412	R O	1711 1716	20	-19	752 728	
22 53 6	- 5 45.8	NGC 7416	O	(2770)	125	157	555	
22 53 49	-37 17.6	NGC 7418	R O O	1510 1440 1469	27	7	363 574 741	

R.A. (1950)	DEC. (1950)	NAME	OBS	HEL VEL (C*Z)	ERR	GAL CORR	REF	COMMENTS
22 53 54	-37 1.0	NGC 7418A	O	2050		9	574	
22 54 0	27 27.0	4 ZW125	O	20040		262	28	
22 54 6	-37 37.0	NGC 7421	R	1832	20	6	885	
22 54 23	35 25.5	4C 35.56	O	35340		276	400	
22 54 24	-36 43.8	IC 1459	O	1645		10	136	
		IC 5265	O	1630			136	
			O	1617			136	
			O	1572			84	
22 54 28	-41 20.4	NGC 7424	O	800	100	-12	844	
			O	904	17		741	
			O	880	100		789	
22 54 32	4 24.6	IC 1460	O	7262	75	196	13	
22 54 37	-43 39.9	IC 5267	O	1715		-24	136	
22 54 39	8 14.3		O	(26980)		210	569	
22 54 42	- 1 18.0	NGC 7428	R	3085	25	174	885	
			O	3050	29		741	
22 54 57	-36 17.7	IC 5269	O	2162	15	12	741	
22 55 18	- 4 3.0	ARP 314NW VV 295	O	3792		163	321	
22 55 18	- 4 3.0	ARP 314SE VV 295	O	3581		163	321	
22 55 48	2 2.0		R	4836	10	187	582	
			O	4797	45		741	
			O	4982	75		582	
22 56 5	14 54.1	MRK 311	O	9300		230	307	
			O	9130	45		382	
22 56 11	-65 27.4	IC 5272	O	3450	200	-128	794	
22 56 12	35 32.0	NGC 7440	R	5665	25	275	582	
			O	5664	15		582	
22 56 34	-65 23.7		O	(11990)	200	-128	794	
22 56 40	-37 58.4	IC 5273	O	1262	37	3	741	
			O	1120	100		844	
22 57 0	-65 20.3		O	(12890)	200	-128	794	

22ʰ57ᵐ

R.A. (1950)	DEC. (1950)	NAME	OBS	HEL VEL (C∗Z)	ERR	GAL CORR	REF	COMMENTS
22 57 30	-13 4.5	NGC 7443	O	(3050)	180	123	555	
22 57 31	-13 6.1	NGC 7444	O	(2920)	130	123	555	
22 57 34	15 42.8	NGC 7448	R	2247	30	232	238	
		ARP 13	R	2196	50		544	
			O	2419	250		13	
22 57 45	25 49.6	B2	O	7040		258	485	
22 57 48	26 32.0	4 ZW128	O	7060		259	262	
22 57 50	16 7.0	MRK 522	O	(9657)		233	447	
22 58 9	7 2.0	NGC 7455 MRK 523	O	(7741)		204	447	
22 58 9	16 5.6	MRK 312	O	9900		233	307	
22 58 12	-13 11.0	NGC 7450 MRK 1126	O	2968		122	921	SEYF
22 58 26	-59 54.2		O	10211	103	-103	613	
22 58 29	-59 54.4		O	10006	72	-103	613	
22 58 37	29 52.7	NGC 7457	O	525	250	265	13	
			O	746	120		555	
22 58 38	16 7.2	NGC 7454	O	2007	9	233	938	
22 58 42	12 27.0		R	2812	10	222	860	
22 58 45	9 20.0	MRK 524	O	4435		212	447	
22 59 0	28 8.0		O	8700		262	534	
22 59 10	1 59.7	NGC 7460	O	3326	28	185	741	
22 59 14	15 48.0		P	1995	10	232	860	
22 59 23	15 42.8	NGC 7463	R	2058		231	744	★
		HOL 802A	R	(2053)			914	
			O	2440	26		148	
22 59 32	15 41.8	NGC 7465	R	1959		231	779	
		MRK 313	O	1815	45		382	
		HOL 802B	O	1994	28		148	
			O	2100			307	
22 59 36	26 47.0	NGC 7466 MRK 1127	O	7151		259	920	★

R.A. (1950)	DEC. (1950)	NAME	OBS	HEL VEL (C*Z)	ERR	GAL CORR	REF	COMMENTS
22 59 43	15 42.3	NGC 7464	O	1872	32	231	148	
		HOL 802C	O	1800			534	
23 0 0	-41 6.0	NGC 7462	O	944	41	-12	789	
23 0 12	32 19.0		O	5965	25	269	582	
23 0 12	38 26.0	MRK 1128	O	4432		278	921	
23 0 23	-18 57.6	PKS	O	38460	180	95	709	
			O	38700			189	
			O	38100			711	
23 0 31	16 20.0	NGC 7468	R	2093		233	480	
		MRK 314	R	2083	12		914	
			O	2030	45		382	
			O	2208			480	
			O	1700			307	
23 0 31	16 20.0	MRK 314A	O	2086	131	233	830	
23 0 31	16 20.0	MRK 314B	O	2133	115	233	830	
23 0 31	16 20.0	MRK 314C	O	2132	129	233	830	
23 0 44	8 36.3	NGC 7469	R	4992		208	779	SEYF
		ARP 298	R	4754			707	
		KDG 575A	O	4861	82		608	
		HOL 803A	O	4807			748	
			O	4832			966	
			O	4875	42		812	
			O	4822	20		473	
			O	4868	20		473	
			O	4846	20		473	
			O	4770			574	
			O	4868			80	
			O	4850			352	
			O	4910			357	
			O	4780	40		13	
			O	4692			13	
23 0 47	8 37.4	IC 5283	O	4875		208	748	★
		ARP 298	O	5012	39		812	
		KDG 575B	O	4875			80	
		HOL 803B						
23 1 36	22 21.2	MRK 315	O	12000		248	307	SEYF
		2 ZW187	O	11620			262	
			O	11620	45		382	
23 2 7	- 8 57.3		O	14241	130	139	761	
23 2 24	16 24.0	VV 738	O	8668		232	774	

R.A. (1950)	DEC. (1950)	NAME	OBS	HEL VEL (C*Z)	ERR	GAL CORR	REF	COMMENTS
23 2 26	12 3.1	NGC 7479	R R O O O O	2382 2358 2340 2492 2421 2425	50 65	219	744 544 904 13 34 13	
23 2 43	-72 54.6		O	35400		-161	675	
23 3 44	39 11.2	4C 39.72	O	61840		277	400	
23 3 57	-44 31.3		O	21000		-31	767	
23 4 6	-43 10.1		O	12266	65	-24	680	
23 4 6	-43 9.7		O	12804	150	-24	680	
23 4 6	9 45.0	MRK 1129	O	4439		211	921	
23 4 12	-43 11.2		O	12499	200	-24	680	
23 4 24	22 40.0	2 ZW188 IC 5285	R O O	5960 6221 6220		248	389 262 389	
23 4 53	3 50.2	4C 03.56	O	45900		190	868	
23 5 0	32 6.0	NGC 7490	R	6214	15	267	752	
23 5 5	22 43.6		R	6239	15	248	829	
23 5 28	-62 28.6		O	8240	200	-116	794	
23 6 30	11 48.0	NGC 7495	R O O O	4894 4895 4200 4889	10 25	217	582 582 624 777	 *
23 6 35	17 54.4	NGC 7497	R R	1705 1716	9	235	914 744	
23 6 59	-43 42.0	NGC 7496	O O	1470 1540		-27	136 574	
23 7 18	8 14.0	3 ZW 94	O	12577		204	262	
23 7 24	7 15.0		O	4893		201	549	
23 7 51	7 18.6	NGC 7499	O O	11595 11916	44 50	201	594 13	
23 7 54	7 21.0	NGC 7501	O	12714	50	201	13	

412

R.A. (1950)	DEC. (1950)	NAME	OBS	HEL. VEL (C*Z)	ERR	GAL CORR	REF	COMMENTS
23 8 0	7 52.0		0	6404		203	549	
23 8 10	7 17.8	NGC 7503	0	13229	65	201	13	
		4C 07.61	0	10836	74		594	
23 8 24	-63 45.8		0	27730	200	-123	794	
23 8 24	-63 45.4		0	(18290)	200	-123	794	
23 8 24	-63 45.1		0	(18290)	200	-123	794	
23 8 54	- 0 27.0	MRK 1130	0	8638		172	921	
23 9 18	9 14.0		0	6573		207	549	
23 9 26	-61 47.4		0	7940	200	-114	794	
23 9 27	-28 48.8	NGC 7507	0	1546	12	44	658	
			0	1637	75		13	
23 9 40	-61 44.1		0	16190	200	-114	794	
23 9 41	-21 46.5		0	31800		78	717	
23 9 56	9 3.2	3CR456	0	69910		206	120	
23 10 12	10 27.0	MRK 526	0	(6684)		211	447	
23 10 18	5 31.0	IC 1474	R	3471	20	194	939	
			0	3370			549	
23 10 18	12 25.0	NGC 7515	R	4467	8	217	914	
23 10 22	5 0.4	3CR458	0	86640	300	192	585	
23 10 36	6 3.0	NGC 7518	R	3531	20	195	939	
		MRK 527	0	(3253)			447	
			0	3568			549	
23 10 36	6 3.0	NGC 7518A	0	643	190	195	594	
23 10 36	6 3.0	NGC 7518B	0	886	78	195	594	
23 10 36	6 8.0		R	4850	20	196	939	
			0	4812			549	
23 11 10	13 45.0	NGC 7525	0	12300		221	307	
		MRK 316						
23 11 20	23 32.9	MRK 317	0	6300		247	307	
23 11 30	8 42.0	NGC 7529	R	4552	20	204	939	
			0	4569			549	

R.A. (1950)	DEC. (1950)	NAME	OBS	HEL VEL (C*Z)	ERR	GAL CORR	REF	COMMENTS
23 11 34	12 54.3	MRK 528	O	26980		218	569	
23 11 48	- 3 0.0	NGC 7532 MRK 529	O	3287		160	447	
23 12 2	4 13.5	NGC 7537 KDG 578 HOL 805B	R R R O O O	2687 2679 2654 2601 2682 2710	13 20 46 30	188	914 939 893 594 148 549	
23 12 3	-43 52.4	NGC 7531	O O O	1604 1710 1876	63 60 60	-30	789 113 555	
23 12 11	4 15.6	NGC 7541 KDG 578 HOL 805A	R R R R O O O O	2678 2712 2675 2665 2685 2392 2672 2662	7 20 10 83 100 30	188	914 893 939 744 936 594 13 915	★
23 12 12	7 26.0		O	4650		200	449	
23 12 36	9 24.0		O	4757		206	549	
23 12 48	18 41.0	NGC 7550A ARP 99	O	4806		234	321	★
23 12 48	18 41.0	NGC 7550SE ARP 99	O	4987		234	321	
23 12 48	18 41.0	NGC 7550NW ARP 99	O	4858		234	321	
23 12 51	-59 19.6		O	13731	24	-104	763	
23 13 0	-38 47.2		O	2102	70	-5	603	
23 13 6	7 1.0	NGC 7564	O	10240		198	449	
23 13 12	6 26.0	NGC 7557	O	3612	55	196	594	
23 13 18	9 14.0		O	12512		205	549	
23 13 24	6 25.0	NGC 7562	O O O	3806 3523 3626	 51	195	13 594 842	
23 13 25	-42 51.5	NGC 7552 IC 5294 PKS	R O O O	1580 1662 1657 1782	 60	-25	671 136 145 113	

R.A. (1950)	DEC. (1950)	NAME	OBS	HEL VEL (C*Z)	ERR	GAL CORR	REF	COMMENTS
23 13 31	-42 53.2	NGC 7552	O	1600		-25	574	
			O	1664			155	
23 13 36	-42 31.8	NGC 7599	O	1535		-24	145	
23 14 2	3 48.9	3CR459	O	65890		186	143	
			O	65980			120	
23 14 12	8 37.0		O	6511		203	549	
23 14 18	29 19.0		O	5100		259	534	
23 14 42	18 25.0	NGC 7578A VV 181 ARP 170	O	11952		233	321	
23 14 48	5 23.0	KDG 579A HOL 807	O	(9648)	170	191	567	
			O	9817	36		812	
			O	10002			549	
23 14 48	5 24.0	KDG 579B HOL 807	O	10064		191	549	
			O	9969	56		812	
			O	10167	59		812	
23 14 48	18 26.0	NGC 7578B VV 181 ARP 170	O	12093		233	321	
23 14 54	- 5 1.0	NGC 7576	O	3481	45	151	72	
			O	3616	50		13	
23 15 6	9 9.0	NGC 7579	O	12310		204	549	
23 15 6	13 43.8	NGC 7580 MRK 318	R	4411		219	779	
			O	4769	45		382	
			O	4800			307	
23 15 18	7 5.0		O	12610		197	549	
23 15 18	7 8.0	NGC 7583	O	3782		197	549	
23 15 24	- 4 56.0	NGC 7585 ARP 223	O	3333	65	151	13	
			O	3385			13	
23 15 24	9 9.0	NGC 7584	O	12910		204	549	
23 15 24	9 23.0	NGC 7587B KDG 580B	O	8303	15	205	908	
23 15 24	9 24.0	NGC 7587A KDG 580A	O	8932		205	549	
			O	8614	40		908	
23 15 39	-42 38.7	NGC 7582 PKS	O	1495	60	-25	761	
			O	1451			136	
			O	1620	12		965	
			O	1631			145	

R.A. (1950)	DEC. (1950)	NAME	OBS	HEL VEL (C∗Z)	ERR	GAL CORR	REF	COMMENTS
23 15 39	-42 38.7	NGC 7582	O	1550		-25	574	
23 15 42	3 54.0		O	(18770)		185	303	
23 15 44	6 18.8	NGC 7591	R	4964	15	194	829	
			R	4953	16		914	
			R	4950	20		939	
			R	4976	20		752	
			O	5001			549	
23 15 54	- 4 42.0	NGC 7592	O	7289	54	152	72	
		VV 731	O	7280			776	
23 16 11	-42 30.7	NGC 7590	O	1733		-25	145	
			O	1520			136	
			O	1340			84	
			O	1440			574	
23 16 12	8 57.0	NGC 7601	O	8205		203	549	
23 16 12	18 25.0	NE	O	11610		232	321	
23 16 12	18 25.0	SW	O	10838		232	321	
23 16 12	24 57.0	MRK 319	O	8051	40	249	567	
		KDG 581A	O	8048	15		908	
			O	8400			307	
23 16 12	24 59.0	KDG 581B	O	8126	100	249	567	
			O	7965	26		908	
23 16 18	- 7 51.0	NGC 7600	O	3391	60	138	13	
23 16 18	6 46.0		O	4947		195	549	
23 16 24	- 0 1.0	NGC 7603	O	8800		170	269	SEYF
		MRK 530	O	8714			447	
		ARP 92	O	8700			388	
			O	8700			592	
23 16 24	6 35.0		R	4181	20	195	939	
			O	4236			549	
23 16 27	- 0 1.5		O	16900		170	269	
			O	17060			878	
23 16 30	- 8 45.6	NGC 7606	O	2096		134	426	
			O	2341	75		13	
23 16 36	-42 31.8	NGC 7599	O	1730	30	-25	741	
			O	1778	63		789	
23 16 36	7 25.0		R	3759	20	198	939	

R.A. (1950)	DEC. (1950)	NAME	OBS	HEL VEL (C*Z)	ERR	GAL CORR	REF	COMMENTS
23 16 36	7 50.0	IC 5309	R	4162	20	199	939	
			O	4264			549	
23 16 42	8 4.0	NGC 7608	R	3480	40	200	939	
			O	3559			549	
23 17 0	9 13.0	NGC 7609 ARP 150 VV 20	O	11927		203	549	
23 17 5	7 47.4	NGC 7611	O	3383	65	199	13	
			O	3400	75		2	
			O	3301			549	
23 17 12	8 17.0	NGC 7612	O	3188		200	549	
23 17 12	9 55.0	NGC 7610	R	3550	10	206	582	
			R	3550	20		939	
			O	3506	60		582	
23 17 35	25 56.4	MRK 322	O	8100		250	307	
			O	8200	200		820	
23 17 36	7 53.0	NGC 7617	O	4072	150	199	13	
			O	3900	100		2	
23 17 37	23 56.7	NGC 7620 MRK 321	O	9600		246	307	
23 17 42	7 43.0		R	2798	20	198	939	
23 17 42	7 55.9	NGC 7619	O	3757	50	199	13	
			O	3800			529	
			O	3755			227	
			O	3800	75		2	
			O	3839			842	
23 17 55	27 2.4	NGC 7624 MRK 323	R	4188		252	610	
			R	4351	25		752	
			O	4500			307	
23 18 0	8 7.0	NGC 7623	O	3463	65	199	13	
			O	3800	125		2	
23 18 0	16 57.0	NGC 7625 3 ZW102 ARP 212 VV 280	R	1641		227	310	
			R	1624	12		914	
			R	1603			610	
			R	1620	20		455	
			R	1654	9		622	
			O	1620			209	
			O	1828			13	
			O	1706	100		13	

R.A. (1950)	DEC. (1950)	NAME	OBS	HEL VEL (C∗Z)	ERR	GAL CORR	REF	COMMENTS
23 18 10	7 56.6	NGC 7626	O	3439		199	842	
		PKS	O	3385	15		658	
			O	3700	100		2	
			O	3465	33		938	
			O	3357	50		13	
			O	3351			227	
			O	3613			281	
23 18 24	-58 21.0		O	3180	120	-101	598	
23 18 36	7 5.0		O	3075		195	549	
23 18 49	25 6.6	IC 5315	R	4437		248	901	
23 18 55	7 56.6	NGC 7631	R	3742	20	198	939	
			R	3760	5		914	
			O	3875	100		789	
			O	3772			549	
23 19 0	23 19.0	3CR460	O	80300		244	717	
			O	(81010)			657	
23 19 6	8 42.0		R	3575	20	201	939	
			O	3602			549	
23 19 10	8 36.8	NGC 7634	O	3190		200	426	
			O	3235			549	
23 19 12	8 47.0		R	2849	20	201	939	
23 19 24	4 45.0	KDG 582	O	4951	20	187	908	
23 19 18	4 43.0	KDG 582	O	5029	25	187	908	
23 19 26	-42 46.2	NGC 7632	O	1600		-27	574	
			O	1553	53		789	
23 19 33	27 16.5	4C 27.50	O	35640		252	400	
23 19 43	40 34.2	NGC 7640	R	360	10	273	402	
			R	377			934	
			R	374	2		944	
			R	360			171	
			R	380	30		216	
			R	370			156	
			R	363	25		183	
			R	373	14		489	
			O	388	28		148	
			O	423			13	
23 20 12	26 51.0		O	8094		251	949	
23 20 30	6 1.0	3 ZW103	O	16480	100	191	820	

R.A. (1950)	DEC. (1950)	NAME	OBS	HEL VEL (C*Z)	ERR	GAL CORR	REF	COMMENTS
23 20 36	12 46.0	3 ZW104	O	12782		213	262	
23 20 43	32 15.1	B2	O	5040		261	485	
23 20 58	16 35.3		O	12981	65	225	902	
23 21 13	16 32.7		O	11527	65	225	902	
23 21 17	16 25.9		O	12456	65	224	902	
23 21 22	9 23.6	NGC 7648	O	3591		202	549	
		IC 1486	O	(3845)			447	
		MRK 531						
23 21 20	16 37.3		O	11998	65	225	902	
23 21 23	16 36.3		O	13128	65	225	902	
23 21 24	16 24.4		O	13335	65	224	902	
23 21 25	16 38.5		O	12136	65	225	902	
23 21 26	16 28.5		O	11155	65	224	902	
23 21 26	16 30.1		O	12144	65	224	902	
			O	12213	65		902	
23 21 28	16 32.2		O	12630	65	224	902	
23 21 35	16 31.0		O	12357	65	224	902	
23 21 38	16 23.9		O	12966	65	224	902	
23 21 45	16 30.6		O	12393	65	224	902	
23 22 0	14 24.0		O	12514	75	218	13	
23 22 12	14 24.0		O	13434	50	218	13	
23 22 18	15 0.0	NGC 7653	R	4265	9	220	914	
		IC 1488	R	6202	20		752	DIS
23 22 30	-58 4.0	NGC 7650	O	10100	160	-100	670	
23 22 31	27 46.8	4C 27.51	O	95600		252	899	
23 22 44	-12 23.9	PKS	O	24630		115	120	
23 22 54	-74 54.0		O	6000		-171	896	
23 23 18	26 45.0	NGC 7660	O	5707	33	250	938	
23 23 48	-32 40.0		R	62	5	20	720	

R.A. (1950)	DEC. (1950)	NAME	OBS	HEL VEL (C*Z)	ERR	GAL CORR	REF	COMMENTS
23 23 48	32 40.0	4 ZW148	0	5225	200	261	820	
23 24 2	17 59.5	MRK 324	0	1500		227	307	
			0	1511	45		382	
23 24 12	24 48.0	NGC 7664	R	3481	15	245	752	
			R	3472	8		914	
			0	3464	10		740	
23 24 22	11 5.0	MRK 532	0	(7589)		206	447	
23 24 48	12 12.0	NGC 7671	0	4126		210	549	
			0	4217	94		594	
			0	4129			13	
23 25 0	12 6.6	NGC 7672	0	4394	93	209	594	
23 25 12	23 18.9	NGC 7673	R	3407	20	241	829	
		4 ZW149	R	3404			480	
		MRK 325	R	3368			389	
		KDG 584A	R	3359			610	
			R	3412	10		914	
			R	3408	29		622	
			0	3391	22		500	
			0	3460			303	
			0	3402			480	
			0	3250	120		856	
			0	3054			262	
			0	3402	70		473	
			0	3300			307	
23 25 24	8 30.0	NGC 7674	0	8707		197	549	
		MRK 533	0	8637			874	
		ARP 182	0	8647			447	
		VV 343						
23 25 30	- 2 28.0	MRK 1131	0	1944		156	921	
23 25 30	8 29.0	NGC 7675	0	8664		197	549	
		VV 343						
23 25 36	23 15.3	NGC 7677	R	3489		241	610	
		MRK 326	R	3540			480	
		KDG 584B	R	3544	20		829	
			R	3554	7		914	
			0	3637	48		500	
			0	3620	30		473	
			0	3900			307	
			0	3583			480	
23 25 36	24 33.0	4 ZW150	0	5046		244	237	
			0	4920			28	

R.A. (1950)	DEC. (1950)	NAME	OBS	HEL VEL (C*Z)	ERR	GAL CORR	REF	COMMENTS
23 25 58	22 8.7	NGC 7678	R	3459	50	238	544	
		ARP 28	R	3487	6		914	
			R	3491			744	
			O	3446	65		13	
23 26 0	-48 18.0		O	16556		-56	515	★
23 26 2	-41 36.7	IC 5325	O	1464	10	-24	741	
			O	1527			728	
23 26 3	14 28.3	A 2326	R	-240		216	216	PEG SYS
		DDO 216	R	-179	10		492	
			R	-185			744	
23 26 12	3 14.0	NGC 7679	R	5181		178	538	
		MRK 534	O	5203	44		594	
		ARP 216	O	5202	40		13	
		VV 329	O	5101			13	
			O	5174			106	
			O	5250			569	
23 26 12	14 25.0	4 ZW152	O	20640		216	303	
23 26 27	17 1.0	NGC 7681A	O	7480	47	224	594	
23 26 27	17 1.0	NGC 7681B	O	6992	68	224	594	
23 26 30	3 15.0	NGC 7682	R	5155	30	178	885	
		ARP 216	O	5168	33		594	
		VV 329						
23 27 33	40 43.1	DDO 217	R	424	10	271	492	
23 27 36	25 15.0	4 ZW153	R	5307	62	244	622	
			O	5326			262	
23 29 30	28 40.3	MRK 930A	O	5319	124	251	830	
23 29 41	2 9.0	MRK 536	O	3450		172	569	
23 29 41	25 35.0	MRK 535	O	(7101)		244	447	
23 29 54	-54 22.0	NGC 7689	O	1450	100	-85	844	
			O	1963	65		741	
23 30 14	-51 58.3	NGC 7690	O	1425		-74	574	
			O	1317	35		741	
23 30 36	-45 17.4	IC 5328	O	3049	48	-43	741	
23 30 48	20 57.0	IC 5329	O	5702	35	232	908	
		KDG 585						

23ʰ31ᵐ

R.A. (1950)	DEC. (1950)	NAME	OBS	HEL VEL (C*Z)	ERR	GAL CORR	REF	COMMENTS
23 31 0	20 51.0	IC 5331 KDG 585	O	5775	36	232	908	
23 31 12	29 46.0	KDG 586 VV 314 ARP 46	O O O	5290 (5224) 4900	18 93	252	908 812 449	
23 31 18	-24 0.2	PKS	O	14300		58	600	
23 31 24	29 47.0	KDG 586 VV 314 ARP 46	O O	3742 (3882)	15 50	252	908 812	
23 31 48	-36 22.6	IC 5332	R	702		-1	671	
23 31 51	4 42.1	MRK 537	O	5850		181	569	
23 32 22	17 57.0	DDO 218	R	1394	15	223	492	
23 32 32	27 5.7	4C27.53A	O	18170		246	595	*
23 32 46	-56 17.4	NGC 7702	O O	3152 6486	34	-95	741 84	DIS
23 33 6	0 55.2		R	2596	10	166	860	
23 33 18	23 20.0	NGC 7712	R	3051	6	237	914	
23 33 41	1 52.7	NGC 7714/5 VV 51	R	2805	40	169	455	
23 33 41	1 52.7	NGC 7714 MRK 538 ARP 284 KDG 587A	R O O O O	2803 2818 2780 2793 2833	15 27	169	914 500 176 748 13	
23 33 48	-38 13.0	NGC 7713	R O O O	686 1010 667 2921	100 30	-10	671 844 148 145	DIS
23 33 49	1 53.0	NGC 7715 KDG 587B ARP 284	O O	2795 2831		169	13 748	
23 33 54	0 1.0	NGC 7716	O	2546	150	162	13	
23 33 56	20 51.4	IC 5338 3C 464	O	16620		231	400	
23 34 0	-20 44.7		O	7933		72	728	*

R.A. (1950)	DEC. (1950)	NAME	OBS	HEL VEL (C*Z)	ERR	GAL CORR	REF	COMMENTS
23 34 50	0 6.9	DDO 219	R	2677	10	162	492	
23 35 9	29 51.2	MRK 328	R	1379		251	779	BRB GAL
			O	1200			307	
			O	1320	30		28	
23 35 24	31 43.0		O	4976		254	777	
			O	4800			534	
23 35 30	31 21.0	MRK 1132	O	8886		254	921	
23 35 40	26 55.5		R	7956	9	244	745	
23 35 44	30 25.8	B2	O	0		252	485	
23 35 48	26 37.0		O	9244		244	659	
23 35 48	26 53.0		O	9271		244	659	
23 36 0	26 33.0		O	9509		243	659	
23 36 0	26 42.0		O	10859		244	659	
23 36 0	26 45.3	NGC 7720N/S	O	8790		244	120	★
23 36 0	26 45.3	NGC 7720N	O	7911		244	659	
		KDG 588B	O	7954			444	
			O	8171	96		812	
23 36 0	26 45.3	NGC 7720S	O	9045		244	659	
		KDG 588A	O	9015	28		444	
		3CR465	O	9140			255	
			O	9004	49		812	
23 36 6	26 42.0		O	9694		244	659	
23 36 12	- 6 48.0	NGC 7721	O	2039	27	133	148	
23 36 12	26 30.0		O	9726		243	659	
23 36 12	26 44.0	IC 5342	O	9218		244	659	
23 36 12	26 56.0		O	9458		244	659	
23 36 18	26 56.0		O	9201		244	659	
23 36 18	26 59.0		O	9240		244	659	
23 36 24	-13 14.0	NGC 7723	O	1823		105	777	
			O	1973			13	
23 36 42	26 50.0		O	7499		244	659	

R.A. (1950)	DEC. (1950)	NAME	OBS	HEL VEL (C*Z)	ERR	GAL CORR	REF	COMMENTS
23 36 48	26 58.0		O	9695		244	659	
23 37 18	-12 34.0	NGC 7727	R	1814	100	108	885	
		ARP 222	O	1877			13	
		VV 67	O	1839	30		13	
23 37 24	26 33.0		O	8657		243	659	
23 37 24	26 52.0	NGC 7728	O	9534		244	659	
		3C 465F	O	9410			595	
23 38 6	- 2 53.0	3 ZW114	O	7040		148	303	
23 38 8	25 57.0	KDG 589A	R	743	10	241	860	
			O	748	14		908	
23 38 12	26 33.0		O	11560		242	659	
23 38 43	- 9 18.2		O	73700		121	717	
23 39 0	3 28.0	NGC 7731	R	2891	7	172	885	
		KDG 590	O	2800	100		594	
23 39 6	3 27.0	NGC 7732	O	2873	65	172	594	
		KDG 590	O	2968	24		500	
23 39 13	- 3 56.8	ARP 295A	O	6860		143	464	
		VV 34	O	6777	60		13	
23 39 18	- 1 36.0	3 ZW116	O	6900	100	153	820	
23 39 27	- 3 53.6	ARP 295B	O	6960		143	464	
		VV 34	O	7016	60		13	
23 39 28	26 32.2	3C 465H	O	28300		242	595	
23 40 29	19 8.8	MRK 330	O	4074	45	223	382	
			O	4200			307	
23 40 48	28 14.0		O	11023		245	321	
23 41 22	25 47.9	NGC 7741	R	786	25	239	183	
		KDG 589B	R	753			744	
			R	749	5		914	
			O	718	15		908	
			O	718	60		555	
			O	729	50		13	
			O	808	11		148	
23 41 36	27 27.0	MRK 1133	O	7077		243	920	★
23 41 43	10 29.3	NGC 7742	O	1629	40	196	13	
			O	1748			13	

R.A. (1950)	DEC. (1950)	NAME	OBS	HEL VEL (C*Z)	ERR	GAL CORR	REF	COMMENTS
23 41 48	9 39.0	NGC 7743	O	1802	65	193	13	
23 42 21	8 54.7	A	O	12380		190	255	
23 42 21	8 54.7	B	O	12140		190	255	
23 42 26	-43 11.4	NGC 7744	O	2990	51	-37	741	
23 42 33	29 26.2	4C 29.70	O	39180		247	400	
23 42 38	6 45.7		O	5354	42	182	741	
23 43 18	-29 49.0	NGC 7749	O	10269	150	25	690	
23 43 36	-28 24.0		O	8353	150	31	690	
23 43 41	33 5.4		R	4929	15	253	829	
23 43 42	-28 17.0		O	7551	150	32	690	
23 44 6	-29 22.0		O	10258	150	27	690	
23 44 6	3 31.2	NGC 7750	R	2934	9	170	914	
			O	2913	59		741	
23 44 16	6 35.0	NGC 7751	R	3261	11	181	914	
23 44 27	- 0 43.4	MRK 540	O	(21287)		153	569	
23 44 28	29 10.7	NGC 7752	O	4980		245	195	
		4 ZW165	O	4989	27		500	
		MRK 1134	O	4585			921	
		ARP 86	O	4902	16		543	
		VV 5	O	4845	77		72	*
		KDG 591	O	4800			195	
			O	5100			534	
23 44 33	29 12.2	NGC 7753	R	5163	11	245	914	
		4 ZW165	O	4868	32		72	*
		ARP 86	O	5206			364	
		VV 5	O	5200	25		543	
		KDG 591	O	5423	90		500	
23 44 36	-28 15.0		O	8424	150	32	690	
23 44 42	-28 23.8		O	8810	150	31	648	
23 44 53	-28 24.7		O	8210	150	31	648	
			O	8116	150		690	
23 44 53	-28 23.0		O	8300	150	31	648	KLE 44
			O	7990	150		690	

425

R.A. (1950)	DEC. (1950)	NAME	OBS	HEL VEL (C*Z)	ERR	GAL CORR	REF	COMMENTS
23 44 56	-28 23.0		O	8300	150	31	648	
			O	8496	150		690	
23 45 0	-28 24.0		O	8301	150	31	690	
23 45 7	-28 27.5		O	8360	150	30	648	
23 45 10	-28 25.2		O	7940	150	31	648	
23 45 12	-28 25.0		O	8554	150	31	690	
			O	8330	150		648	
23 45 18	-30 48.0	NGC 7755	O	3148	74	19	203	
			O	2930			574	
			O	2898	35		148	
23 45 18	-28 26.0		O	7972	150	30	690	
23 45 24	-28 36.0		O	10095	150	30	690	
23 45 25	-28 25.2		O	7320	150	30	648	
23 45 48	-28 30.0		O	8660	150	30	690	
23 45 48	-28 23.0		O	8923	150	30	690	
23 45 48	-28 21.0		O	10091	150	31	690	
23 45 54	-28 33.0		O	8117	150	30	690	
23 45 57	18 27.5	3C 467	O	189360		218	584	
23 46 12	3 53.6	NGC 7756	O	3148	23	170	741	
23 46 13	3 53.7	NGC 7757	R	2953	8	170	914	
		ARP 68	R	2950			744	
			R	2955	10		752	
23 46 17	25 56.4	DDO 220	R	799	10	237	492	
23 46 24	5 51.6		O	3848	30	177	741	
23 47 21	-61 51.5		O	13490	200	-123	794	
23 47 21	-61 48.9		O	9140	200	-123	794	
23 47 24	-28 13.0		O	8750	150	30	690	
22 47 26	11 20.8		O	7985		222	847	
23 47 52	30 13.3	B2	O	116900		246	899	
23 48 0	26 37.7		O	8175	65	238	902	

R.A. (1950)	DEC. (1950)	NAME	OBS	HEL VEL (C*Z)	ERR	GAL CORR	REF	COMMENTS
23 48 0	28 43.0	MRK 1135	O	8448		242	921	
23 48 3	26 55.4		O	7573	65	239	902	
23 48 16	27 0.60		O	7967	65	239	902	
			O	7944			949	
23 48 18	-41 1.0	NGC 7764	O	1632	24	-30	724	
			O	1759	70		603	
			O	1640	100		844	
			O	1640	42		789	
			O	1759			293	
23 48 20	26 53.29	NGC 7765	O	7483	40	238	651	
			O	7525			949	
			O	7567	65		902	
23 48 24	26 48.52	NGC 7767 IC 1511	O	8055	65	238	902	
23 48 24	26 50.91	NGC 7766	O	7938	50	238	651	
			O	8031	65		902	
23 48 26	26 52.17	NGC 7768	O	7831	65	238	902	
			O	7940			255	
			O	8218			777	
			O	8214			949	
			O	7143	30		651	
23 48 27	26 57.78		O	8142	65	238	902	
23 48 28	26 56.47		O	7468	65	238	902	
23 48 29	26 58.80		O	8448	65	238	902	
			O	8634			949	
23 48 29	27 0.66		O	9063	65	239	902	
23 48 31	19 52.3	NGC 7769 KDG 592A	R	4204	14	221	914	
			O	4344			748	
			O	4349			13	
23 48 32	0 46.4		O	8244	21	157	741	
23 48 46	20 18.0	KDG 593A	O	5323		222	748	
			O	5256	50		812	
23 48 47	27 0.93		O	7915	65	238	902	
23 48 48	19 48.0	NGC 7770	O	4338		220	13	
23 48 52	19 50.0	NGC 7771 KDG 592B	O	4276		220	13	
			O	4278			748	
			O	4293	57		72	

427

23ʰ48ᵐ

R.A. (1950)	DEC. (1950)	NAME	OBS	HEL VEL (C*Z)	ERR	GAL CORR	REF	COMMENTS
23 48 54	20 18.5	MRK 331	O	5400		222	307	
		KDG 593B	O	5378			748	
			O	5244	30		812	
			O	5323	47		500	
			O	5363	45		382	
			O	5312	30		567	
23 48 54	35 23.0	MRK 1136	O	7835		255	921	
23 48 58	26 57.5		O	8220	65	238	902	
23 49 3	26 55.4		O	7771	65	238	902	
23 49 6	-28 39.0		O	7799	150	28	690	
23 49 15	27 0.6		R	8002	6	238	745	
23 49 19	-39 18.1		O	12695	30	-22	613	
23 49 22	-39 17.6		O	12567	58	-22	613	
23 49 22	- 1 25.9	4C-01.61	O	52200		148	189	SEYF
23 49 24	-28 14.0		O	8899	150	29	690	
23 49 24	-28 12.0		O	8794	150	30	690	
23 49 28	27 3.0		O	7974	65	238	902	
23 49 42	-28 44.0		O	8814	150	27	690	
23 49 54	-28 37.0		O	(7836)	150	27	690	
23 50 12	26 51.0		O	8046	65	237	902	
23 50 44	-41 5.1		O	9082	140	-31	613	
23 50 47	-41 5.3	NGC 7764A	O	9162	61	-51	613	
			O	8860	58		789	
			O	9080	40		844	
			O	9101	70		603	
23 50 47	-41 5.3	NGC 7764E	O	7830	40	-31	844	*
			O	7839	70		603	
23 50 50	-41 5.9		O	9312	64	-31	613	
23 50 59	7 50.6	NGC 7780	O	5155	58	182	741	
23 51 20	7 41.5	NGC 7782	R	5377	9	181	914	
			O	5368	20		741	
			O	(5960)	150		555	

428

R.A. (1950)	DEC. (1950)	NAME	OBS	HEL VEL (C*Z)	ERR	GAL CORR	REF	COMMENTS
23 51 30	-10 42.53		O	23352	120	108	393	
23 51 30	-10 36.72		O	21152	120	108	393	
23 51 32	-10 42.18		O	22592	120	108	393	
23 51 36	-10 46.55		O	22242	120	108	393	
23 51 36	0 6.0	NGC 7783A/B ARP 323 VV 208	O	7779		153	274	
23 51 36	0 6.0	NGC 7783A KDG 595A	O	7012	34	153	594	
23 51 36	0 6.0	NGC 7783B KDG 595B	O	6702	65	153	594	
23 51 39	-10 41.83		O O	23092 23105	120	108	393 444	
23 51 39	-10 39.86		O	23562	120	108	393	
23 51 42	-10 31.62		O	22791	120	109	393	
23 51 45	-10 49.32		O	21323	120	107	393	
23 51 47	-10 41.89		O	23292	120	108	393	
23 51 53	-10 38.95		O	20142	120	108	393	
23 52 38	49 33.4	DA 611	O	71100		268	586	
23 52 46	5 38.3	NGC 7785	O O	3820 3846	65	173	842 13	
23 53 13	10 11.1		R	1739	10	188	860	
23 53 28	7 14.6	MRK 541	R O	11716 (11814)		178	707 447	SEYF
23 53 30	- 1 15.8	IC 1515 KDG 597	O O	6739 6714	43 27	146	741 908	
23 53 33	- 1 11.7	IC 1516 KDG 597	O O	7390 7262	43 28	146	741 908	
23 54 12	16 32.0	KDG 598A ARP 262 VV 255	O	1828	22	208	812	
23 54 12	16 33.0	KDG 598B ARP 262 VV 255	O	1723	13	208	812	

23^h54^m

R.A. (1950)	DEC. (1950)	NAME	OBS	HEL VEL (C*Z)	ERR	GAL CORR	REF	COMMENTS
23 54 26	-35 11.3	PKS	O	14600	30	-5	355	
23 54 26	- 2 22.0	MRK 542	O	7354		141	447	
23 54 36	12 11.0	IC 1519	O	8573	236	194	908	
		KDG 599	O	8771	50		812	
23 54 36	12 12.0	IC 1518	O	11191	100	194	908	
		KDG 599	O	11270	63		812	
23 54 42	47 38.0		O	13077		266	321	
23 54 48	47 37.0		O	12886		266	321	
23 54 54	47 36.0		O	13187		266	321	
23 55 15	-32 52.1	NGC 7793	R	227		5	671	
			R	205	12		216	
			R	205			171	
			O	177			13	
			O	221			892	
			O	200			574	
			O	215			619	
			O	286	200		13	
23 56 0	10 26.0	NGC 7794	R	5280	32	188	914	
23 56 3	27 38.0	4C 27.54	O	123500		236	899	
23 56 23	-55 44.1	NGC 7796	O	3411	100	-100	741	
			O	3511			84	
			O	3300			574	
23 56 29	-61 11.7	PKS	O	28890	30	-123	355	
			O	28870	40		350	
			O	28800			627	
23 56 48	46 37.0	IC 1525	R	5031	15	265	752	
			R	5018	10		829	
23 56 52	20 28.5	NGC 7798	R	2407	8	218	914	
		MRK 332	O	2700			307	
23 57 27	0 25.3	PKS	O	25150	120	150	350	
23 57 28	46 59.7		R	5017	10	265	829	
23 57 30	- 0 21.0	KDG 600	O	7028	32	147	812	
			O	5400			475	DIS
23 57 36	- 0 19.0	KDG 600	O	7006	36	147	812	
			O	7200			475	

R.A. (1950)	DEC. (1950)	NAME	OBS	HEL VEL (C*Z)	ERR	GAL CORR	REF	COMMENTS
23 57 48	39 12.9		R	334	10	256	860	
23 58 0	26 3.0	MRK 1137	O	7239		231	921	
23 58 6	22 30.0		O	8080		223	814	
23 58 36	4 13.0	IC 5374 KDG 601	O	8932	20	164	908	
23 58 36	4 15.0	IC 5375 KDG 601	O	9073	55	164	908	
23 58 48	31 9.0	NGC 7805 ARP 112 VV 226 KDG 602	O	4816	35	242	812	
23 58 54	31 10.0	NGC 7806 ARP 112 VV 226 KDG 602	O	4633	73	242	812	
23 58 46	12 50.0	NGC 7803	R	5401	25	194	829	
23 59 5	23 12.4	3 ZW125A KDG 603 VV 254	O O O O	4367 4379 4579 4581	10 59	224	76 908 748 500	
23 59 9	23 13.1	3 ZW125B KDG 603 VV 254	O O O O	4361 4436 4585 4313	16 20	224	76 908 748 500	
23 59 23	-15 44.1	DDO 221 A 2359	R R R O O	-123 -115 -130 -76 -78	10 5 20	82	492 934 156 11 13	WLM GAL
23 59 53	3 4.5	MRK 543	O	7641		159	569	SEYF

CHAPTER 4

Notes to the Catalogue

00 07 57 +10 41.8   IIIZW2. POSSIBLE QSO.

00 24 00 +16 53.4   PROBABLY OBSERVED BY OKE (REF.292), WHO ASSIGNS A VALUE Z=0.38+/-0.01 TO A GALAXY OF ZW CL 0024+1654.

00 31 33 +39 07.6   IN THE TABLE THE VALUE VH=CZ IS PRESENTED; THE RELATIVISTIC RECESSION VELOCITY HOWEVER IS 184500 KM/S.

00 32 05 +30 09.5   FORMERLY IDENTIFIED WITH THE RADIOSOURCE B2 0032+30 ACCURATE MEASUREMENTS BY GRUEFF ET AL. (ASTRON.AND AP. 1981, 44, 241) SHOW THE IDENTIFICATION TO BE UNFOUNDED. THE RADIOSOURCE B2 0032+30 IS NOW CONSIDERED AN EMPTY FIELD.

00 35 14 -33 59.4   RING GALAXY, OPTICAL MEASUREMENTS REFER TO THE RING AND WERE OBTAINED AVERAGING SEVEN SPECTRA OF EMISSION REGIONS.

00 45 08 -25 33.7   NGC253. THE AUTHORS OF REF.245 CLAIM THAT BECAUSE OF THE PRESENCE OF AN EXPANSION OUTFLOW FROM THE NUCLEUS THE REAL VELOCITY IS VH=250 KM/S. ACCORDING TO REF.66 THE REDSHIFT IN REF.13 WHICH REFERS TO AN "EMISSION PATCH 2'.7 N OF THE NUCLEUS" IS WRONG. THERE APPEAR TO BE NO EMISSION PATCHES IN THE VICINITY. PROBABLY NGC253 HAS BEEN CONFUSED WITH NGC300 IN TABLE I OF REF.13. MEASUREMENT OF REF.755 IS OBTAINED AVERAGING VELOCITIES OF PAIRS OF HII REGIONS IN THE SPIRAL ARMS DIAMETRICALLY OPPOSED WITH RESPECT TO THE NUCLEUS.

00 48 30 -07 20.0   NGC275. IN REF.744 THE AUTHORS POINT OUT THAT THEIR MEASUREMENT COULD HAVE BEEN CONFUSED BY THE PRESENCE OF NGC274.

00 51 00 +12 25.0   1ZW1. THE HIGHEST VH VALUE OF REF.825 REFERS TO EMISSION LINES (BOTH PERMITTED AND FORBIDDEN), THE LOWEST ONLY TO FORBIDDEN ONES.

00 53 42 -14 33.0   POSSIBLE SEYFERT.

00 55 11 -63 41.5   SEE NOTE FOR 00 55 15 -63 42.0.

00 55 15 -63 42.0   DOUBLET ESO 079-IG-13. COORDINATES IN REF.680 ARE DIFFERENT FROM THOSE GIVEN HERE TAKEN FROM REF.896. THERE IS NO DOUBT HOWEVER THAT THE OBJECTS DESCRIBED ARE THE SAME.

00 55 40 +26 35.5   NGC326. THIS GALAXY HAS TWO COMPONENTS WITH ZN = 0.0471 AND ZS = 0.0495 RESPECTIVELY.

01 06 06 +72 56.0   THE IDENTIFICATION OF THIS GALAXY WITH THE

RADIOSOURCE 3CR31.3 IS DOUBTFUL,(SEE JENKINS,C.J., POOLEY,G.G., RILEY,G.M., 1977 MEM.RAS, 84, 61).

01 13 19 +32 49.5   MRK1. ALL THE VELOCITIES LISTED (REFS.192,280,707,723 AND 779) REFER TO THE GALAXY MRK1 WHICH ACCORDING TO 2RCBG IS NGC449 BUT OTHER AUTHORS IDENTIFY IT WITH IC1661 (SEE KOJOIAN ET AL., 1978, A.J., 82, 1545).

01 17 42 −41 28.0   AGUERO 6. THE REDSHIFT QUOTED FROM REF.719 IS THE AVERAGE OF THE MEASUREMENTS ON SIX DIFFERENT CONDENSATIONS.

01 21 51 −59 03.9   FAIRALL 9. DOUBLET OF GALAXIES. REF.761 REFERS TO THE MAIN GALAXY; REF.695 TO ITS COMPANION.

01 22 00 +03 32.0   NGC520. FOR THIS PECULIAR GALAXY WE HAVE LISTED THE HIGHEST PEAK VELOCITIES FOR RADIO MEASUREMENTS AND THE VHS OF THE BRIGHTEST CONDENSATIONS FOR OPTICAL MEASUREMENTS WHEN SPECIFIED IN THE REFERENCE. FOR A DETAILED DISCUSSION OF THIS OBJECT SEE REF.951. THE VH FROM REF.951 REFERS TO THE MAIN CONDENSATION OF A COMPLEX OPTICAL STRUCTURE.

01 32 06 +34 48.0   POSSIBLE SEYFERT.

01 45 06 +27 10.7   NGC672. RADIO MEASUREMENTS POSSIBLY CONFUSED BY IC1727 (01 44 42 +27 05.1).

01 54 24 +32 00.1   THE RADIO GALAXY 4C31.06B IS ASSOCIATED WITH A CLUSTER OF GALAXIES AND, AT PRESENT, IT IS IMPOSSIBLE TO IDENTIFY WHICH SPECIFIC GALAXY IN THE CLUSTER IS THE RADIOSOURCE. REF.595 DOES NOT MENTION TO WHICH GALAXY OF THE CLUSTER THE VH LISTED REFERS.

01 56 06 −01 42.0   IT IS NOT CLEAR WHETHER THE GALAXY OBSERVED IN REF.475 IS THE BRIGHTEST COMPONENT OF THE DOUBLET KDG48 OR NOT. FROM THE QUOTED MAG=14.9 WE IDENTIFIED IT WITH KDG48B (REF.908). HOWEVER ITS VELOCITY IS VERY SIMILAR TO THE "A" COMPONENT.

02 05 14 +02 28.5   MRK586. POSSIBLE QSO.

02 10 00 +86 06.1   REF.391 GIVES RADIAL VELOCITIES OF FOUR GALAXIES (01 10 00 +86 06.1; 02 11 01 +86 05.3; 02 11 02 +86 06.7; 02 12 08 +86 06.4) IN THE FIELD OF THE RADIOSOURCE 3C61.1. KRISTIAN ET AL.(1978, AP.J., 219, 803) HOWEVER ARGUE THAT THE CORRECT IDENTIFICATION IS ANOTHER GALAXY, FURTHER AWAY FOR WHICH NO VH IS, AS YET, AVAILABLE.

02 11 01 +86 05.3   SEE NOTE FOR 02 10 00 +86 06.1.

02 11 02 +86 06.7   SEE NOTE FOR 02 10 00 +86 06.1.

02 11 28 +04 56.5   MRK1027. GALAXY WITH DOUBLE NUCLEUS. VH FOR THE
                     TWO NUCLEI ARE IDENTIFIED WITH A AND B.

02 12 08 +86 06.4   SEE NOTE FOR 02 10 00 +86 06.1.

02 40 22 +30 07.8   NGC1058. THE MEASUREMENT OF REF.13 IS RECOGNIZED
                     AS WRONG IN REF.97.

02 56 00 -36 54.0   THE COORDINATES LISTED FOR THIS OBJECT IN REF.728
                     ARE NOT ACCURATE ENOUGH TO SAY WHETHER THIS
                     OBJECT IS THE SAME AS 02 56 56 -36 48.6 OR NOT
                     (REF.680).

02 56 56 -36 48.6   SEE NOTE FOR 02 56 00 -36 54.0.

02 58 43 +35 38.5   4C35.06. THE REDSHIFT GIVEN IN REF.255 IS THE
                     AVERAGE BETWEEN THE REDSHIFTS OF TWO COMPONENTS
                     OF THE DOUBLE GALAXY SYSTEM.

03 07 46 -20 47.4   NGC1232A. IN REF.789 IT IS SUGGESTED THAT VH
                     MEASURED IN REF.528 IS INCORRECT.

03 17 39 -66 40.7   NGC1313-I. MEASUREMENTS OF REFS.49 AND 431 ARE
                     CORRECTED IN REFS.488 AND 474 RESPECTIVELY.
                     NGC1313-I IS ERRONEOUSLY CALLED IC1313 IN REF.49.

03 54 06 -42 31.0   NGC1487. THE RADIAL VELOCITY FROM REF.719 IS
                     OBTAINED AVERAGING MEASUREMENTS ALONG FOUR
                     CONDENSATIONS.

04 30 31 +05 15.0   THE MEASUREMENT OF REF.880 REFERS TO THE STELLAR
                     POPULATION OF THE NEBULOSITY SURROUNDING THE
                     SEYFERT NUCLEUS.

05 53 06 +03 23.1   2ZW40. VH REPORTED IN REF.262 IS IN ERROR. A
                     REMEASUREMENT OF THE SAME SPECTRUM BY SARGENT
                     (AP.J., 1970, 162, L155) YIELDS VH=750 KM/S.

07 22 33 -09 33.6   NGC2377. THIS GALAXY IS NOT, AS PREVIOUSLY
                     BELIEVED, THE OPTICAL COUNTERPART OF 3C178 (SEE
                     HASCHICK ET AL., 1980, AP.J., 239, 774).

07 23 37 +69 19.1   MRK71 IS AN HII REGION IN THIS DWARF IRREGULAR.

07 25 48 +33 58.0   NGC2389. IN REF.744 VH IS CONFUSED BY THE
                     PRESENCE OF NGC2388, 3'.4 W.

07 38 00 +73 54.0   VERY APPROXIMATE COORDINATES FOR TB0738+73.

07 43 31 +39 08.4   VV117 C,D,ETC. IN COL. 3 REFER TO EMISSION
                     REGIONS AS INDICATED IN PHOTOGRAPH FROM REF.19.

07 44 36 +55 57.0   4C56.16. ONE OF TWO POSSIBLE IDENTIFICATIONS
                     TOGETHER WITH A GALAXY AT 07 46 00 +56 03.0. VH
                     MEASURED FROM CENTRAL COMPONENT. A JET FROM
                     NUCLEUS HAS VH=7237+/-7 KM/S.

07 46 00 +56 03.0   SEE NOTE FOR 07 44 36 +55 57.0.

07 54 00 +73 12.0   VERY APPROXIMATE COORDINATES FOR TB0754+73.

08 20 48 +21 38.0   ERRONEOUSLY IDENTIFIED WITH NGC2341 BY REF.321.

08 26 18 +66 04.0   CLUSTER ABELL 665. NUMBERS OF PRESENT AND FOLLOWING GALAXIES REFER TO THE FINDING CHART PUBLISHED IN REF.633.

08 30 06 +57 45.0   ALSO OBSERVED IN REF.521 BUT MEASUREMENT CONFUSED BY A NEARBY STAR. THIS UNCERTAINTY IS ALSO CONFIRMED IN REF.908.

08 32 11 +46 40.0   A AND B IN COL.3 INDICATE CONDENSATIONS (COMPANION GALAXIES?) NORTH OF 08 32 11 +46 39.9 (MKN92).

08 32 12 +46 40.0   SEE NOTE FOR 08 32 11 +46 40.0.

08 55 06 +03 23.0   A COMPOSITE SPECTRUM BY OKE (REF.292) OF BOTH COMPONENTS SUGGESTS Z=0.20+/-0.01.

08 55 18 +37 16.0   SMALL COMPANION OF II-HZ4 IS INDICATED BY B.

08 58 02 +60 20.9   IN REF.261 (TABLE I) A VH FOR MRK18 IS GIVEN BUT ACTUALLY IT REFERS TO NGC2726.

09 03 36 +60 40.0   NGC2742. THE VALUE OF 1276 KM/S (REF.688) IS AN AVERAGE OF 3 INDEPENDENT MEASUREMENTS.

09 13 13 +53 39.1   MRK104 GALAXY WITH A DOUBLE NUCLEUS. VH FOR THE TWO NUCLEI ARE IDENTIFIED WITH A AND B.

09 17 30 +01 15.0   POOR CLUSTER MKW1S. NUMBERS 1, 2, AND 3 REFER TO IDENTIFICATIONS IN REF.750. NO FINDING CHARTS AVAILABLE AS FAR AS WE ARE AWARE.

09 23 48 +74 48.0   APPROXIMATE COORDINATES OF THE BRIGHTEST MEMBER OF A787. THIS GALAXY HAS A DOUBLE NUCLEUS.

09 26 00 +74 00.0   VERY APPROXIMATE COORDINATES FOR TB0926+74.

09 29 20 +21 43.2   NGC2903. THE VALUE OF 589 KM/S (REF.33) IS AN AVERAGE OF 4 INDEPENDENT MEASUREMENTS.

09 42 43 +73 13.2   OUR IDENTIFICATION OF THE OBJECT OBSERVED BY REF.214 WITH COMPONENT A OF MRK121 IS TENTATIVE.

09 44 00 -04 38.0   VERY APPROXIMATE POSITION OF "PECULIAR COMPANION ABOUT 87'NE OF NGC2974" (09 40 00 -03 28.1)(REF.360).

09 47 18 +00 51.1   MRK1236 SATELLITE OF NGC3023 (09 47 18 +00 51.3) PROJECTED ON A SPIRAL ARM.

09 47 18 +00 51.3   THE MEASUREMENT OF REF.776 PUBLISHED AS SE KNOT,

PROBABLY REFERS TO MRK1236.

09 51 27 +69 17.8   MEASUREMENT REFERRED TO NGC3031 PLUS NGC3034 PLUS NGC3037.

09 51 43 +69 55.0   NGC3034.   REF.470   MEASURED   HI   ABSORBTION; REF.621   H92   ALPHA   RECOMBINATION   LINE. MEASUREMENT IN REF.216 POSSIBLY CONFUSED BY THE GALAXY.

09 52 12 +09 30.0   NGC3049=MRK710. GALAXY WITH A DOUBLE NUCLEUS. VH FOR THE TWO NUCLEI ARE IDENTIFIED WITH A AND B.

09 54 00 +15 53.0   MRK712. GALAXY WITH A DOUBLE NUCLEUS.   VH FOR THE TWO NUCLEI ARE IDENTIFIED WITH A AND B.

09 57 52 +72 21.9   NGC3066 POSSIBLE SEYFERT.

10 07 30 -66 48.0   IC2554 VH FROM REF.719 IS AN AVERAGE OF 3 MEASUREMENTS ON 3 CONDENSATIONS.

10 20 48 +11 13.0   MRK721=IC606 GALAXY WITH A DOUBLE NUCLEUS.   VH FOR THE TWO NUCLEI ARE IDENTIFIED WITH A AND B.

10 25 42 -43 38.9   NGC3256.VH FROM REF.719 IS AN AVERAGE OF 3 MEASUREMENTS ON 3 CONDENSATIONS.

10 47 02 +33 14.7   NGC3395. VH FROM REF.744 IS CONFUSED BY NGC3396 IN THE BEAM.

10 47 08 +33 15.3   NGC3396. IN REF.242 THE MEASURMENT WAS DONE ASSUMING THE CENTER OF THIS IRREGULAR GALAXY AS THE BRIGHTEST EASTWARD CONCENTRATION.   OTHER AUTHORS HAVE MEASURED VH OF THIS OBJECT WHERE A KNOT IS PRESENT.

10 49 24 +33 13.0   NGC3430. IN REF.693 POSSIBLE CONFUSION CAUSED BY NGC3424 (10 49 00 +33 10.0) IN THE BEAM.

10 51 12 +54 34.0   DWARF COMPANION OF NGC3448 (10 51 40 +54 34.5).

10 51 40 +54 34.5   NGC3448. OBJECT WITH TWO NUCLEAR COMPONENTS HERE CALLED E AND W.

10 52 16 +40 43.5   THIS ENTRY REFERS TO A DIFFERENT GALAXY FROM THE FOLLOWING ONE ALTHOUGH COORDINATES ARE THE SAME.

11 10 06 +09 20.0   MRK731=IC676. GALAXY WITH A DOUBLE NUCLEUS.   VH FOR THE TWO NUCLEI ARE IDENTIFIED WITH A AND B.

11 13 53 +29 31.2   4C29.41. COMPLEX FIELD FORMED BY TWO GALAXIES AND A SOUTHERN KNOT (11 13 55 +29 31.0).   THE REDSHIFT OF REF.485 REFERS TO THE IDENTIFICATION OF THE 4C RADIOSOURCE WITH THE NORTHERN GALAXY.

11 18 04 +23 44.3   GALAXY ERRONEOUSLY IDENTIFIED WITH 3CR256 (SEE

REF.585 AND REF.946).

11 29 07 +71 05.1   ARP329. VH GIVEN IN REF.77 IS AN AVERAGE OF
                     VELOCITIES   OF   ALL   GALAXIES   IN   THE   VV172
                     (A,B,C,D,E,)GROUP.

11 29 54 +53 13.5   MRK176. THE VELOCITY FROM REF.308 COULD REFER TO
                     THE   ENTIRE   GROUP   OF   INTERACTING   GALAXIES   VV150
                     RATHER THAN TO MRK176 ALONE. FURTHERMORE MRK176
                     APPEARS TO BE THE C COMPONENT OF THE VV GROUP AS
                     LISTED IN THE 2RCBG AND NOT THE   D   COMPONENT   AS
                     STATED   IN   THE   NOTES   OF   THE   SAME   2RCBG.
                     IDENTIFICATION WITH MEMBERS MEASURED IN   REFS.360
                     AND 40 IS TENTATIVE.

11 40 36 +20 01.0   GALAXY 97-78 IN CGCG. GUDHEUS (1976, AP.J., 208,
                     267) ERRONEOUSLY IDENTIFIES THIS OBJECT   WITH
                     IC2951, (SEE REF.753).

11 41 50 +37 25.3   THIS AND THE FOLLOWING GALAXY ARE BOTH ASSOCIATED
                     WITH THE RADIOSOURCE B2 1141+37A,B.

11 42 00 +57 47.0   SHB123.   THE NUMBERS REFER TO THE   FINDING   CHART
                     PUBLISHED   IN   BAIER   ET AL., (1974, ASTROFIZIKA,
                     10, 327).

11 54 07 +48 36.8   NGC3985. IN REF.789 IT IS NOT SPECIFIED   WHETHER
                     VH   REFERS TO THE GALAXY ALONE OR WHETHER IT ALSO
                     INCLUDES THE CLOSE COMPANION KDG310A   (11   54   00
                     +48 36.0).

11 55 00 +53 39.0   NGC3992. IT APPEARS THAT   THE   AUTHORS   OF   THIS
                     REF.   HAVE   MISIDENTIFIED   ONE   OF   THE   TWO
                     COMPONENTS OF NGC3991 WITH NGC3992 (SEE 11 54   56
                     +32 36.80).

12 08 00 +39 41.0   NGC4151. VH OF REF.411 REFERS   TO   THE   NUCLEUS.
                     IN   THE   SAME   REF.   VH FOR VARIOUS ZONES OF THE
                     SAME GALAXY ARE ALSO GIVEN. AUTHORS OF   REF.693
                     WARN THAT THERE MAY HAVE BEEN CONFUSION CAUSED BY
                     PRESENCE OF NGC4156 IN THE BEAM.

12 08 04 +70 38.8   MRK199. VH FROM REF.341 IS A   MEASUREMENT   ALONG
                     THE   MINOR   AXIS   OF   THE   MRK199   GROUP   OF   3
                     INTERACTING OBJECTS.

12 15 06 +12 40.0   IC3105. PROBABLY A COMPANION GALAXY IS PROJECTED
                     CLOSE   TO   THE DISK OF THIS GALAXY.   THE EMISSION
                     LINE   SPECTRUM   EXHIBITS   SEVERAL   COMPONENTS   AT
                     DIFFERENT VELOCITIES.

12 16 42 -01 33.0   MRK1320. POSSIBLY A QSO (SEE REF.920).

12 16 42 +14 09.6   OBJECT N.58 IN THE FIELD 12 27   +13   (FAINT   BLUE
                     OBJECTS   OF HARO-LUYTEN) COULD BE SOMETHING OTHER
                     THAN A GALAXY BUT PROBABLY IS   A   DWARF   IN   THE
                     VIRGO CLUSTER.

12 17 24 +05 37.0   NGC4273=HOL366A. ERRONEOUSLY CALLED HOL368C IN 2RCBG.

12 18 45 +11 47.3   NGC4294. THE RADIO MEASUREMENTS OF REF.744 PROBABLY CONFUSED BY NGC4299 (12 19 08 +11 46.8).

12 19 32 +75 35.2   MRK205. VH IN REF.854 IS AN AVERAGE BETWEEN EMISSION AND ABSORPTION.

12 23 56 +09 08.9   NGC4411A. IN REF.789 IT IS ARGUED THAT THE MEASUREMENTS REPORTED IN REF.694 ARE IN ERROR BECAUSE A STAR WAS MEASURED IN PLACE OF THE GALAXY.

12 25 04 +26 30.4   THE IDENTIFICATION WITH THE RADIOSOURCE B2 1225 +26 IS DOUBTFUL (SEE GRUEFF ET AL., 1981, ASTRON.AND AP.SUPPL., 44, 241).

12 28 10 +41 54.9   NGC4490. VH IN REF.693 PROBABLY CONFUSED BY THE PRESENCE OF NGC4485 (12 28 06 +41 58.6) IN THE BEAM.

12 28 18 +12 40.0   NGC4486. VH IN REF.444 IS AN AVERAGE OF 5 MEASUREMENTS. VH=1040 KM/S (REF.165) IS IN EMISSION VH=1260 KM/S (REF.165) IS IN ABSORPTION.

12 34 18 +13 26.4   NGC4569. THE POSITIVE VH VALUE FROM REF.13 IS IN ERROR (SEE REF.273 AND RODGERS,A.W., AND FREEMAN,K.C., 1970, AP.J., 161, L109). REF.117 ALSO GIVES A SIMILAR VALUE. IT IS POSSIBLE THAT BOTH VHS REFER TO IC3583 (12 34 12 +13 32.0) CLOSE COMPANION OF NGC4569.

12 38 24 +28 15.0   VH FROM REF.408 PROBABLY REFERS TO THE A COMPONENT OF THE DOUBLET HARO31A/B.

12 41 32 +32 26.5   NGC4656. IN REF.693 THE MEASUREMENT WAS PROBABLY CONFUSED BY NGC4657 IN THE BEAM.

12 42 06 +37 23.0   NGC4662. MISIDENTIFIED WITH NGC4462 IN REF.582.

12 43 42 +31 00.0   NGC4676A. RADIAL VELOCITIES ALONG THE TAIL ARE REPORTED BY THEYS ET AL. (1972, PASP, 84, 851).

12 45 14 +04 36.6   NGC4698. VH FROM REF.153 IS OUR AVERAGE OF 2 DIFFERENT VALUES OBTAINED FROM 2 DIFFERENT LINES.

12 50 48 +37 05.0   NGC4774A AND B. VH FOR NGC4774A REFERS TO THE COMPACT COMPONENT, VH FOR NGC4774B REFERS TO THE RING OF THE GALAXY.

12 52 36 +08 20.0   NGC4795A. IN REF.694 VH IS GIVEN FOR NGC4795A WHICH COULD INDICATE THE CLOSE COMPANION OF NGC4795 I.E. NGC4796 (12 52 36 +08 20.0).

12 54 05 +57 08.6   MRK231. ABSORPTION LINE REDSHIFT. THE HIGHER VALUE IN REF.304 ORIGINATES FROM EMISSION LINES.

12 56 39 +35 06.9   NGC4861. VH FROM REF.13 AND REF.77 COULD REFER
                      TO IC3961 (12 56 40 +35 07.9) THE WESTERN
                      COMPONENT OF VV797; (SEE REF.430 AND NOTE TO
                      TABLE 5 IN REF.13).

12 57 06 +28 19.0   THIS ENTRY REFERS TO A DIFFERENT GALAXY FROM THE
                      FOLLOWING ONE ALTHOUGH COORDINATES ARE THE SAME.

12 57 18 +28 14.0   THIS ENTRY REFERS TO A DIFFERENT GALAXY FROM THE
                      FOLLOWING ONE ALTHOUGH COORDINATES ARE THE SAME.

12 57 36 +28 13.0   THIS GALAXY (RB68) AND THE FOLLOWING (RB71) HAVE
                      THE SAME COORDINATES (SEE ROOD,H.J., BAUM,W.A.,
                      1967, A.J., 72, 398).

13 00 43 +36 07.8   BRACCESI OBJECT B234, A GALAXY AND NOT A QSO.

13 06 36 +28 18.0   THIS ENTRY REFERS TO A DIFFERENT GALAXY FROM THE
                      FOLLOWING ONE ALTHOUGH COORDINATES ARE THE SAME.

13 14 24 +12 49.0   NGC5058=MRK786. SYSTEM FORMED BY A LARGE GALAXY
                      (NGC5058A) AND A FAINT COMPANION (NGC5058B).
                      NGC5058A HAS A DOUBLE NUCLEUS. VH FOR THE TWO
                      NUCLEI ARE IDENTIFIED AS MRK786A AND MRK786B.

13 22 32 -42 45.5   NGC5128. IN REF.62 AN ABSORPTION SPECTRUM IS
                      ALSO GIVEN FROM WHICH VH=450+/-50 KM/S IS
                      DERIVED.

13 27 47 +47 27.3   NGC5194. IN REF.693 POSSIBLE CONFUSION CAUSED BY
                      NGC5195 (13 27 52 +47 31.8) IN THE BEAM.

13 31 45 -45 17.6   COORDINATES GIVEN IN REFS.378 AND 865 ARE SEVERAL
                      ARC MIN. OFF, FOR CORRECT POSITION DISCUSSION
                      SEE REF.701.

13 37 06 -31 23.4   NGC5253. VH IN REF.145 IS THE AVERAGE OF VH=285
                      AND VH=817 MEASURED IN TWO DIFFERENT PARTS OF THE
                      GALAXY.

13 47 48 -48 48.5   CAPITAL LETTERS IN THIS AND THE TWO FOLLOWING
                      ENTRIES REFER TO IDENTIFICATIONS IN FIG.1 OF
                      REF.765.

13 59 09 +59 34.1   NGC5430=MRK799. GALAXY WITH A DOUBLE NUCLEUS.
                      VH FOR THE TWO NUCLEI ARE IDENTIFIED WITH A AND
                      B.

14 00 48 -05 49.8   NGC5426. IN REF.744 POSSIBLE CONFUSION CAUSED BY
                      NGC5427 (14 00 49 -05 47.5) IN THE BEAM.

14 01 12 +41 50.6   MRK283. THE COORDINATES GIVEN IN COL.1 AND 2
                      REFER TO THE BRIGHTEST GALAXY OF A SMALL GROUP.
                      IT HAS NOT BEEN POSSIBLE TO IDENTIFY WHICH OBJECT
                      WAS CLASSIFIED AS MRK283 (SEE PETERSON, S.D.,
                      1973 A.J., 78, 811).

14 01 28 +54 35.6 MEASUREMENT REFERRING TO VV344 (NGC5457 PLUS NGC5474).

14 15 18 -43 09.7 NGC5530. THE ZERO VELOCITY GIVEN IN REF.574 IS HIGHLY DISCREPANT FROM THE VELOCITIES GIVEN BY OTHERS. THE COORDINATES IN REF.574 (14 17 00 -43 16) ARE ALSO QUITE DIFFERENT FROM THE COORDINATES OF NGC5530. THERE ARE THEREFORE SERIOUS DOUBTS WHETHER THE MEASUREMENT OF REF.574 REFERS TO NGC5530 AT ALL.

14 24 48 +63 31.0 VERY APPROXIMATE COORDINATES OF THE FIRST MEMBER OF A1918.

14 24 54 +20 03.0 MRK813. POSSIBLE QSO.

14 29 48 +05 24.0 ALSO TOLOLO 1429.8 +064. THERE IS A DISCREPANCY BETWEEN COORDINATES GIVEN AND COORDINATES IN THE NAME (SEE REF.683).

14 33 55 +55 20.9 4C55.29. OUR ACCURATE POSITION OF THE OPTICAL COUNTERPART OF THE 4C RADIOSOURCE IN A1940 ARE: 14 33 55.03+/-0.10 55 20 53.6+/-0.9 IN GOOD AGREEMENT WITH THE POSITION GIVEN BY PAULINY-TOTH ET AL., (1978, A.J., 83, 451).

14 39 00 +53 42.4 INTERACTING OBJECT WITH MRK477 (14 39 03 +53 42.9).

14 47 18 +27 59.2 OPTICAL COUNTERPART OF A COMPONENT OF B2 1447+27. RADIOCOMPONENT B APPEARS TO BE UNRELATED TO A.

14 49 00 +09 32.0 ARP173. FROM REF.49 IT IS UNCLEAR WHETHER THE MEASUREMENTS ON ARP173 REFER TO THE DOUBLET KDG439 OR TO ONLY ONE COMPONENT.

15 02 47 +26 12.6 3CR310. UNCLEAR FROM REF.824 IF Z GIVEN REFERS TO A AND B COMPONENTS OF OPTICAL COUNTERPART OF THIS 3CR RADIOSOURCE (CLOSE DOUBLET OF GALAXIES) OR TO THE NW ONE (A) ONLY.

15 06 05 +34 34.7 THE 21CM VH OF REF.678 IS IN ABSORPTION.

15 14 06 +07 10.0 TWO ANONIMOUS GALAXIES IN THE SHANE AND WIRTANEN CLOUD AT THE SAME FORMAL COORDINATES (SEE SHANE,C.D., WIRTANEN,C.A., 1954, A.J., 59, 285).

15 14 30 +07 12.0 GALAXY PAIR N 125 IN REF.594. UNFORTUNATELY THE INFORMATION AVAILABLE IN THIS REFERENCE IS NOT SUFFICIENT TO IDENTIFY TO WHICH GALAXY IN A2052 THEY ARE REFERRING.

15 16 19 +42 55.6 1ZW107A/B. COORDINATES GIVEN IN REF.77 ARE QUITE DIFFERENT FROM THE ONES GIVEN HERE. HOWEVER, THERE IS VERY LITTLE DOUBT THAT THE MEASUREMENT REFERS TO THE GALAXY DOUBLET 1ZW107A/B.

15 17 51 +28 45.4   POSSIBLE SEYFERT.

15 57 01 +20 54.2   SEYFERT   SEXTET.   NGC6027/A/B/C/,   THE
                     IDENTIFICATION LETTERS ATTACHED TO THE NGC NUMBER
                     FOLLOW THOSE IN THE FINDING CHART IN REF.274

16 00 00 +16 09.0   POSSIBLY ASSOCIATED WITH THE RADIOSOURCE 1559
                     +16W1 (15 59 57 +16 09.51).

16 02 18 +17 51.1   NGC6041C.   GROUP FORMED OF 3 GALAXIES, TWO
                     (NGC6041 A AND B) NORTH OF THE THIRD WHICH
                     ARBITRARILY WE HAVE CALLED C.

16 02 54 +17 33.0   IC1173. FROM TABLE 2 OF REF.857 IT IS NOT CLEAR
                     WHETHER OR NOT VH REFERS TO IC 1173. HOWEVER
                     COORDINATES GIVEN TO THE MEASUREMENT ARE THE SAME
                     AS COORDINATES OF IC1173 IN THE GCGC CATALOGUE
                     AND THE AUTHORS STATE IN THE TEXT THAT GCGC
                     COORDINATES ARE USED.

16 03 09 +18 16.6   IN REF.23 THE GALAXY IDENTIFIED AS NGC6055IN
                     REF.277 WAS REOBSERVED.   IT IS CLAIMED THAT
                     COORDINATES ARE QUITE DIFFERENT FROM THOSE OF
                     NGC6055 (16   03 18 +18 17.7) THEREFORE WE REPORT
                     BOTH OBSERVATIONS AS IF THEY WERE TWO SEPARATE
                     OBJECTS.

16 11 25 -60 47.6   OPTICAL POSITION FROM TAB.3 IN REF.627 PROBABLY
                     MISPRINTED IN THEIR FIG.10.

16 27 00 +39 43.5   GALAXY N.99 IN ROOD AND SASTRY (1972, A.J., 77,
                     451) OF A2199. THE AUTHORS OF REF.56 HAVE
                     ADMITTED THEIR VH IS INCORRECT (SEE PRIVATE
                     COMMUNICATION IN REF.186)

16 47 58 +45 32.5   ARP103A-C. ERRONEOUSLY CALLED ZWICKY TRIPLET IN
                     2RCBG PAGE 52

17 20 27 +24 48.0   ALSO V395HER BELIEVED TO BE A STAR, LATER PROVEN
                     TO BE A GALAXY.

17 21 12 -64 54.1   NGC6328. THIS NGC NUMBER IS LISTED AS "NOT
                     EXISTENT" IN THE NRGC.

17 48 55 +68 42.8   POSSIBLE SEYFERT.

18 09 24 -85 26.0   NGC6438.   PECULIAR   SYSTEM   CONSISTING   OF:
                     LENTICULAR COMPONENT (A), RING SHAPED IRREGULAR
                     WITH NUCLEUS (B) AND ARMS (C).   IDENTIFICATIONS
                     HERE   PROPOSED   FOR   REF.857   ARE TENTATIVE.
                     MEASUREMENTS OF REF.316 ARE INCORRECTLY REPORTED
                     IN TAB.1 OF REF.830.

18 22 06 +66 35.0   BRIGHT   AND   COMPACT   OBJECT,   PROBABLY   A
                     SUPERASSOCIATION IN THE DOUBLE GALAXY NGC 6636.

18 36 13 +17 09.1   3CR386. ALSO OBSERVED IN REF.120; HOWEVER THAT

MEASUREMENT TURNED OUT TO BE UNRELIABLE (SEE REF.291).

19 00 00 +40 41.0   AVERAGE VH OBTAINED FROM 5 MEASUREMENTS ON KNOTS.

19 09 26 +52 08.0   ALSO V1102CYG. BELIEVED TO BE A STAR, LATER PROVEN TO BE A GALAXY.

19 40 23 +50 30.9   3CR402. IN THE FIELD OF THE 3CR RADIOSOURCE THERE ARE 4 GALAXIES HERE DESIGNATED A,B,C,AND D.

19 42 07 −14 55.7   NGC6822. MEASUREMENTS IN REF.13 REFER TO REGION 5 OF NGC6822 IN HUBBLE (1925, AP.J., 62, 402). MEASUREMENTS IN REF.72 REFER TO REGION 6 AND 7 OF THE SAME PAPER BY HUBBLE AND COULD BE GLOBULAR CLUSTERS OF NGC6822. MEASUREMENTS IN REF.650 REFER TO REGION 1 AND 3. THESE REGIONS PROBABLY REFER TO HUBBLE'S PAPER ALSO BUT IT IS NOT SPECIFICALLY SO STATED.

19 57 22 −47 12.5   NGC6845. DESIGNATION OF REGIONS A,B,AND C COMES FROM REF.380.

21 04 45 +76 21.1   3CR427.1. THE QUOTED REDSHIFT IMPLIES VH=CZ OF COLUMN 5. THE RELATIVISTIC RECESSION VELOCITY IS VH=195320 KM/S.

21 45 21 −35 11.2   IC5135. ERRONEOUSLY CALLED NGC7135 IN REF.148.

22 36 19 +33 48.5   NGC7343. MATERNE AND TAMMANN (1974, ASTRON.AND AP., 35, 441) POINT OUT THAT VH IN REF.13 IS IN ERROR.

22 44 00 −65 19.3   THE EAST COMPONENT OF IC5250 HAS TWO NUCLEI, A NW AND A SW ONE. THE DISCREPANCY IN VELOCITIES MEASURED FOR THIS DOUBLE NUCLEUS COULD BE A TYPING MISTAKE IN THE ORIGINAL REF.794 (PAG.354) (Z=0.0115 FOR NW, Z=0.0015 FOR SE).

22 47 25 +11 20.6   NGC7385. VH FROM REF.350 PROBABLY REFERS TO NGC7386 (22 47 32 +11 26.0) (SEE REF.595 NOTE 8 AND TABLE 2).

22 52 34 +12 57.2   NGC7413 GALAXY PREVIOUSLY IDENTIFIED WITH 3CR455. NOW THIS RADIOSOURCE IS BELIEVED TO BE A QSO (SEE REF.946).

22 59 23 +15 42.8   NGC7463. THE RADIO MEASUREMENTS OF REF.744 WERE CONFUSED BY THE PRESENCE OF NGC7465 (22 59 32 +15 41.8) IN THE BEAM.

22 59 36 +26 47.0   POSSIBLE SEYFERT.

23 00 47 +08 37.4   IC5283. THE VELOCITY FIELD OF IC5283 IS ALSO STUDIED IN REF.80.

23 06 30 +11 48.0   NGC7495. VH FROM REF.624 WAS MEASURED FROM

SUPERNOVA 1973N.

23 12 11 +04 15.6   NGC7541. IN REF.744 MEASUREMENTS POSSIBLY CONFUSED BY THE PRESENCE OF NGC7537 (23 12 02 +04 13.5) IN THE BEAM.

23 12 48 +18 41.0   THE PRESENT GALAXY AND THE TWO FOLLOWING ONES HERE IDENTIFIED AS NGC7550A AND NGC7550NW/SE ARE IDENTIFIED AS NGC7549, NGC7547 AND NGC7550 IN THE 2RCBG.

23 26 00 −40 18.0   APPROXIMATE COORDINATES FOR TOLOLO 2326-403.

23 32 32 +27 05.7   ALSO 3CR465A-B.

23 34 00 −20 44.7   QUOTED COORDINATES ARE ONLY INDICATIVE (SEE REF.728).

23 36 00 +26 45.3   NGC7720. GALAXY WITH A DOUBLE NUCLEUS. THE RADIOSOURCE 3C465 IS ASSOCIATED WITH THE SOUTHERN COMPONENT. REF.120 REFERS TO THE FINDING CHART PUBLISHED BY GRIFFIN (1963, A.J., 68, 421). IT IS UNCLEAR TO WHICH OF THE NUCLEAR COMPONENTS THE MEASUREMENTS OF REF.120 REFER, SINCE IN GRIFFINS' PAPER THE GALAXY IS MENTIONED AS NGC7720 ONLY. THE VH IN REF.4 HAS A VALUE INTERMEDIATE BETWEEN VALUES GIVEN BY OTHER AUTHORS FOR THE TWO SEPARATE COMPONENTS, WE HAVE ASSUMED HERE THAT IT REFERS TO THE WHOLE NUCLEAR PART OF NGC7720 WITH BOTH COMPONENTS IN THE SLIT.

23 41 36 +27 27.0   MRK1133. POSSIBLE SEYFERT.

23 44 28 +29 10.7   REF.72 ERRONEOUSLY ATTRIBUTED THE VH OF NGC7753 TO NGC7752 AND VICEVERSA (SEE REF.364).

23 44 33 +29 12.2   NGC7753. SEE NOTE FOR 23 44 28 +29 10.7.

23 50 47 −41 05.3   NGC7764E. REF.844 REFERS TO THIS GALAXY AS TO THE ELLIPTICAL COMPANION OF NGC7764A. IT IS NOT CLEAR WHETHER IT SHOULD BE IDENTIFIED WITH OBJECT "B" OF REF.613, SINCE IN REF.844 NO COORDINATES ARE GIVEN. FURTHERMORE, SINCE VH ARE ALSO DISCREPANT NGC7764E HAS BEEN CONSIDERED AS A FOURTH GALAXY OF THE 293-IG08 GROUP DESCRIBED IN REF.613. COORDINATES ARE THE SAME AS FOR NGC7764A.

CHAPTER 5

References in Chronological Order

1   STROMBERG,G., 1925, AP.J., 61, 352

2   HUMASON,M.L., 1931, AP.J., 71, 35

3   MOORE,J.H., 1932, PUB. LICK OBS. N.18

4   BABCOCK,H.W., 1939, LICK OBS.BULL. N.498, P.41

5   STRUVE,O., LINKE,W., 1940, PASP, 52, 139

6   MAYALL,N.U., ALLER,L.H., 1942, AP.J., 95, 5

7   EVANS,D.S., 1952, OBSERVATORY, 72, 164

8   PAGE,T., 1952, AP.J., 116, 63

9   BAADE,W., MINKOWSKY,R., 1954, AP.J., 119, 206

10   KERR,F.J., HINDMAN,J.V., ROBINSON,B.J., 1954,
     AUSTRAL.J.PHYS., 7, 297

11   HUMASON,M.L., WAHLQUIST,H.D., 1955, A.J., 60, 254

12   EVANS,D.S., 1956, MNRAS, 116, 659

13   HUMASON,M.L., MAYALL,N.U., SANDAGE,A.R., 1956, A.J.,
     61, 97

14   MAYALL,N.U., 1956, QUOTED IN REF.83

15   RAIMOND,E., VOLDERS,L.M.J.S., 1957, BAN, 14, 19

16   VAN DE HULST,H.C., RAIMOND,E., VAN WOERDEN,H., 1957,
     BAN, 14, 1

17   EVANS,D.S., WAYMAN,P.A., 1958, MNASSA, 17, 137

18   BURBIDGE,E.M., BURBIDGE,G.R., 1959, AP.J., 129, 271

19   BURBIDGE,E.M., BURBIDGE,G.R., 1959, AP.J., 130, 12

20   BURBIDGE,E.M., BURBIDGE,G.R., PRENDERGAST,K.H., 1959,
     AP.J., 130, 26

21   BURBIDGE,E.M., BURBIDGE,G.R., PRENDERGAST,K.H., 1959
     AP.J., 130, 739

22   BURBIDGE,G.R., BURBIDGE,E.M., 1959, AP.J., 130, 629

23   MINKOWSKY,R., 1959, AP.J., 130, 1028

24   MINKOWSKY,R., OSTERBROCK,D., 1959, AP.J., 129, 583

25   MUNCH,G., 1959, PASP, 71, 101

26   VOLDERS,L., 1959, BAN, 14, 323

27   VOLDERS,L., VAN DE HULST, H.C., 1959, IAU SYMP. 9, 423

28   BARBON,R., 1960, MEM.SOC.ASTR.ITALIANA, 40, 211

29   BURBIDGE,E.M., BURBIDGE,G.R., 1960, AP.J., 132, 30

30   BURBIDGE,E.M., BURBIDGE,G.R., PRENDERGAST,K.H., 1960,
      AP.J., 131, 549

31   BURBIDGE,E.M., BURBIDGE,G.R., PRENDERGAST,K.H., 1960,
      AP.J., 131, 282

32   BURBIDGE,E.M., BURBIDGE,G.R., PRENDERGAST,K.H., 1960,
      AP.J., 132, 661

33   BURBIDGE,E.M., BURBIDGE,G.R., PRENDERGAST,K.H., 1960,
      AP.J., 132, 640

34   BURBIDGE,E.M., BURBIDGE,G.R., PRENDERGAST,K.H., 1960,
      AP.J., 132, 654

35   DUFLOT-AUGARDE,R., 1960, PUB.OBS. HAUTE PROVENCE,
      5, N.14

36   MINKOWSKI,R., 1960, ANN. ASTROPHYS., 23, 385

37   MINKOWSKI,R., 1960, AP.J., 132, 908

38   ZWICKY,F., HUMASON,M.L., 1960, AP.J., 132, 627

39   ZWICKY,F., HUMASON,M.L., 1960, QUOTED BY HUMASON,M.L.,
      GATES,H.S., 1960 PASP, 72, 208

40   BURBIDGE,E.M., BURBIDGE,G.R., 1961, A.J., 66, 541

41   BURBIDGE,E.M., BURBIDGE,G.R., 1961, AP.J., 133, 726

42   BURBIDGE,E.M., BURBIDGE,G.R., 1961, AP.J., 134, 244

43   BURBIDGE,E.M., BURBIDGE,G.R., 1961, AP.J., 134, 248

44   BURBIDGE,E.M., BURBIDGE,G.R., PRENDERGAST,K.H., 1961,
        AP.J., 134, 874

45   BURBIDGE,E.M., BURBIDGE,G.R., PRENDERGAST,K.H., 1961,
        AP.J., 134, 237

46   BURBIDGE,E.M., BURBIDGE,G.R., PRENDERGAST,K.H., 1961,
        AP.J., 133, 814

47   BURBIDGE,E.M., BURBIDGE,G.R., PRENDERGAST,K.H., 1961,
        AP.J., 134, 232

48   DE VAUCOULEURS,G., 1961, AP.J., 133, 405

49   DE VAUCOULEURS,G., DE VAUCOULEURS,A., 1961, MEM. RAS,
        68, 69

50   DUFLOT-AUGARDE,R., 1961, COMPTES RENDUS AC. SCI.
        PARIS, SER B, 253, 224

51   EVANS,D.S., HARDIG.G.A., 1961, MNASSA, 20, 64

52   GREENSTEIN,J.L., 1961, AP.J., 133, 335

53   HEIDMANN,J., 1961, BAN, 15, 314

54   HUMASON,M.L., GOMES,A.M., KEARNS,C.E., 1961, PASP,
        73, 175

55   LOVASICH,J.L., MAYALL,N.U., NEYMAN,N.J., SCOTT,E.,
        1961, PROC. IV BERKELEY SYMP. ON MATH. STAT. AND
        PROB., VOL.III, P.187

56   MINKOWSKY,R., 1961, A.J., 66, 558

57   PAGE,T., 1961, PROC. IV BERKELEY SYMP. ON MATH.
        STAT. AND PROB., VOL III, PAG.277

58   VOLDERS,L., HOGBOM,J.A., 1961, BAN, 15, 307

59   ZWICKY,F., HUMASON,M.L., 1961, AP.J., 133, 794

60   ZWICKY,F., QUOTED BY HUMASON,M.L., 1961, PASP, 73,
        185

61   BURBIDGE,E.M., BURBIDGE,G.R., 1962, AP.J., 135, 366

62   BURBIDGE,E.M., BURBIDGE,G.R., 1962, NATURE, 194, 367

63   BURBIDGE,E.M., BURBIDGE,G.R., PRENDERGAST,K.H., 1962,
        AP.J., 136, 704

64   BURBIDGE,E.M., BURBIDGE,G.R., PRENDERGAST,K.H., 1962,
        AP.J., 136, 119

65   BURBIDGE,E.M., BURBIDGE,G.R., PRENDERGAST,K.H., 1962,
        AP.J., 136, 128

66   BURBIDGE,E.M., BURBIDGE,G.R., PRENDERGAST,K.H., 1962,
        AP.J., 136, 339

67   DIETER,N.H., 1962, A.J., 67, 217

68   DIETER,N.H., 1962, A.J., 67, 313

69   DIETER,N.H., 1962, A.J., 67, 317

70   GREENSTEIN,J.L., 1962, AP.J., 135, 679

71   GREENSTEIN,J.L., ZWICKY,F., 1962, PASP, 74, 35

72   MAYALL,N.U., DE VAUCOULEURS,A., 1962, A.J., 67, 363

73   ROBERTS,M.S., 1962, A.J., 67, 431

74   ROBERTS,M.S., 1962, A.J., 67, 437

75   BURBIDGE,E.M., 1963, QUOTED BY SANDAGE,A., 1963,
        AP.J., 138, 863

76   BURBIDGE,E.M., BURBIDGE,G.R., 1963, AP.J., 138, 1306

77   BURBIDGE,E.M., BURBIDGE,G.R., HOYLE,F., 1963, AP.J.,
        138, 873

78   BURBIDGE,E.M., BURBIDGE,G.R., PRENDERGAST,K.H., 1963,
        AP.J., 137, 376

79   BURBIDGE,E.M., BURBIDGE,G.R., PRENDERGAST,K.H., 1963,
        AP.J., 138, 375

80   BURBIDGE,E.M., BURBIDGE,G.R., PRENDERGAST,K.H., 1963,
        AP.J., 137, 1022

81   BURKE,B.F., TURNER,K.C., TUVE,M.A., 1963, CARNEGIE
        INSTIT. YEAR BOOK, 62, 293

82   BURLEY,J., 1963, A.J., 68, 274

83   DE VAUCOULEURS,G., DE VAUCOULEURS,A., 1963, AP.J.,

137, 363

84   EVANS,D.S., 1963, MNASSA, 22, 140

85   WAYMAN, 1963, QUOTED IN REF.84

86   BURBIDGE,E.M., BURBIDGE,G.R., 1964, AP.J., 140, 1307

87   BURBIDGE,E.M., BURBIDGE,G.R., 1964, AP.J., 140, 1445

88   BURBIDGE,E.M., BURBIDGE,G.R., CRAMPIN,D.J., RUBIN,V.C.,
        PRENDERGAST,K.H., 1964, AP.J., 139, 1058

89   BURBIDGE,E.M., BURBIDGE,G.R., CRAMPIN,D.J., RUBIN,V.C.,
        PRENDERGAST,K.H., 1964, AP.J., 139, 539

90   BURBIDGE,E.M., BURBIDGE,G.R., PRENDERGAST,K.H., 1964,
        AP.J., 140, 1617

91   BURBIDGE,E.M., BURBIDGE,G.R., RUBIN,V.C., 1964, AP.J.,
        140, 942

92   BURBIDGE,E.M., BURBIDGE,G.R., RUBIN,V.C.,PRENDERGAST,K.H.,
        1964, AP.J., 140, 85

93   EPSTEIN,E.E., 1964, A.J., 69, 490

94   RUBIN,V.C., BURBIDGE,E.M., BURBIDGE,G.R., 1964, AP.J.,
        140, 94

95   RUBIN,V.C., BURBIDGE,E.M., BURBIDGE,G.R., 1964, AP.J.,
        140, 1304

96   RUBIN,V.C., BURBIDGE,E.M., BURBIDGE,G.R., PRENDERGAST,K.H.,
        1964, AP.J., 140, 80

97   ZWICKY,F., 1964, AP.J., 139, 514

98   ZWICKY,F., 1964, AP.J., 140, 1467

99   ZWICKY,F., HUMASON,M.L., 1964, AP.J., 139, 269

100   ARGYLE,E., 1965, AP.J., 141, 750

101   ARP,H., 1965, AP.J., 142, 402

102   BERTOLA,F., 1965, ANN. ASTROPHYS., 28, 574

103   BERTOLA,F., 1965, CONTR. OSS. ASTROF. ASIAGO N.172,
        P. 35.

104   BRANDT,J.C., 1965, MNRAS, 129, 309

105   BURBIDGE,E.M., BURBIDGE,G.R., 1965, AP.J., 142, 1351

106   BURBIDGE,E.M., BURBIDGE,G.R., 1965, AP.J., 142, 634

107   BURBIDGE,E.M., BURBIDGE,G.R., PRENDERGAST,K.H., 1965,
        AP.J., 142, 641

108   BURBIDGE,E.M., BURBIDGE,G.R., PRENDERGAST,K.H., 1965,
        AP.J., 142, 154

109   BURBIDGE,E.M., BURBIDGE,G.R., PRENDERGAST,K.H., 1965,
        AP.J., 142, 649

110   DEMOULIN,M.H., 1965, COMPTES RENDUS AC.SCI. PARIS,
        SER.B, 260, 3287

111   DEMOULIN,M.H., 1965, PUB.OBS. HAUTE PROVENCE, 8, N 1

112   DUFLOT,R., 1965, PUB.OBS. HAUTE PROVENCE, 8, N 16

113   EVANS,D.S., MALIN,S.R., 1965, MNASSA, 24, 32

114   HOGLUND,B., ROBERTS,M.S., 1965, AP.J., 142, 1366

115   KINMAN,T.D., 1965, AP.J., 142, 1241

116   ROBERTS,M.S., 1965, AP.J., 142, 148

117   ROBINSON,B.J., KOEHLER,J.A., 1965, NATURE, 208, 993

118   RUBIN,V.C., BURBIDGE,E.M., BURBIDGE,G.R., CRAMPIN,D.J.,
        PRENDERGAST,K.H., 1965, AP.J., 141, 759

119   RUBIN,V.C., BURBIDGE,E.M., BURBIDGE,G.R., PRENDERGAST,K.H.,
        1965, AP.J., 141, 885

120   SCHMIDT,M., 1965, AP.J., 141, 1

121   SEIELSTAD,G.A., WHITEOAK,J.B., 1965, AP.J., 142, 616

122   ZWICKY,F., 1965, AP.J., 142, 1293

123   ZWICKY,F., 1965, AP.J., 143, 192

124   ZWICKY,F., KARPOWICZ,M., KOWAL,C.T., 1965, CATALOGUE
        OF GALAXIES AND OF CLUSTERS OF GALAXIES, CALIFORNIA

125   BERTOLA,F., 1966, MEM. SOC. ASTR. ITALIANA, 37, 433

126   BOTTINELLI,L., GOUGUENHEIM,L., HEIDMANN,J., HEIDMANN,N.,
        WELIACEW,L., 1966, COMPTES RENDUS AC.SCI.
        PARIS, SER.B, 263, 223

127   BRUNDAGE,W.D., KRAUS, J.D., 1966, SCIENCE, 153, 411

128   BURBIDGE,E.M., BURBIDGE,G.R., 1966, AP.J., 145, 661

129   GOTTESMAN,S.T., DAVIES,R.D., REDDISH,V.C., 1966,
        MNRAS, 133, 359

130   HODGE,P.W., MERCHANT,A.E., 1966, AP.J., 144, 875

131   MENG,S.Y., KRAUS,J.D., 1966, A.J., 71, 170

132   ROBERTS,M.S., 1966, AP.J., 144, 639

133   ROBINSON,B.J., VAN DAMME, K.J., 1966, AUSTRAL.J.PHYS.,
        19, 111

134   SANDAGE,A., 1966, AP.J., 145, 1

135   SERSIC,J.L., 1966, ZEIT. FUR ASTR., 64, 202

136   SHOBBROOK,R.R., 1966, MNRAS, 131, 293

137   VAN DAMME,K.J., 1966, AUSTRAL.J.PHYS., 19, 687

138   WESTERLUND,B.E., SMITH,L.F., 1966, AUSTRAL.J.PHYS.,
        19, 181

139   WESTERLUND,B.E., STOKES,N.R., 1966, AP.J., 145, 354

140   ANDRILLAT,Y., SOUFFRIN,S., 1967, COMPTES RENDUS AC.
        SCI. PARIS, SER.B, 264, 89

141   ARP,H., 1967, AP.J., 148, 321

142   BERTOLA,F., 1967, ATTI DEL X CONVEGNO DELLA SOC.ASTR.
        ITALIANA,P. 209

143   BURBIDGE,E.M., 1967, AP.J., 149, L51

144   BURBIDGE,E.M., BURBIDGE,G.R., SHELTON,J.W., 1967,
        AP.J., 150, 783

145   CARRANZA,G.J.,1967, OBSERVATORY, 87, 38

146   CHINCARINI,G., WALKER,M.F., 1967, AP.J., 147, 407

147   CHINCARINI,G., WALKER,M.F., 1967, AP.J., 149, 487

148   DE VAUCOULEURS,A., DE VAUCOULEURS,G., 1967, A.J.,
        72, 730

149   DE VAUCOULEURS,G., 1967, IAU SYMP., N.30, P.9

150   DE VAUCOULEURS,G., 1967, IAU SYMP., N.30, P.91

151   DELANNOY,J., GUELIN,M., WELIASCHEW,W.L., 1967,
        QUOTED IN REF.183

152   EVANS,D.S., 1967, OBSERVATORY, 87, 224

153   GATES,H.S., ZWICKY,F., BERTOLA,F., CIATTI,F.,
        RUDICKI,K., 1967, A.J. 72, 912

154   PAGE,T., 1967, A.J., 72, 821

155   PASTORIZA,M., 1967, OBSERVATORY, 87, 225

156   ROGSTAD,D.H., ROUGOOR,G.W., WHITEOAK,J.B., 1967,
        AP.J., 150, 9

157   RUBIN,V.C., FORD,W.K.JR, 1967, PASP, 79, 322

158   RUBIN,V.C., MOORE,S., BERTIAU,F.C., 1967, A.J., 72, 59

159   SANDAGE,A., 1967, AP.J., 150, L145

160   SANDAGE,A., 1967, AP.J., 150, L9

161   SARGENT,W.L.W., 1967, PASP, 79, 369

162   SHOBBROOK,R.R., ROBINSON,B.J., 1967, AUSTRAL.J.PHYS.,
        20, 131

163   VORONTSOV-VELYAMINOV,B.A., 1967, ASTR.CIRC.(USSR),
        445, 6

164   WALKER,M.F., CHINCARINI,G., 1967, AP.J., 147, 416

165   WALKER,M.F., HAYES,S., 1967, AP.J., 149, 481

166   WILLS,D., 1967, AP.J., 148, L57

167   ANDRILLAT,Y., 1968, A.J., 73, 862

168   ARP,H., 1968, AP.J., 152, 1101

169   ARP,H., 1968, PASP 80, 129

170   BERTOLA,F., 1968, MEM. SOC. ASTR. ITALIANA, 39, 453

171   BOTTINELLI,L., GOUGUENHEIM,L., HEIDMANN,J., HEIDMANN,N.,
        1968, ANN. ASTROPHYS., 31, 205

172   BRACCESI,A., LYNDS,R., SANDAGE,A., 1968, AP.J., 152,
        L105

173   BURBIDGE,E.M., BURBIDGE,G.R., 1968, AP.J., 151, 99

174   BURBIDGE,E.M., BURBIDGE,G.R., 1968, AP.J., 154, 857

175   CARRANZA,C.J., 1968, BOL.ASSOC.ARG.ASTRON.,N.14, P.38

176   DEMOULIN,M.H., BURBIDGE,E.M., BURBIDGE,G.R., 1968,
        AP.J., 153, 31

177   DIBAI,E.A., ESIPOV,V.F., 1968, SOVIET ASTRONOMY, 12, 561

178   DU PUY,D.L., 1968, PASP, 80, 29

179   FORD,W.K.JR., PURGATHOFER,A.T., RUBIN,V.C., 1968,
        AP.J., 153, L39

180   FORD,W.K.JR., RUBIN,V.C., 1968, PASP, 80, 466

181   GORDON,K.J., REMAGE,N.H., ROBERTS,M.S., 1968, AP.J.,
        154, 845

182   LEWIS,B.M., 1968, PROC. AUSTRAL. ASTR. SOC., 1, 104

183   ROBERTS,M.S., 1968, A.J., 73, 945

184   ROBERTS,M.S., 1968, AP.J., 151, 117

185   RUBIN,V.C., FORD,W.K.JR., 1968, AP.J., 154, 431

186   SARGENT,W.L.W., 1968, A.J., 73, 893

187   SARGENT,W.L.W., 1968, AP.J., 152, L31

188   SARGENT,W.L.W., 1968, AP.J., 153, L135

189   SEARLE,L., BOLTON,J.G., 1968, AP.J., 154, L101

190   SERSIC,J.L., PASTORIZA,M., CARRANZA,G., 1968, ASTROPHYS.
        LETT., 2, 45

191   WALKER,M.F., 1968, AP.J., 151, 71

192   WEEDMAN,D.W., KHACHIKYAN,E.E., 1968, ASTROFIZIKA,
        4, 587 (ASTROPHYSICS 4, 243)

193   ALLOIN,D., ANDRILLAT,Y., 1969, COMPTES RENDUS AC.SCI.
        PARIS, SER.B, 268, 139

194   ANDERSON,K.S., KRAFT,R.P., 1969, AP.J., 158, 859

195   ARP,H., 1969, ASTRON.AND AP., 3, 418

196   BACHALL,J.M., SCHMIDT,M., GUNN,J.E., 1969, AP.J.,
        157, L77

197   BARBON,R., 1969, MEM.SOC.ASTR.ITALIANA, 40, 559

198   BEALE,J.S., DAVIES,R.D., 1969, NATURE, 221, 531

199   BERTOLA,F., D'ODORICO,S., FORD,W.K.JR, RUBIN,V.C.,
        1969, AP.J., 157, L27

200   BURBIDGE,E.M., DEMOULIN,M.H., 1969, AP.J., 157, L155

201   BURBIDGE,E.M., DEMOULIN,M.H., 1969, ASTROPHYS. LETT.,
        4., 89

202   CARRANZA,G.J., CRILLON,R., MONNET,G., 1969, ASTRON.
        AP., 1, 479

203   CATCHPOLE,R.M., EVANS,D.S., JONES,D.H.P., 1969,
        OBSERVATORY,89, 21

204   CRILLON,R., MONNET,G., 1969, ASTRON.AND AP., 1, 449

205   CRILLON,R., MONNET,G., 1969, ASTRON.AND AP., 2, 1

206   DE VENY,J.B., LYNDS,C.R., 1969, PASP, 81, 535

207   DEHARVENG,J.M., PELLET,A., 1969, ASTRON.AND AP.,
        1, 208

208   DEMOULIN,M.H., 1969, AP.J., 156, 325

209   DEMOULIN,M.H., 1969, AP.J., 157, 69

210   DEMOULIN,M.H., 1969, AP.J., 157, 75

211   DEMOULIN,M.H., 1969, AP.J., 157, 81

212   DEMOULIN,M.H., TUNG CHAN,Y.W., 1969, AP.J., 156, 501

213   DU PUY,D.L., DE VENY,J.B., 1969, PASP, 81, 637

214   FAIRALL,A.P., ANGIONE,R.J., 1969, PASP, 81, 685

215   FORD,W.K.JR., 1969, CARNEGIE INSTIT. YEAR BOOK 67, 35

216   GOUGUENHEIM,L., 1969, ASTRON.AND AP., 3, 281

217   GUELIN,M., HUCHTMEIER,W., 1969, QUOTED IN REF.245

218   GUELIN,M., WELIACHEW,L., 1969, ASTRON.AND AP., 1, 10

219   LEWIS,B.M., 1969, PROC. AUSTRAL. ASTRON. SOC., 1, 288

220   RUDNICKI,K., TARRARO,I., 1969, ACTA ASTRON., 19, 171

221   SERSIC,J.L., 1969, NATURE, 224, 253

222   SERSIC,J.L., CARRANZA,G., 1969, INF.BUL. FOR THE
        SOUTHERN HEMISPHERE N.14, P.32

223   VAN DEN BERGH,S., 1969, AP.J. SUPPL. 19, 145

224   WEEDMAN,D.W., KHACHIKYAN,E.E., 1969, ASTROFIZIKA,
        5, 113 (ASTROPHYSICS, 5, 51)

225   WELCH,G.A., WALLERSTEIN,G., 1969, PASP, 81, 23

226   WELIACHEW,L., 1969, ASTRON.AND AP., 3, 402

227   WESTERLUND,B.E., WALL,J.V., 1969, A.J., 74, 335

228   WILLS,D., BOLTON,J.G., 1969, AUSTRAL.J.PHYS., 22, 775

229   ZWICKY,F., SARGENT,W.L.W., KOWAL,C., 1969, PASP, 81, 224

230   ALLEN,R.J., 1970, ASTRON.AND AP., 7, 330

231   ARAKELYAN,M.A., DIBAI,E.A., ESIPOV,V.F., 1970, ASTROFIZIKA, 6, 39 (ASTROPHYSICS, 6, 14)

232   ARAKELYAN,M.A., DIBAI,E.A., ESIPOV,V.F., MARKARIAN,B.E., 1970, ASTR. CIRC.(USSR), N.586, 1

233   ARAKELYAN,M.A., DIBAI,E.A., ESIPOV,V.F., MARKARYAN,B.E., 1970, ASTROFIZIKA, 6, 357 (ASTROPHYSICS, 6, 189)

234   ARP,H., 1970, ASTROPHYS. LETT., 5, 257

235   ARP,H., BERTOLA,F., 1970, ASTROPHYS. LETT., 6, 65

236   ARP,H., VISVANATHAN,N., 1970, ASTROPHYS. LETT., 5, 73

237   BARBON,R., 1970, MEM.SOC.ASTR.ITALIANA, 41, 129

238   BOTTINELLI,L., CHAMARAUX,P., GOUGUENHEIM,L., LAQUE',R., 1970, ASTRON.AND AP., 6, 453

239   BOULESTEIX,J., DUBOUT-CRILLON,R., MONNET,G., 1970, ASTRON.AND AP., 8, 204

240   BURBIDGE,E.M., 1970, AP.J., 160, L33

241   CHAMARAUX,P., HEIDMANN,J., LAUQUE',R., 1970, ASTRON. AND AP., 8, 424

242   D'ODORICO,S., 1970, AP.J., 160, 3

243   DEHERVENG,J.M., PELLET,A., 1970, ASTRON.AND AP., 7, 210

244   DEMOULIN,M.H., 1970, AP.J., 160, L79

245   DEMOULIN,M.H., BURBIDGE,E.M., 1970, AP.J., 159, 799

246   DU PUY,D.L., 1970, A.J., 75, 1143

247   GOTTESMAN,S.T., DAVIES,R.D., 1970, MNRAS, 149, 263

248   GUELIN,M., WELIACHEW,L., 1970, ASTRON.AND AP., 7, 141

249   GUELIN,M., WELIACHEW,L., 1970, ASTRON.AND AP., 9, 155

250   LEWIS,B.M., 1970, OBSERVATORY, 90, 264

251   MC CUTCHEON,W.H., DAVIES,R.D., 1970, MNRAS, 150, 337

252   MONNET,G., 1970, IAU SYMP. N.38, P.73

253   PAGE,T., 1970, AP.J., 159, 791

254   PASTORIZA,M., 1970, BOL.ASSOC.ARG.ASTRON.,N.15, 1

255   PETERSON,B.A., 1970, A.J., 75, 695

256   ROBERTS,M.S., 1970, AP.J., 161, L9

257   ROBERTS,M.S., WARREN, J.L., 1970, ASTRON.AND AP.,
        6, 165

258   RODGERS,A.W., FREEMAN, K.C., 1970, AP.J., 161, L109

259   RUBIN,V.C., FORD,W.K.JR., 1970, AP.J., 159, 379

260   RUBIN,V.C., FORD,W.K.JR., D'ODORICO,S., 1970, AP.J.,
        160, 801

261   SARGENT,W.L.W., 1970, AP.J., 159, 765

262   SARGENT,W.L.W., 1970, AP.J., 160, 405

263   VAN DEN BERGH,S., 1970, PASP, 82, 1374

264   WEEDMAN,D.W., 1970, AP.J., 159, 405

265   WEEDMAN,D.W., 1970, AP.J., 161, L113

266   ZWICKY,F., OKE,J.B., NEUGEBAUER,G., SARGENT,W.L.W.,
        FAIRALL,A.P., 1970, PASP, 82, 93

267   ALLEN,R.J., DARCHY,B.F., LAUQUE',R., 1971, ASTRON.
        AND AP., 10, 198

268   ARAKELYAN,M.A., DIBAI,E.A., ESIPOV,V.F., 1971,
        ASTROFIZIKA, 7, 177 (ASTROPHYSICS, 7, 102)

269   ARP,H., 1971, ASTROPHYS. LETT., 7, 221

270   ARP,H., 1971, ASTROPHYS. LETT., 9, 1

271   BOTTINELLI,L., CHAMARAUX,P., GERARD,E., GOUGUENHEIM.L.,
        HEIDMANN,J., KAZES,I., LAUQUE',R., 1971, ASTRON.
        AND AP., 12, 264
272   BURBIDGE,E.M., BURBIDGE,G.R., SOLOMON,P.M.,
        STRITTMATTER,P.A., 1971, AP.J. 170, 233

273   BURBIDGE,E.M., HODGE,P.M., 1971, AP.J., 166, 1

274   BURBIDGE,E.M., SARGENT,W.L.W., 1971, NUCLEI OF
        GALAXIES., ED. O'CONNEL, PONTIF.ACC.SCI.SCRIPTA VARIA
        N. 35, P.351

275   BURNS,W.R., ROBERTS,M.S., 1971, AP.J., 166, 265

276   CHINCARINI,G., ROOD,H.J., 1971, AP.J., 168,321

277   DE JAGER,G., DAVIES, R.D., 1971, MNRAS, 153, 9

278   DENISYUK,E.K., 1971, ASTR.CIRC.(USSR), 615, 4

279   DENISYUK,E.K., 1971, ASTR.CIRC.(USSR), 621, 7

280   DENISYUK,E.K., 1971, ASTR.CIRC.(USSR), 624, 1

281   DISNEY,M.J., CROMWELL,R.H., 1971, AP.J., 164, L35

282   FAIRALL,A.P., 1971, MNRAS, 153, 383

283   FORD,W.K.JR., RUBIN,V.C., ROBERTS,M.S., 1971, A.J.,
        76, 22

284   GORDON,K.J., 1971, AP.J., 169, 235

285   GREEN,F.,AND ROBERTS,M.S., 1971, QUOTED IN REF.283

286   GUNN,J.E., 1971, AP.J., 164, L113

287   KHACHIKYAN,E.E., WEEDMAN,D.W., 1971, ASTROFIZIKA 7,
        389 (ASTROPHYSICS, 7, 231)

288   KINTNER,E.C., 1971, A.J., 76, 409

289   KODAIRA,K., 1971, PUB.ASTR.SOC. JAPAN, 23, 589

290   KUNKEL,W.E., BRADT,H.V., 1971, AP.J., 170, L7

291   LYNDS,R., 1971, AP.J., 168, L87

292   OKE,J.B., 1971, AP.J., 170, 193

293   PASTORIZA,M., AGUERO,E., 1971, BOL.ASOC.ARG.ASTRON.,
        N.16, 3

294   ROBERTS,M.S., 1971, QUOTED IN REF.513

295   ROGSTAD,D.H., 1971, ASTRON.AND AP., 13, 108

296   ROGSTAD,D.H., SHOSTAK,G.S., 1971, ASTRON.AND AP.,
        13, 99

297   SHOSTAK,G.S., WELIACHEW,L., 1971, AP.J., 169, L71

298   SPINRAD,H., SARGENT,W.L.W., OKE,J.B., NEUGEBAUER,G.,
        LANDAU,R., KING,I.R., GUNN,J.E., GARMIRE,G.,
        DIETER,N.H., 1971, AP.J., 163, L25

299   TIFFT,W.G., GREGORY,S.A., 1971, PASP, 83,810

300   ULRICH,M.H., 1971, AP.J., 163, 441

301   WELIACHEW,L., 1971, PASP, 83, 609

302   WHITEOAK,J.B., GARDNER,F.F., 1971, ASTROPHYS. LETT.,
        8, 57

303   ZWICKY,F., ZWICKY,M.A., 1971, CATALOGUE OF SELECTED
        COMPACT GALAXIES AND OF POST ERUPTIVE GALAXIES
        (ZURICH)

304   ADAMS,T.F., WEEDMAN,D.W., 1972, AP.J., 173, L109

305   ARAKELYAN,M.A., DIBAI,E.A., ESIPOV,V.F., 1972, ASTR.
        CIRC.(USSR), 717, 7

306   ARAKELYAN,M.A., DIBAI,E.A., ESIPOV,V.F., 1972,
        ASTROFIZIKA, 8, 329 (ASTROPHYSICS, 8, 197)

307   ARAKELYAN,M.A., DIBAI,E.A., ESIPOV,V.F., 1972,
        ASTROFIZIKA, 8, 177 (ASTROPHYSICS, 8, 106)

308   ARAKELYAN,M.A., DIBAI,E.A., ESIPOV,V.F., 1972,
        ASTROFIZIKA, 8, 33 (ASTROPHYSICS, 8, 17)

309   ARP,H., BURBIDGE,E.M., MACKAY,C.D., STRITTMATTER,P.A.,
        1972, AP.J., 171, L41

310   BALKOWSKI,C., BOTTINELLI,L., GOUGUENHEIM,L., HEIDMANN,J.,
        1972, ASTRON.AND AP., 21, 303

311   BARBON,R., 1972, MEM.SOC.ASTR.ITALIANA, 43, 313

312   BAUTZ,L.P., 1972, A.J., 77, 331

313   BOHUSKI,T.J., BURBIDGE,E.M., BURBIDGE,G.R., SMITH,M.G.,
        1972, AP.J., 175, 329

314   BOTTINELLI,L., GOUGUENHEIM,L., HEIDMANN,J., 1972,
        ASTRON.AND AP., 17, 445

315   BRACCESI,A., FORMIGGINI,L., GIOIA,I., SARGENT,W.L.W.,
        1972, PASP, 84, 592

316   BURBIDGE,E.M., BURBIDGE,G.R., 1972, AP.J., 171, 253

317   BURBIDGE,E.M., BURBIDGE,G.R., 1972, AP.J., 172, 37

318   BURBIDGE,E.M., STRITTMATTER,P.A., 1972, AP.J., 172,
        L37

319   BURBIDGE,E.M., STRITTMATTER,P.A., SMITH,H.E., SPINRAD,H.,
        1972, AP.J., 178, L43

320   BURBIDGE,G.R., O'DELL,S.L., STRITTMATTER,P.A., 1972,
        AP.J., 175, 601

321   CHINCARINI,G., ROOD,H.J., 1972, A.J., 77, 4

322   CHINCARINI,G., ROOD,H.J., 1972, A.J., 77, 448

323   DANZIGER,I.J., CHROMEY,F.R., 1972, ASTROPHYS. LETT.,
        10, 99

324   GOTTESMAN,S.T., WELIACHEW,L., 1972, ASTROPHYS. LETT.,
        12, 63

325   HECKATHORN,H.M., 1972, AP.J., 173, 501

326   HUCHTMEIER,W.K., 1972, ASTRON.AND AP., 17, 207

327   KARACHENTSEV,I.D., 1972, QUOTED IN REF.428

328   KAROJI,H., KODAIRA,K., 1972, PUB.ASTR.SOC. JAPAN,
        24, 239

329   KHACHIKYAN,E.E., PANOSSIAN,H.A., 1972, ASTR.CIRC.
        (USSR), 698, 1

330   LEWIS,B.M., 1972, AUSTRAL.J.PHYS., 25, 315

331   LOVE,R., 1972, NATURE PHYS. SCI., 235, 53

332   LYNDS,R., 1972, IAU SYMP. N. 44, P.376

333   LYNDS,R., 1972, QUOTED BY SANDAGE A., 1972, AP.J.,
        178, 25

334   MARGON,B., SPINRAD,H., HEILES,C., TOVMASSIAN,H.,
        HARLAN,E., BOWYER,S., LAMPTON,M., 1972, AP.J., 178,
        L77

335   MORTON,D.C., CHEVALIER,R.A., 1972, AP.J., 174, 489

336   O'CONNELL,R.W., KRAFT,R.P., 1972, AP.J., 175, 335

337   OEMLER,A.JR., GUNN,J.E., OKE,J.B., 1972, AP.J., 176,
        L47

338   PETERSON,B.A., BOLTON,J.G., 1972, AP.J., 173, L19

339   ROBINSON,L.B., WAMPLER,E.J., 1972, AP.J., 171, L83

340   ROGSTAD,D.H., SHOSTAK,G.S., 1972, AP.J., 176, 315

341   SARGENT,W.L.W., 1972, AP.J., 173, 7

342   SARGENT,W.L.W., 1972, AP.J., 176, 581

343   SERSIC,J.L., AGUERO,E.L., 1972, AP. AND SP.SCI., 19,
        387

344   SERSIC,J.L., CARRANZA,G., PASTORIZA,M., 1972, AP.
        AND SP.SCI., 19, 469

345   SHIELDS,G.A., OKE,J.B., SARGENT,W.L.W., 1972, AP.J.,
        176, 75

346   SIMKIN,S.M., 1972, NATURE, 239, 43

347   STOCKTON,A., 1972, AP.J., 173, 247

348   TAKADA,M., KODAIRA,K., 1972, PUB.ASTR.SOC. JAPAN,
        24, 525

349   TIFFT,W.G., 1972, AP.J., 175, 613

350   TRITTON,K.P., 1972, MNRAS, 158, 277

351   ULRICH,M.H., 1972, AP.J., 171, L35

352   ULRICH,M.H., 1972, AP.J., 171, L37

353   ULRICH,M.H., 1972, AP.J., 178, 113

354   WHITEHURST,R.N., ROBERTS,M.S., 1972, AP.J., 175, 347

355   WHITEOAK,J.B., 1972, AUSTRAL.J.PHYS., 25, 233

356   ALLEN,R.J., 1973, QUOTED IN REF.468

357   ANDERSON,K.S., 1973, AP.J., 182, 369

358   ARAKELYAN,M.A., DIBAI,E.A., ESIPOV,V.F., 1973,
        ASTROFIZIKA, 9, 319 (ASTROPHYSICS, 9, 180)

359   ARAKELYAN,M.A., DIBAI,E.A., ESIPOV,V.F., 1973,
        ASTROFIZIKA, 9, 325 (ASTROPHYSICS, 9, 183)

360   ARP,H., 1973, AP.J., 185, 797

361   ARP,H., KHACHIKYAN,E.E., 1973, ASTROFIZIKA, 9, 509
        (ASTROPHYSICS, 9, 308)

362   BALKOWSKI,C., BOTTINELLI,L., CHAMARAUX.P., GOUGUENHEIM,L.,
        HEIDMANN,J., 1973, ASTRON.AND AP., 25, 319

363   BALKOWSKI,C., BOTTINELLI,L., GOUGUENHEIM,L., HEIDMANN,J.,
        1973, ASTRON.AND AP., 23, 139

364   BERTOLA,F., D'ODORICO,S., 1973, ASTROPHYS. LETT.,
        13, 161

365   BOND,H.E., SARGENT,W.L.W., 1973, AP.J., 185, L109

366   BOTTINELLI,L., CHAMARAUX,P., GOUGUENHEIM,L., HEIDMANN,J.,
        1973, ASTRON.AND AP., 29, 217

367   BOTTINELLI,L., GOUGUENHEIM,L., 1973, ASTRON.AND AP.,
        29, 425

368   BOTTINELLI,L., GOUGUENHEIM,L., HEIDMANN,J., 1973,
        ASTRON.AND AP., 22, 281

369   CHINCARINI,G., HECKATHORN,H.M., 1973, PASP, 85, 568

370   CHROMEY,F.R., 1973, ASTRON.AND AP., 29, 77

371   DAVIES,R.D., 1973, MNRAS, 161, 25P

372   DAVIES,R.D., LEWIS,B.M., 1973, MNRAS, 165, 231

373   DE VAUCOULEURS,G., DE VAUCOULEURS,A., 1973, A.J.,
        78, 377

374   DENISYUK,E.K., LIPOVETSKY,V.A., 1973, ASTR.CIRC.
        (USSR), 798, 2

375   DENISYUK,E.K., PAVLOVA,N.N., 1973, ASTR.CIRC.(USSR),
        N 797, 1

376   DISNEY,M.J., 1973, AP.J., 181, L55

377   DISNEY,M.J., ROGERS,A., 1973, QUART.J.RAS, 14, 438

378   DOTTORI,H.A., FOURCADE,C.R., 1973, ASTRON.AND AP.,
        23, 405

379   GOTTESMAN,S.T., WRIGHT,M.C.H., 1973, AP.J., 184, 71

380   GRAHAM,J.A., RUBIN,V.C., 1973, AP.J., 183, 19

381   GREGORY,S.A., CONNOLLY,L.P., 1973, AP.J., 182, 351

382   HUCHRA,J., SARGENT,W.L.W., 1973, AP.J., 186, 433

383   HUCHTMEIER,W.K., 1973, ASTRON.AND AP., 22, 27

384   HUCHTMEIER,W.K., 1973, ASTRON.AND AP., 22, 91

385   HUCHTMEIER,W.K., 1973, ASTRON.AND AP., 23, 93

386   KHACHIKYAN,E.E., 1973, ASTROFIZIKA, 8, 529
        (ASTROPHYSICS, 8, 311)

387   KHACHIKYAN,E.E., 1973, ASTROFIZIKA, 9, 157,
        (ASTROPHYSICS, 9, 87)

388   KOPILOV,I.M., LIPOVETSKY,V.A., PRONIK,V.I., CHUVAEV,K.K.,
        1973, ASTR. CIRC.(USSR), 755, 1

389   LAUQUE',R., 1973, ASTRON.AND AP., 23, 253

390   LEWIS,B.M., DAVIES,R.D., 1973, MNRAS, 165, 213

391   MILLER,J.S., ROBISON,L.B., WAMPLER,E.J., 1973, AP.J.,

179, L83

392   MORTON,D.C., CHEVALIER,R.A., 1973, AP.J., 179, 55

393   OEMLER,A.JR., 1973, AP.J., 180, 11

394   PENSTON,M.V., PENSTON,M.J., 1973, MNRAS, 162, 109

395   ROBINSON,L.B., WAMPLER,E.J., 1973, AP.J., 179, L135

396   ROBINSON,L.B., WAMPLER,E.J., 1973, AP.J., 179, L79

397   ROGSTAD,D.H., SHOSTAK,G.S., ROTS,A.H., 1973, ASTRON.
        AND AP., 22, 111

398   SAKKA,K., OKA,S., WAKAMATSU,K., 1973, PUB. ASTR.SOC.
        JAPAN, 25, 317

399   SAKKA,K., OKA,S., WAKAMATSU,K., 1973, PUB.ASTR.SOC.
        JAPAN, 25, 153

400   SARGENT,W.L.W., 1973, AP.J., 182, L13

401   SARGENT,W.L.W., 1973, PASP, 85, 281

402   SEIELSTAD,G.A., WRIGHT,M.C.H., 1973, AP.J., 184, 343

403   SHOSTAK,G.S., 1973, BULL. AM. ASTR. SOC., 5, 430

404   SHOSTAK,G.S., ROGSTAD,D.H., 1973, ASTRON.AND AP.,
        24, 405

405   SPINRAD,H., SMITH,H.E., 1973, AP.J., 179, L71

406   STOCKTON,A., 1973, NATURE PHYS. SCI. 246, 25

407   TIFFT,W.G., 1973, AP.J., 179, 29

408   TIFFT,W.G., GREGORY,S.A., 1973, AP.J., 181, 15

409   TIFFT,W.G., JEWSBURY,C.P., SARGENT,T.A., 1973, AP.J.,
        185, 115

410   TRITTON,K.P., SCHILIZZI,R.T., 1973, MNRAS, 165, 245

411   ULRICH,M.H., 1973, AP.J., 181, 51

412   VAN DER KRUIT,P.C., 1973, AP.J., 186, 807

413   VAN DER KRUIT,P.C., 1973, ASTRON.AND AP., 61, 171

414   WARNER,P.J., WRIGHT,M.C.H., BALDWIN,J.E., 1973,
        MNRAS, 163, 163

415   WELIACHEW,L., GOTTESMAN,S.T., 1973, ASTRON.AND AP.,
        24, 59

416   WHITEOAK,J.B., GARDNER,F.F., 1973, ASTROPHYS. LETT.,
        15, 211

417   WRIGHT,M.C.H., SEIELSTAD,G.A., 1973, ASTROPHYS. LETT.,
        13, 1

418   ALLEN,R.J., GOSS,W.M., SANCISI,R., SULLIVAN,W.T.III,
        VAN WOERDEN,H., 1974, IAU SYMP. N. 58, P.425

419   ANDERSON,K.S., 1974, AP.J., 189, 195

420   ARP,H., HEIDMANN,J., KHACHIKYAN,E.E., 1974, ASTROFIZIKA,
        10, 7 (ASTROPHYSICS, 10, 2)

421   ARP,H., KHACHIKIAN,E.E., 1974, ASTROFIZIKA, 10, 173,
        (ASTROPHYSICS, 10, 106)

422   ARP,H., KHACHIKIAN,E.E., ANDREASYAN,N.K., 1974,
        ASTROFIZIKA, 10, 625 (ASTROPHYSICS, 10, 399)

423   ARP,H., SARGENT,W.L.W., KHACHIKYAN,E.E., ANDREASYAN,N.K.,
        1974, ASTROFIZIKA, 10, 298 (ASTROPHYSICS, 10, 179)

424   BALKOWSKI,C., BOTTINELLI, L., CHAMARAUX,P., GOUGUENHEIM,L.,
        HEIDMANN,J., 1974, ASTRON.AND AP., 34, 43

425   BARBIERI,C., BERTOLA,F., DI TULLIO,G., 1974, ASTRON.
        AND AP., 35, 463

426   BARBON,R., CAPACCIOLI,M., 1974, ASTRON.AND AP., 35,
        151

427   BOND,H.E., TIFFT,W.G., 1974, PASP, 86, 981

428   BORCHKHADZE,T.M., 1974, ASTROFIZIKA, 10, 493,
        (ASTROPHYSICS, 10, 311)

429   CAROZZI,N., CHAMARAUX,P., DUFLOT,R., 1974, ASTRON.
        AND AP., 33, 113

430   CAROZZI,N., CHAMARAUX,P., DUFLOT-AUGARDE,R., 1974,
        ASTRON.AND AP., 30, 21

431   CARRANZA,G.J., AGUERO,E., 1974, OBSERVATORY, 94, 7

432   CHROMEY,F.R., 1974, ASTRON.AND AP., 31, 165

433   CHROMEY,F.R., 1974, ASTRON.AND AP., 37, 7

434   DANZIGER,I.J., SCHUSTER,H.E., 1974, ASTRON.AND AP., 34, 301

435   DENISYUK,E.K., 1974, ASTR.CIRC.(USSR), 809, 1

436   DENISYUK,E.K., 1974, ASTR.CIRC.(USSR), 809, 2

437   DENISYUK,E.K., BABKIN,I.G., SINYAEVA,N.V., 1974, ASTR.CIRC.(USSR), N. 837, 2

438   DENISYUK,E.K., LIPOVETSKY,V.A., 1974, ASTROFIZIKA, 10, 315 (ASTROPHYSICS, 10,195)

439   DISNEY,M.J.,,1974, AP.J., 193, L103

440   FREEMAN,K.C., DE VAUCOULEURS,G., 1974, AP.J., 194, 569

441   GOAD,J.W., 1974, AP.J., 192, 311

442   GRAHAM,J.A., 1974, OBSERVATORY, 94, 290

443   HODGE,P.W., 1974, PASP, 86, 645

444   JENNER,D.C., 1974, AP.J., 191, 55

445   KAZARYAN,M.A., KHACHIKYAN,E.E., 1974, ASTROFIZIKA, 10, 477 (ASTROPHYSICS 10, 299)

446   KHACHIKYAN,E.E., ANDREASYAN,N.K., SARGENT,W.L.W., 1974, ASTROFIZIKA, 10, 297.(ASTROPHYSICS, 10, 177)

447   KOPILOV,I.M., LIPOVETSKY,V.A., PRONIK,V.I., CHUVAEV,K.K., 1974, ASTROFIZIKA, 10, 483 (ASTROPHYSICS, 10, 305)

448   KORMENDY,J., SARGENT,W.L.W., 1974, AP.J., 193, 19

449   KOWAL,C.T., ZWICKY,F., SARGENT,W.L.W., SEARLE,L., 1974, PASP, 86, 516

450   LYNDS,R., 1974, QUOTED BY MATERNE,J., TAMMANN,G.A., 1974, ASTRON.AND AP., 35, 441

451   LYNDS,R., 1974, QUOTED IN REF.444

452   LYNDS,R., 1974, QUOTED IN REF.428

453   OSMER,P.S., SMITH,M.G., WEEDMAN,D.W., 1974, AP.J.,

192, 279

454   OSMER,P.S., SMITH,M.G., WEEDMAN,D.W., 1974, AP.J.,
       189, 187

455   PETERSON,S.D., SHOSTAK,G.S., 1974, A.J., 79, 767

456   ROGSTAD,D.H., LOCKART,I.A., WRIGHT,M.C.H., 1974,
       AP.J., 193, 309

457   ROTS,A.H., SHANE,W.W., ASTRON.AND AP., 1974, 31, 245

458   RUBIN,V.C., 1974, AP.J., 191, 645

459   SCHMIDT,M.,1974, QUOTED BY WILLIS,A.G., STROM,R.C.,
       WILSON,A.S., 1974, NATURE, 250, 625

460   SHOSTAK,G.S., 1974, AP.J., 187, 19

461   SHOSTAK,G.S., 1974, AP.J., 189, L1

462   SHOSTAK,G.S., 1974, ASTRON.AND AP., 31, 97

463   STOCKTON,A., 1974, AP.J., 187, 219

464   STOCKTON,A., 1974, AP.J., 190, L47

465   STRITTMATTER,P.A., CARSWELL,R.F., GILBERT,G.,
       BURBIDGE,E.M., 1974, AP.J., 190, 509

466   TIFFT,W.G., 1974, AP.J., 188, 221

467   TULLY,R.B., 1974, AP.J. SUPPL., 27, 437

468   VAN DER KRUIT,P.C., 1974, AP.J., 188, 3

469   VAN DER KRUIT,P.C., 1974, AP.J., 192, 1

470   WELIACHEW,L., 1974, AP.J., 191, 639

471   WILLS,D., WILLS,B.J., 1974, AP.J., 190, 271

472   ZWICKY,F., 1974, IN: SUPERNOVAE AND SUPERNOVA REMNANTS,
       COSMOVICI ED. REIDEL, P.1

473   AFANASIEV,V.L., KARACHENTSEV,I.D., NOTNI,P., 1975,
       ASTRON. NACHR., 296, 233

474   AGUERO,E., CARRANZA,G.J., 1975, OBSERVATORY, 95, 179

475   ARAKELYAN,M.A., DIBAI,E.A., ESIPOV,V.F., 1975,
        ASTROFIZIKA, 11, 377 (ASTROPHYSICS, 11, 254)

476   ARAKELYAN,M.A., DIBAI,E.A., ESIPOV,V.F., 1975,
        ASTROFIZIKA, 11, 15 (ASTROPHYSICS, 11, 8)

477   ARP,H., O'CONNELL,R.W., 1975, AP.J., 197, 291

478   BENVENUTI,P., CAPACCIOLI,M., D'ODORICO,S., 1975,
        ASTRON.AND AP., 41, 91

479   BERTOLA,F., CAPACCIOLI,M., 1975, AP.J., 200, 439

480   BOTTINELLI,L., DUFLOT,R., GOUGUENHEIM,L., HEIDMANN,J.,
        1975, ASTRON.AND AP., 41, 61

481   BOTTINELLI,L., GOUGUENHEIM,L., 1975, ASTRON.AND AP.,
        39, 341

482   BURBIDGE,E.M., SMITH,H.E., BURBIDGE,G.R., 1975, AP.J.,
        199, L137

483   CHERIGUENE,M.F., 1975, IN LA DINAMIQUE DES GALAXIES
        SPIRALES, COLLOQUE CNRS, 241, 439

484   CHINCARINI,G., MARTINS,D., 1975, AP.J., 196, 335

485   COLLA,G., FANTI,C., FANTI,R., GIOIA,I., LARI,C.,
        LEQUEUX,J., LUCAS,R., ULRICH,M.H., 1975, ASTRON.AND
        AP. SUPPL., 20, 1

486   DE VAUCOULEURS,G., 1975, AP.J. SUPPL., 29, 193

487   DE VAUCOULEURS,G., DE VAUCOULEURS,A., 1975, AP.J.,
        197, L1

488   DE VAUCOULEURS,G., DE VAUCOULEURS,A., 1975,
        OBSERVATORY, 95, 178

489   DEAN,J.F., DAVIES,R.D., 1975, MNRAS, 170, 503

490   DEHARVENG,J.M., PELLET,A., 1975, ASTRON.AND AP.,
        38, 15

491   DOROSHENKO,V.T., TEREBIZH,V.YU., 1975, ASTROFIZIKA,
        11, 631 (ASTROPHYSICS, 11, 422)

492   FISHER,J.R., TULLY,R.B., 1975, ASTRON.AND AP., 44,
        151

493   GOTTESMAN,S.T., WELIACHEW,L., 1975, AP.J., 195, 23

494   GRAHAM,J.A., SARGENT,W.L.W., 1975, QUOTED BY

SCHILIZZI,R.T., 1975, MEM.RAS, 79, 75

495   GREGORY,S.A., 1975, AP.J., 199, 1

496   GREGORY,S.A., 1975, PASP, 87, 833

497   HUCHTMEIER,W.K., BOHNENSTENGEL,H.D., 1975, ASTRON.
        AND AP., 44, 479

498   HUCHTMEIER,W.K., BOHNENSTENGEL,H.D., 1975, ASTRON.
        AND AP., 41, 477

499   HUCHTMEIER,W.K., TAMMANN,G.A., WENDKER,H.J., 1975,
        ASTRON.AND AP., 42, 205

500   KARACHENTSEV,I.D., PRONIK,V.I., CHUVAEV,K.K., 1975,
        ASTRON.AND AP., 41, 375

501   MATHEWSON,D.S., FORD,V.L., MURRAY,J.D., 1975,
        OBSERVATORY, 95, 176

502   METLOV,V.G., 1975, SOVIET ASTRON. LETT. 1, 220

503   MINKOWSKI,R., OORT,J.H., VAN HOUTEN,G.J., DAVIS,M.M.,
        1975, QUOTED IN REF.530

504   MIRZOYAN,L.V., MILLER,J.S., OSTERBROCK,D.E., 1975,
        AP.J., 196, 687

505   QUINTANA, H., MELNICK,J., 1975, PASP, 87, 863

506   RICKARD,J.J., 1975, ASTRON.AND AP., 40, 339

507   RUBIN,V.C., FORD,W.K.JR, PETERSON,C.J., 1975, AP.J.,
        199, 39

508   RUBIN,V.C., THONNARD,N., FORD,W.K.JR, 1975, AP.J.,
        199, 31

509   SANCISI,R., ALLEN,R.J., VAN ALBADA,T.S., 1975, IN
        LA DINAMIQUE DES GALAXIES SPIRALES, COLLOQUE CNRS,
        241, 295

510   SANDAGE,A., 1975, AP.J., 202, 563

511   SARGENT,W.L.W., 1975, QUOTED BY SCHILIZZI,R.T., 1975,
        MEM. RAS, 79, 75

512   SIEFERT, P.T., GOTTESMAN, S.T., WRIGHT, M.C.H., 1975,
        IN LA DINAMIQUE DES GALAXIES SPIRALES, COLLOQUE
        CNRS, 241,425.

513   SIMKIN,S.M., 1975, AP.J., 195, 293

514   SIMKIN,S.M., 1975, AP.J., 200, 567

515   SMITH,G.M., 1975, AP.J., 202, 591

516   SPINRAD,H., 1975, AP.J., 199, L1

517   SPINRAD,H., 1975, AP.J., 199, L3 AND 1978, AP.J.,
        220, L135

518   SPINRAD,H., SMITH,H.E., HUNSTEAD,R., RYLE,M., 1975,
        AP.J., 198, 7

519   TIFFT,W.G., HILSMAN,K.A., CORRADO,L.C., 1975, AP.J.,
        199, 16

520   TIFFT,W.G., TARENGHI,M., 1975, AP.J., 199, 10

521   ULRICH,M.H., 1975, ASTRON.AND AP., 40, 337

522   ULRICH,M.H., KINMAN,T.D., LYNDS,C.R., RIEKE,G.H.,
        EKERS,R.D., 1975, AP.J. 198, 261

523   VAN ALBADA,G.D., SHANE,W.W., 1975, ASTRON.AND AP.,
        42, 433

524   VAN DER KRUIT,P.C., 1975, AP.J., 195, 611

525   VIDAL,N.V., 1975, ASTRON.AND AP., 42, 145

526   VIDAL,N.V., 1975, PASP, 87, 625

527   VIDAL,N.V., PETERSON,B.A., 1975, AP.J., 196, L95

528   WELCH,G.A., CHINCARINI,G., ROOD,H.J., 1975, A.J.,
        80, 77

529   WESTPHAL,J.A., KRISTIAN,J., SANDAGE,A., 1975, AP.J.,
        197, L95

530   WILLIAMS,T.B., 1975, AP.J., 199, 586

531   WINTER,A.J.B., 1975, MNRAS, 172, 1

532   ALLEN,D.A., WRIGHT,A.E., GOSS,W.M., 1976, MNRAS, 1
        77, 91

533   ARAKELYAN,M.A., DIBAI,E.A., ESIPOV,V.F., 1976,
        ASTROFIZIKA, 12, 195 (ASTROPHYSICS, 12, 122)

534   ARAKELYAN,M.A., DIBAI,E.A., ESIPOV,V.F., 1976,
        ASTROFIZIKA, 12, 683 (ASTROPHYSICS, 12, 456)

535   ARKHIPOVA,V.P., ESIPOV,V.F., SAVEL'EVA,M.V., 1976,
        SOVIET ASTRON. LETT., 20, 521

536   ARP,H., 1976, AP.J., 210, L59

537   BAARS,J.W.M., WENDKER,H.J., 1976, ASTRON.AND AP.,
        48, 405

538   BALICK,B., FABER,S.M., GALLAGHER,J.S., 1976, AP.J.,
        209, 710

539   BARBIERI,C., ROMANO,G., 1976, ASTRON.AND AP., 50,
        15

540   BARBON,R., CAPACCIOLI,M., 1976, ASTRON.AND AP., 49,
        125

541   NO REFERENCE

542   BENVENUTI,P., CAPACCIOLI,M., D'ODORICO,S., 1976,
        ASTRON.AND AP., 53, 141

543   BENVENUTI,P., D'ODORICO,S., VETTOLANI,G., 1976,
        PROC. THIRD EUROPEAN ASTRON. MEETING, TBILISI, P. 393

544   BOTTINELLI,L., GOUGUENHEIM,L., 1976, ASTRON.AND AP.,
        47, 381

545   BRIDLE,A.H., FOMALONT,E.B., 1976, ASTRON.AND AP.,
        52, 107

546   CAROZZI,N., 1976, ASTRON.AND AP., 49, 425

547   CAROZZI,N., 1976, ASTRON.AND AP., 49, 431

548   CHINCARINI,G., ROOD,H.J., 1976, AP.J., 206, 30

549   CHINCARINI,G., ROOD,H.J., 1976, PASP, 88, 388

550   COLEMAN,G.D., HINTZEN,P., SCOTT,J.S., TARENGHI,M.,
        1976, NATURE, 262, 476

551   COTTRELL,G.A., 1976, MNRAS, 174, 455

552   COTTRELL,G.A., 1976, MNRAS, 177, 463

553   CRAINE,E.R., WARNER,J.W., 1976, AP.J., 206, 359

554   DE VAUCOULEURS,A., SHOBBROOK,R.R., STROBEL,A., 1976,
        A.J., 81, 219

555   DE VAUCOULEURS,G., DE VAUCOULEURS,A., 1976, A.J.,
        81, 595

556   DENISYUK,E.K., DOSTAL,V.A., 1976, ASTR.CIRC.(USSR),
        N. 931, 7

557   DENISYUK,E.K., LIPOVETSKY,V.A., AFANASIEV,V.L., 1976,
        ASTROFIZIKA, 12, 665 (ASTROPHYSICS, 12, 442)

558   DIBAI,E.A., DOROSHENKO,V.T., TEREBIZH,V.YU., 1976,
        ASTROFIZIKA, 12, 689 (ASTROPHYSICS, 12, 459)

559   DICKENS,R,J; MOSS,C., 1976, MNRAS, 174, 47

560   DUFLOT,R., 1976, ASTRON.AND AP., 48, 437

561   FISHER,J.R., TULLY,R.B., 1976, ASTRON.AND AP., 53,
        397

562   FRICKE,K.J., KAUFMANN,J.P., 1976, MITT.ASTRON.GES.,
        38, 102

563   GRANDI,S.A., OSTERBROCK,D.E., PHILLIPS,M.M., 1976,
        ASTRON.AND AP., 51, 323

564   GUDEHUS,D.H., 1976, AP.J., 208, 267

565   HUCHRA,J., 1976, QUOTED BY DE BRUYN,A.G., 1977,
        ASTRON.AND AP., 58, 221

566   HUCHTMEIER,W.K., TAMMANN,G.A., WENDKER,H.J., 1976,
        ASTRON.AND AP., 46, 381

567   KARACHENTSEV,I.D., PRONIK,V.I., CHUVAEV,K.K., 1976,
        ASTRON.AND AP., 51, 185

568   KHACHIKYAN,E.E., 1976, ASTRON. NACHR. 297, 287

569   KOPYLOV,I.M., LIPOVETSKY,V.A., PRONIK,V.I., CHUVAEV,K.K.,
        1976, ASTROFIZIKA, 12, 189 (ASTROPHYSICS, 12, 119)

570   KOSKI,A., 1976 PH.D. THESIS, UNIVERSITY OF CALIFORNIA,
        SANTA CRUZ

571   KOWAL,C.T., HUCHRA,J., SARGENT,W.L.W., 1976, PASP,
        88, 521

572   KRUMM,N., SALPETER,E.E., 1976, AP.J., 208, L7

573   LYNDS,R., TOOMRE,A., 1976, AP.J., 209, 382

574   MARTIN,W.L., 1976, MNRAS, 175, 633

575   OSTERBROCK,D.E., KOSKI,A.T., PHILLIPS,M.M., 1976,
        AP.J., 206, 898

576   PASTORIZA,M., AGUERO,E., 1976, AP. AND SP.SCI., 39,

577   PETERSON,B.A., JAUNCEY,D.L., WRIGHT,A.E., CONDON,J.J.,
       1976, AP.J., 207,L5

578   PETERSON,C.J., RUBIN,V.C., FORD,W.K.JR., THONNARD,N.,
       1976, AP.J., 208, 662

579   PHILLIPS,M.M., 1976, AP.J., 208, 37

580   ROGSTAD,D.H., WRIGHT,M.C.H., LOCKHART,I.A., 1976,
       AP.J., 204, 703

581   ROOD,H.J., DICKEL,J.R., 1976, AP.J., 205, 346

582   RUBIN,V.C., FORD,W.K.JR, THONNARD,N., ROBERTS,M.S.,
       GRAHAM,J.A., 1976, A.J., 81, 687

583   SANDAGE,A., KRISTIAN,J., WESTPHAL,J.A., 1976, AP.J.,
       205, 688

584   SMITH,H.E., SPINRAD,H., HUNSTEAD,R., 1976, AP.J.,
       206, 345

585   SMITH,H.E., SPINRAD,H., SMITH,E.C., 1976, PASP, 88,
       621

586   SPINRAD,H., 1976, PASP, 88, 565

587   SPINRAD,H., BACHALL,N.A., 1976, PASP, 88, 660

588   SPINRAD,H., LIEBERT,J., SMITH,H.E., HUNSTEAD,R., 1
       976, AP.J., 206, L79

589   SPINRAD,H., SMITH,H.E., 1976, AP.J., 206, 355

590   TARENGHI,M., SCOTT,J.S., 1976, AP.J., 207, L9

591   THEYS,J.C., SPIEGEL,E.A., 1976, AP.J., 208, 650

592   THOLINE,J.E., OSTERBROCK,D.E., 1976, AP.J., 210, L
       117

593   TIFFT,W.G., GREGORY,S.A., 1976, AP.J., 205, 696

594   TURNER,E.L., 1976, AP.J., 208, 20

595   ULRICH,M.H., 1976, AP.J., 206, 364

596   VAN DER KRUIT,P.C., 1976, ASTRON.AND AP., 49, 161

597   VAN DER KRUIT,P.C., 1976, ASTRON.AND AP., 52, 85

598   WEST,R.M., 1976, ASTRON.AND AP., 46, 327

599   WEST,R.M., 1976, ASTRON.AND AP., 53, 435

600   WILLS,D., WILLS,B.J., 1976, AP.J. SUPPL., 31, 143

601   WILSON,A.S., PENSTON,M.V., FOSBURY,R.A.E., BOKSENBERG,A.,
        1976, MNRAS, 177, 673

602   ADAMS,M.T., RUDNICK,L., 1977, A.J., 82, 857

603   AGUERO,E., CARRANZA,G.J., 1977, OBSERVATORY, 97, 241

604   ANDRILLAT,Y., SWINGS,J.P., 1977, ASTROPHYS. LETT.,
        18, 151

605   ARP,H., 1977, QUOTED IN REF.614

606   BACHALL,N.A., SARGENT,W.L.W., 1977, AP.J., 217, L19

607   BALDWIN,J.A., WAMPLER,E.J., BURBIDGE,E.M., O'DELL,S.L.,
        SMITH,H.E., HAZARD,C.,NORDISIECK,K.H., POOLE,Y.G.,
        STEIN,W.A., 1977,   AP.J., 215, 408

608   BARBIERI,C., DI SEREGO ALIGHIERI,S., ZAMBON,M., 19
        77, ASTRON.AND AP., 57, 353

609   BERTOLA,F., CAPACCIOLI,M., 1977, AP.J., 211, 697

610   BIEGING,J.H., BIERMANN,P., 1977, ASTRON.AND AP., 60,
        361

611   BLACKMAN,C.P., 1977, MNRAS, 178, 15

612   BOLTON,J.G., SAVAGE,A., 1977, AUSTRAL.J.PHYS.SUPPL.,
        44, 21

613   BORCHKHADZE,T.M., BREYSACHER,J., LAUSTSEN,S., SCHUSTER,H.E.,
        WEST, R.M., 1977, ASTRON.AND AP. SUPPL., 30, 35

614   BOSMA,A., EKERS,R.D., LEQUEUX,J., 1977, ASTRON.AND
        AP., 57, 97

615   BOSMA,A., VAN DER HULST,J.M., SULLIVAN,W.T.III, 1977,
        ASTRON.AND AP., 57, 373

616   BOTTINELLI,L., GOUGUENHEIM,L., 1977, ASTRON.AND AP.,
        54, 641

617   BOTTINELLI,L., GOUGUENHEIM,L., 1977, ASTRON.AND AP.,

60, L23

618   CAROZZI,N., 1977, ASTRON.AND AP., 55, 261

619   CARRANZA,G.J., AGUERO,E.L., 1977, ASTROPYS. AND SPACE
        SCI., 47, 397

        ..BARSKY,D.A., FALGARONE,E.G., LEQUEUX,J., 1977,
        ASTRON.AND AP., 59, L5

621   CHAISSON,E.J., RODRIGUEZ,L.F., 1977, AP.J., 214, L111

622   CHAMARAUX,P., 1977, ASTRON.AND AP., 60, 67

623   CHINCARINI,G., ROOD,H.J., 1977, AP.J., 214, 351

624   CIATTI,F., ROSINO,L., 1977, ASTRON.AND AP., 57, 73

625   COMBES,F., GOTTESMAN,S.T., WELIACHEW,L., 1977,
        ASTRON.AND AP., 59, 181

626   COOKE,J.A., EMERSON,D., NANDY,K., REDDISH,V.C.,
        SMITH,M.G., 1977, MNRAS, 178, 687

627   DANZIGER,I.J., 1977, QUOTED BY CHRISTIANSEN,W.N.,
        ET AL., 1977, MNRAS, 181, 183

628   DANZIGER,I.J., FOSBURY,R.A.E., PENSTON,M.V., 1977,
        MNRAS, 179, 41P

629   DAWE,J.A., DICKENS,R.J., PETERSON,B.A., 1977, MNRAS,
        178, 675

630   DENISYUK,E.K., LIPOVETSKY,V.A., 1977, SOVIET ASTRON.
        LETT., 3, 3

631   DOSTAL',V.A., 1977, SOVIET ASTRON. LETT., 3, 30

632   FABER,S.M., BALIK,B., GALLAGHER,J.S., KNAPP,G.R.,
        1977, AP.J., 214, 383

633   FABER,S.M., DRESSLER,A., 1977, A.J., 82, 187

634   FOSBURY,R.A.E., HAWARDEEN,T.G., 1977, MNRAS, 178,
        473

635   FOSBURY,R.A.E., MEBOLD,U., GOSS,W.M., VAN WOERDEN,H.,
        1977, MNRAS, 179, 89

636   FREEMAN,K.C., KARLSSON,B., LYNGA,G., BURRELL,J.F.,
        VAN WOERDEN,H., GOSS,W.M., MEBOLD,U., 1977, ASTRON.
        AND AP., 55, 445

637   GALLAGHER,J.S., KNAPP,G.R., FABER,S.M., BALIK,B.,
        1977, AP.J., 215, 463

638   GRANDI,S.A., 1977, AP.J., 215, 446

639   HINTZEN,P., SCOTT,J.S., TARENGHI,M., 1977, AP.J.,
        212, 8

640   HUCHRA,J., HOESSEL,J., ELIAS,J., 1977, A.J., 82, 6
        74

641   HUCHRA,J., THUAN,T.X., 1977, AP.J., 216, 694

642   HUCHTMEIER,W.K., TAMMANN,G.A., WENDKER,H.J., 1977,
        ASTRON.AND AP., 57, 313

643   KINMAN,T.D., RUBIN,V.C., THONNARD,N., FORD,W.K.JR,
        PETERSON,C.J., 1977, A.J., 82, 879

644   KIRSHNER,R.P., 1977, AP.J., 212, 319

645   KNAPP,G.R., GALLAGHER,J.S., FABER,S.M., BALICK,B.,
        1977, A.J., 82, 106

646   KRON,R.G., SPINRAD,H., KING,I.R., 1977, AP.J., 217,
        951

647   KRUMM,N., SALPETER,E.E., 1977, ASTRON.AND AP., 56,
        465

648   MACCACARO,T., COOKE,B.A., WARD,M.J., PENSTON,M.V.,
        HAYNES,R.F., 1977, MNRAS, 180, 465

649   MEBOLD,U., GOSS,W.M., FOSBURY,R.A.E., 1977, MNRAS,
        180, 11P

650   MELNIK,J., 1977, AP.J., 213, 15

651   MELNIK,J., SARGENT,W.L.W., 1977, AP.J., 215, 401

652   MITTON,S., HAZARD,C., WHELAN,J.A.J., 1977, MNRAS,
        179, 569

653   MOSS,C., DICKENS,R.J., 1977, MNRAS, 178, 701

654   OSTERBROCK,D.E., PHILLIPS,M.M., 1977, PASP, 89, 251

655   PENSTON,M.V, FOSBURY,R.A.E., WARD,M.J., WILSON,A.S.,
        1977, MNRAS, 180, 19

656   RUBIN,V.C., THONNARD,N., FORD,W.K.JR, 1977, AP.J.,
        217, L1

657   SARGENT,W.L.W., 1977, AP.J., 212, L105

658   SARGENT,W.L.W., SCHECHTER,P.L., BOKSENBERG,A.,

SHORTRIDGE,K., 1977, AP.J., 212, 326

659   SCOTT,J.S., ROBERTSON,J.W., TARENGHI,M., 1977,
ASTRON.AND AP., 59, 23

660   SERSIC,J.L., BAJAJA,E., COLOMB,R., 1977, ASTRON.AND
AP., 59, 19

661   SMITH,H.E., BURBIDGE,E.M., BALDWIN, J.A., TOHLINE,J.E.,
WAMPLER, E.J., HAZARD,C., MURDOCH,H.S., 1977.
AP.J., 215, 427

662   SPINRAD,H., 1977, PASP, 89, 116

663   TIFFT,W.G., TARENGHI,M., 1977, AP.J., 217, 944

664   VAN ALBADA,G.D., 1977, ASTRON.AND AP., 61, 297

665   VAN DER KRUIT,P.C., 1977, ASTRON.AND AP., 61, 171

666   VAN WOERDEN,H., MEBOLD,U, GOSS,W.M., SIEGMAN,G., 1
977, QUOTED IN REF.741

667   VIDAL,N.V., WICKRAMASINGHE,D.T., 1977, MNRAS, 180,
305

668   WAGGETT,P.C., WARNER,P.J., BALDWIN,J.E., 1977,
MNRAS, 181, 465

669   WARD,M.J., WILSON,A.S., DISNEY,M.J., ELVIS,M.,
MACCACARO,T., 1977, ASTRON.AND AP., 59, L19

670   WEST,R.M., 1977, ASTRON.AND AP.SUPPL., 27, 73

671   WHITEOAK,J.B., GARDNER,F.F., 1977, AUSTRAL.J.PHYS.,
30, 187

672   WRIGHT,A.E., JAUNCEY,D.L., PETERSON,B.A., CONDON,J.J.,
1977, AP.J. 211, L115.

673   WYCKOFF,S., WEHINGER,P.A., 1977, ASTROPHYS. AND SPACE
SCI., 48, 421

674   ABELL,G.O., EASTMOND,T.S., JENNER,D.C., 1978, AP.J.,
221, L1

675   ALLEN,D.A., LONGMORE,A.J., HAWARDEN,T.G., CANNON,R.D.,
ALLEN,C.J., 1978, MNRAS, 184, 303

676   ALLEN,R.J., VAN DER HULST,J.M., GOSS,W.M., HUCHTMEIER,W.,
1978, ASTRON. AND AP., 64, 359

677   ALLSOPP,N.J., 1978, MNRAS, 184, 397

678   BAAN,W.A., HASCHICK,A.D., GREENFIELD,P.E., 1978,
AP.J., 222, L7

679   BALKOWSKI,C., CHAMARAUX,P., WELIACHEW,L., 1978,
        ASTRON.AND AP., 69, 263

680   BERGWALL,N.A.S., EKMAN,A.B.G., LAUBERTS,A., WESTERLUND,B.E.,
        BORCHKHADZE,T.M., BREYSACHER,J., LAUSTSEN,S.,
        MULLER,A.B., SCHUSTER,H.E., SURDEJ,J., WEST,R.M.,
        1978, ASTRON.AND AP. SUPPL., 33, 243

681   BERTOLA,F., CAPACCIOLI,M., 1978, AP.J., 219, 404

682   BIEGING,J.H., 1978, ASTRON.AND AP., 64, 23

683   BOHUSKI,T.J., FAIRALL,A.P., WEEDMAN,D.W., 1978, AP.J.,
        221, 776

684   BOTTINELLI,L., DUFLOT,R., GOUGUENHEIM,L., 1978,
        ASTRON.AND AP., 63, 363

685   BOTTINELLI,L., GOUGUENHEIM,L., 1978, ASTRON.AND AP.,
        64, L3

686   BRADT,H.V., BURKE,B.F., CANIZARES,C.R., GREENFIELD,P.E.,
        KELLY,R.L., MCCLINTOCK,J.E., VAN PARADIJS,J.,
        KOSKI,A.T., 1978, AP.J., 226, L111

687   BURBIDGE,E.M., SMITH,H.E., BURBIDGE,G.R., 1978, AP.J.,
        219, 400

688   CAROZZI-MEYSONNIER,N., 1978, ASTRON.AND AP. SUPPL.,
        33, 411

689   CAROZZI-MEYSSONNIER,N., 1978, ASTRON.AND AP., 63,
        415

690   CHINCARINI,G., TARENGHI,M., BETTIS,C., 1978, AP.J.,
        221, 34

691   DANZIGER,I.J., GOSS,W.H., FRATER,R.H., 1978, MNRAS,
        184, 341

692   DAVIS,M., HUCHRA,J., TONRY,J., LATHAM,D., 1978, IAU
        CIRC., N 3202

693   DICKEL,J.R., ROOD,H.J., 1978, AP.J., 223, 391

694   EASTMOND,T.S., ABELL,G.O., 1978, PASP, 90, 367

695   FAIRALL,A.P., 1978, MNASSA, 37, 72

696   FANTI,R., GIOIA,I., LARI,C., ULRICH,M.H., 1978,
        ASTRON.AND AP. SUPPL., 34, 341

697   FELDMAN,F.R., MARSHALL,F.E., PHILLIPS,M.M., 1978,
        IAU CIRC. N.3293

698   FISHER,J.R., TULLY,R.B., 1978, QUOTED IN REF.741

699   FOSBURY,R.A.E., 1978, QUOTED IN REF.783

700   FOSBURY,R.A.E., MEBOLD,U., GOSS,W.M., DOPITA,M.A.,
         1978, MNRAS, 183, 549

701   GRAHAM,J.A., 1978, PASP, 90, 237

702   GREEN,R.F., WILLIAMS,T.B., MORTON,D.C., 1978, AP.J.,
         226, 729

703   GREGORY,S.A., THOMPSON,L.A., 1978, AP.J., 222, 784

704   HARTWICK,F.D.A., SARGENT,W.L.W., 1978, AP.J., 221,
         512

705   HASCHICK,A.D., BAAN,W.A., BURKE,B.F., 1978, AP.J.,
         225, 343

706   HAVLEN,R.J., QUINTANA,H., 1978, AP.J., 220, 14

707   HECKMAN,T.M., BALICK,B., SULLIVAN,W.T.III, 1978,
         AP.J., 224, 745

708   HINTZEN,P., OEGERLE,W.R., SCOTT,J.S., 1978, A.J.,
         83, 478

709   HUNSTEAD,R.W., MURDOCK,H.S., SHOBBROCK,R.R., 1978,
         MNRAS, 185, 149

710   JAFFE,W.J., PEROLA,G.C., TARENGHI,M., 1978, AP.J.,
         224, 808

711   JAUNCEY,D.L., WRIGHT,A.E., PETERSON,B.A., CONDON,J.J.,
         1978, AP.J., 219, L1

712   JENNER,D., 1978, QUOTED IN REF.707

713   KARACHENTSEV,I.D., TIFFT,W.G., 1978, ASTRON.AND AP.,
         63, 411

714   KIRSHNER,R.P., OEMLER,A.JR, SCHECHTER,P.L., 1978,
         A.J., 83, 1549

715   KNAPP,G.R., FABER,S.M., GALLAGER,J.S., 1978, A.J.,
         83, 11

716   KNAPP,G.R., GALLAGHER,J.S., FABER,S.M., 1978, A.J.,
         83, 139

717   KRISTIAN,J., SANDAGE,A., WESTPHAL,J.A., 1978, AP.J.,
         221, 383

718   LAING,R.A., LONGAIR,M.S., RILEY,J.N., KIBBLEWHITE,E.J.,
         GUNN,J.E., 1978, MNRAS, 183, 547

719   LAUBERTS,A., BERGVALL,N.A.S., EKMAN,A.B.G., WESTERLUND,B.E.,
       1978, ASTRON.AND AP.SUPPL., 35, 55

720   LONGMORE,A.J., HAWARDEN,T.G., WEBSTER,B.L., GOSS,W.M.,
       MEBOLD,U., 1978, MNRAS, 183, 97P

721   MILLS,B.Y., HUNSTEAD,R.W., SKELLERN,D.J., 1978,
       MNRAS, 185, 51P

722   OKE,J.B., 1978, AP.J., 219, L97

723   OSTERBROCK,D.E., 1978, QUOTED IN REF.707

724   PEDREROS,M., 1978, PASP, 90, 14

725   PENCE,W.D., 1978, PUB. IN ASTR. UNIV. TEXAS, N.14

726   PETERSON,B.M., 1978, AP.J., 223, 740

727   PETERSON,B.M., CRAINE,E.R., STRITTMATTER,P.A., 1978,
       PASP, 90, 386

728   PETERSON,C.J., 1978, PASP, 90, 10

729   PETERSON,C.J., RUBIN,V.C., FORD,W.K.JR., ROBERTS,M.S.,
       1978, AP.J., 226, 770

730   PETERSON,C.J., RUBIN,V.C., FORD,W.K.JR., THONNARD,N.,
       1978, AP.J., 219, 31

731   PINEDA,F.J., DELVAILLE,J.P., HUCHRA,J., DAVIS,M.,
       1978, IAU CIRC., N 3202

732   PRABHU,T.P., 1978, KODAIKANAL OBS. BULL., SER A, 2,
       105

733   REIF,K., MEBOLD,U., GOSS,W.M., 1978, ASTRON.AND AP.,
       67, L1

734   ROBERTS,M.S., 1978, QUOTED IN REF.829

735   RODGERS,A.W., PETERSON,B.A., HARDING,P., 1978, AP.J.,
       225, 768

736   ROTS,A.H., 1978, A.J. 83, 219.

737   RUBIN,V.C., 1978, AP.J., 224, L55

738   RUBIN,V.C., FORD,W.K.JR, STROM,K.M., STROM,S.E.,
       ROMANISHIN,W., 1978, AP.J., 224, 782

739   RUBIN,V.C., FORD,W.K.JR., PETERSON,C.J., LYNDS,C.R.,
       1978, AP.J., 37, 235

740   RUBIN,V.C., FORD,W.K.JR., THONNARD,N., 1978, AP.J.,
        225, L107

741   SANDAGE,A., 1978, A.J., 83, 904

742   SARGENT,W.L.W., YOUNG,P.J., BOKSEMBERG,A., SHORTRIDGE,K.,
        LYNDS,C.R., HARTWICK,F.D.A., 1978, AP.J., 221, 731

743   SCHNOPPER,H.W., DAVIS,M., DELVAILLE,J.P., GELLER,M.J.,
        HUCHRA,J., 1978, NATURE, 275, 719

744   SHOSTAK,G.S., 1978, ASTRON.AND AP., 68, 321

745   SILVERGLATE,P.R., KRUMM,N., 1978, AP.J., 224, L99

746   SPINRAD,H., SMITH,H.E., 1978, QUOTED IN REF.717

747   STAUFFER,J., SPINRAD,H., 1978, PASP, 90, 20

748   STOKE,J.T., TIFFT,W.G., KAFTAN-KASSIM,M.A., 1978,
        A.J., 83, 322

749   SULLIVAN,W.T.III, JOHNSON,P.E., 1978, AP.J., 225,
        751

750   THOMAS,J.C., BATCHELOR,D., 1978, A.J., 83, 1160

751   THOMPSON,L.A., WALKER,W.J.JR., GREGORY,S.A., 1978,
        PASP, 90, 644

752   THONNARD,N., RUBIN,V.C., FORD,W.K.JR, ROBERTS,M.S.,
        1978, A.J., 83, 1564

753   TIFFT,W.G., 1978, AP.J., 222, 54

754   TULLY,R.B., BOTTINELLI,L., FISHER,J.R., GOUGHENHEIM,L.,
        SANCISI,R., VAN WOERDEN,H., 1978, ASTRON.AND AP.,63, 37

755   ULRICH,M.H., 1978, AP.J., 219, 424

756   ULRICH,M.H., 1978, AP.J., 221, 422

757   ULRICH,M.H., 1978, AP.J., 222, L3

758   ULRICH,M.H., 1978, QUOTED IN REF. 762

759   VAN DER KRUIT,P.C., BOSMA,A., 1978 ASTRON.AND AP.
        SUPPL., 34, 259

760   WARD,M.J., WILSON,A.S., 1978, ASTRON.AND AP., 70,

761   WARD,M.J., WILSON,A.S., PENSTON,M.V., ELVIS,M.,
      MACCACARO,T., TRITTON,K.P., 1978, AP.J., 223, 788

762   WELIACHEW,L., SANCISI,R., GUELIN,M., 1978, ASTRON.
      AND AP., 65, 37

763   WEST,R.M., BORCHKHADZE,T.M., BREYSACHER,J., LAUSTSEN,S.,
      SCHUSTER,H.E., 1978, ASTRON.AND AP.SUPPL., 31, 55

764   WEST,R.M., DANKS,A.C., ALCAINO,G., 1978, ASTRON.AND
      AP., 65, 151

765   WESTERLUND,B.E., BERGWALL,N.A.S., EKMAN,A.B.G.,
      LAUBERTS,A., 1978, ASTRON.AND AP.SUPPL., 31, 427

766   WESTPHAL,J.A., KRISTIAN,J., QUOTED BY KRISTIAN J.,
      SANDAGE,A. KATEM B., 1978, AP.J. 219, 803

767   ZEALEY,W., 1978, IAU CIRC. N.3293

768   AFANASIEV,V.L., KARACHENTSEV,I.D., LIPOVETSKY,V.A.,
      1979, ASTRON. NACHR., 300, 77

769   ALLEN,R.J., SHOSTAK,G.S., 1979, ASTRON.AND AP.SUPPL.,
      35, 163

770   ALLSOPP,N.J., 1979, MNRAS, 186, 343

771   ALLSOPP,N.J., 1979, MNRAS, 187, 537

772   ALLSOPP,N.J., 1979, MNRAS, 188, 371

773   ALLSOPP,N.J., 1979, MNRAS, 188, 765

774   ARKHIPOVA,V.P., ESIPOV,V.F., 1979, SOVIET ASTRON.
      LETT., 5, 140

775   BALKOWSKI,C., 1979, ASTRON.AND AP., 78, 190

776   BARBIERI,C., CASINI,C., HEIDMANN,J., DI SEREGO,S.,
      ZAMBON,M., 1979, ASTRON.AND AP.SUPPL., 37, 559

777   BARBON,R., CAPACCIOLI,M., TIFFT,W.G., 1979, ASTRON.
      AND AP.SUPPL., 36, 129

778   BECK,S.C., LACY,J.H., GEBALLE,T.R., 1979, AP.J., 231,
      28

779   BIERMANN,P., CLARKE,J.N., FRICKE,K.J., 1979, ASTRON.
      AND AP., 75, 19

780   BIERMANN,P., CLARKE,J.N., FRICKE,K.J., 1979, ASTRON.
      AND AP., 75, 7

781   BOTTINELLI,L., GOUGUENHEIM,L., 1979, ASTRON.AND AP., 76, 176

782   BOTTINELLI,L., GOUGUENHEIM,L., 1979, ASTRON.AND AP., 74, 172

783   BURBIDGE,G.R., CROWNE,A.H., 1979, AP.J. SUPPL., 40, 583

784   CAROZZI-MEYSSONIER,N., 1979, ASTRON.AND AP. SUPPL., 37, 529

785   CASINI,C., HEIDMANN,J., TARENGHI,M., 1979, ASTRON. AND AP., 73, 216

786   CHINCARINI,G., GIOVANELLI,R., HAYNES,M.P., 1979, A.J., 84, 1500

787   COHEN,R.J., 1979, MNRAS, 187, 839

788   COMTE,G., MONNET,G., ROSADO,M., 1979, ASTRON.AND AP., 72, 73

789   DE VAUCOULEURS,G., DE VAUCOULEURS,A., NIETO,J.L., 1979, A.J., 84, 1811

790   DENNEFELD,M., LAUSTSEN,S., MATERNE,J., 1979, ASTRON. AND AP., 74, 123

791   FAIRALL,A.P., 1979, MNASSA, 38, 18

792   FAIRALL,A.P., 1979, MNASSA, 38, 24

793   FAIRALL,A.P., 1979, MNRAS, 188, 343

794   FAIRALL,A.P., 1979, MNRAS, 188, 349

795   FAIRALL,A.P., 1979, NATURE, 279, 140

796   GIOVANELLI,R., 1979, AP.J., 227, L125

797   GOAD,J.W., DE VENY,J.B., GOAD,L.E., 1979, AP.J. SUPPL., 39, 439

798   GORDON,K.J., GORDON,C.P., 1979, ASTROPHYS. LETTERS, 20, 9

799   GRAHAM,J.A., 1979, AP.J., 232, 60

800   GRIERSMITH,D., VISVANATHAN,N., 1979, ASTRON.AND AP., 79, 329

801   HAWARDEN,T.G., LONGMORE,A.J., CANNON,R.D., ALLEN,D.A., 1979, MNRAS, 186, 495

802   HAWARDEN,T.G., VAN WOERDEN,H., MEBOLD,U., GOSS,W.M.,
       PETERSON,B.A., 1979, ASTRON.AND AP. 76, 230

803   HAYNES,M.P., 1979, A.J., 84, 1830

804   HAYNES,M.P., GIOVANELLI,R., ROBERTS,M.S., 1979, AP.J.,
       229, 83

805   HELOU,G., SALPETER,E.E., KRUMM,N., 1979, AP.J., 228,
       L1

806   HINTZEN,P., 1979, PASP, 91, 426

807   HINTZEN,P., SCOTT,J.S., 1979, ASTRON.AND AP., 74,
       116

808   HUCHRA,J., DAVIS,M., 1979, QUOTED IN REF.931

809   HUCHTMEIER,W.K., 1979, ASTRON.AND AP., 75, 170

810   JOEVEER,M., KAASIK,A., EINASTO,J., 1979, ASTROFIZIKA,
       15, 19 (ASTROPHYSICS, 15, 16)

811   KARACHENTSEV,I.D., KARACHENTSEVA,V.E., SHCHERBANOVSKII,A.L.,
       1979, SOVIET ASTRON. LETT., 4, 261

812   KARACHENTSEV,I.D., SARGENT,W.L.W., ZIMMERMANN,B, 1979
       ASTROFIZIKA, 15 25, (ASTROPHYSICS, 15, 19)

813   KARACHENTSEVA,V.E., KARACHENTSEV,I.D., 1979,
       ASTROFIZIKA, 15, 589 (ASTROPHYSICS, 15, 396)

814   KAZARYAN,M.A., 1979, ASTROFIZIKA, 15, 193 (ASTROPHYSICS,
       15, 117)

815   KNAPP,G.R., KERR,F.J., HANDERSON,A.P., 1979, AP.J.,
       234, 448

816   KODAIRA,K., IYE,M., 1979, PUB.ASTR.SOC. JAPAN, 31,
       647

817   KRUMM,N., SALPETER,E.E., 1979, A.J. 84, 1138

818   KRUMM,N., SALPETER,E.E., 1979, AP.J., 228, 64

819   KUNTH,D., SARGENT,W.L.W., 1979, ASTRON.AND AP., 76,
       50

820   KUNTH,D., SARGENT,W.L.W., 1979, ASTRON.AND AP.SUPPL.,
       36, 259

821   LONGMORE,A.J., HAWARDEN,T.G., CANNON,R.D., ALLEN,D.A.,
       MEBOLD,U., GOSS,W.M., REIF,K., 1979, MNRAS, 188, 285

822   MCALPINE,G.M., WILLIAMS,G.A., LEWIS,D.W., 1979, PASP,

91, 746

823   MEBOLD,U., GOSS,W.M., VAN WOERDEN,H., HAWARDEN,T.G.,
       SIEGMAN,B., 1979, ASTRON.AND AP., 74, 100

824   MILEY,G.K., OSTERBROCK,D.E., 1979, PASP, 91, 257

825   OKE,J.B., LAUER,T.R., 1979, AP.J., 230, 360

826   OSTERBROCK,D.E., GRANDI,S.A., 1979, AP.J., 228, L59

827   PASTORIZA,M., 1979, AP.J. 234, 837.

828   PETERSON,B.A., WRIGHT,A.E., JAUNCEY,D.L., CONDON,J.J.,
       1979, AP.J., 232, 400

829   PETERSON,S.D., 1979, AP.J. SUPPL., 40, 527

830   PETROSYAN,A.R., SAAKYAN,K.A., KHACHIKYAN,E.E., 1979,
       ASTROFIZIKA, 15, 209 (ASTROPHYSICS, 15, 142)

831   PETROSYAN,A.R., SAAKYAN,K.A., KHACHIKYAN,E.E., 1979,
       ASTROFIZIKA, 15, 373 (ASTROPHYSICS, 15, 250)

832   PHILLIPS,M.M., FELDMAN,F.R., MARSHALL,F.E., WAMSTEKER,W.,
       1979, ASTRON.AND AP., 76, L14

833   REAKES,M., 1979, MNRAS, 187, 509

834   REAKES,M., 1979, MNRAS, 187, 525

835   ROGSTAD,D.H., CRUTCHER,R.M., CHU,K., 1979, AP.J.,
       229, 509

836   ROSE,J.A., GRAHAM,J.A., 1979, AP.J., 231, 320

837   ROTS,A.H., 1979, ASTRON.AND AP., 80, 255

838   RUBIN,V.C., FORD,W.K.JR., ROBERTS,M.S., 1979, AP.J.,
       230, 35

839   SANCISI,R., ALLEN,R.J., 1979, ASTRON.AND AP., 74, 73

840   SANCISI,R., ALLEN,R.J., SULLIVAN,W.T.III, 1979, ASTRON.
       AND AP., 78, 217

841   SARGENT,W.L.W., 1979, QUOTED IN REF.780

842   SCHECHTER,P.L., GUNN,J.E., 1979, AP.J., 229, 472

843   SCHILD,R., DAVIS,M., 1979, A.J., 84, 311

844   SERSIC,J.L., ARIAS,J.C., ARAUJO,A., 1979, OBSERVATORY,
        99, 130

845   SERSIC,J.L., CERRUTI,M.A., 1979, OBSERVATORY, 99,
        150

846   SIMKIN,S.M., 1979, AP.J., 234, 56

847   SIMKIN,S.M., EKERS,R.D., 1979, A.J., 84, 56

848   SMITH,H.E., 1979, QUOTED IN REF.783

849   SMITH,H.E., JUNKKARINEN,V.T., SPINRAD,H., GRUEFF,G.,
        VIGOTTI,M., 1979, AP.J., 231, 307

850   SPINRAD,H., 1979, QUOTED IN REF.783

851   SPINRAD,H., KRON,R., HUNSTEAD,R.W., 1979, QUOTED IN
        REF.783

852   SPINRAD,H., STAUFFER,J., HARLAN,E., 1979, PASP, 91,
        619

853   STAUFFER,J., SPINRAD,H., SARGENT,W.L.W., 1979, AP.J.,
        228, 379

854   STOCKTON,A., WYCKOFF,S., WEHINGER,P.A., 1979, AP.J.,
        231, 673

855   SULENTIC,J.W., ARP,H., LORRE,J., 1979, AP.J., 233,
        44

856   TAMURA,S., HASEGAWA,M., 1979, PUB.ASTR.SOC. JAPAN,
        31, 329

857   TARENGHI,M., TIFFT,W.G., CHINCARINI,G., ROOD,H.J.,
        THOMPSON,L.A.,1979, AP.J. 234, 793.

858   TARTER,J.C., WRIGHT,M.C.H., 1979, ASTRON.AND AP.,
        76, 127

859   THUAN,T.X., MARTIN,G.E., 1979, AP.J., 232, L11

860   THUAN,T.X., SEITZER,P.O., 1979, AP.J., 231, 327

861   VAN DEN HULST,J.M., 1979, ASTRON.AND AP., 71, 131

862   VAN DER HULST,J.M., HUCHTMEIER,W.K., 1979, ASTRON.
        AND AP., 78, 82

863   VENNIK,J., 1979, PUBL. TARTU ASTR.OBS., 47, 163

864   WALSH,D., WILLS,B.J., WILLS,D., 1979, MNRAS, 189,
        667.

865   WEBSTER,B.L., GOSS,W.M., HAWARDEN,T.G., LONGMORE,A.J.,
        MEBOLD,U., 1979, MNRAS, 186, 31

866   WEST,R.M., 1979, ASTRON.AND AP., 71, 262

867   WEST,R.M., 1979, IAU CIRC. N.3415

868   WILLS,B.J., WILLS,D., 1979, AP.J. SUPPL., 41, 689

869   WILSON,A.S., PENSTON,M.V., 1979, AP.J., 232, 389

870   WRIGHT,A.E., PETERSON,B.A., JAUNCEY,D.L., CONDON,J.J.,
        1979, AP.J., 229, 73

871   ZASOV,A.V., KARACHENTSEV,I.D., 1979, SOVIET ASTRON.
        LETT., 5, 126

872   AARONSON,M., MOULD,J., HUCHRA,J., SULLIVAN,W.T.III,
        SCHOMMER,R.A., BOTHUN,G.D., 1980, AP.J., 239, 12

873   AFANASIEV,V.L., KARACHENTSEV,I.D., ARKHIPOVA,V.P.,
        DOSTAL,V.A., METLOV, V.G., 1980, ASRON.AND AP., 91
        302

874   AFANASIEV,V.L., LIPOVETSKY,V.A., MARKARIAN,B.E.,
        STEPANIAN,J.A., 1980, ASTROFIZIKA, 16, 193
        (ASTROPHYSICS, 16, 119)

875   ALLEN,R.J., SULLIVAN,W.T.III, 1980, ASTRON.AND AP.,
        1980, 84, 181

876   APPENZELLER,I., MOLLENHOFF,C., 1980, ASTRON.AND AP.,
        81, 54

877   ARP,H., 1980, AP.J., 236, 63

878   ARP,H., 1980, AP.J., 239, 469

879   ARP,H., 1980, AP.J., 240, 415

880   BALDWIN,J.A., CARSWELL,R.F., WAMPLER,E.J., SMITH,H.E.,
        BURBIDGE,E.M., BOKSENBERG,A., 1980, AP.J., 236, 388

881   BLACKMAN,C.P., 1980, MNRAS, 190, 459

882   BLACKMAN,C.P., 1980, MNRAS, 191, 123

883   BOSMA,A., CASINI,C., HEIDMANN,J., VAN DER HULST,J.M.,
        VAN WOERDEN,H., 1980, ASTRON. AND AP., 89, 345

884   BOTTINELLI,L., GOUGUENHEIM,L., 1980, ASTRON.AND AP.,
        88, 108

885   BOTTINELLI,L., GOUGUENHEIM,L., PATUREL,G., 1980,
        ASTRON.AND AP., 88, 32

886   BRIGGS,F.H., WOLFE,A.M., KRUMM,N., SALPETER,E.E.,
        1980, AP.J., 238, 510

887   BRIGGS,S.A., BOKSENBERG,A., CARSWELL,R.F., SARGENT.W.L.W.,
        1980, MNRAS, 191, 665

888   CAROZZI-MEYSSONIER,N., 1980, ASTRON.AND AP., 92, 189

889   CHAMARAUX,P., BALKOWSKI,C., GERARD,E., 1980, ASTRON.
        AND AP., 83, 38

890   COMBES,F., FOY,F.C., GOTTESMAN,S.T., WELIACEW,L.,
        1980, ASTRON.AND AP., 84, 85

891   DAVIES,R.D., DAVIDSON,G.P., JOHNSON,S.C., 1980,
        MNRAS, 191, 253

892   DAVOUST,E., DE VAUCOULEURS,G., 1980, AP.J., 242, 30

893   DICKEL,J.R., ROOD,H.J., 1980, A.J., 85, 1003

894   FAIRALL,A.P., 1980, MNASSA, 39, 55

895   FAIRALL,A.P., 1980, MNRAS, 191, 391

896   FAIRALL,A.P., 1980, MNRAS, 192, 389

897   GOSS,W.M., DANZIGER,I.J., FOSBURY,R.A.E., BOKSENBERG,A.,
        1980, MNRAS, 190, 23P

898   GOTTESMAN,S.T., 1980, A.J., 85, 824

899   GRUEFF,G., VIGOTTI,M., SPINRAD,H., 1980, ASTRON.AND
        AP., 86, 50

900   HART,L., DAVIES,R.D., JOHNSON,S.C., 1980, MNRAS,
        191, 269

901   HAYNES,M.P., GIOVANELLI,R., 1980, AP.J., 240, L87

902   HINTZEN,P., 1980, A.J., 85, 626

903   HINTZEN,P., SCOTT,J.S., 1980, AP.J., 239, 765

904   HUA,C.T., DONAS,J., DOAN,N.H., 1980, ASTRON.AND AP.,

90, 8

905   HUCHTMEIER,W.K., SEIRADAKIS,J.H., MATERNE,J., 1980,
       ASTRON.AND AP., 91, 341

906   HUCHTMEIER,W.K., SEIRADAKIS,J.H., TAMMANN,G.A., 1980,
       ASTRON.AND AP. 89, 95

907   JONES,J.E., JONES,B.J.T., 1980, MNRAS, 191, 685

908   KARACHENTSEV,I.D., 1980, AP.J. SUPPL., 44, 137

909   KARACHENTSEV,I.D., KOPYLOV,A.I., 1980, MNRAS, 192,
       109

910   KELTON,P.W., 1980, A.J., 85, 89

911   KIRSHNER,R.P., MALUMUTH,E.M., 1980, AP.J., 236, 366

912   KRISS,G.A., CANIZARES,C.R., MC CLINTOK,J.E.,
       FEIGELSON,E.D., 1980, AP.J., 235, L61

913   KRISTIAN,J., SANDAGE,A., WESTPHAL,J., 1980, QUOTED
       IN REF.946

914   KRUMM,N., SALPETER,E.E., 1980, A.J., 85, 1312

915   KYAZUMOV,G.A., 1980, SOVIET ASTRON. LETT., 6, 220

916   KYAZUMOV,G.A., BARABANOV,A.V., 1980, SOVIET ASTRON.
       LETT., 6, 181

917   LAUSTSEN,S., WEST,R.M., 1980, J. ASTROPHYS. ASTRON.,
       1, 177

918   LEQUEUX,J., VIALLEFOND,F., 1980, ASTRON.AND AP.,
       91, 269

919   MARCELIN,M., COMTE,G., COURTES,G., GEORGELIN,Y.P.,
       1980, PASP, 92, 38

920   MARKARIAN,B.E., LIPOVETSKY,V.A., STEPANIAN,J.A.,
       1980, ASTROFIZIKA, 16, 609 (ASTROPHYSICS, 16, 353)

921   MARKARIAN,B.E., LIPOVETSKY,V.A., STEPANIAN,J.A.,
       1980, ASTROFIZIKA, 16, 5 (ASTROPHYSICS, 16, 1)

922   METLOV,V.G., 1980, SOVIET ASTRON. LETT., 6, 110

923   NEWTON,K., 1980, MNRAS, 191, 169

924   PETERSON,C.J., 1980, A.J., 85, 226

925   PETERSON,C.J., 1980, PASP, 92, 397

926   PETROSYAN,A.R., SAAKIAN,K.A., KHACHIKIAN,E.E., 1980,
        ASTROFIZIKA, 16, 589

927   PETROSYAN,A.R., SAAKYAN,K.A., KHACHIKIAN,E.E., 1980,
        ASTROFIZIKA, 16, 621 (ASTROPHYSICS, 16, 360)

928   PETROSYAN,A.R., SAAKYAN,K.A., KHACHIKYAN,E.E., 1980,
        SOVIET ASTRON. LETT., 6, 144

929   PETROSYAN,A.R., SAAKYAN,K.A., KHACHIKYAN,E.E., 1980,
        SOVIET ASTRON. LETT. 6, 288

930   PHILLIPS,M.M., 1980, AP.J., 236, L45

931   PINEDA,F.J., DELVAILLE,J.P., GRINDLAY,J.E., SCHNOPPER,H.W.,
        1980, AP.J., 237, 414

932   PRABHU,T.P., 1980, J.ASTROPHYS.ASTR., 1, 129

933   REAKES,M., 1980, MNRAS, 192, 297

934   ROTS,A.H., 1980, ASTRON.AND AP.SUPPL., 41, 189

935   RUBIN,V.C., 1980, AP.J., 238, 808

936   RUBIN,V.C., FORD,W.K.JR., THONNARD,N., 1980, AP.J.,
        238, 471

937   RUBIN,V.C., PETERSON,C.J., FORD,W.K.JR., 1980, AP.J.,
        239, 50

938   SCHECHTER,P.L., 1980, A.J., 85, 801

939   SCHOMMER, SULLIVAN, BOTHUM, 1980, QUOTED IN REF.872

940   SCHWARTZ,D.A., DAVIS,M., DOXSEY,R.E., GRIFFITHS,R.E.,
        HUCHRA,J., JOHNSTON,M.D., MUSHOTZKY,R.F., SWANK,J.J.,
        TONRY,J., 1980, AP.J., 238, L53

941   SCHWEIZER,F., 1980, AP.J., 237, 303

942   SHANE,W.W., 1980, ASTRON.AND AP., 82, 314

943   SHANE,W.W., KRUMM,N., NORMAN,C., 1980, QUOTED IN
        REF.556

944   SHOSTAK,G.S., ALLEN,R.J., 1980, ASTRON.AND AP., 81,
        167

945   SHUDER,J.M., 1980, AP.J., 240, 32

946   SMITH,H.E., SPINRAD,H., 1980, PASP, 92, 553

947   SPINRAD,H., STAUFFER,J., BUTCHER,H., 1980, QUOTED
         IN REF.946

948   STAUFFER,J., 1980, QUOTED IN REF.946

949   STAUFFER,J., 1980, QUOTED IN REF.902

950   STAUFFER,J., SPINRAD,H., 1980, AP.J., 235, 347

951   STOCKTON,A., BERTOLA,F., 1980, AP.J., 235, 37

952   STOUGHTON,R., OSTERBROCK,D.E., 1980, PASP, 92, 117

953   SULENTIC,J.W., 1980, AP.J., 241, 67

954   SULENTIC,J.W., 1980, ASTRON.AND AP., 88, 94

955   SULLIVAN,W.T.,III,SCHOMMER,R.A.,BOTHUM,G.D.,BATES,B.A.,
         1980, QUOTED IN REF.872

956   TULLY,R.B., 1980, AP.J., 237, 390

957   TULLY,R.B., 1980, QUOTED IN REF.907

958   ULRICH,M.H., BUTCHER,H., MAYER,D.L., 1980, NATURE,
         288, 459

959   ULRICH,M.H., PEQUIGNOT,D., 1980, AP.J., 238, 45

960   VAN ALBADA,G.D., 1980, ASTRON.AND AP., 90, 123

961   VIALLEFOND,F., ALLEN,R.J., DE BOER,J.A., 1980,
         ASTRON.AND AP., 82, 207

962   VIALLEFOND,F., ALLEN,R.J., DE BOER,J.A., 1980,
         ASTRON.AND AP., 82, 207

963   VORONTSOV VELYAMINOV,B.A., DOSTAL,V.A., METLOV,V.G.,
         1980, SOVIET ASTRON. LETT., 6, 217

964   VORONTSOV VELYAMINOV,B.A., METLOV,V.G., 1980, SOVIET
         ASTRON. LETT., 6, 109

965   WARD,M.J., PENSTON,M.V., BLADES,J.C., TURTLE,A.J.,
         1980, MNRAS, 193, 563

966   WESTIN,A.M., 1980, ASTRON.AND AP., 89, L11

967   WILKERSON,M.S., 1980, AP.J., 240, L115

CHAPTER 6

References in Alphabetical Order

872   AARONSON,M., MOULD,J., HUCHRA,J., SULLIVAN,W.T.III, SCHOMMER,R.A., BOTHUN,G.D., 1980, AP.J., 239, 12

674   ABELL,G.O., EASTMOND,T.S., JENNER,D.C., 1978, AP.J., 221, L1

602   ADAMS,M.T., RUDNICK,L., 1977, A.J., 82, 857

304   ADAMS,T.F., WEEDMAN,D.W., 1972, AP.J., 173, L109

873   AFANASIEV,V.L., KARACHENTSEV,I.D., ARKHIPOVA,V.P., DOSTAL,V.A., METLOV, V.G., 1980, ASRON.AND AP., 91, 302

768   AFANASIEV,V.L., KARACHENTSEV,I.D., LIPOVETSKY,V.A., 1979, ASTRON. NACHR., 300, 77

473   AFANASIEV,V.L., KARACHENTSEV,I.D., NOTNI,P., 1975, ASTRON. NACHR., 296, 233

874   AFANASIEV,V.L., LIPOVETSKY,V.A., MARKARIAN,B.E., STEPANIAN,J.A., 1980, ASTROFIZIKA, 16, 193 (ASTROPHYSICS, 16, 119)

474   AGUERO,E., CARRANZA,G.J., 1975, OBSERVATORY, 95, 179

603   AGUERO,E., CARRANZA,G.J., 1977, OBSERVATORY, 97, 241

675   ALLEN,D.A., LONGMORE,A.J., HAWARDEN,T.G., CANNON,R.D., ALLEN,C.J., 1978, MNRAS, 184, 303

532   ALLEN,D.A., WRIGHT,A.E., GOSS,W.M., 1976, MNRAS, 177, 91

230   ALLEN,R.J., 1970, ASTRON.AND AP., 7, 330

356   ALLEN,R.J., 1973, QUOTED IN REF.468

267   ALLEN,R.J., DARCHY,B.F., LAUQUE',R., 1971, ASTRON. AND AP., 10, 198

418   ALLEN,R.J., GOSS,W.M., SANCISI,R., SULLIVAN,W.T.III, VAN WOERDEN,H., 1974, IAU SYMP. N. 58, P.425

769   ALLEN,R.J., SHOSTAK,G.S., 1979, ASTRON.AND AP.SUPPL., 35, 163

875   ALLEN,R.J., SULLIVAN,W.T.III, 1980, ASTRON.AND AP., 1980, 84, 181

676   ALLEN,R.J., VAN DER HULST,J.M., GOSS,W.M., HUCHTMEIER,W., 1978, ASTRON. AND AP., 64, 359

193   ALLOIN,D., ANDRILLAT,Y., 1969, COMPTES RENDUS AC.SCI. SER.B, 268, 139

677   ALLSOPP,N.J., 1978, MNRAS, 184, 397

770   ALLSOPP,N.J., 1979, MNRAS, 186, 343

771   ALLSOPP,N.J., 1979, MNRAS, 187, 537

772   ALLSOPP,N.J., 1979, MNRAS, 188, 371

773   ALLSOPP,N.J., 1979, MNRAS, 188, 765

357   ANDERSON,K.S., 1973, AP.J., 182, 369

419   ANDERSON,K.S., 1974, AP.J., 189, 195

194   ANDERSON,K.S., KRAFT,R.P., 1969, AP.J., 158, 859

167   ANDRILLAT,Y., 1968, A.J., 73, 862

140   ANDRILLAT,Y., SOUFFRIN,S., 1967, COMPTES RENDUS AC.
        SCI. PARIS, SER.B, 264, 89

604   ANDRILLAT,Y., SWINGS,J.P., 1977, ASTROPHYS. LETT.,
        18, 151

876   APPENZELLER,I., MOLLENHOFF,C., 1980, ASTRON.AND AP.,
        81, 54

231   ARAKELYAN,M.A., DIBAI,E.A., ESIPOV,V.F., 1970,
        ASTROFIZIKA, 6, 39 (ASTROPHYSICS, 6, 14)

268   ARAKELYAN,M.A., DIBAI,E.A., ESIPOV,V.F., 1971,
        ASTROFIZIKA, 7, 177 (ASTROPHYSICS, 7, 102)

305   ARAKELYAN,M.A., DIBAI,E.A., ESIPOV,V.F., 1972,
        ASTR.CIRC.(USSR), 717, 7

307   ARAKELYAN,M.A., DIBAI,E.A., ESIPOV,V.F., 1972,
        ASTROFIZIKA, 8, 177 (ASTROPHYSICS, 8, 106)

306   ARAKELYAN,M.A., DIBAI,E.A., ESIPOV,V.F., 1972,
        ASTROFIZIKA, 8, 329 (ASTROPHYSICS, 8, 197)

308   ARAKELYAN,M.A., DIBAI,E.A., ESIPOV,V.F., 1972,
        ASTROFIZIKA, 8, 33 (ASTROPHYSICS, 8, 17)

358   ARAKELYAN,M.A., DIBAI,E.A., ESIPOV,V.F., 1973,
        ASTROFIZIKA, 9, 319 (ASTROPHYSICS, 9, 180)

359   ARAKELYAN,M.A., DIBAI,E.A., ESIPOV,V.F., 1973,
        ASTROFIZIKA, 9, 325 (ASTROPHYSICS, 9, 183)

476   ARAKELYAN,M.A., DIBAI,E.A., ESIPOV,V.F., 1975,
        ASTROFIZIKA, 11, 15 (ASTROPHYSICS, 11, 8)

475   ARAKELYAN,M.A., DIBAI,E.A., ESIPOV,V.F., 1975,
          ASTROFIZIKA, 11, 377 (ASTROPHYSICS, 11, 254)

533   ARAKELYAN,M.A., DIBAI,E.A., ESIPOV,V.F., 1976,
          ASTROFIZIKA, 12, 195 (ASTROPHYSICS, 12, 122)

534   ARAKELYAN,M.A., DIBAI,E.A., ESIPOV,V.F., 1976,
          ASTROFIZIKA, 12, 683 (ASTROPHYSICS, 12, 456)

232   ARAKELYAN,M.A., DIBAI,E.A., ESIPOV,V.F., MARKARIAN,B.E.,
          1970, ASTR. CIRC.(USSR), N.586, 1

233   ARAKELYAN,M.A., DIBAI,E.A., ESIPOV,V.F., MARKARYAN,B.E.,
          1970, ASTROFIZIKA, 6, 357 (ASTROPHYSICS, 6, 189)

100   ARGYLE,E., 1965, AP.J., 141, 750

774   ARKHIPOVA,V.P., ESIPOV,V.F., 1979, SOVIET ASTRON.
          LETT., 5, 140

535   ARKHIPOVA,V.P., ESIPOV,V.F., SAVEL'EVA,M.V., 1976,
          SOVIET ASTRON. LETT., 20, 521

101   ARP,H., 1965, AP.J., 142, 402

141   ARP,H., 1967, AP.J., 148, 321

168   ARP,H., 1968, AP.J., 152, 1101

169   ARP,H., 1968, PASP 80, 129

195   ARP,H., 1969, ASTRON.AND AP., 3, 418

234   ARP,H., 1970, ASTROPHYS. LETT., 5, 257

269   ARP,H., 1971, ASTROPHYS. LETT., 7, 221

270   ARP,H., 1971, ASTROPHYS. LETT., 9, 1

360   ARP,H., 1973, AP.J., 185, 797

536   ARP,H., 1976, AP.J., 210, L59

605   ARP,H., 1977, QUOTED IN REF.614

877   ARP,H., 1980, AP.J., 236, 63

878   ARP,H., 1980, AP.J., 239, 469

879   ARP,H., 1980, AP.J., 240, 415

235   ARP,H., BERTOLA,F., 1970, ASTROPHYS. LETT., 6, 65

309   ARP,H., BURBIDGE,E.M., MACKAY,C.D., STRITTMATTER,P.A.,
          1972, AP.J., 171, L41

420   ARP,H., HEIDMANN,J., KHACHIKYAN,E.E., 1974,
          ASTROFIZIKA, 10, 7 (ASTROPHYSICS, 10, 2)

421   ARP,H., KHACHIKIAN,E.E., 1974, ASTROFIZIKA, 10, 173,
          (ASTROPHYSICS, 10, 106)

422   ARP,H., KHACHIKIAN,E.E., ANDREASYAN,N.K., 1974,
          ASTROFIZIKA, 10, 625 (ASTROPHYSICS, 10, 399)

361   ARP,H., KHACHIKYAN,E.E., 1973, ASTROFIZIKA, 9, 509
          (ASTROPHYSICS, 9, 308)

477   ARP,H., O'CONNELL,R.W., 1975, AP.J., 197, 291

423   ARP,H., SARGENT,W.L.W., KHACHIKYAN,E.E., ANDREASYAN,N.K.,
          1974, ASTROFIZIKA, 10, 298 (ASTROPHYSICS, 10, 179)

236   ARP,H., VISVANATHAN,N., 1970, ASTROPHYS. LETT., 5, 73

9   BAADE,W., MINKOWSKY,R., 1954, AP.J., 119, 206

678   BAAN,W.A., HASCHICK,A.D., GREENFIELD,P.E., 1978,
          AP.J., 222, L7

537   BAARS,J.W.M., WENDKER,H.J., 1976, ASTRON.AND AP.,
          48, 405

4   BABCOCK,H.W., 1939, LICK OBS.BULL. N.498, P.41

196   BACHALL,J.M., SCHMIDT,M., GUNN,J.E., 1969, AP.J.,
          157, L77

606   BACHALL,N.A., SARGENT,W.L.W., 1977, AP.J., 217, L19

880   BALDWIN,J.A., CARSWELL,R.F., WAMPLER,E.J., SMITH,H.E.,
          BURBIDGE,E.M., BOKSENBERG,A., 1980, AP.J., 236, 388

607   BALDWIN,J.A., WAMPLER,E.J., BURBIDGE,E.M., O'DELL,S.L.,
          SMITH,H.E., HAZARD,C.,NORDISIECK,K.H., POOLEY,G.,
          STEIN,W.A., 1977, AP.J., 215, 408

538   BALICK,B., FABER,S.M., GALLAGHER,J.S., 1976, AP.J.,
          209, 710

775   BALKOWSKI,C., 1979, ASTRON.AND AP., 78, 190

501

424   BALKOWSKI,C., BOTTINELLI, L., CHAMARAUX,P., GOUGUENHEIM,L.,
        HEIDMANN,J., 1974, ASTRON.AND AP., 34, 43

362   BALKOWSKI,C., BOTTINELLI,L., CHAMARAUX.P., GOUGUENHEIM,L.,
        HEIDMANN,J., 1973, ASTRON.AND AP., 25, 319

310   BALKOWSKI,C., BOTTINELLI,L., GOUGUENHEIM,L., HEIDMANN,J.,
        1972, ASTRON.AND AP., 21, 303

363   BALKOWSKI,C., BOTTINELLI,L., GOUGUENHEIM,L., HEIDMANN,J.,
        1973, ASTRON.AND AP., 23, 139

679   BALKOWSKI,C., CHAMARAUX,P., WELIACHEW,L., 1978,
        ASTRON.AND AP., 69, 263

425   BARBIERI,C., BERTOLA,F., DI TULLIO,G., 1974, ASTRON.
        AND AP., 35, 463

776   BARBIERI,C., CASINI,C., HEIDMANN,J., DI SEREGO,S.,
        ZAMBON,M., 1979, ASTRON.AND AP.SUPPL., 37, 559

608   BARBIERI,C., DI SEREGO ALIGHIERI,S., ZAMBON,M., 1977,
        ASTRON.AND AP., 57, 353

539   BARBIERI,C., ROMANO,G., 1976, ASTRON.AND AP., 50, 15

 28   BARBON,R., 1960, MEM.SOC.ASTR.ITALIANA, 40, 211

197   BARBON,R., 1969, MEM.SOC.ASTR.ITALIANA, 40, 559

237   BARBON,R., 1970, MEM.SOC.ASTR.ITALIANA, 41, 129

311   BARBON,R., 1972, MEM.SOC.ASTR.ITALIANA, 43, 313

426   BARBON,R., CAPACCIOLI,M., 1974, ASTRON.AND AP., 35,
        151

540   BARBON,R., CAPACCIOLI,M., 1976, ASTRON.AND AP., 49,

777   BARBON,R., CAPACCIOLI,M., TIFFT,W.G., 1979, ASTRON.
        AND AP.SUPPL., 36, 129

312   BAUTZ,L.P., 1972, A.J., 77, 331

198   BEALE,J.S., DAVIES,R.D., 1969, NATURE, 221, 531

778   BECK,S.C., LACY,J.H., GEBALLE,T.R., 1979, AP.J., 231,
        28

478   BENVENUTI,P., CAPACCIOLI,M., D'ODORICO,S., 1975,
        ASTRON.AND AP., 41, 91

542   BENVENUTI,P., CAPACCIOLI,M., D'ODORICO,S., 1976,

ASTRON.AND AP., 53, 141

543   BENVENUTI,P., D'ODORICO,S., VETTOLANI,G., 1976,
        PROC. THIRD EUROPEAN ASTRON. MEETING, TBILISI, P. 393

680   BERGWALL,N.A.S., EKMAN,A.B.G., LAUBERTS,A., WESTERLUND,B.E.,
        BORCHKHADZE,T.M., BREYSACHER,J., LAUSTSEN,S.,
        MULLER,A.B., SCHUSTER,H.E., SURDEJ,J., WEST,R.M.,
        1978, ASTRON.AND AP. SUPPL., 33, 243

102   BERTOLA,F., 1965, ANN. ASTROPHYS., 28, 574

103   BERTOLA,F., 1965, CONTR. OSS. ASTROF. ASIAGO N.172,
        P. 35.

125   BERTOLA,F., 1966, MEM. SOC. ASTR. ITALIANA, 37, 433

142   BERTOLA,F., 1967, ATTI DEL X CONVEGNO DELLA SOC.ASTR.
        ITALIANA,P. 209

170   BERTOLA,F., 1968, MEM. SOC. ASTR. ITALIANA, 39, 453

479   BERTOLA,F., CAPACCIOLI,M., 1975, AP.J., 200, 439

609   BERTOLA,F., CAPACCIOLI,M., 1977, AP.J., 211, 697

681   BERTOLA,F., CAPACCIOLI,M., 1978, AP.J., 219, 404

364   BERTOLA,F., D'ODORICO,S., 1973, ASTROPHYS. LETT.,
        13, 161

199   BERTOLA,F., D'ODORICO,S., FORD,W.K.JR, RUBIN,V.C.,
        1969, AP.J., 157, L27

682   BIEGING,J.H., 1978, ASTRON.AND AP., 64, 23

610   BIEGING,J.H., BIERMANN,P., 1977, ASTRON.AND AP., 60,
        361

779   BIERMANN,P., CLARKE,J.N., FRICKE,K.J., 1979, ASTRON.
        AND AP., 75, 19

780   BIERMANN,P., CLARKE,J.N., FRICKE,K.J., 1979, ASTRON.
        AND AP., 75, 7

611   BLACKMAN,C.P., 1977, MNRAS, 178, 15

881   BLACKMAN,C.P., 1980, MNRAS, 190, 459

882   BLACKMAN,C.P., 1980, MNRAS, 191, 123

313   BOHUSKI,T.J., BURBIDGE,E.M., BURBIDGE,G.R., SMITH,M.G.,

1972, AP.J., 175, 329

683   BOHUSKI,T.J., FAIRALL,A.P., WEEDMAN,D.W., 1978, AP.J.,
       221, 776

612   BOLTON,J.G., SAVAGE,A., 1977, AUSTRAL.J.PHYS.SUPPL.,
       44, 21

365   BOND,H.E., SARGENT,W.L.W., 1973, AP.J., 185, L109

427   BOND,H.E., TIFFT,W.G., 1974, PASP, 86, 981

428   BORCHKHADZE,T.M., 1974, ASTROFIZIKA, 10, 493,
       (ASTROPHYSICS, 10, 311)

613   BORCHKHADZE,T.M., BREYSACHER,J., LAUSTSEN,S., SCHUSTER,H.E.,
       WEST, R.M., 1977, ASTRON.AND AP. SUPPL., 30, 35

883   BOSMA,A., CASINI,C., HEIDMANN,J., VAN DER HULST,J.M.,
       VAN WOERDEN,H., 1980, ASTRON. AND AP., 89, 345

614   BOSMA,A., EKERS,R.D., LEQUEUX,J., 1977, ASTRON.AND
       AP., 57, 97

615   BOSMA,A., VAN DER HULST,J.M., SULLIVAN,W.T.III, 1977,
       ASTRON.AND AP., 57, 373

271   BOTTINELLI,L., CHAMARAUX,P., GERARD,E., GOUGUENHEIM,L.,
       HEIDMANN,J., KAZES,I., LAUQUE',R., 1971, ASTRON.
       AND AP., 12, 264

366   BOTTINELLI,L., CHAMARAUX,P., GOUGUENHEIM,L., HEIDMANN,J.,
       1973, ASTRON.AND AP., 29, 217

238   BOTTINELLI,L., CHAMARAUX,P., GOUGUENHEIM,L., LAQUE',R.,
       1970, ASTRON.AND AP., 6, 453

684   BOTTINELLI,L., DUFLOT,R., GOUGUENHEIM,L., 1978,
       ASTRON.AND AP., 63, 363

480   BOTTINELLI,L., DUFLOT,R., GOUGUENHEIM,L., HEIDMANN,J.,
       1975, ASTRON.AND AP., 41, 61

367   BOTTINELLI,L., GOUGUENHEIM,L., 1973, ASTRON.AND AP.,
       29, 425

481   BOTTINELLI,L., GOUGUENHEIM,L., 1975, ASTRON.AND AP.,
       39, 341

544   BOTTINELLI,L., GOUGUENHEIM,L., 1976, ASTRON.AND AP.,
       47, 381

616   BOTTINELLI,L., GOUGUENHEIM,L., 1977, ASTRON.AND AP.,
       54, 641

617   BOTTINELLI,L., GOUGUENHEIM,L., 1977, ASTRON.AND AP.,
       60, L23

685   BOTTINELLI,L., GOUGUENHEIM,L., 1978, ASTRON.AND AP.,
       64, L3

782   BOTTINELLI,L., GOUGUENHEIM,L., 1979, ASTRON.AND AP.,
         74, 172

781   BOTTINELLI,L., GOUGUENHEIM,L., 1979, ASTRON.AND AP.,
         76, 176

884   BOTTINELLI,L., GOUGUENHEIM,L., 1980, ASTRON.AND AP.,
         88, 108

314   BOTTINELLI,L., GOUGUENHEIM,L., HEIDMANN,J., 1972,
         ASTRON.AND AP., 17, 445

368   BOTTINELLI,L., GOUGUENHEIM,L., HEIDMANN,J., 1973,
         ASTRON.AND AP., 22, 281

171   BOTTINELLI,L., GOUGUENHEIM,L., HEIDMANN,J., HEIDMANN,N.,
         1968, ANN. ASTROPHYS., 31, 205

126   BOTTINELLI,L., GOUGUENHEIM,L., HEIDMANN,J., HEIDMANN,N.,
         WELIACEW,L., 1966, COMPTES RENDUS AC.SCI. PARIS,
         SER.B, 263, 223

885   BUTTINELLI,L., GOUGUENHEIM,L., PATUREL,G., 1980,
         ASTRON.AND AP., 88, 32

239   BOULESTEIX,J., DUBOUT-CRILLON,R., MONNET,G., 1970,
         ASTRON.AND AP., 8, 204

315   BRACCESI,A., FORMIGGINI,L., GIOIA,I., SARGENT,W.L.W.,
         1972, PASP, 84, 592

172   BRACCESI,A., LYNDS,R., SANDAGE,A., 1968, AP.J., 152,
         L105

686   BRADT,H.V., BURKE,B.F., CANIZARES,C.R., GREENFIELD,P.E.,
         KELLY,R.L., MCCLINTOCK,J.E., VAN PARADIJS,J.,
         KOSKI,A.T., 1978, AP.J., 226, L111

104   BRANDT,J.C., 1965, MNRAS, 129, 309

545   BRIDLE,A.H., FOMALONT,E.B., 1976, ASTRON.AND AP.,
         52, 107

886   BRIGGS,F.H., WOLFE,A.M., KRUMM,N., SALPETER,E.E.,
         1980, AP.J., 238, 510

887   BRIGGS,S.A., BOKSENBERG,A., CARSWELL,R.F., SARGENT,W.L.W.,
         1980, MNRAS, 191, 665

127   BRUNDAGE,W.D., KRAUS, J.D., 1966, SCIENCE, 153, 411

 75   BURBIDGE,E.M., 1963, QUOTED BY SANDAGE,A., 1963,
         AP.J., 138, 863

143   BURBIDGE,E.M., 1967, AP.J., 149, L51

240   BURBIDGE,E.M., 1970, AP.J., 160, L33

18   BURBIDGE,E.M., BURBIDGE,G.R., 1959, AP.J., 129, 271

19   BURBIDGE,E.M., BURBIDGE,G.R., 1959, AP.J., 130, 12

29   BURBIDGE,E.M., BURBIDGE,G.R., 1960, AP.J., 132, 30

40   BURBIDGE,E.M., BURBIDGE,G.R., 1961, A.J., 66, 541

41   BURBIDGE,E.M., BURBIDGE,G.R., 1961, AP.J., 133, 726

42   BURBIDGE,E.M., BURBIDGE,G.R., 1961, AP.J., 134, 244

43   BURBIDGE,E.M., BURBIDGE,G.R., 1961, AP.J., 134, 248

61   BURBIDGE,E.M., BURBIDGE,G.R., 1962, AP.J., 135, 366

62   BURBIDGE,E.M., BURBIDGE,G.R., 1962, NATURE, 194, 367

76   BURBIDGE,E.M., BURBIDGE,G.R., 1963, AP.J., 138, 1306

86   BURBIDGE,E.M., BURBIDGE,G.R., 1964, AP.J., 140, 1307

87   BURBIDGE,E.M., BURBIDGE,G.R., 1964, AP.J., 140, 1445

105   BURBIDGE,E.M., BURBIDGE,G.R., 1965, AP.J., 142, 1351

106   BURBIDGE,E.M., BURBIDGE,G.R., 1965, AP.J., 142, 634

128   BURBIDGE,E.M., BURBIDGE,G.R., 1966, AP.J., 145, 661

173   BURBIDGE,E.M., BURBIDGE,G.R., 1968, AP.J., 151, 99

174   BURBIDGE,E.M., BURBIDGE,G.R., 1968, AP.J., 154, 857

316   BURBIDGE,E.M., BURBIDGE,G.R., 1972, AP.J., 171, 253

317   BURBIDGE,E.M., BURBIDGE,G.R., 1972, AP.J., 172, 37

89   BURBIDGE,E.M., BURBIDGE,G.R., CRAMPIN,D.J., RUBIN,V.C.,
        PRENDERGAST,K.H., 1964, AP.J., 139, 539

88   BURBIDGE,E.M., BURBIDGE,G.R., CRAMPIN,D.J., RUBIN,V.C.,

PRENDERGAST,K.H., 1964, AP.J., 139, 1058

77   BURBIDGE,E.M., BURBIDGE,G.R., HOYLE,F., 1963, AP.J.,
        138, 873

20   BURBIDGE,E.M., BURBIDGE,G.R., PRENDERGAST,K.H., 1959,
        AP.J., 130, 26

21   BURBIDGE,E.M., BURBIDGE,G.R., PRENDERGAST,K.H., 1959,
        AP.J., 130, 739

31   BURBIDGE,E.M., BURBIDGE,G.R., PRENDERGAST,K.H., 1960,
        AP.J., 131, 282

30   BURBIDGE,E.M., BURBIDGE,G.R., PRENDERGAST,K.H., 1960,
        AP.J., 131, 549

33   BURBIDGE,E.M., BURBIDGE,G.R., PRENDERGAST,K.H., 1960,
        AP.J., 132, 640

34   BURBIDGE,E.M., BURBIDGE,G.R., PRENDERGAST,K.H., 1960,
        AP.J., 132, 654

32   BURBIDGE,E.M., BURBIDGE,G.R., PRENDERGAST,K.H., 1960,
        AP.J., 132, 661

46   BURBIDGE,E.M., BURBIDGE,G.R., PRENDERGAST,K.H., 1961,
        AP.J., 133, 814

47   BURBIDGE,E.M., BURBIDGE,G.R., PRENDERGAST,K.H., 1961,
        AP.J., 134, 232

45   BURBIDGE,E.M., BURBIDGE,G.R., PRENDERGAST,K.H., 1961,
        AP.J., 134, 237

44   BURBIDGE,E.M., BURBIDGE,G.R., PRENDERGAST,K.H., 1961,
        AP.J., 134, 874

64   BURBIDGE,E.M., BURBIDGE,G.R., PRENDERGAST,K.H., 1962,
        AP.J., 136, 119

65   BURBIDGE,E.M., BURBIDGE,G.R., PRENDERGAST,K.H., 1962,
        AP.J., 136, 128

66   BURBIDGE,E.M., BURBIDGE,G.R., PRENDERGAST,K.H., 1962,
        AP.J., 136, 339

63   BURBIDGE,E.M., BURBIDGE,G.R., PRENDERGAST,K.H., 1962,
        AP.J., 136, 704

80   BURBIDGE,E.M., BURBIDGE,G.R., PRENDERGAST,K.H., 1963,
        AP.J., 137, 1022

78   BURBIDGE,E.M., BURBIDGE,G.R., PRENDERGAST,K.H., 1963,
        AP.J., 137, 376

79   BURBIDGE,E.M., BURBIDGE,G.R., PRENDERGAST,K.H., 1963,
        AP.J., 138, 375

90   BURBIDGE,E.M., BURBIDGE,G.R., PRENDERGAST,K.H., 1964,
        AP.J., 140, 1617

108   BURBIDGE,E.M., BURBIDGE,G.R., PRENDERGAST,K.H., 1965,
        AP.J., 142, 154

107   BURBIDGE,E.M., BURBIDGE,G.R., PRENDERGAST,K.H., 1965,
        AP.J., 142, 641

109   BURBIDGE,E.M., BURBIDGE,G.R., PRENDERGAST,K.H., 1965,
        AP.J., 142, 649

 91   BURBIDGE,E.M., BURBIDGE,G.R., RUBIN,V.C., 1964, AP.J.,
        140, 942

 92   BURBIDGE,E.M., BURBIDGE,G.R., RUBIN,V.C.,PRENDERGAST,K.H.,
        1964, AP.J., 140, 85

144   BURBIDGE,E.M., BURBIDGE,G.R., SHELTON,J.W., 1967,
        AP.J., 150, 783

272   BURBIDGE,E.M., BURBIDGE,G.R., SOLOMON,P.M., STRITTMATTER,P.A.,
        1971, AP.J. 170, 233

200   BURBIDGE,E.M., DEMOULIN,M.H., 1969, AP.J., 157, L155

201   BURBIDGE,E.M., DEMOULIN,M.H., 1969, ASTROPHYS. LETT.,
        4., 89

273   BURBIDGE,E.M., HODGE,P.M., 1971, AP.J., 166, 1

274   BURBIDGE,E.M., SARGENT,W.L.W., 1971, NUCLEI OF GALAXIES.,
        ED. O'CONNEL, PONTIF.ACC.SCI.SCRIPTA VARIA
        N. 35, P.351

482   BURBIDGE,E.M., SMITH,H.E., BURBIDGE,G.R., 1975, AP.J.,
        199, L137

687   BURBIDGE,E.M., SMITH,H.E., BURBIDGE,G.R., 1978, AP.J.,
        219, 400

318   BURBIDGE,E.M., STRITTMATTER,P.A., 1972, AP.J., 172,
        L37

319   BURBIDGE,E.M., STRITTMATTER,P.A., SMITH,H.E., SPINRAD,H.,
        1972, AP.J., 178, L43

 22   BURBIDGE,G.R., BURBIDGE,E.M., 1959, AP.J., 130, 629

783   BURBIDGE,G.R., CROWNE,A.H., 1979, AP.J. SUPPL., 40,
        583

320   BURBIDGE,G.R., O'DELL,S.L., STRITTMATTER,P.A., 1972,
        AP.J., 175, 601

 81   BURKE,B.F., TURNER,K.C., TUVE,M.A., 1963, CARNEGIE
        INSTIT. YEAR BOOK, 62, 293

 82   BURLEY,J., 1963, A.J., 68, 274

275   BURNS,W.R., ROBERTS,M.S., 1971, AP.J., 166, 265

546   CAROZZI,N., 1976, ASTRON.AND AP., 49, 425

547   CAROZZI,N., 1976, ASTRON.AND AP., 49, 431

618   CAROZZI,N., 1977, ASTRON.AND AP., 55, 261

429   CAROZZI,N., CHAMARAUX,P., DUFLOT,R., 1974, ASTRON.
         AND AP., 33, 113

430   CAROZZI,N., CHAMARAUX,P., DUFLOT-AUGARDE,R., 1974,
         ASTRON.AND AP., 30, 21

688   CAROZZI-MEYSONNIER,N., 1978, ASTRON.AND AP. SUPPL.,
         33, 411

784   CAROZZI-MEYSSONIER,N., 1979, ASTRON.AND AP. SUPPL.,
         37, 529

888   CAROZZI-MEYSSONIER,N., 1980, ASTRON.AND AP., 92, 189

689   CAROZZI-MEYSSONNIER,N., 1978, ASTRON.AND AP., 63, 415

175   CARRANZA,C.J., 1968, BOL.ASSOC.ARG.ASTRON.,N.14, P.38

431   CARRANZA,G.J., AGUERO,E., 1974, OBSERVATORY, 94, 7

619   CARRANZA,G.J., AGUERO,E.L., 1977, ASTROPYS. AND SPACE
         SCI., 47, 397

202   CARRANZA,G.J., CRILLON,R., MONNET,G., 1969, ASTRON.
         AND AP., 1, 479

145   CARRANZA,G.J.,1967, OBSERVATORY, 87, 38

785   CASINI,C., HEIDMANN,J., TARENGHI,M., 1979, ASTRON.
         AND AP., 73, 216

203   CATCHPOLE,R.M., EVANS,D.S., JONES,D.H.P., 1969, OB
         SERVATORY,89, 21

620   CESARSKY,D.A., FALGARONE,E.G., LEQUEUX,J., 1977,
         ASTRON.AND AP., 59, L5

621   CHAISSON,E.J., RODRIGUEZ,L.F., 1977, AP.J., 214, L111

622   CHAMARAUX,P., 1977, ASTRON.AND AP., 60, 67

889   CHAMARAUX,P., BALKOWSKI,C., GERARD,E., 1980, ASTRON.
         AND AP., 83, 38

241   CHAMARAUX,P., HEIDMANN,J., LAUQUE',R., 1970, ASTRON.
        AND AP., 8, 424

483   CHERIGUENE,M.F., 1975, IN LA DINAMIQUE DES GALAXIE
        S SPIRALES, COLLOQUE CNRS, 241, 439

786   CHINCARINI,G., GIOVANELLI,R., HAYNES,M.P., 1979, A.J.,
        84, 1500

369   CHINCARINI,G., HECKATHORN,H.M., 1973, PASP, 85, 568

484   CHINCARINI,G., MARTINS,D., 1975, AP.J., 196, 335

276   CHINCARINI,G., ROOD,H.J., 1971, AP.J., 168,321

321   CHINCARINI,G., ROOD,H.J., 1972, A.J., 77, 4

322   CHINCARINI,G., ROOD,H.J., 1972, A.J., 77, 448

548   CHINCARINI,G., ROOD,H.J., 1976, AP.J., 206, 30

549   CHINCARINI,G., ROOD,H.J., 1976, PASP, 88, 388

623   CHINCARINI,G., ROOD,H.J., 1977, AP.J., 214, 351

690   CHINCARINI,G., TARENGHI,M., BETTIS,C., 1978, AP.J.,
        221, 34

146   CHINCARINI,G., WALKER,M.F., 1967, AP.J., 147, 407

147   CHINCARINI,G., WALKER,M.F., 1967, AP.J., 149, 487

370   CHROMEY,F.R., 1973, ASTRON.AND AP., 29, 77

432   CHROMEY,F.R., 1974, ASTRON.AND AP., 31, 165

433   CHROMEY,F.R., 1974, ASTRON.AND AP., 37, 7

624   CIATTI,F., ROSINO,L., 1977, ASTRON.AND AP., 57, 73

787   COHEN,R.J., 1979, MNRAS, 187, 839

550   COLEMAN,G.D., HINTZEN,P., SCOTT,J.S., TARENGHI,M.,
        1976, NATURE, 262, 476

485   COLLA,G., FANTI,C., FANTI,R., GIOIA,I., LARI,C.,
        LEQUEUX,J., LUCAS,R., ULRICH,M.H., 1975, ASTRON.AND

AP. SUPPL., 20, 1

890 COMBES,F., FOY,F.C., GOTTESMAN,S.T., WELIACEW,L., 1980, ASTRON.AND AP., 84, 85

625 COMBES,F., GOTTESMAN,S.T., WELIACHEW,L., 1977, ASTRON. AND AP., 59, 181

788 COMTE,G., MONNET,G., ROSADO,M., 1979, ASTRON.AND AP., 72, 73

626 COOKE,J.A., EMERSON,D., NANDY,K., REDDISH,V.C., SMITH,M.G., 1977, MNRAS, 178, 687

551 COTTRELL,G.A., 1976, MNRAS, 174, 455

552 COTTRELL,G.A., 1976, MNRAS, 177, 463

553 CRAINE,E.R., WARNER,J.W., 1976, AP.J., 206, 359

204 CRILLON,R., MONNET,G., 1969, ASTRON.AND AP., 1, 449

205 CRILLON,R., MONNET,G., 1969, ASTRON.AND AP., 2, 1

242 D'ODORICO,S., 1970, AP.J., 160, 3

627 DANZIGER,I.J., 1977, QUOTED BY CHRISTIANSEN,W.N., ET AL., 1977, MNRAS, 181, 183

323 DANZIGER,I.J., CHROMEY,F.R., 1972, ASTROPHYS. LETT., 10, 99

628 DANZIGER,I.J., FOSBURY,R.A.E., PENSTON,M.V., 1977, MNRAS, 179, 41P

691 DANZIGER,I.J., GOSS,W.H., FRATER,R.H., 1978, MNRAS, 184, 341

434 DANZIGER,I.J., SCHUSTER,H.E., 1974, ASTRON.AND AP., 34, 301

371 DAVIES,R.D., 1973, MNRAS, 161, 25P

891 DAVIES,R.D., DAVIDSON,G.P., JOHNSON,S.C., 1980, MNRAS, 191, 253

372 DAVIES,R.D., LEWIS,B.M., 1973, MNRAS, 165, 231

692 DAVIS,M., HUCHRA,J., TONRY,J., LATHAM,D., 1978, IAU CIRC., N 3202

892 DAVOUST,E., DE VAUCOULEURS,G., 1980, AP.J., 242, 30

629   DAWE,J.A., DICKENS,R.J., PETERSON,B.A., 1977, MNRAS, 178, 675

277   DE JAGER,G., DAVIES, R.D., 1971, MNRAS, 153, 9

148   DE VAUCOULEURS,A., DE VAUCOULEURS,G., 1967, A.J., 72, 730

554   DE VAUCOULEURS,A., SHOBBROOK,R.R., STROBEL,A., 1976, A.J., 81, 219

48   DE VAUCOULEURS,G., 1961, AP.J., 133, 405

149   DE VAUCOULEURS,G., 1967, IAU SYMP., N.30, P.9

150   DE VAUCOULEURS,G., 1967, IAU SYMP., N.30, P.91

486   DE VAUCOULEURS,G., 1975, AP.J. SUPPL., 29, 193

49   DE VAUCOULEURS,G., DE VAUCOULEURS,A., 1961, MEM. RAS, 68, 69

83   DE VAUCOULEURS,G., DE VAUCOULEURS,A., 1963, AP.J., 137, 363

373   DE VAUCOULEURS,G., DE VAUCOULEURS,A., 1973, A.J., 78, 377

487   DE VAUCOULEURS,G., DE VAUCOULEURS,A., 1975, AP.J., 197, L1

488   DE VAUCOULEURS,G., DE VAUCOULEURS,A., 1975, OBSERVATORY, 95, 178

555   DE VAUCOULEURS,G., DE VAUCOULEURS,A., 1976, A.J., 81, 595

789   DE VAUCOULEURS,G., DE VAUCOULEURS,A., NIETO,J.L., 1979, A.J., 84, 1811

206   DE VENY,J.B., LYNDS,C.R., 1969, PASP, 81, 535

489   DEAN,J.F., DAVIES,R.D., 1975, MNRAS, 170, 503

207   DEHARVENG,J.M., PELLET,A., 1969, ASTRON.AND AP., 1, 208

490   DEHARVENG,J.M., PELLET,A., 1975, ASTRON.AND AP., 38, 15

243   DEHERVENG,J.M., PELLET,A., 1970, ASTRON.AND AP., 7, 210

151   DELANNOY,J., GUELIN,M., WELIASCHEW,W.L., 1967, QUOTED IN REF.183

110   DEMOULIN,M.H., 1965, COMPTES RENDUS AC.SCI. PARIS,
        SER.B, 260, 3287

111   DEMOULIN,M.H., 1965, PUB.OBS. HAUTE PROVENCE, 8, N 1

208   DEMOULIN,M.H., 1969, AP.J., 156, 325

209   DEMOULIN,M.H., 1969, AP.J., 157, 69

210   DEMOULIN,M.H., 1969, AP.J., 157, 75

211   DEMOULIN,M.H., 1969, AP.J., 157, 81

244   DEMOULIN,M.H., 1970, AP.J., 160, L79

245   DEMOULIN,M.H., BURBIDGE,E.M., 1970, AP.J., 159, 799

176   DEMOULIN,M.H., BURBIDGE,E.M., BURBIDGE,G.R., 1968,
        AP.J., 153, 31

212   DEMOULIN,M.H., TUNG CHAN,Y.W., 1969, AP.J., 156, 501

278   DENISYUK,E.K., 1971, ASTR.CIRC.(USSR), 615, 4

279   DENISYUK,E.K., 1971, ASTR.CIRC.(USSR), 621, 7

280   DENISYUK,E.K., 1971, ASTR.CIRC.(USSR), 624, 1

435   DENISYUK,E.K., 1974, ASTR.CIRC.(USSR), 809, 1

436   DENISYUK,E.K., 1974, ASTR.CIRC.(USSR), 809, 2

437   DENISYUK,E.K., BABKIN,I.G., SINYAEVA,N.V., 1974,
        ASTR.CIRC.(USSR), N. 837, 2

556   DENISYUK,E.K., DOSTAL,V.A., 1976, ASTR.CIRC.(USSR),
        N. 931, 7

374   DENISYUK,E.K., LIPOVETSKY,V.A., 1973, ASTR.CIRC.
        (USSR), 798, 2

438   DENISYUK,E.K., LIPOVETSKY,V.A., 1974, ASTROFIZIKA,
        10, 315 (ASTROPHYSICS, 10,195)

630   DENISYUK,E.K., LIPOVETSKY,V.A., 1977, SOVIET ASTRON.
        LETT., 3, 3

557   DENISYUK,E.K., LIPOVETSKY,V.A., AFANASIEV,V.L., 1976,

ASTROFIZIKA, 12, 665 (ASTROPHYSICS, 12, 442)

375   DENISYUK,E.K., PAVLOVA,N.N., 1973, ASTR.CIRC.(USSR),
       N 797, 1

790   DENNEFELD,M., LAUSTSEN,S., MATERNE,J., 1979, ASTRON.
       AND AP., 74, 123

558   DIBAI,E.A., DOROSHENKO,V.T., TEREBIZH,V.YU., 1976,
       ASTROFIZIKA, 12, 689 (ASTROPHYSICS, 12, 459)

177   DIBAI,E.A., ESIPOV,V.F., 1968, SOVIET ASTRONOMY,
       12, 561

693   DICKEL,J.R., ROOD,H.J., 1978, AP.J., 223, 391

893   DICKEL,J.R., ROOD,H.J., 1980, A.J., 85, 1003

559   DICKENS,R,J, MOSS,C., 1976, MNRAS, 174, 47

 67   DIETER,N.H., 1962, A.J., 67, 217

 68   DIETER,N.H., 1962, A.J., 67, 313

 69   DIETER,N.H., 1962, A.J., 67, 317

376   DISNEY,M.J., 1973, AP.J., 181, L55

281   DISNEY,M.J., CROMWELL,R.H., 1971, AP.J., 164, L35

377   DISNEY,M.J., ROGERS,A., 1973, QUART.J.RAS, 14, 438

439   DISNEY,M.J.,,1974, AP.J., 193, L103

491   DOROSHENKO,V.T., TEREBIZH,V.YU., 1975, ASTROFIZIKA,
       11, 631 (ASTROPHYSICS, 11, 422)

631   DOSTAL',V.A., 1977, SOVIET ASTRON. LETT., 3, 30

378   DOTTORI,H.A., FOURCADE,C.R., 1973, ASTRON.AND AP.,
       23, 405

178   DU PUY,D.L., 1968, PASP, 80, 29

246   DU PUY,D.L., 1970, A.J., 75, 1143

213   DU PUY,D.L., DE VENY,J.B., 1969, PASP, 81, 637

112   DUFLOT,R., 1965, PUB.OBS. HAUTE PROVENCE, 8, N 16

560   DUFLOT,R., 1976, ASTRON.AND AP., 48, 437

35   DUFLOT-AUGARDE,R., 1960, PUB.OBS. HAUTE PROVENCE,
        5, N.14

50   DUFLOT-AUGARDE,R., 1961, COMPTES RENDUS AC. SCI.
        PARIS, SER B, 253, 224

694   EASTMOND,T.S., ABELL,G.O., 1978, PASP, 90, 367

93   EPSTEIN,E.E., 1964, A.J., 69, 490

7   EVANS,D.S., 1952, OBSERVATORY, 72, 164

12   EVANS,D.S., 1956, MNRAS, 116, 659

84   EVANS,D.S., 1963, MNASSA, 22, 140

152   EVANS,D.S., 1967, OBSERVATORY, 87, 224

51   EVANS,D.S., HARDIG.G.A., 1961, MNASSA, 20, 64

113   EVANS,D.S., MALIN,S.R., 1965, MNASSA, 24, 32

17   EVANS,D.S., WAYMAN,P.A., 1958, MNASSA, 17, 137

632   FABER,S.M., BALIK,B., GALLAGHER,J.S., KNAPP,G.R.,
        1977, AP.J., 214, 383

633   FABER,S.M., DRESSLER,A., 1977, A.J., 82, 187

282   FAIRALL,A.P., 1971, MNRAS, 153, 383

695   FAIRALL,A.P., 1978, MNASSA, 37, 72

791   FAIRALL,A.P., 1979, MNASSA, 38, 18

792   FAIRALL,A.P., 1979, MNASSA, 38, 24

793   FAIRALL,A.P., 1979, MNRAS, 188, 343

794   FAIRALL,A.P., 1979, MNRAS, 188, 349

795   FAIRALL,A.P., 1979, NATURE, 279, 140

894   FAIRALL,A.P., 1980, MNASSA, 39, 55

895   FAIRALL,A.P., 1980, MNRAS, 191, 391

896   FAIRALL,A.P., 1980, MNRAS, 192, 389

214   FAIRALL,A.P., ANGIONE,R.J., 1969, PASP, 81, 685

696   FANTI,R., GIOIA,I., LARI,C., ULRICH,M.H., 1978,
        ASTRON.AND AP. SUPPL., 34, 341

697   FELDMAN,F.R., MARSHALL,F.E., PHILLIPS,M.M., 1978,
        IAU CIRC. N.3293

492   FISHER,J.R., TULLY,R.B., 1975, ASTRON.AND AP.,
        44, 151

561   FISHER,J.R., TULLY,R.B., 1976, ASTRON.AND AP.,
        53, 397

698   FISHER,J.R., TULLY,R.B., 1978, QUOTED IN REF.741

215   FORD,W.K.JR., 1969, CARNEGIE INSTIT. YEAR BOOK 67,
        35

179   FORD,W.K.JR., PURGATHOFER,A.T., RUBIN,V.C., 1968,
        AP.J., 153, L39

180   FORD,W.K.JR., RUBIN,V.C., 1968, PASP, 80, 466

283   FORD,W.K.JR., RUBIN,V.C., ROBERTS,M.S., 1971, A.J.,
        76, 22

699   FOSBURY,R.A.E., 1978, QUOTED IN REF.783

634   FOSBURY,R.A.E., HAWARDEEN,T.G., 1977, MNRAS, 178,
        473

700   FOSBURY,R.A.E., MEBOLD,U., GOSS,W.M., DOPITA,M.A.,
        1978, MNRAS, 183, 549

635   FOSBURY,R.A.E., MEBOLD,U., GOSS,W.M., VAN WOERDEN,H.,
        1977, MNRAS, 179, 89

440   FREEMAN,K.C., DE VAUCOULEURS,G., 1974, AP.J., 194,
        569

636   FREEMAN,K.C., KARLSSON,B., LYNGA,G., BURRELL,J.F.,
        VAN WOERDEN,H., GOSS,W.M., MEBOLD,U., 1977, ASTRON.
        AND AP., 55, 445

562   FRICKE,K.J., KAUFMANN,J.P., 1976, MITT.ASTRON.GES.,
             38, 102

637   GALLAGHER,J.S., KNAPP,G.R., FABER,S.M., BALIK,B.,
             1977, AP.J., 215, 463

153   GATES,H.S., ZWICKY,F., BERTOLA,F., CIATTI,F.,
             RUDNICKI,K., 1967, A.J. 72, 912

796   GIOVANELLI,R., 1979, AP.J., 227, L125

441   GOAD,J.W., 1974, AP.J., 192, 311

797   GOAD,J.W., DE VENY,J.B., GOAD,L.E., 1979, AP.J.
             SUPPL., 39, 439

284   GORDON,K.J., 1971, AP.J., 169, 235

798   GORDON,K.J., GORDON,C.P., 1979, ASTROPHYS. LETTERS,
             20, 9

181   GORDON,K.J., REMAGE,N.H., ROBERTS,M.S., 1968, AP.J.,
             154, 845

897   GOSS,W.M., DANZIGER,I.J., FOSBURY,R.A.E., BOKSENBERG,A.,
             1980, MNRAS, 190, 23P

898   GOTTESMAN,S.T., 1980, A.J., 85, 824

247   GOTTESMAN,S.T., DAVIES,R.D., 1970, MNRAS, 149, 263

129   GOTTESMAN,S.T., DAVIES,R.D., REDDISH,V.C., 1966,
             MNRAS, 133, 359

324   GOTTESMAN,S.T., WELIACHEW,L., 1972, ASTROPHYS. LETT.,
             12, 63

493   GOTTESMAN,S.T., WELIACHEW,L., 1975, AP.J., 195, 23

379   GOTTESMAN,S.T., WRIGHT,M.C.H., 1973, AP.J., 184, 71

216   GOUGUENHEIM,L., 1969, ASTRON.AND AP., 3, 281

442   GRAHAM,J.A., 1974, OBSERVATORY, 94, 290

701   GRAHAM,J.A., 1978, PASP, 90, 237

799   GRAHAM,J.A., 1979, AP.J., 232, 60

380   GRAHAM,J.A., RUBIN,V.C., 1973, AP.J., 183, 19

494   GRAHAM,J.A., SARGENT,W.L.W., 1975, QUOTED BY
          SCHILIZZI,R.T., 1975, MEM.RAS, 79, 75

638   GRANDI,S.A., 1977, AP.J., 215, 446

563   GRANDI,S.A., OSTERBROCK,D.E., PHILLIPS,M.M., 1976,
          ASTRON.AND AP., 51, 323

285   GREEN,F.,AND ROBERTS,M.S., 1971, QUOTED IN REF.283

702   GREEN,R.F., WILLIAMS,T.B., MORTON,D.C., 1978, AP.J.,
          226, 729

 52   GREENSTEIN,J.L., 1961, AP.J., 133, 335

 70   GREENSTEIN,J.L., 1962, AP.J., 135, 679

 71   GREENSTEIN,J.L., ZWICKY,F., 1962, PASP, 74, 35

495   GREGORY,S.A., 1975, AP.J., 199, 1

496   GREGORY,S.A., 1975, PASP, 87, 833

381   GREGORY,S.A., CONNOLLY,L.P., 1973, AP.J., 182, 351

703   GREGORY,S.A., THOMPSON,L.A., 1978, AP.J., 222, 784

800   GRIERSMITH,D., VISVANATHAN,N., 1979, ASTRON.AND AP.,
          79, 329

899   GRUEFF,G., VIGOTTI,M., SPINRAD,H., 1980, ASTRON.AND
          AP., 86, 50

564   GUDEHUS,D.H., 1976, AP.J., 208, 267

217   GUELIN,M., HUCHTMEIER,W., 1969, QUOTED IN REF.245

218   GUELIN,M., WELIACHEW,L., 1969, ASTRON.AND AP., 1, 10

248   GUELIN,M., WELIACHEW,L., 1970, ASTRON.AND AP., 7, 141

249   GUELIN,M., WELIACHEW,L., 1970, ASTRON.AND AP., 9, 155

286   GUNN,J.E., 1971, AP.J., 164, L113

900   HART,L., DAVIES,R.D., JOHNSON,S.C., 1980, MNRAS,

191, 269

704   HARTWICK,F.D.A., SARGENT,W.L.W., 1978, AP.J., 221, 512

705   HASCHICK,A.D., BAAN,W.A., BURKE,B.F., 1978, AP.J., 225, 343

706   HAVLEN,R.J., QUINTANA,H., 1978, AP.J., 220, 14

801   HAWARDEN,T.G., LONGMORE,A.J., CANNON,R.D., ALLEN,D.A., 1979, MNRAS, 186, 495

802   HAWARDEN,T.G., VAN WOERDEN,H., MEBOLD,U., GOSS,W.M., PETERSON,B.A., 1979, ASTRON.AND AP. 76, 230

803   HAYNES,M.P., 1979, A.J., 84, 1830

901   HAYNES,M.P., GIOVANELLI,R., 1980, AP.J., 240, L87

804   HAYNES,M.P., GIOVANELLI,R., ROBERTS,M.S., 1979, AP.J., 229, 83

325   HECKATHORN,H.M., 1972, AP.J., 173, 501

707   HECKMAN,T.M., BALICK,B., SULLIVAN,W.T.III, 1978, AP.J., 224, 745

53   HEIDMANN,J., 1961, BAN, 15, 314

805   HELOU,G., SALPETER,E.E., KRUMM,N., 1979, AP.J., 228, L1

806   HINTZEN,P., 1979, PASP, 91, 426

902   HINTZEN,P., 1980, A.J., 85, 626

708   HINTZEN,P., OEGERLE,W.R., SCOTT,J.S., 1978, A.J., 83, 478

807   HINTZEN,P., SCOTT,J.S., 1979, ASTRON.AND AP., 74, 116

903   HINTZEN,P., SCOTT,J.S., 1980, AP.J., 239, 765

639   HINTZEN,P., SCOTT,J.S., TARENGHI,M., 1977, AP.J., 212, 8

443   HODGE,P.W., 1974, PASP, 86, 645

130   HODGE,P.W., MERCHANT,A.E., 1966, AP.J., 144, 875

114   HOGLUND,B., ROBERTS,M.S., 1965, AP.J., 142, 1366

904   HUA,C.T., DONAS,J., DOAN,N.H., 1980, ASTRON.AND AP.,
         90, 8

565   HUCHRA,J., 1976, QUOTED BY DE BRUYN,A.G., 1977,
         ASTRON.AND AP., 58, 221

808   HUCHRA,J., DAVIS,M., 1979, QUOTED IN REF.931

640   HUCHRA,J., HOESSEL,J., ELIAS,J., 1977, A.J., 82,
         674

382   HUCHRA,J., SARGENT,W.L.W., 1973, AP.J., 186, 433

641   HUCHRA,J., THUAN,T.X., 1977, AP.J., 216, 694

326   HUCHTMEIER,W.K., 1972, ASTRON.AND AP., 17, 207

383   HUCHTMEIER,W.K., 1973, ASTRON.AND AP., 22, 27

384   HUCHTMEIER,W.K., 1973, ASTRON.AND AP., 22, 91

385   HUCHTMEIER,W.K., 1973, ASTRON.AND AP., 23, 93

809   HUCHTMEIER,W.K., 1979, ASTRON.AND AP., 75, 170

498   HUCHTMEIER,W.K., BOHNENSTENGEL,H.D., 1975, ASTRON.
         AND AP., 41, 477

497   HUCHTMEIER,W.K., BOHNENSTENGEL,H.D., 1975, ASTRON.
         AND AP., 44, 479

905   HUCHTMEIER,W.K., SEIRADAKIS,J.H., MATERNE,J., 1980,
         ASTRON.AND AP., 91, 341

906   HUCHTMEIER,W.K., SEIRADAKIS,J.H., TAMMANN,G.A., 1980,
         ASTRON.AND AP. 89, 95

499   HUCHTMEIER,W.K., TAMMANN,G.A., WENDKER,H.J., 1975,
         ASTRON.AND AP., 42, 205

566   HUCHTMEIER,W.K., TAMMANN,G.A., WENDKER,H.J., 1976,
         ASTRON.AND AP., 46, 381

642   HUCHTMEIER,W.K., TAMMANN,G.A., WENDKER,H.J., 1977,
         ASTRON.AND AP., 57, 313

  2   HUMASON,M.L., 1931, AP.J., 71, 35

 54   HUMASON,M.L., GOMES,A.M., KEARNS,C.E., 1961, PASP,
         73, 175

13   HUMASON,M.L., MAYALL,N.U., SANDAGE,A.R., 1956, A.J.,
        61, 97

11   HUMASON,M.L., WAHLQUIST,H.D., 1955, A.J., 60, 254

709   HUNSTEAD,R.W., MURDOCK,H.S., SHOBBROCK,R.R., 1978,
        MNRAS, 185, 149

710   JAFFE,W.J., PEROLA,G.C., TARENGHI,M., 1978, AP.J.,
        224, 808

711   JAUNCEY,D.L., WRIGHT,A.E., PETERSON,B.A., CONDON,J.J.,
        1978, AP.J., 219, L1

712   JENNER,D., 1978, QUOTED IN REF.707

444   JENNER,D.C., 1974, AP.J., 191, 55

810   JOEVEER,M., KAASIK,A., EINASTO,J., 1979, ASTROFIZIKA,
        15, 19 (ASTROPHYSICS, 15, 16)

907   JONES,J.E., JONES,B.J.T., 1980, MNRAS, 191, 685

327   KARACHENTSEV,I.D., 1972, QUOTED IN REF.428

908   KARACHENTSEV,I.D., 1980, AP.J. SUPPL., 44, 137

811   KARACHENTSEV,I.D., KARACHENTSEVA,V.E., SHCHERBANOVSKII,
        1979, SOVIET ASTRON. LETT., 4, 261

909   KARACHENTSEV,I.D., KOPYLOV,A.I., 1980, MNRAS, 192,
        109

500   KARACHENTSEV,I.D., PRONIK,V.I., CHUVAEV,K.K., 1975,
        ASTRON.AND AP., 41, 375

567   KARACHENTSEV,I.D., PRONIK,V.I., CHUVAEV,K.K., 1976,
        ASTRON.AND AP., 51, 185

812   KARACHENTSEV,I.D., SARGENT,W.L.W., ZIMMERMANN,B,
        1979, ASTROFIZIKA, 15 25, (ASTROPHYSICS, 15, 19)

713   KARACHENTSEV,I.D., TIFFT,W.G., 1978, ASTRON.AND AP.,
        63, 411

813   KARACHENTSEVA,V.E., KARACHENTSEV,I.D., 1979,
        ASTROFIZIKA, 15, 589 (ASTROPHYSICS, 15, 396)

328   KAROJI,H., KODAIRA,K., 1972, PUB.ASTR.SOC. JAPAN,
        24, 239

814   KAZARYAN,M.A., 1979, ASTROFIZIKA, 15, 193
        (ASTROPHYSICS, 15, 117)

445   KAZARYAN,M.A., KHACHIKYAN,E.E., 1974, ASTROFIZIKA,

10, 477 (ASTROPHYSICS 10, 299)

910   KELTON,P.W., 1980, A.J., 85, 89

10   KERR,F.J., HINDMAN,J.V., ROBINSON,B.J., 1954,
         AUSTRAL.J.PHYS., 7, 297

386   KHACHIKYAN,E.E., 1973, ASTROFIZIKA, 8, 529
         (ASTROPHYSICS, 8, 311)

387   KHACHIKYAN,E.E., 1973, ASTROFIZIKA, 9, 157,
         (ASTROPHYSICS, 9, 87)

568   KHACHIKYAN,E.E., 1976, ASTRON. NACHR. 297, 287

446   KHACHIKYAN,E.E., ANDREASYAN,N.K., SARGENT,W.L.W.,
         1974, ASTROFIZIKA, 10, 297.(ASTROPHYSICS, 10, 177)

329   KHACHIKYAN,E.E., PANOSSIAN,H.A., 1972, ASTR.CIRC.
         (USSR), 698, 1

287   KHACHIKYAN,E.E., WEEDMAN,D.W., 1971, ASTROFIZIKA 7,
         389 (ASTROPHYSICS, 7, 231)

115   KINMAN,T.D., 1965, AP.J., 142, 1241

643   KINMAN,T.D., RUBIN,V.C., THONNARD,N., FORD,W.K.JR,
         PETERSON,C.J., 1977, A.J., 82, 879

288   KINTNER,E.C., 1971, A.J., 76, 409

644   KIRSHNER,R.P., 1977, AP.J., 212, 319

911   KIRSHNER,R.P., MALUMUTH,E.M., 1980, AP.J., 236, 366

714   KIRSHNER,R.P., OEMLER,A.JR, SCHECHTER,P.L., 1978,
         A.J., 83, 1549

715   KNAPP,G.R., FABER,S.M., GALLAGER,J.S., 1978, A.J.,
         83, 11

716   KNAPP,G.R., GALLAGHER,J.S., FABER,S.M., 1978, A.J.,
         83, 139

645   KNAPP,G.R., GALLAGHER,J.S., FABER,S.M., BALICK,B.,
         1977, A.J., 82, 106

815   KNAPP,G.R., KERR,F.J., HANDERSON,A.P., 1979, AP.J.,
         234, 448

289   KODAIRA,K., 1971, PUB.ASTR.SOC. JAPAN, 23, 589

816   KODAIRA,K., IYE,M., 1979, PUB.ASTR.SOC. JAPAN, 31,
         647

388   KOPILOV,I.M., LIPOVETSKY,V.A., PRONIK,V.I., CHUVAEV,K.K.,
        1973, ASTR. CIRC.(USSR), 755, 1

447   KOPILOV,I.M., LIPOVETSKY,V.A., PRONIK,V.I., CHUVAEV,K.K.,
        1974, ASTROFIZIKA, 10, 483 (ASTROPHYSICS, 10, 305)

569   KOPYLOV,I.M., LIPOVETSKY,V.A., PRONIK,V.I., CHUVAEV,K.K.,
        1976, ASTROFIZIKA, 12, 189 (ASTROPHYSICS, 12,

448   KORMENDY,J., SARGENT,W.L.W., 1974, AP.J., 193, 19

570   KOSKI,A., 1976 PH.D. THESIS, UNIVERSITY OF CALIFORNIA,
        SANTA CRUZ

571   KOWAL,C.T., HUCHRA,J., SARGENT,W.L.W., 1976, PASP,
        88, 521

449   KOWAL,C.T., ZWICKY,F., SARGENT,W.L.W., SEARLE,L.,
        1974, PASP, 86, 516

912   KRISS,G.A., CANIZARES,C.R., MC CLINTOK,J.E., FEIGELSON,E.D.,
        1980, AP.J., 235, L61

913   KRISTIAN,J., SANDAGE,A., WESTPHAL,J., 1980, QUOTED
        IN REF.946

717   KRISTIAN,J., SANDAGE,A., WESTPHAL,J.A., 1978, AP.J.,
        221, 383

646   KRON,R.G., SPINRAD,H., KING,I.R., 1977, AP.J., 217,
        951

572   KRUMM,N., SALPETER,E.E., 1976, AP.J., 208, L7

647   KRUMM,N., SALPETER,E.E., 1977, ASTRON.AND AP., 56,
        465

817   KRUMM,N., SALPETER,E.E., 1979, A.J. 84, 1138

818   KRUMM,N., SALPETER,E.E., 1979, AP.J., 228, 64

914   KRUMM,N., SALPETER,E.E., 1980, A.J., 85, 1312

290   KUNKEL,W.E., BRADT,H.V., 1971, AP.J., 170, L7

819   KUNTH,D., SARGENT,W.L.W., 1979, ASTRON.AND AP., 76,
        50

820   KUNTH,D., SARGENT,W.L.W., 1979, ASTRON.AND AP.SUPPL.,
        36, 259

915   KYAZUMOV,G.A., 1980, SOVIET ASTRON. LETT., 6, 220

916   KYAZUMOV,G.A., BARABANOV,A.V., 1980, SOVIET ASTRON.
        LETT., 6, 181

718   LAING,R.A., LONGAIR,M.S., RILEY,J.N., KIBBLEWHITE,E.J.,
         GUNN,J.E., 1978, MNRAS, 183, 547

719   LAUBERTS,A., BERGVALL,N.A.S., EKMAN,A.B.G., WESTERLUND,B.E.,
         1978, ASTRON.AND AP.SUPPL., 35, 55

389   LAUQUE',R., 1973, ASTRON.AND AP., 23, 253

917   LAUSTSEN,S., WEST,R.M., 1980, J. ASTROPHYS. ASTRON.,
         1, 177

918   LEQUEUX,J., VIALLEFOND,F., 1980, ASTRON.AND AP.,
         91, 269

182   LEWIS,B.M., 1968, PROC. AUSTRAL. ASTR. SOC., 1, 104

219   LEWIS,B.M., 1969, PROC. AUSTRAL. ASTRON. SOC., 1, 288

250   LEWIS,B.M., 1970, OBSERVATORY, 90, 264

330   LEWIS,B.M., 1972, AUSTRAL.J.PHYS., 25, 315

390   LEWIS,B.M., DAVIES,R.D., 1973, MNRAS, 165, 213

821   LONGMORE,A.J., HAWARDEN,T.G., CANNON,R.D., ALLEN,D.A.,
         MEBOLD,U., GOSS,W.M., REIF,K., 1979, MNRAS, 188, 285

720   LONGMORE,A.J., HAWARDEN,T.G., WEBSTER,B.L., GOSS,W.M.,
         MEBOLD,U., 1978, MNRAS, 183, 97P

 55   LOVASICH,J.L., MAYALL,N.U., NEYMAN,N.J., SCOTT,E.,
         1961, PROC. IV BERKELEY SYMP. ON MATH. STAT. AND
         PROB., VOL.III, P.187

331   LOVE,R., 1972, NATURE PHYS. SCI., 235, 53

291   LYNDS,R., 1971, AP.J., 168, L87

332   LYNDS,R., 1972, IAU SYMP. N. 44, P.376

333   LYNDS,R., 1972, QUOTED BY SANDAGE A., 1972, AP.J.,
         178, 25

450   LYNDS,R., 1974, QUOTED BY MATERNE,J., TAMMANN,G.A.,
         1974, ASTRON.AND AP., 35, 441

452   LYNDS,R., 1974, QUOTED IN REF.428

451   LYNDS,R., 1974, QUOTED IN REF.444

573   LYNDS,R., TOOMRE,A., 1976, AP.J., 209, 382

648   MACCACARO,T., COOKE,B.A., WARD,M.J., PENSTON,M.V.,
        HAYNES,R.F., 1977, MNRAS, 180, 465

919   MARCELIN,M., COMTE,G., COURTES,G., GEORGELIN,Y.P.,
        1980, PASP, 92, 38

334   MARGON,B., SPINRAD,H., HEILES,C., TOVMASSIAN,H.,
        HARLAN,E., BOWYER,S., LAMPTON,M., 1972, AP.J., 178,
        L77

921   MARKARIAN,B.E., LIPOVETSKY,V.A., STEPANIAN,J.A.,
        1980, ASTROFIZIKA, 16, 5 (ASTROPHYSICS, 16, 1)

920   MARKARIAN,B.E., LIPOVETSKY,V.A., STEPANIAN,J.A.,
        1980, ASTROFIZIKA, 16, 609 (ASTROPHYSICS, 16, 353)

574   MARTIN,W.L., 1976, MNRAS, 175, 633

501   MATHEWSON,D.S., FORD,V.L., MURRAY,J.D., 1975,
        OBSERVATORY, 95, 176

14   MAYALL,N.U., 1956, QUOTED IN REF.83

6   MAYALL,N.U., ALLER,L.H., 1942, AP.J., 95, 5

72   MAYALL,N.U., DE VAUCOULEURS,A., 1962, A.J., 67, 363

251   MC CUTCHEON,W.H., DAVIES,R.D., 1970, MNRAS, 150, 337

822   MCALPINE,G.M., WILLIAMS,G.A., LEWIS,D.W., 1979,
        PASP, 91, 746

649   MEBOLD,U., GOSS,W.M., FOSBURY,R.A.E., 1977, MNRAS,
        180, 11P

823   MEBOLD,U., GOSS,W.M., VAN WOERDEN,H., HAWARDEN,T.G.,
        SIEGMAN,B., 1979, ASTRON.AND AP., 74, 100

650   MELNIK,J., 1977, AP.J., 213, 15

651   MELNIK,J., SARGENT,W.L.W., 1977, AP.J., 215, 401

131   MENG,S.Y., KRAUS,J.D., 1966, A.J., 71, 170

502   METLOV,V.G., 1975, SOVIET ASTRON. LETT. 1, 220

922   METLOV,V.G., 1980, SOVIET ASTRON. LETT., 6, 110

824   MILEY,G.K., OSTERBROCK,D.E., 1979, PASP, 91, 257

391   MILLER,J.S., ROBISON,L.B., WAMPLER,E.J., 1973,
        AP.J., 179, L83

721   MILLS,B.Y., HUNSTEAD,R.W., SKELLERN,D.J., 1978,
        MNRAS, 185, 51P

 36   MINKOWSKI,R., 1960, ANN. ASTROPHYS., 23, 385

 37   MINKOWSKI,R., 1960, AP.J., 132, 908

503   MINKOWSKI,R., OORT,J.H., VAN HOUTEN,G.J., DAVIS,M.M.,
        1975, QUOTED IN REF.530

 23   MINKOWSKY,R., 1959, AP.J., 130, 1028

 56   MINKOWSKY,R., 1961, A.J., 66, 558

 24   MINKOWSKY,R., OSTERBROCK,D., 1959, AP.J., 129, 583

504   MIRZOYAN,L.V., MILLER,J.S., OSTERBROCK,D.E., 1975.
        AP.J., 196, 687

652   MITTON,S., HAZARD,C., WHELAN,J.A.J., 1977, MNRAS,
        179, 569

252   MONNET,G., 1970, IAU SYMP. N.38, P.73

  3   MOORE,J.H., 1932, PUB. LICK OBS. N.18

335   MORTON,D.C., CHEVALIER,R.A., 1972, AP.J., 174, 489

392   MORTON,D.C., CHEVALIER,R.A., 1973, AP.J., 179, 55

653   MOSS,C., DICKENS,R.J., 1977, MNRAS, 178, 701

 25   MUNCH,G., 1959, PASP, 71, 101

923   NEWTON,K., 1980, MNRAS, 191, 169

336   O'CONNELL,R.W., KRAFT,R.P., 1972, AP.J., 175, 335

393   OEMLER,A.JR., 1973, AP.J., 180, 11

337   OEMLER,A.JR., GUNN,J.E., OKE,J.B., 1972, AP.J.,
        176, L47

292   OKE,J.B., 1971, AP.J., 170, 193

722   OKE,J.B., 1978, AP.J., 219, L97

825   OKE,J.B., LAUER,T.R., 1979, AP.J., 230, 360

454   OSMER,P.S., SMITH,M.G., WEEDMAN,D.W., 1974, AP.J.,
        189, 187

453   OSMER,P.S., SMITH,M.G., WEEDMAN,D.W., 1974, AP.J.,
        192, 279

723   OSTERBROCK,D.E., 1978, QUOTED IN REF.707

826   OSTERBROCK,D.E., GRANDI,S.A., 1979, AP.J., 228, L59

575   OSTERBROCK,D.E., KOSKI,A.T., PHILLIPS,M.M., 1976,
        AP.J., 206, 898

654   OSTERBROCK,D.E., PHILLIPS,M.M., 1977, PASP, 89, 251

  8   PAGE,T., 1952, AP.J., 116, 63

 57   PAGE,T., 1961, PROC. IV BERKELEY SYMP. ON MATH.
        STAT. AND PROB., VOL III, PAG.277

154   PAGE,T., 1967, A.J., 72, 821

253   PAGE,T., 1970, AP.J., 159, 791

155   PASTORIZA,M., 1967, OBSERVATORY, 87, 225

254   PASTORIZA,M., 1970, BOL.ASSOC.ARG.ASTRON.,N.15, 1

827   PASTORIZA,M., 1979, AP.J. 234, 837.

293   PASTORIZA,M., AGUERO,E., 1971, BOL.ASOC.ARG.ASTRON.,
        N.16, 3

576   PASTORIZA,M., AGUERO,E., 1976, AP. AND SP.SCI., 39,
        201

724   PEDREROS,M., 1978, PASP, 90, 14

725   PENCE,W.D., 1978, PUB. IN ASTR. UNIV. TEXAS, N.14

655   PENSTON,M.V, FOSBURY,R.A.E., WARD,M.J., WILSON,A.S.,
        1977, MNRAS, 180, 19

394   PENSTON,M.V., PENSTON,M.J., 1973, MNRAS, 162, 109

255   PETERSON,B.A., 1970, A.J., 75, 695

338   PETERSON,B.A., BOLTON,J.G., 1972, AP.J., 173, L19

577   PETERSON,B.A., JAUNCEY,D.L., WRIGHT,A.E., CONDON,J.J.,
          1976, AP.J., 207,L5

828   PETERSON,B.A., WRIGHT,A.E., JAUNCEY,D.L., CONDON,J.J.,
          1979, AP.J., 232, 400

726   PETERSON,B.M., 1978, AP.J., 223, 740

727   PETERSON,B.M., CRAINE,E.R., STRITTMATTER,P.A., 1978,
          PASP, 90, 386

728   PETERSON,C.J., 1978, PASP, 90, 10

924   PETERSON,C.J., 1980, A.J., 85, 226

925   PETERSON,C.J., 1980, PASP, 92, 397

729   PETERSON,C.J., RUBIN,V.C., FORD,W.K.JR., ROBERTS,M.S.,
          1978, AP.J., 226, 770

578   PETERSON,C.J., RUBIN,V.C., FORD,W.K.JR., THONNARD,N.,
          1976, AP.J., 208, 662

730   PETERSON,C.J., RUBIN,V.C., FORD,W.K.JR., THONNARD,N.,
          1978, AP.J., 219, 31

829   PETERSON,S.D., 1979, AP.J. SUPPL., 40, 527

455   PETERSON,S.D., SHOSTAK,G.S., 1974, A.J., 79, 767

926   PETROSYAN,A.R., SAAKIAN,K.A., KHACHIKIAN,E.E., 1980,
          ASTROFIZIKA, 16, 589

927   PETROSYAN,A.R., SAAKYAN,K.A., KHACHIKIAN,E.E., 1980,
          ASTROFIZIKA, 16, 621 (ASTROPHYSICS, 16, 360)

830   PETROSYAN,A.R., SAAKYAN,K.A., KHACHIKYAN,E.E., 1979
          ASTROFIZIKA, 15, 209 (ASTROPHYSICS, 15, 142)

831   PETROSYAN,A.R., SAAKYAN,K.A., KHACHIKYAN,E.E., 1979,
          ASTROFIZIKA, 15, 373 (ASTROPHYSICS, 15, 250)

929   PETROSYAN,A.R., SAAKYAN,K.A., KHACHIKYAN,E.E., 1980,
          SOVIET ASTRON. LETT. 6, 288

928   PETROSYAN,A.R., SAAKYAN,K.A., KHACHIKYAN,E.E., 1980,

SOVIET ASTRON. LETT., 6, 144

579   PHILLIPS,M.M., 1976, AP.J., 208, 37

930   PHILLIPS,M.M., 1980, AP.J., 236, L45

832   PHILLIPS,M.M., FELDMAN,F.R., MARSHALL,F.E., WAMSTEKER,W.,
         1979, ASTRON.AND AP., 76, L14

931   PINEDA,F.J., DELVAILLE,J.P., GRINDLAY,J.E., SCHNOPPER,H.W.,
         1980, AP.J., 237, 414

731   PINEDA,F.J., DELVAILLE,J.P., HUCHRA,J., DAVIS,M.,
         1978, IAU CIRC., N 3202

732   PRABHU,T.P., 1978, KODAIKANAL OBS. BULL., SER A,
         2, 105

932   PRABHU,T.P., 1980, J.ASTROPHYS.ASTR., 1, 129

505   QUINTANA, H., MELNICK,J., 1975, PASP, 87, 863

15   RAIMOND,E., VOLDERS,L.M.J.S., 1957, BAN, 14, 19

833   REAKES,M., 1979, MNRAS, 187, 509

834   REAKES,M., 1979, MNRAS, 187, 525

933   REAKES,M., 1980, MNRAS, 192, 297

733   REIF,K., MEBOLD,U., GOSS,W.M., 1978, ASTRON.AND AP.,
         67, L1

506   RICKARD,J.J., 1975, ASTRON.AND AP., 40, 339

73   ROBERTS,M.S., 1962, A.J., 67, 431

74   ROBERTS,M.S., 1962, A.J., 67, 437

116   ROBERTS,M.S., 1965, AP.J., 142, 148

132   ROBERTS,M.S., 1966, AP.J., 144, 639

183   ROBERTS,M.S., 1968, A.J., 73, 945

184   ROBERTS,M.S., 1968, AP.J., 151, 117

256   ROBERTS,M.S., 1970, AP.J., 161, L9

294   ROBERTS,M.S., 1971, QUOTED IN REF.513

734   ROBERTS,M.S., 1978, QUOTED IN REF.829

257   ROBERTS,M.S., WARREN, J.L., 1970, ASTRON.AND AP.,
          6, 165

117   ROBINSON,B.J., KOEHLER,J.A., 1965, NATURE, 208, 993

133   ROBINSON,B.J., VAN DAMME, K.J., 1966, AUSTRAL.J.PHYS.,
          19, 111

339   ROBINSON,L.B., WAMPLER,E.J., 1972, AP.J., 171, L83

395   ROBINSON,L.B., WAMPLER,E.J., 1973, AP.J., 179, L135

396   ROBINSON,L.B., WAMPLER,E.J., 1973, AP.J., 179, L79

258   RODGERS,A.W., FREEMAN, K.C., 1970, AP.J., 161, L109

735   RODGERS,A.W., PETERSON,B.A., HARDING,P., 1978, AP.J.,
          225, 768

295   ROGSTAD,D.H., 1971, ASTRON.AND AP., 13, 108

835   ROGSTAD,D.H., CRUTCHER,R.M., CHU,K., 1979, AP.J.,
          229, 509

456   ROGSTAD,D.H., LOCKART,I.A., WRIGHT,M.C.H., 1974,
          AP.J., 193, 309

156   ROGSTAD,D.H., ROUGOOR,G.W., WHITEOAK,J.B., 1967,
          AP.J., 150, 9

296   ROGSTAD,D.H., SHOSTAK,G.S., 1971, ASTRON.AND AP.,
          13, 99

340   ROGSTAD,D.H., SHOSTAK,G.S., 1972, AP.J., 176, 315

397   ROGSTAD,D.H., SHOSTAK,G.S., ROTS,A.H., 1973, ASTRON.
          AND AP., 22, 111

580   ROGSTAD,D.H., WRIGHT,M.C.H., LOCKHART,I.A., 1976,
          AP.J., 204, 703

581   ROOD,H.J., DICKEL,J.R., 1976, AP.J., 205, 346

836   ROSE,J.A., GRAHAM,J.A., 1979, AP.J., 231, 320

736   ROTS,A.H., 1978, A.J. 83, 219.

837   ROTS,A.H., 1979, ASTRON.AND AP., 80, 255

934   ROTS,A.H., 1980, ASTRON.AND AP.SUPPL., 41, 189

457   ROTS,A.H., SHANE,W.W., ASTRON.AND AP., 1974, 31, 245

458   RUBIN,V.C., 1974, AP.J., 191, 645

737   RUBIN,V.C., 1978, AP.J., 224, L55

935   RUBIN,V.C., 1980, AP.J., 238, 808

95   RUBIN,V.C., BURBIDGE,E.M., BURBIDGE,G.R., 1964,
      AP.J., 140, 1304

94   RUBIN,V.C., BURBIDGE,E.M., BURBIDGE,G.R., 1964,
      AP.J., 140, 94

118   RUBIN,V.C., BURBIDGE,E.M., BURBIDGE,G.R., CRAMPIN,D.J.,
      PRENDERGAST,K.H., 1965, AP.J., 141, 759

96   RUBIN,V.C., BURBIDGE,E.M., BURBIDGE,G.R., PRENDERGAST,K.H.,
      1964, AP.J., 140, 80

119   RUBIN,V.C., BURBIDGE,E.M., BURBIDGE,G.R., PRENDERGAST,K.H.,
      1965, AP.J., 141, 885

157   RUBIN,V.C., FORD,W.K.JR, 1967, PASP, 79, 322

507   RUBIN,V.C., FORD,W.K.JR, PETERSON,C.J., 1975, AP.J.,
      199, 39

738   RUBIN,V.C., FORD,W.K.JR, STROM,K.M., STROM,S.E.,
      ROMANISHIN,W., 1978, AP.J., 224, 782

582   RUBIN,V.C., FORD,W.K.JR, THONNARD,N., ROBERTS,M.S.,
      GRAHAM,J.A., 1976, A.J., 81, 687

185   RUBIN,V.C., FORD,W.K.JR., 1968, AP.J., 154, 431

259   RUBIN,V.C., FORD,W.K.JR., 1970, AP.J., 159, 379

260   RUBIN,V.C., FORD,W.K.JR., D'ODORICO,S., 1970, AP.J.,
      160, 801

739   RUBIN,V.C., FORD,W.K.JR., PETERSON,C.J., LYNDS,C.R.,
      1978, AP.J., 37, 235

838   RUBIN,V.C., FORD,W.K.JR., ROBERTS,M.S., 1979, AP.J.,

230, 35

740   RUBIN,V.C., FORD,W.K.JR., THONNARD,N., 1978, AP.J.,
225, L107

936   RUBIN,V.C., FORD,W.K.JR., THONNARD,N., 1980, AP.J.,
238, 471

158   RUBIN,V.C., MOORE,S., BERTIAU,F.C., 1967, A.J., 72, 59

937   RUBIN,V.C., PETERSON,C.J., FORD,W.K.JR., 1980,
AP.J., 239, 50

508   RUBIN,V.C., THONNARD,N., FORD,W.K.JR, 1975, AP.J.,
199, 31

656   RUBIN,V.C., THONNARD,N., FORD,W.K.JR, 1977, AP.J.,
217, L1

220   RUDNICKI,K., TARRARO,I., 1969, ACTA ASTRON., 19, 171

398   SAKKA,K., OKA,S., WAKAMATSU,K., 1973, PUB. ASTR.SOC.
JAPAN, 25, 317

399   SAKKA,K., OKA,S., WAKAMATSU,K., 1973, PUB.ASTR.SOC.
JAPAN, 25, 153

839   SANCISI,R., ALLEN,R.J., 1979, ASTRON.AND AP., 74,
73

840   SANCISI,R., ALLEN,R.J., SULLIVAN,W.T.III, 1979,
ASTRON.AND AP., 78, 217

509   SANCISI,R., ALLEN,R.J., VAN ALBADA,T.S., 1975, IN
LA DINAMIQUE DES GALAXIES SPIRALES, COLLOQUE CNRS,
241, 295

134   SANDAGE,A., 1966, AP.J., 145, 1

159   SANDAGE,A., 1967, AP.J., 150, L145

160   SANDAGE,A., 1967, AP.J., 150, L9

510   SANDAGE,A., 1975, AP.J., 202, 563

741   SANDAGE,A., 1978, A.J., 83, 904

583   SANDAGE,A., KRISTIAN,J., WESTPHAL,J.A., 1976, AP.J.,
205, 688

161   SARGENT,W.L.W., 1967, PASP, 79, 369

186   SARGENT,W.L.W., 1968, A.J., 73, 893

187   SARGENT,W.L.W., 1968, AP.J., 152, L31

188   SARGENT,W.L.W., 1968, AP.J., 153, L135

261   SARGENT,W.L.W., 1970, AP.J., 159, 765

262   SARGENT,W.L.W., 1970, AP.J., 160, 405

341   SARGENT,W.L.W., 1972, AP.J., 173, 7

342   SARGENT,W.L.W., 1972, AP.J., 176, 581

400   SARGENT,W.L.W., 1973, AP.J., 182, L13

401   SARGENT,W.L.W., 1973, PASP, 85, 281

511   SARGENT,W.L.W., 1975, QUOTED BY SCHILIZZI,R.T., 1975,
        MEM. RAS, 79, 75

657   SARGENT,W.L.W., 1977, AP.J., 212, L105

841   SARGENT,W.L.W., 1979, QUOTED IN REF.780

658   SARGENT,W.L.W., SCHECHTER,P.L., BOKSENBERG,A.,
        SHORTRIDGE,K., 1977, AP.J., 212, 326

742   SARGENT,W.L.W., YOUNG,P.J., BOKSEMBERG,A., SHORTRIDGE,K.,
        LYNDS,C.R., HARTWICK,F.D.A., 1978, AP.J., 221, 731

938   SCHECHTER,P.L., 1980, A.J., 85, 801

842   SCHECHTER,P.L., GUNN,J.E., 1979, AP.J., 229, 472

843   SCHILD,R., DAVIS,M., 1979, A.J., 84, 311

120   SCHMIDT,M., 1965, AP.J., 141, 1

459   SCHMIDT,M.,1974, QUOTED BY WILLIS,A.G., STROM,R.C.,
        WILSON,A.S., 1974, NATURE, 250, 625

743   SCHNOPPER,H.W., DAVIS,M., DELVAILLE,J.P., GELLER,M.J.,
        HUCHRA,J., 1978, NATURE, 275, 719

939   SCHOMMER, SULLIVAN, BOTHUM, 1980, QUOTED IN REF.872

940   SCHWARTZ,D.A., DAVIS,M., DOXSEY,R.E., GRIFFITHS,R.E.,

HUCHRA,J., JOHNSTON,M.D., MUSHOTZKY,R.F., SWANK,J.J.,
TONRY,J., 1980, AP.J., 238, L53

941   SCHWEIZER,F., 1980, AP.J., 237, 303

659   SCOTT,J.S., ROBERTSON,J.W., TARENGHI,M., 1977,
          ASTRON.AND AP., 59, 23

189   SEARLE,L., BOLTON,J.G., 1968, AP.J., 154, L101

121   SEIELSTAD,G.A., WHITEOAK,J.B., 1965, AP.J., 142, 616

402   SEIELSTAD,G.A., WRIGHT,M.C.H., 1973, AP.J., 184, 343

135   SERSIC,J.L., 1966, ZEIT. FUR ASTR., 64, 202

221   SERSIC,J.L., 1969, NATURE, 224, 253

343   SERSIC,J.L., AGUERO,E.L., 1972, AP. AND SP.SCI.,
          19, 387

844   SERSIC,J.L., ARIAS,J.C., ARAUJO,A., 1979, OBSERVATORY,
          99, 130

660   SERSIC,J.L., BAJAJA,E., COLOMB,R., 1977, ASTRON.AND
          AP., 59, 19

222   SERSIC,J.L., CARRANZA,G., 1969, INF.BUL. FOR THE
          SOUTHERN HEMISPHERE N.14, P.32

344   SERSIC,J.L., CARRANZA,G., PASTORIZA,M., 1972, AP.
          AND SP.SCI., 19, 469

845   SERSIC,J.L., CERRUTI,M.A., 1979, OBSERVATORY, 99,
          150

190   SERSIC,J.L., PASTORIZA,M., CARRANZA,G., 1968,
          ASTROPHYS. LETT., 2, 45

942   SHANE,W.W., 1980, ASTRON.AND AP., 82, 314

943   SHANE,W.W., KRUMM,N., NORMAN,C., 1980, QUOTED IN
          REF.556

345   SHIELDS,G.A., OKE,J.B., SARGENT,W.L.W., 1972, AP.J.,
          176, 75

136   SHOBBROOK,R.R., 1966, MNRAS, 131, 293

162   SHOBBROOK,R.R., ROBINSON,B.J., 1967, AUSTRAL.J.PHYS.,
          20, 131

403   SHOSTAK,G.S., 1973, BULL. AM. ASTR. SOC., 5, 430

460   SHOSTAK,G.S., 1974, AP.J., 187, 19

461   SHOSTAK,G.S., 1974, AP.J., 189, L1

462   SHOSTAK,G.S., 1974, ASTRON.AND AP., 31, 97

744   SHOSTAK,G.S., 1978, ASTRON.AND AP., 68, 321

944   SHOSTAK,G.S., ALLEN,R.J., 1980, ASTRON.AND AP., 81,
        167

404   SHOSTAK,G.S., ROGSTAD,D.H., 1973, ASTRON.AND AP.,
        24, 405

297   SHOSTAK,G.S., WELIACHEW,L., 1971, AP.J., 169, L71

945   SHUDER,J.M., 1980, AP.J., 240, 32

512   SIEFERT, P.T., GOTTESMAN, S.T., WRIGHT, M.C.H., 1975,
        IN LA DINAMIQUE DES GALAXIES SPIRALES, COLLOQUE
        CNRS, 241,425.

745   SILVERGLATE,P.R., KRUMM,N., 1978, AP.J., 224, L99

346   SIMKIN,S.M., 1972, NATURE, 239, 43

513   SIMKIN,S.M., 1975, AP.J., 195, 293

514   SIMKIN,S.M., 1975, AP.J., 200, 567

846   SIMKIN,S.M., 1979, AP.J., 234, 56

847   SIMKIN,S.M., EKERS,R.D., 1979, A.J., 84, 56

515   SMITH,G.M., 1975, AP.J., 202, 591

848   SMITH,H.E., 1979, QUOTED IN REF.783

661   SMITH,H.E., BURBIDGE,E.M., BALDWIN, J.A., TOHLINE,J.E.,
        WAMPLER, E.J., HAZARD,C., MURDOCH,H.S., 1977,
        AP.J., 215, 427

849   SMITH,H.E., JUNKKARINEN,V.T., SPINRAD,H., GRUEFF,G.,
        VIGOTTI,M., 1979, AP.J., 231, 307

946   SMITH,H.E., SPINRAD,H., 1980, PASP, 92, 553

584   SMITH,H.E., SPINRAD,H., HUNSTEAD,R., 1976, AP.J.,
206, 345

585   SMITH,H.E., SPINRAD,H., SMITH,E.C., 1976, PASP, 88,
621

516   SPINRAD,H., 1975, AP.J., 199, L1

517   SPINRAD,H., 1975, AP.J., 199, L3 AND 1978, AP.J.,
220, L135

586   SPINRAD,H., 1976, PASP, 88, 565

662   SPINRAD,H., 1977, PASP, 89, 116

850   SPINRAD,H., 1979, QUOTED IN REF.783

587   SPINRAD,H., BACHALL,N.A., 1976, PASP, 88, 660

851   SPINRAD,H., KRON,R., HUNSTEAD,R.W., 1979, QUOTED IN
REF.783

588   SPINRAD,H., LIEBERT,J., SMITH,H.E., HUNSTEAD,R.,
1976, AP.J., 206, L79

298   SPINRAD,H., SARGENT,W.L.W., OKE,J.B., NEUGEBAUER,G.,
LANDAU,R., KING,I.R., GUNN,J.E., GARMIRE,G., DIETER,N.H.,
1971, AP.J., 163, L25

405   SPINRAD,H., SMITH,H.E., 1973, AP.J., 179, L71

589   SPINRAD,H., SMITH,H.E., 1976, AP.J., 206, 355

746   SPINRAD,H., SMITH,H.E., 1978, QUOTED IN REF.717

518   SPINRAD,H., SMITH,H.E., HUNSTEAD,R., RYLE,M., 1975,
AP.J., 198, 7

947   SPINRAD,H., STAUFFER,J., BUTCHER,H., 1980, QUOTED
IN REF.946

852   SPINRAD,H., STAUFFER,J., HARLAN,E., 1979, PASP, 91,
619

949   STAUFFER,J., 1980, QUOTED IN REF.902

948   STAUFFER,J., 1980, QUOTED IN REF.946

747   STAUFFER,J., SPINRAD,H., 1978, PASP, 90, 20

950   STAUFFER,J., SPINRAD,H., 1980, AP.J., 235, 347

853   STAUFFER,J., SPINRAD,H., SARGENT,W.L.W., 1979, AP.J.,
        228, 379

347   STOCKTON,A., 1972, AP.J., 173, 247

406   STOCKTON,A., 1973, NATURE PHYS. SCI. 246, 25

463   STOCKTON,A., 1974, AP.J., 187, 219

464   STOCKTON,A., 1974, AP.J., 190, L47

951   STOCKTON,A., BERTOLA,F., 1980, AP.J., 235, 37

854   STOCKTON,A., WYCKOFF,S., WEHINGER,P.A., 1979, AP.J.,
        231, 673

748   STOKE,J.T., TIFFT,W.G., KAFTAN-KASSIM,M.A., 1978,
        A.J., 83, 322

952   STOUGHTON,R., OSTERBROCK,D.E., 1980, PASP, 92, 117

465   STRITTMATTER,P.A., CARSWELL,R.F., GILBERT,G., BURBIDGE,E.M.,
        1974, AP.J., 190, 509

  1   STROMBERG,G., 1925, AP.J., 61, 352

  5   STRUVE,O., LINKE,W., 1940, PASP, 52, 139

953   SULENTIC,J.W., 1980, AP.J., 241, 67

954   SULENTIC,J.W., 1980, ASTRON.AND AP., 88, 94

855   SULENTIC,J.W., ARP,H., LORRE,J., 1979, AP.J., 233, 44

955   SULLIVAN,W.T.,III,SCHOMMER,R.A.,BOTHUM,G.D.,BATES,B.A.,
        1980, QUOTED IN REF.872

749   SULLIVAN,W.T.III, JOHNSON,P.E., 1978, AP.J., 225, 751

348   TAKADA,M., KODAIRA,K., 1972, PUB.ASTR.SOC. JAPAN,
        24, 525

856   TAMURA,S., HASEGAWA,M., 1979, PUB.ASTR.SOC. JAPAN,
        31, 329

590   TARENGHI,M., SCOTT,J.S., 1976, AP.J., 207, L9

857   TARENGHI,M., TIFFT,W.G., CHINCARINI,G., ROOD,H.J.,
         THOMPSON,L.A.,1979, AP.J. 234, 793.

858   TARTER,J.C., WRIGHT,M.C.H., 1979, ASTRON.AND AP.,
         76, 127

591   THEYS,J.C., SPIEGEL,E.A., 1976, AP.J., 208, 650

592   THOLINE,J.E., OSTERBROCK,D.E., 1976, AP.J., 210, L117

750   THOMAS,J.C., BATCHELOR,D., 1978, A.J., 83, 1160

751   THOMPSON,L.A., WALKER,W.J.JR., GREGORY,S.A., 1978,
         PASP, 90, 644

752   THONNARD,N., RUBIN,V.C., FORD,W.K.JR, ROBERTS,M.S.,
         1978, A.J., 83, 1564

859   THUAN,T.X., MARTIN,G.E., 1979, AP.J., 232, L11

860   THUAN,T.X., SEITZER,P.O., 1979, AP.J., 231, 327

349   TIFFT,W.G., 1972, AP.J., 175, 613

407   TIFFT,W.G., 1973, AP.J., 179, 29

466   TIFFT,W.G., 1974, AP.J., 188, 221

753   TIFFT,W.G., 1978, AP.J., 222, 54

299   TIFFT,W.G., GREGORY,S.A., 1971, PASP, 83,810

408   TIFFT,W.G., GREGORY,S.A., 1973, AP.J., 181, 15

593   TIFFT,W.G., GREGORY,S.A., 1976, AP.J., 205, 696

519   TIFFT,W.G., HILSMAN,K.A., CORRADO,L.C., 1975, AP.J.,
         199, 16

409   TIFFT,W.G., JEWSBURY,C.P., SARGENT,T.A., 1973, AP.J.,
         185, 115

520   TIFFT,W.G., TARENGHI,M., 1975, AP.J., 199, 10

663   TIFFT,W.G., TARENGHI,M., 1977, AP.J., 217, 944

350   TRITTON,K.P., 1972, MNRAS, 158, 277

410   TRITTON,K.P., SCHILIZZI,R.T., 1973, MNRAS, 165, 245

467   TULLY,R.B., 1974, AP.J. SUPPL., 27, 437

956   TULLY,R.B., 1980, AP.J., 237, 390

957   TULLY,R.B., 1980, QUOTED IN REF.907

754   TULLY,R.B., BOTTINELLI,L., FISHER,J.R., GOUGHENHEIM,L., SANCISI,R., VAN WOERDEN,H., 1978, ASTRON.AND AP., 63, 37

594   TURNER,E.L., 1976, AP.J., 208, 20

300   ULRICH,M.H., 1971, AP.J., 163, 441

351   ULRICH,M.H., 1972, AP.J., 171, L35

352   ULRICH,M.H., 1972, AP.J., 171, L37

353   ULRICH,M.H., 1972, AP.J., 178, 113

411   ULRICH,M.H., 1973, AP.J., 181, 51

521   ULRICH,M.H., 1975, ASTRON.AND AP., 40, 337

595   ULRICH,M.H., 1976, AP.J., 206, 364

755   ULRICH,M.H., 1978, AP.J., 219, 424

756   ULRICH,M.H., 1978, AP.J., 221, 422

757   ULRICH,M.H., 1978, AP.J., 222, L3

758   ULRICH,M.H., 1978, QUOTED IN REF. 762

958   ULRICH,M.H., BUTCHER,H., MAYER,D.L., 1980, NATURE, 288, 459

522   ULRICH,M.H., KINMAN,T.D., LYNDS,C.R., RIEKE,G.H., EKERS,R.D., 1975, AP.J. 198, 261

959   ULRICH,M.H., PEQUIGNOT,D., 1980, AP.J., 238, 45

664   VAN ALBADA,G.D., 1977, ASTRON.AND AP., 61, 297

960   VAN ALBADA,G.D., 1980, ASTRON.AND AP., 90, 123

523   VAN ALBADA,G.D., SHANE,W.W., 1975, ASTRON.AND AP.,
         42, 433

137   VAN DAMME,K.J., 1966, AUSTRAL.J.PHYS., 19, 687

 16   VAN DE HULST,H.C., RAIMOND,E., VAN WOERDEN,H., 1957,
         BAN, 14, 1

223   VAN DEN BERGH,S., 1969, AP.J. SUPPL. 19, 145

263   VAN DEN BERGH,S., 1970, PASP, 82, 1374

861   VAN DEN HULST,J.M., 1979, ASTRON.AND AP., 71, 131

862   VAN DER HULST,J.M., HUCHTMEIER,W.K., 1979, ASTRON.
         AND AP., 78, 82

412   VAN DER KRUIT,P.C., 1973, AP.J., 186, 807

413   VAN DER KRUIT,P.C., 1973, ASTRON.AND AP., 61, 171

468   VAN DER KRUIT,P.C., 1974, AP.J., 188, 3

469   VAN DER KRUIT,P.C., 1974, AP.J., 192, 1

524   VAN DER KRUIT,P.C., 1975, AP.J., 195, 611

596   VAN DER KRUIT,P.C., 1976, ASTRON.AND AP., 49, 161

597   VAN DER KRUIT,P.C., 1976, ASTRON.AND AP., 52, 85

665   VAN DER KRUIT,P.C., 1977, ASTRON.AND AP., 61, 171

759   VAN DER KRUIT,P.C., BOSMA,A., 1978 ASTRON.AND AP.
         SUPPL., 34, 259

666   VAN WOERDEN,H., MEBOLD,U, GOSS,W.M., SIEGMAN,G.,
         1977, QUOTED IN REF.741

863   VENNIK,J., 1979, PUBL. TARTU ASTR.OBS., 47, 163

961   VIALLEFOND,F., ALLEN,R.J., DE BOER,J.A., 1980, ASTRON.

AND AP., 82, 207

962   VIALLEFOND,F., ALLEN,R.J., DE BOER,J.A., 1980, ASTRON.
        AND AP., 82, 207

525   VIDAL,N.V., 1975, ASTRON.AND AP., 42, 145

526   VIDAL,N.V., 1975, PASP, 87, 625

527   VIDAL,N.V., PETERSON,B.A., 1975, AP.J., 196, L95

667   VIDAL,N.V., WICKRAMASINGHE,D.T., 1977, MNRAS, 180,
        305

26   VOLDERS,L., 1959, BAN, 14, 323

58   VOLDERS,L., HOGBOM,J.A., 1961, BAN, 15, 307

27   VOLDERS,L., VAN DE HULST, H.C., 1959, IAU SYMP. 9,
        423

963   VORONTSOV VELYAMINOV,B.A., DOSTAL,V.A., METLOV,V.G.,
        1980, SOVIET ASTRON. LETT., 6, 217

964   VORONTSOV VELYAMINOV,B.A., METLOV,V.G., 1980, SOVIET
        ASTRON. LETT., 6, 109

163   VORONTSOV-VELYAMINOV,B.A., 1967, ASTR.CIRC.(USSR),
        445, 6

668   WAGGETT,P.C., WARNER,P.J., BALDWIN,J.E., 1977, MNRAS,
        181, 465

191   WALKER,M.F., 1968, AP.J., 151, 71

164   WALKER,M.F., CHINCARINI,G., 1967, AP.J., 147, 416

165   WALKER,M.F., HAYES,S., 1967, AP.J., 149, 481

864   WALSH,D., WILLS,B.J., WILLS,D., 1979, MNRAS, 189,
        667.

965   WARD,M.J., PENSTON,M.V., BLADES,J.C., TURTLE,A.J.,
        1980, MNRAS, 193, 563

760   WARD,M.J., WILSON,A.S., 1978, ASTRON.AND AP., 70,
        L79

669   WARD,M.J., WILSON,A.S., DISNEY,M.J., ELVIS,M., MACCACARO,T.,
        1977, ASTRON.AND AP., 59, L19

761   WARD,M.J., WILSON,A.S., PENSTON,M.V., ELVIS,M., MACCACARO,T.,
        TRITTON,K.P., 1978, AP.J., 223, 788

414   WARNER,P.J., WRIGHT,M.C.H., BALDWIN,J.E., 1973, MNRAS, 163, 163

85   WAYMAN, 1963, QUOTED IN REF.84

865   WEBSTER,B.L., GOSS,W.M., HAWARDEN,T.G., LONGMORE,A.J., MEBOLD,U., 1979, MNRAS, 186, 31

264   WEEDMAN,D.W., 1970, AP.J., 159, 405

265   WEEDMAN,D.W., 1970, AP.J., 161, L113

192   WEEDMAN,D.W., KHACHIKYAN,E.E., 1968, ASTROFIZIKA, 4, 587 (ASTROPHYSICS 4, 243)

224   WEEDMAN,D.W., KHACHIKYAN,E.E., 1969, ASTROFIZIKA, 5, 113 (ASTROPHYSICS, 5, 51)

528   WELCH,G.A., CHINCARINI,G., ROOD,H.J., 1975, A.J., 80, 77

225   WELCH,G.A., WALLERSTEIN,G., 1969, PASP, 81, 23

226   WELIACHEW,L., 1969, ASTRON.AND AP., 3, 402

301   WELIACHEW,L., 1971, PASP, 83, 609

470   WELIACHEW,L., 1974, AP.J., 191, 639

415   WELIACHEW,L., GOTTESMAN,S.T., 1973, ASTRON.AND AP., 24, 59

762   WELIACHEW,L., SANCISI,R., GUELIN,M., 1978, ASTRON. AND AP., 65, 37

598   WEST,R.M., 1976, ASTRON.AND AP., 46, 327

599   WEST,R.M., 1976, ASTRON.AND AP., 53, 435

670   WEST,R.M., 1977, ASTRON.AND AP.SUPPL., 27, 73

866   WEST,R.M., 1979, ASTRON.AND AP., 71, 262

867   WEST,R.M., 1979, IAU CIRC. N.3415

763   WEST,R.M., BORCHKHADZE,T.M., BREYSACHER,J., LAUSTSEN,S., SCHUSTER,H.E., 1978, ASTRON.AND AP.SUPPL., 31, 55

764   WEST,R.M., DANKS,A.C., ALCAINO,G., 1978, ASTRON.AND.

AP., 65, 151

765   WESTERLUND,B.E., BERGWALL,N.A.S., EKMAN,A.B.G.,
       LAUBERTS,A., 1978, ASTRON.AND AP.SUPPL., 31, 427

138   WESTERLUND,B.E., SMITH,L.F., 1966, AUSTRAL.J.PHYS.,
       19, 181

139   WESTERLUND,B.E., STOKES,N.R., 1966, AP.J., 145, 354

227   WESTERLUND,B.E., WALL,J.V., 1969, A.J., 74, 335

966   WESTIN,A.M., 1980, ASTRON.AND AP., 89, L11

766   WESTPHAL,J.A., KRISTIAN,J., QUOTED BY KRISTIAN J.,
       SANDAGE,A. KATEM B., 1978, AP.J. 219, 803

529   WESTPHAL,J.A., KRISTIAN,J., SANDAGE,A., 1975, AP.J.,
       197, L95

354   WHITEHURST,R.N., ROBERTS,M.S., 1972, AP.J., 175, 347

355   WHITEOAK,J.B., 1972, AUSTRAL.J.PHYS., 25, 233

302   WHITEOAK,J.B., GARDNER,F.F., 1971, ASTROPHYS. LETT.,
       8, 57

416   WHITEOAK,J.B., GARDNER,F.F., 1973, ASTROPHYS. LETT.,
       15, 211

671   WHITEOAK,J.B., GARDNER,F.F., 1977, AUSTRAL.J.PHYS.,
       30, 187

967   WILKERSON,M.S., 1980, AP.J., 240, L115

530   WILLIAMS,T.B., 1975, AP.J., 199, 586

868   WILLS,B.J., WILLS,D., 1979, AP.J. SUPPL., 41, 689

166   WILLS,D., 1967, AP.J., 148, L57

228   WILLS,D., BOLTON,J.G., 1969, AUSTRAL.J.PHYS., 22,
       775

471   WILLS,D., WILLS,B.J., 1974, AP.J., 190, 271

600   WILLS,D., WILLS,B.J., 1976, AP.J. SUPPL., 31, 143

869   WILSON,A.S., PENSTON,M.V., 1979, AP.J., 232, 389

601   WILSON,A.S., PENSTON,M.V., FOSBURY,R.A.E., BOKSENBERG,A.,
        1976, MNRAS, 177, 673

531   WINTER,A.J.B., 1975, MNRAS, 172, 1

672   WRIGHT,A.E., JAUNCEY,D.L., PETERSON,B.A., CONDON,J.J.,
        1977, AP.J. 211, L115.

870   WRIGHT,A.E., PETERSON,B.A., JAUNCEY,D.L., CONDON,J.J.,
        1979, AP.J., 229, 73

417   WRIGHT,M.C.H., SEIELSTAD,G.A., 1973, ASTROPHYS. LETT.,
        13, 1

673   WYCKOFF,S., WEHINGER,P.A., 1977, ASTROPHYS. AND SPACE
        SCI., 48, 421

871   ZASOV,A.V., KARACHENTSEV,I.D., 1979, SOVIET ASTRON.
        LETT., 5, 126

767   ZEALEY,W., 1978, IAU CIRC. N.3293

 97   ZWICKY,F., 1964, AP.J., 139, 514

 98   ZWICKY,F., 1964, AP.J., 140, 1467

122   ZWICKY,F., 1965, AP.J., 142, 1293

123   ZWICKY,F., 1965, AP.J., 143, 192

472   ZWICKY,F., 1974, IN: SUPERNOVAE AND SUPERNOVA REMNANTS,
        COSMOVICI ED. REIDEL, P.1

 38   ZWICKY,F., HUMASON,M.L., 1960, AP.J., 132, 627

 39   ZWICKY,F., HUMASON,M.L., 1960, QUOTED BY HUMASON,M.L.,
        GATES,H.S., 1960 PASP, 72, 208

 59   ZWICKY,F., HUMASON,M.L., 1961, AP.J., 133, 794

 99   ZWICKY,F., HUMASON,M.L., 1964, AP.J., 139, 269

124   ZWICKY,F., KARPOWICZ,M., KOWAL,C.T., 1965, CATALOGUE
        OF GALAXIES AND OF CLUSTERS OF GALAXIES, CALIFORNIA
        INSTITUTE OF TECHNOLOGY VOL. V, FIELD 385, PAG.52

266   ZWICKY,F., OKE,J.B., NEUGEBAUER,G., SARGENT,W.L.W.,
        FAIRALL,A.P., 1970, PASP, 82, 93

 60   ZWICKY,F., QUOTED BY HUMASON,M.L., 1961, PASP, 73,
        185

229   ZWICKY,F., SARGENT,W.L.W., KOWAL,C., 1969, PASP, 81,

224

303   ZWICKY,F., ZWICKY,M.A., 1971, CATALOGUE OF SELECTED
       COMPACT GALAXIES AND OF POST ERUPTIVE GALAXIES
       (ZURICH)

CHAPTER 7

Notes to the References

8   No coordinates are given for the companion of NGC3769 and
     therefore no radial velocity is reported in the catalogue
     since we were unable to identify it. The components of
     HOL485 (NGC4782 and NGC4783) have been exchanged in tab.1 of
     this ref. i.e.the object with a higher redshift is NGC4783
     and not NGC4782.

11   We were unable to identify NGC224'.

27   The VH values for NGC598 and NGC5457 are reported under
     ref.26.

40   As far as we understand it the only new measurements reported
     in this ref. are the redshifts of the members of VV150 and
     therefore only those appear in the catalogue.

59   Redshifts for galaxies members of the Wild triplet were
     published in ref.13.

77   Several measurements reported in this ref. were also
     previously published elsewhere by the same authors and
     therefore are not repeated.

81   Results from the same measurements are also reported in an
     abstract in A.J., 1963, 68, 295 where the VH of M33 (NGC598)
     is given as 179+/-1 where, obviously, a minus sign has been
     omitted.

99   Tab.1, galaxy N.11, read NGC548 and not NGC598.

112 Redshifts for NGC4631 and NGC4656 are also published in
     ref.50.

232 Spectra of MRK291 and MRK298, also published in ref.268 are
     listed under the latter ref.

241 In Tab.1 read 1ZW17=1ZW18 and 1ZW114=1ZW115.

261 In tab.1 NGC2726 is misidentified with MRK18.

262 O'Konnell,R.W.and Kraft,R.P.,(1972, Ap.J., 175, 335) point
     out that all objects in the 1ZW catalog with running number N
     greater than 4 are incorrectly given in ref.262 with running
     number N - 1 (e.g. 1ZW128 should be 1ZW129). The radial
     velocity of 1ZW1 was already published in ref 396.

268 Some of the specta contained in this ref. were previously
     published by the same authors in ref.232.

321 The spectra of galaxies in the Perseus cluster listed in this
     ref. were previously published in ref.276 and therefore

catalogued only once. Two galaxies are listed at 22 52 00 +32 13.0 and identified as NW and SE. However one is led to believe (2RCBG) that SE is 4ZW123A (22 51 48 +32 14) which actually is NW in the pair.

340   The radial velocity of IC342 was also published in ref.397 and is listed under that ref.

379   The value of VH given for this reference is quoted in the paper as Wright 1973 in preparation. However we were unable to find the paper by Wright in the literature.

401   In this ref.measurements of redshifts for distant clusters of galaxies are given. In the catalogue we have reported only those measurements for which there was clear indication of which galaxy had been observed.

403   Only VH of NGC7318B is a new measurement. Other redshifts are contained in ref.180.

426   The authors state that their radial velocities are:"heliocentric velocities corrected for galactic rotation" We ASSUMED a correction was applied.

448   VH for galaxy G in VV166 is also given in the paper but we were unable to find its coordinates.

471   In tab.3 of this ref. the radiosource OX169 is considered a galaxy whilst it is generally considered a QSO.

485   In tab.1 NGC708 is listed as NGC703 (R.Fanti private communication)

492   In tab.2 of this ref. read DDO135=NGC4523 and not DDO135=NGC4522.

510   Despite Sandage's statement (ref.741 page 907) according to which measurements were supposed to supersede previous ones, 3 galaxies in the list of this ref. were missing in the aforementioned paper and therefore listed.

522   Only NGC6454 has a new VH measurement. The other two galaxies in this ref. are listed also in ref.485 and therefore appear under that ref.

548   This paper apperently is a compilation of redshifts some of which to be published. However all VH values labelled CR3 in table 1, as far as we know have never been published and therefore are listed in the present CRVG.

581   Partially revised in ref.693. Data unrevised are reported. This ref. also contains data from Dickel and Rood (1975, A.J., 80, 584).

582   NGC4662 is misidentified as NGC4462 (object 120 in tab.1) Furthermore for objects N 131, 134, 154, 167, no coordinates are given in tab.1 although redshifts are given in tab.2.

583   Measurements referring to A520 are also published in ref.529.

586   The radial velocity of the radiosource H2147+14 (Z=0.196) has not been included because of lack of coordinates.

623   In tab.1 the galaxy listed as NGC598 is in fact NGC548.

633   No coordinates but finding charts only are given in this ref. Object N35 in Abell2029 (tab.1) is missing in the finding chart and therefore unidentifiable among the hundreds of galaxies of the cluster. The remarks for redshifts are B for background, F for foreground but no caption is provided for remark A at the end of the tables.

641   In tab. 1 NGC4051 is incorrectly labelled NGC4951.

662   The measurements defined helicentric in the text (see Tab.1a) are compared to some measurements from ref.13 which are corrected (see Tab.1b). It could therefore be the case that the measurements presented are in fact corrected for galactic rotation.

663   In tab.1 of this ref. read 3C 277.3 instead of 3C 377.3.

670   The radial velocities of seven galaxies are not listed under this ref. since they are alredy published in ref. 598 and listed there.

683   This reference lists also the radial velocities of 12 emission line Tololo galaxies with unpublished coordinates, which therefore were not included in the catalogue.

690   No velocities are reported in the catalogue for galaxy B1 and B2 (tab.1) because these galaxies are not marked on the finding charts and no coordinates are given.

693   Data in this reference revise part of those in ref. 581. The data unrevised is reported here. Data from Dickel and Rood (1975, A.J., 80, 584) also revised in this ref. except for one galaxy which was listed also in ref.581.

708   Coordinates for 3CR66 given in this reference are wrong.

717   Coordinates referring to optical identification of 3C460 are incorrect. Instead of 23 19 00 +23 30 22 read 23 19 00 +23 19.00 as from ref.946.

740   Some of the redshifts in tab.1 were also published in other references by the same authors and therefore listed under the other references (ref.729, 738, 936).

741   The object called NGC4470 has the coordinates of NGC4472 and VH given in Tab.4 is compatable with VH of NGC4472 so in this catalog the measu-rements of this ref. have been attributed to NGC4472 and not to NGC4470 as suggested. The coordinates of A509 are 05 09 25 -14 51.0 and not 05 09 25 -14 15.1 In tab.5 the radiosource B2 1113+29 has a misprint in the coordinates, read 11 hours instead of 10 hours in R.A.

744   This ref. contains the radio measurements presented in Ap.J., 1975, 198, 527.

750   Out of 12 radial velocities measured for cD suspected galaxies in this ref. only 4 are reported in the catalogue because of lack of coordinates.

756   In this ref. 76 radial velocities for galaxies members of 4 clusters are listed; unfortunately no coordinates are given only finding charts. We were able to find coordinates for 20 galaxies in 1615+35 and one galaxy in 1621+38. For lach of positions all other velocities are missing from the catalogue.

779   See note to ref.780.

780   For some objects the authors confirm the RCBG2 value and therefore do not report their own measurements and state that their values are equal to the RCBG2 values within the errors.

783   Both this ref. and ref.946 report the same value for the redshift of 3C300 i.e. the Carnegie Institute Yearbooks of 1972. However we must point out that the abovementioned Yearbook does not contain such a value; actually there is no mention of 3C300 at all. Furthermore the radiosource identification is attributed to their ref.48 which unfortunately is missing from their reference list. The value of the radial velocity of 3C300, although listed here because present in two authoritative sources should therefore be taken with caution.

789   Coordinates of NGC4235 have wrong declination. The correct value is +07 28 and not +16 28.

793   The radial velocities of two galaxies (18 42 -63 23 and 18 43 -63 12) were published also in ref.801 and therefore listed here only under that ref.

806   The radial velocities given in this ref. were corrected for
        relativistic effects. This correction has been removed.

853   Galaxy N.27 does not appear in the finding charts provided by
        the authors and we were unable to find its coordinates.
        Therefore its redshift is not listed here.

858   Two galaxies are indicated in this reference as A1127 and
        A1214.   Their names should read A 1127+22 and A1214+29 and
        must not be confused with A1127 A1214 as in 1RCBG.

899   The radial velocity of the optical counterpart of 3CR200 has
        been listed only once under ref.946.

903   Galaxies N.34(14W119) and 36(14W115) (Tab.1)   were   not
        included because no coordinates are given and the finding
        charts turned out to be radio maps (see Harris et al., 1979,
        Astron. and Ap. Suppl., 39, 215) with crosses for optical
        objects.

908   The coordinates of KDG281 given in page 143 are: NGC3646 (11
        19 00 +20 36) ;   NGC3649 (11 19 36 +20 28). The correct
        declination should be +20 26.7 for NGC3646 and +20 29.0 for
        NGC3649 respectively.

921   Radial velocities of MRK1127 and MRK1133 were also published
        by the same authors in ref.920 and listed here under that
        ref.

936   Redshifts for NGC801, NGC3672 and NGC7664 in tab.1 where also
        published in ref.740 and therefore are listed under that ref.

APPENDIX A

Peculiar Names

ABB.	NAMES	RA	DEC
AMB KNT	AMBARTSUMIAN KNOT	11 08 31	+28 57.8
AND NAEB	ANDROMEDA NEBULA = NGC224	00 40 00	+41 00.1
BAD A	BAADE A	00 47 06	+42 19.
BAD B	BAADE B	00 47 06	+42 20.
BRB GAL	BARBON GALAXY	23 35 09	+29 51.2
CEN A	CENTAURUS A = NGC1258	13 22 32	−42 45.5
CIR GAL	CIRCINUS GALAXY	14 09 18	−65 06.3
COP SEP	COPELAND SEPTET = NGC3745−54	11 35 12	+22 15.
CRW GAL	CARTWEEL GALAXY = A0035	00 35 14	−33 59.4
DRA SYS	DRACO SYSTEM = DDO208	17 19 24	+57 58.1
FAI9	FAIRALL 9	01 21 51	−59 03.9
FAT 703	FATH 703	15 11 00	−15 17.
FFO	FOURCADE FIGUEROA OBJECT	13 31 48	−48 17.6
FOR A	FORNAX A = NGC1316	03 20 47	−37 23.1
FOR SYS	FORNAX SYSTEM	02 38 06	−34 41.
HOL I	HOLMBERG I = DDO63	09 36 00	+71 24.9
HOL II	HOLMBERG II = DDO50	08 13 43	+70 52.3
HOL IIX	HOLMBERG VIII = DDO166	13 10 59	+36 28.6
HOL IV	HOLMBERG VI = DDO185	13 52 56	+54 08.8
HOL V	HOLMBERG V	13 38 42	+54 35.0
HOL VII	HOLMBERG VII = DDO137	12 32 12	+06 34.3
HYA A	HYDRA A	09 15 41	−11 53.0
KLE 44	KLEMOLA 44 GROUP	23 44 54	−28 25.
LEO TRI	LEO TRIPLET = HOL246A−C	11 16 18	+13 22.
LMC	LARGE MAGELLANIC CLOUD	05 24 00	−69 48.
MAF I	MAFFEI I	02 32 36	+59 26.
MAF II	MAFFEI II	02 38 08	+59 23.4
PEG SYS	PEGASUS SYSTEM = DDO216	23 26 03	+14 28.3
PER A	PERSEUS A	03 16 30	+41 19.8
PIN	PINWHEEL = NGC5457 = M101	14 01 28	+54 35.6
REI 80	REINMUTH 80 = NGC4517A	12 29 54	+00 39.9
SCP SYS	SCULPTOR SYSTEM	00 57 36	−33 58.
SEX A	SEXTANS A = DDO75	10 08 34	−04 27.7
SEX B	SEXTANS B = DDO70	09 57 23	+05 34.1
SEY SEX	SEYFERT SEXTET = NGC 6027A−D	11 57 00	+20 54.
SMC	SMALL MAGELLANIC CLOUD	00 51 00	−73 06.
SOM	SOMBRERO = NGC4594 = M104	12 37 23	−11 21.0
STP QUI	STEPHAN QUINTET = NGC7317−20	22 33 36	+33 41.0
UMI DWF	URSA MINOR DWARF = DDO199	15 08 25	+67 27.9
WHIR	WHIRLPOOL = NGC5194 = M51	13 25 47	+47 27.3
WLD TRI	WILD TRIPLET = ARP248A−C	11 44 12	−03 33.
WLM GAL	WOLF LUNDMARK MELOTTE = DDO221	23 59 23	−15 44.1
X COM	X COMAE	12 57 58	+28 40.2
ZW SEY	ZWICKY SEYFERT	00 39 30	+40 03.0
ZW SEY	ZWICKY SEYFERT	09 34 30	+01 20.
ZW TRI	ZWICKY TRIPLET = ARP103A−C	16 47 58	+45 32.5
ZWO N.2	ZWICKY OBJECT N.2 = DDO105	11 55 54	+38 21.

APPENDIX B

Messier's Galaxies

MESSIER'S GALAXIES

M	NGC	R.A.	DEC.
M 31	NGC 224	00 04 00	+41 00.1
M 32	NGC 221	00 39 58	+40 35.5
M 33	NGC 598	01 31 03	+30 23.9
M 49	NGC4472	12 27 14	+08 16.7
M 51	NGC5194	13 27 47	+47 27.3
M 58	NGC4579	12 35 13	+12 05.7
M 59	NGC4621	12 39 31	+11 55.2
M 60	NGC4649	12 41 09	+11 49.4
M 61	NGC4303	12 19 22	+04 45.1
M 63	NGC5055	13 13 35	+42 17.9
M 64	NGC4826	12 54 17	+21 57.1
M 65	NGC3623	11 16 18	+13 22.0
M 66	NGC3627	11 17 38	+13 15.8
M 74	NGC 628	01 34 00	+15 32.0
M 77	NGC1068	02 40 06	−00 14.0
M 81	NGC3031	09 51 27	+69 17.8
M 82	NGC3034	09 51 43	+69 55.0
M 83	NGC5236	13 34 10	−29 36.8
M 84	NGC4374	12 22 32	+13 09.9
M 85	NGC4382	12 22 53	+18 28.0
M 86	NGC4406	12 23 42	+13 13.5
M 87	NGC4486	12 28 18	+12 40.0
M 88	NGC4501	12 29 27	+14 41.7
M 89	NGC4552	12 33 06	+12 50.0
M 90	NGC4569	12 34 18	+13 26.4
M 91	NGC4548	12 32 54	+14 46.0
M 94	NGC4736	12 48 32	+41 23.5
M 95	NGC3351	10 41 18	+11 58.0
M 96	NGC3368	10 44 08	+12 05.1
M 98	NGC4192	12 11 12	+15 10.0
M 99	NGC4254	12 16 18	+14 42.0
M100	NGC4321	12 20 23	+16 06.0
M101	NGC5457	14 01 28	+54 35.6
M102		NOT EXISTENT	*
M104	NGC4594	12 37 18	−11 21.0
M105	NGC3379	10 45 12	+12 51.0
M106	NGC4258	12 16 30	+47 34.9
M108	NGC3556	11 08 37	+55 56.7
M109	NGC3992	11 55 00	+53 39.0
M110	NGC 205	00 37 38	+41 24.9

* See Mallas,J.H., and Kreimer,E., The Messier Album and
Gingerich's Preface for discussion. The name M102 in the
2RCBG is attributed to NGC5866.

APPENDIX C

Arp's Peculiar Objects

ARP	RA	DEC	ARP	RA	DEC
1	9 21.2	49 34.	103	16 48.0	45 33.
2	16 14.8	47 10.	104	13 30.5	63 1.
3	22 34.0	-3 10.	105	11 8.5	28 59.
4	1 46.0	-12 38.	106	12 13.1	28 27.
5	11 21.8	3 36.	111	13 59.5	34 4.
6	8 9.7	46 9.	112	23 58.9	31 10.
8	1 19.9	-1 8.	113	0 15.9	29 46.
9	8 9.2	73 45.	116	12 41.0	11 51.
12	8 32.2	28 39.	117	14 7.6	17 56.
13	22 57.6	15 43.	118	2 52.7	0 23.
14	22 33.0	-26 19.	119	1 16.8	12 11.
15	22 49.0	-5 50.	120	12 25.2	13 17.
16	11 17.6	13 16.	122	16 2.2	17 53.
18	12 3.0	50 49.	123	5 20.3	-11 33.
19	0 29.2	-5 26.	125	16 36.7	42 2.
22	11 57.0	-18 59.	126	1 55.6	2 50.
23	12 39.1	41 25.	127	0 36.5	-9 17.
24	10 51.6	57 15.	129	9 36.4	32 36.
26	14 1.5	54 36.	133	1 23.2	-1 37.
27	11 18.2	53 27.	134	12 27.2	8 17.
28	23 26.0	22 9.	135	2 37.2	38 51.
29	20 33.8	59 59.	136	14 57.2	54 5.
30	17 22.3	62 13.	137	9 31.4	10 20.
31	1 48.4	21 40.	138	11 56.2	25 19.
32	17 12.6	59 24.	139	13 5.0	26 59.
37	2 40.1	0 14.	140	0 48.5	-7 20.
38	17 31.4	75 44.	141	7 8.2	73 34.
41	3 7.8	-20 47.	142	9 35.1	2 58.
42	15 0.5	23 32.	143	7 43.5	39 9.
46	23 31.4	29 47.	144	0 3.4	-13 41.
49	14 30.0	8 18.	145	2 20.0	41 9.
53	10 32.6	-16 54.	146	0 4.2	-6 55.
55	9 12.6	44 31.	147	3 8.8	1 8.
58	8 29.2	19 22.	148	11 1.1	41 7.
63	9 36.3	32 32.	150	23 17.0	9 13.
65	0 19.2	22 7.	151	11 22.7	54 39.
68	23 46.2	3 54.	152	12 28.3	12 40.
71	16 2.9	17 54.	153	13 22.5	-42 46.
72	11 44.7	18 2.	154	3 20.8	-37 23.
73	16 33.7	46 19.	155	11 20.8	54 7.
75	1 48.8	-4 18.	157	1 22.0	3 32.
76	12 34.3	13 26.	158	1 22.5	33 46.
77	2 44.2	-30 29.	159	12 49.3	26 3.
78	1 56.6	18 46.	160	12 11.7	54 48.
80	8 42.6	74 17.	162	10 48.6	28 15.
81	18 13.2	68 20.	163	12 42.8	27 24.
82	8 8.2	25 21.	166	1 54.6	32 58.
83	11 37.5	15 36.	167	8 46.5	19 16.
84	13 56.4	37 42.	168	0 40.0	40 36.
85	13 27.8	47 27.	169	22 12.3	13 36.
86	23 44.5	29 11.	170	23 14.8	18 26.
87	11 38.1	22 42.	172	16 3.3	17 44.
90	15 24.4	41 51.	173	14 49.0	9 32.
91	15 32.3	15 22.	174	9 55.8	29 7.
92	23 16.4	0 1.	175	12 30.6	11 37.
94	10 20.8	20 6.	176	13 1.3	-11 14.
97	12 3.2	31 22.	178	14 22.0	35 5.
98	1 29.3	31 51.	182	23 25.4	8 30.
99	23 12.8	18 41.	184	5 36.6	69 21.
102	17 18.0	49 2.	185	16 35.0	78 18.

ARP OBJECTS

ARP	RA	DEC	ARP	RA	DEC
186	4 31.6	-8 41.	280	11 35.1	48 10.
189	12 41.2	16 40.	281	12 39.7	32 49.
191	11 4.7	18 42.	282	0 34.2	23 43.
192	10 34.4	18 24.	283	9 14.2	42 13.
193	13 18.3	34 24.	284	23 33.8	1 53.
194	11 55.4	36 40.	286	14 17.6	4 13.
197	11 28.2	20 45.	287	8 59.7	26 8.
199	14 15.0	36 48.	294	11 37.1	32 11.
202	8 57.1	35 55.	295	23 39.2	-3 57.
205	10 51.7	54 35.	298	23 0.7	8 36.
206	10 49.7	36 53.	299	11 25.7	58 50.
209	16 3.0	20 41.	300	9 23.8	68 40.
210	4 26.1	64 44.	301	11 7.2	24 32.
212	23 18.0	16 57.	302	14 54.8	24 48.
213	4 2.6	69 41.	303	9 43.8	3 18.
214	11 29.8	53 21.	304	3 8.8	-9 7.
215	9 10.9	40 19.	307	9 23.0	11 39.
216	23 26.2	3 14.	308	1 23.4	-1 36.
217	10 35.7	53 46.	310	10 15.4	22 5.
220	15 32.8	23 40.	311	17 26.6	58 35.
221	13 46.1	25 26.	312	16 48.3	46 48.
222	23 37.3	-12 34.	313	11 55.0	32 34.
223	23 15.4	-4 56.	314	22 55.3	-4 3.
224	11 48.5	55 22.	315	9 16.7	33 58.
225	8 49.2	78 25.	316	10 15.7	22 9.
226	22 18.0	-24 56.	317	11 16.3	13 22.
227	1 17.2	3 9.	318	2 7.2	-10 23.
229	1 20.9	33 1.	319	22 33.7	33 42.
232	9 31.0	10 22.	320	11 35.3	22 13.
233	10 29.4	54 39.	321	9 36.4	-4 37.
234	11 33.1	54 48.	322	11 29.9	53 14.
235	0 6.2	15 32.	323	23 51.6	0 6.
238	13 13.6	62 24.	324	15 59.9	16 2.
239	13 39.9	55 56.	327	5 19.1	6 38.
240	13 37.4	1 5.	329	11 29.1	71 6.
242	12 43.7	31 0.	330	16 48.0	53 30.
243	8 35.4	25 56.	331	1 4.7	32 3.
244	11 59.3	-18 36.	333	2 36.5	10 38.
245	9 43.4	-14 8.	335	11 1.8	5 6.
247	8 20.7	21 29.	336	8 51.7	58 56.
248	11 44.2	-3 34.	337	9 51.7	69 55.
253	9 40.9	-5 4.			
258	2 36.4	18 9.			
259	4 59.1	-4 20.			
260	12 11.1	16 24.			
261	14 46.8	-9 58.			
262	23 54.2	16 33.			
263	10 22.4	17 25.			
264	10 0.8	40 58.			
266	12 56.7	35 7.			
267	10 33.9	31 48.			
268	8 13.7	70 52.			
269	12 28.1	41 59.			
270	10 47.1	33 15.			
271	14 0.8	-5 50.			
272	16 3.1	17 54.			
273	2 18.4	39 9.			
274	14 32.7	5 35.			
277	12 52.3	2 56.			
279	3 11.9	-2 59.			

APPENDIX D

Vorontsov-Velyaminov's Interacting Galaxies

VV OBJECTS

VV	RA	DEC	VV	RA	DEC
1	13 27.9	47 32.	117	7 43.5	39 8.
3	11 28.2	20 45.	118	11 25.7	58 50.
5	23 44.5	29 12.	119	10 0.8	40 58.
7	15 0.5	23 32.	120	13 59.5	34 4.
8	11 53.6	-19 37.	122	1 55.6	2 50.
9	8 8.2	25 21.	123	7 8.2	73 34.
10	17 18.1	49 6.	126	11 55.4	36 40.
11	10 49.7	36 53.	127	12 42.0	34 40.
13	12 3.2	31 20.	128	12 11.1	16 24.
14	10 51.6	57 15.	138	8 9.7	46 9.
16	15 44.7	18 2.	140	14 46.8	-9 58.
19	13 39.8	55 55.	141	7 24.9	72 37.
20	23 17.0	9 13.	143	2 36.4	18 9.
21	14 0.8	-5 50.	144	11 22.7	54 39.
22	11 20.8	54 70.	147	12 9.7	18 24.
30	12 28.1	41 56.	150	11 29.9	53 14.
31	11 48.5	55 22.	159	15 59.9	16 2.
32	11 1.1	41 7.	166	0 15.8	29 49.
33	13 30.5	63 1.	169	5 19.1	6 38.
34	23 39.5	-3 54.	172	11 29.1	71 5.
35	11 44.0	-3 35.	175	1 53.8	5 23.
40	8 59.7	26 8.	179	12 1.6	20 30.
43	12 30.6	11 37.	181	23 14.7	18 25.
48	13 56.4	37 42.	188	12 25.2	13 17.
50	9 14.3	42 12.	189	1 54.6	32 58.
51	23 33.7	1 53.	190	13 46.1	25 26.
52	9 40.9	-5 4.	192	16 13.4	19 35.
54	1 46.0	10 14.	193	1 4.6	32 8.
55	13 37.4	1 5.	194	16 3.3	17 44.
56	12 41.2	16 40.	197	16 48.3	46 48.
58	9 25.5	76 42.	199	12 13.1	28 27.
65	10 25.7	-43 39.	201	12 52.0	-12 17.
66	11 57.0	-18 59.	206	12 41.2	11 49.
67	23 37.3	-12 34.	207	1 20.9	32 60.
68	22 49.0	-5 49.	208	23 51.6	0 6.
70	14 11.2	7 54.	209	10 20.7	20 9.
71	10 34.4	18 24.	210	14 14.9	36 48.
73	12 39.1	41 26.	211	13 34.5	10 56.
75	11 1.8	5 6.	212	16 2.2	17 53.
76	12 29.1	4 13.	213	16 2.3	17 51.
77	14 22.0	35 5.	215	16 4.2	15 49.
78	3 54.1	-42 31.	216	11 56.2	25 19.
79	8 35.4	25 56.	219	12 34.0	11 32.
80	0 6.2	15 32.	220	16 3.4	17 56.
81	0 48.5	-7 20.	224	12 43.7	31 0.
82	9 36.3	32 32.	226	23 58.9	31 10.
83	9 36.4	32 36.	228	11 37.1	32 11.
86	16 3.0	20 41.	229	11 7.2	24 32.
88	13 28.6	19 42.	230	11 55.5	28 9.
89	17 12.6	59 24.	231	1 22.0	3 32.
90	15 59.8	15 50.	232	17 22.3	62 13.
95	10 22.4	17 25.	237	11 8.5	28 59.
96	2 32.5	37 25.	239	11 4.7	18 42.
101	17 26.5	58 35.	244	15 32.2	15 22.
104	12 11.2	36 55.	245	11 59.3	-18 36.
106	8 23.9	68 39.	246	10 47.0	33 15.
109	14 43.8	8 42.	247	18 13.2	68 20.
112	10 40.2	13 43.	249	11 55.2	32 34.
115	15 57.0	20 54.	250	13 13.7	62 23.
116	9 36.4	-4 37.	251	11 21.8	3 36.

VV OBJECTS

VV	RA	DEC	VV	RA	DEC
252	10 50.8	17 2.	530	14 21.5	-16 30.
253	13 49.8	2 20.	533	10 6.9	54 45.
254	23 59.2	23 13.	538	10 46.7	33 1.
255	23 54.2	16 33.	539	7 30.4	74 34.
261	12 11.7	54 48.	541	9 2.8	25 38.
272	0 3.4	-13 41.	543	13 40.1	30 4.
273	11 48.1	56 44.	544	11 38.5	-6 13.
280	23 18.0	16 57.	553	9 30.4	23 21.
282	11 35.1	22 17.	555	4 27.5	6 50.
285	2 32.2	-9 0.	557	14 16.4	22 3.
288	22 34.0	33 44.	558	12 56.2	14 29.
291	15 47.9	69 38.	561	12 33.6	19 36.
294	10 26.6	70 18.	563	12 30.2	14 20.
295	22 55.3	-4 3.	565	4 59.2	-4 20.
297	20 11.7	-70 54.	569	18 16.0	30 38.
300	11 38.1	22 44.	587	3 11.6	-3 0.
304	19 14.2	-60 36.	594	11 25.5	22 16.
307	10 15.0	22 8.	596	6 5.0	34 16.
308	11 17.7	13 52.	601	5 52.3	48 32.
312	10 20.6	53 21.	611	16 1.4	39 46.
313	12 52.3	2 56.	612	9 1.8	18 40.
314	23 31.2	29 46.	614	12 16.5	9 8.
316	9 35.1	2 58.	615	14 15.0	-7 11.
317	13 44.9	34 8.	617	16 50.5	2 30.
323	2 18.5	39 8.	620	9 47.3	0 51.
324	14 48.9	35 47.	621	20 30.5	-2 11.
326	13 28.3	31 32.	622	0 23.1	25 27.
328	14 2.4	12 58.	625	16 31.8	29 4.
329	23 26.2	3 14.	628	10 52.4	58 9.
331	2 52.6	0 23.	633	13 17.8	33 20.
334	3 8.8	-9 7.	644	7 23.6	72 14.
335	13 53.6	17 45.	645	9 1.7	14 42.
338	1 44.7	27 5.	654	7 43.3	55 4.
339	13 57.8	13 12.	665	4 43.6	3 25.
340	14 54.8	24 48.	672	18 33.5	67 4.
343	23 25.4	8 30.	699	4 57.0	-11 12.
344	14 3.3	53 54.	713	14 44.0	51 47.
347	1 16.8	12 12.	716	9 25.4	50 59.
350	11 37.5	15 36.	727	10 44.3	26 48.
366	12 50.7	4 44.	731	23 15.9	-4 42.
367	11 47.5	25 13.	738	23 2.4	16 24.
371	14 25.6	21 32.	739	14 22.0	27 55.
384	12 0.3	18 19.	761	8 51.9	57 45.
394	14 2.7	54 38.	792	13 56.0	57 14.
426	17 44.6	30 43.	793	3 56.3	78 9.
432	12 15.1	12 40.	794	10 39.8	34 43.
442	16 56.3	20 7.	797	12 56.7	35 8.
454	12 6.8	44 22.	828	13 32.7	51 54.
457	11 50.4	-4 8.	842	14 43.3	31 39.
458	14 32.7	5 35.	851	13 42.8	56 8.
472	16 25.8	51 40.			
486	1 0.7	-7 16.			
493	12 41.8	45 17.			
497	12 10.5	52 32.			
498	11 26.7	20 50.			
517	14 43.3	51 35.			
519	8 42.6	74 17.			
520	12 12.0	6 5.			
523	11 54.9	32 37.			
529	10 41.6	7 1.			

APPENDIX E

Markarian Galaxies

MARKARIAN GALAXIES

MRK	RA			DEC		MRK	RA			DEC	
1	1	13	19	32	49.5	66	13	23	19	57	31.0
2	1	51	56	36	40.2	67	13	39	56	30	46.3
3	6	9	48	71	3.0	68	13	42	48	27	22.2
4	6	21	28	74	20.0	69	13	43	28	29	53.0
5	6	35	24	75	40.2	71	7	23	24	69	19.1
6	6	45	43	74	29.1	72	7	25	43	56	57.2
7	7	22	19	72	40.4	73	7	27	19	63	20.9
8	7	23	38	72	13.8	74	7	27	38	55	21.6
9	7	32	42	58	53.0	75	7	28	42	55	18.2
10	7	43	7	61	3.4	78	7	37	7	65	17.7
12	7	44	41	74	29.1	79	7	38	41	49	55.8
13	7	51	57	60	26.0	82	7	42	57	62	30.3
14	8	5	22	72	57.0	83	7	44	22	54	20.2
15	8	28	49	75	18.6	84	7	51	49	55	50.1
16	8	47	58	73	22.7	86	8	9	58	46	9.0
17	8	47	49	57	18.0	87	8	15	49	74	8.9
18	8	58	2	60	20.9	88	8	24	2	55	52.0
19	9	12	54	59	58.9	89	8	25	54	52	14.6
20	9	17	0	71	45.3	90	8	26	0	52	51.9
21	9	44	57	58	12.2	91	8	28	57	52	46.6
22	9	46	3	55	49.0	92	8	32	3	46	39.9
23	9	53	27	60	12.3	93	8	32	27	66	24.3
24	9	56	0	54	45.1	94	8	34	0	51	48.9
25	10	0	22	59	40.7	95	8	44	22	70	20.9
26	10	8	26	59	8.3	96	8	45	26	46	26.1
27	10	8	32	58	58.9	97	8	46	32	65	49.5
28	10	9	0	58	38.7	98	8	46	0	72	0.1
29	10	14	28	60	18.5	99	8	47	28	61	12.6
30	10	16	19	57	40.2	100	8	54	19	66	39.8
31	10	16	24	57	40.3	101	9	1	24	51	48.8
32	10	23	48	56	31.5	102	9	8	48	46	50.5
33	10	29	22	54	39.4	103	9	11	22	67	58.0
34	10	30	52	60	17.3	104	9	13	52	53	39.1
35	10	42	16	56	13.4	105	9	15	16	71	37.0
36	11	2	16	29	24.6	106	9	16	16	55	34.3
37	11	13	51	29	3.0	107	9	17	51	71	45.3
38	11	15	26	54	1.3	108	9	17	26	64	27.1
39	11	15	30	54	1.4	109	9	19	30	47	27.5
40	11	22	42	54	39.4	110	9	21	42	52	30.2
41	11	33	58	55	7.4	111	9	23	58	68	37.7
42	11	51	5	46	29.0	114	9	26	5	56	4.3
43	12	0	27	39	42.1	115	9	27	27	49	28.7
44	12	0	51	39	5.4	116	9	30	51	55	27.8
45	12	1	2	60	48.5	118	9	39	2	76	34.8
46	12	13	26	41	12.2	119	9	40	26	66	12.4
47	12	13	29	40	51.0	120	9	42	29	72	40.9
48	12	14	1	59	9.4	121	9	42	1	73	13.1
49	12	16	36	4	8.1	122	9	43	36	73	11.8
50	12	20	51	2	57.3	123	9	43	51	56	20.2
52	12	23	9	0	50.9	124	9	45	9	50	43.3
53	12	53	42	27	57.0	125	9	47	42	46	11.6
54	12	54	32	32	43.1	126	9	49	32	52	27.6
55	12	55	0	27	40.6	127	9	51	0	51	28.8
56	12	56	10	27	32.0	128	9	53	10	60	19.5
57	12	56	12	27	26.7	129	9	53	12	46	41.8
58	12	56	42	27	55.0	130	9	56	42	47	32.7
59	12	56	39	35	6.9	131	9	57	39	55	51.7
60	12	57	45	28	8.1	133	9	57	45	72	21.9
62	13	2	56	30	32.6	134	9	59	56	43	25.7
65	13	13	6	59	22.0	135	10	2	6	53	57.5

566

MARKARIAN GALAXIES

MRK	RA			DEC		MRK	RA			DEC	
136	10	3	42	77	9.1	202	12	15	42	58	56.1
138	10	9	18	67	39.5	203	12	15	18	44	27.0
139	10	12	46	44	2.2	205	12	19	46	75	35.2
140	10	13	25	45	34.3	206	12	21	25	67	43.0
141	10	15	39	64	13.2	207	12	22	39	54	46.9
142	10	22	23	51	55.7	209	12	23	23	48	46.1
143	10	23	28	62	35.4	210	12	24	28	48	33.1
144	10	23	54	44	15.7	211	12	25	54	44	46.1
145	10	31	44	64	42.4	212	12	25	44	44	43.0
146	10	32	5	46	49.1	212	12	25	5	44	43.0
147	10	32	26	63	47.7	212	12	25	26	44	43.0
148	10	32	37	44	34.4	213	12	29	37	58	14.3
149	10	34	39	64	31.5	214	12	29	39	66	2.3
150	10	35	40	44	46.9	215	12	30	40	46	2.5
151	10	39	15	48	1.7	216	12	30	15	52	3.1
152	10	45	54	50	18.2	219	12	36	54	56	12.1
153	10	46	4	52	35.8	220	12	41	4	55	10.2
154	10	47	49	50	26.0	220	12	41	49	55	10.2
155	10	48	24	44	50.1	221	12	41	24	55	10.8
156	10	50	11	50	33.0	222	12	43	11	47	22.5
157	10	52	6	49	59.6	223	12	43	6	71	35.5
158	10	56	2	61	47.7	224	12	44	2	48	30.5
159	10	55	25	72	54.7	225	12	44	25	47	26.0
161	10	59	7	45	29.8	226	12	45	7	72	11.2
162	11	2	18	45	1.0	229	12	46	18	47	59.2
163	11	3	35	48	54.3	230	12	53	35	64	32.0
164	11	9	34	51	54.4	231	12	54	34	57	8.6
165	11	15	37	63	33.1	232	12	55	37	59	20.2
166	11	16	29	62	45.4	233	12	56	29	59	24.2
168	11	23	1	47	16.4	234	12	58	1	64	42.8
169	11	23	53	59	25.8	235	12	57	53	33	42.3
170	11	23	56	64	24.8	236	12	58	56	61	55.5
171	11	25	43	58	50.4	237	12	59	43	48	19.8
172	11	26	51	22	3.5	238	12	59	51	65	16.1
173	11	27	45	48	23.0	241	13	3	45	33	14.3
174	11	28	32	56	24.7	242	13	3	32	53	45.5
175	11	29	38	62	47.0	243	13	8	38	60	51.3
176	11	29	54	53	13.5	244	13	10	54	50	39.5
177	11	30	37	55	20.9	245	13	10	37	67	46.0
178	11	30	45	49	30.7	246	13	11	45	56	21.7
179	11	30	51	62	9.9	247	13	12	51	55	3.8
181	11	34	18	20	14.0	248	13	13	18	44	40.0
182	11	34	18	20	12.2	250	13	15	18	44	4.4
183	11	36	0	68	49.2	251	13	17	0	52	18.8
185	11	38	36	47	58.2	253	13	18	36	56	41.9
186	11	43	17	50	28.7	254	13	20	17	51	59.9
187	11	43	46	71	54.2	255	13	20	46	53	13.7
188	11	44	55	56	14.9	256	13	21	55	70	46.4
190	11	49	10	48	57.7	257	13	25	10	55	44.7
191	11	50	29	70	42.7	258	13	26	29	53	42.1
192	11	52	44	51	24.3	259	13	26	44	44	11.4
193	11	52	52	57	56.4	260	13	27	52	45	15.7
194	11	59	36	66	40.7	261	13	29	36	75	49.7
195	12	0	3	64	39.3	262	13	29	3	75	49.8
196	12	1	48	66	52.6	263	13	31	48	69	7.0
197	12	5	18	67	39.8	264	13	22	18	58	8.7
198	12	6	43	47	20.1	265	13	35	43	28	1.4
199	12	8	4	70	38.8	266	13	36	4	48	31.8
200	12	8	52	48	48.6	267	13	37	52	43	18.3
201	12	11	40	54	48.3	268	13	38	40	30	37.8

MARKARIAN GALAXIES

MRK	RA			DEC		MRK	RA			DEC	
269	13	39	19	66	4.5	340	0	25	19	30	52.7
270	13	39	41	67	55.5	341	0	34	41	23	42.6
271	13	39	52	55	55.5	342	0	35	52	13	15.5
271	13	39	48	55	55.3	343	0	35	48	14	45.9
272	13	39	53	43	0.8	345	0	37	53	24	45.1
273	13	42	51	56	8.3	346	0	42	51	27	10.6
275	13	46	25	31	42.5	347	0	45	25	22	6.0
277	13	50	25	64	37.1	348	0	46	25	31	41.0
278	13	51	22	72	58.7	349	0	50	22	21	14.4
279	13	51	52	69	33.2	350	0	54	52	23	37.1
280	13	55	5	29	2.1	352	0	57	5	31	33.5
281	13	55	0	42	5.6	353	1	0	0	22	4.4
282	14	1	3	69	43.3	354	1	1	3	20	9.9
283	14	1	12	41	50.6	355	1	19	12	26	36.4
284	14	4	9	69	22.5	356	1	19	9	26	36.3
285	14	7	40	71	54.2	357	1	19	40	22	54.5
286	14	18	46	71	48.8	358	1	23	46	31	21.2
288	14	50	50	74	1.7	359	1	24	50	18	55.1
289	15	31	23	58	3.0	360	1	41	23	16	48.8
290	15	34	45	58	4.0	361	1	42	45	16	51.4
291	15	52	54	19	20.3	363	1	48	54	21	45.0
292	15	53	45	19	2.3	364	1	54	45	27	37.3
294	15	59	49	18	57.2	365	2	1	49	28	25.1
295	16	1	13	19	19.0	366	2	8	13	13	40.9
296	16	1	13	19	17.9	367	2	10	13	16	51.0
297	16	3	0	20	40.5	368	2	30	0	20	25.5
298	16	3	22	17	56.1	369	2	34	22	20	55.4
299	16	3	31	17	26.4	370	2	37	31	19	5.0
300	16	4	0	18	19.0	372	2	46	0	19	5.9
303	22	14	1	16	13.2	373	6	50	1	50	25.0
304	22	14	45	13	59.5	374	6	55	45	54	15.9
305	22	29	24	19	26.4	375	7	2	24	67	45.6
306	22	29	26	19	26.1	376	7	10	26	45	47.1
307	22	33	31	20	3.9	378	7	13	31	49	47.0
308	22	39	30	19	59.9	382	7	52	30	39	19.1
309	22	50	10	24	27.9	383	7	57	10	39	58.5
311	22	56	5	14	54.1	384	8	0	5	23	32.0
312	22	58	9	16	5.6	385	8	0	9	25	14.6
313	22	59	32	15	41.8	386	8	16	32	22	11.0
314	23	0	31	16	20.0	387	8	21	31	17	29.7
315	23	1	36	22	21.2	388	8	25	36	25	30.0
316	23	11	10	13	45.0	389	8	29	10	22	44.0
317	23	11	20	23	32.9	390	8	32	20	30	42.3
318	23	15	6	13	43.8	391	8	51	6	39	43.7
319	23	16	12	24	57.0	394	9	16	12	26	28.8
321	23	17	37	23	56.7	397	9	21	37	18	1.4
322	23	17	35	25	56.4	398	9	21	35	17	52.6
323	23	17	55	27	2.4	399	9	23	55	35	6.8
324	23	24	2	17	59.5	400	9	23	2	19	36.1
325	23	25	12	23	19.0	401	9	27	12	29	45.5
326	23	25	36	23	15.3	402	9	32	36	30	38.0
328	23	35	9	29	51.2	403	9	37	9	21	27.4
330	23	40	29	19	8.8	404	9	39	29	32	4.6
331	23	48	54	20	18.5	405	9	40	54	32	12.5
332	23	56	52	20	28.5	406	9	41	52	29	50.1
334	0	0	36	21	40.9	407	9	44	36	39	19.0
335	0	3	45	19	55.5	408	9	45	45	33	6.9
336	0	5	26	32	47.5	409	9	46	26	32	27.0
338	0	21	13	14	24.6	410	9	50	13	37	59.0
339	0	22	7	14	32.7	411	9	54	7	33	51.0

568

MARKARIAN GALAXIES

MRK	RA	DEC	MRK	RA	DEC
412	9 55 4	32 28.0	487	15 35 4	55 25.6
413	9 56 21	31 56.3	489	15 42 21	41 15.0
414	10 10 11	35 31.6	490	15 44 11	49 9.1
415	10 25 47	40 6.0	491	15 49 47	43 34.0
416	10 40 24	20 40.9	492	15 56 24	26 57.3
417	10 46 48	23 13.8	493	15 57 48	35 10.0
418	10 50 21	34 10.6	494	15 58 21	30 30.8
419	10 55 48	24 28.0	496	16 10 48	52 35.0
420	11 0 53	38 11.9	497	16 15 53	52 8.1
421	11 1 40	38 28.6	499	16 47 40	48 47.6
422	11 8 14	28 49.0	500	16 47 14	48 48.0
423	11 24 8	35 31.0	501	16 52 8	39 50.4
424	11 27 47	37 0.7	503	16 55 47	63 19.2
426	11 38 12	35 28.0	504	16 59 12	29 28.8
427	11 40 48	36 23.6	506	17 20 48	30 55.5
428	11 41 33	37 27.8	507	17 48 33	68 42.8
429	11 43 49	35 7.8	509	20 41 49	-10 54.3
430	11 48 29	55 21.5	512	21 9 29	-1 35.0
432	11 55 31	28 9.0	513	21 16 31	2 3.0
434	11 56 55	35 10.3	515	21 20 55	-7 57.7
435	12 9 25	40 55.7	516	21 53 25	7 8.0
438	12 20 22	22 43.3	518	21 56 22	11 48.0
439	12 22 8	39 39.7	520	21 58 8	10 18.7
440	12 25 2	36 58.0	522	22 57 2	16 7.0
441	12 42 6	41 1.0	523	22 58 6	7 2.0
442	12 44 59	35 37.5	524	22 58 59	9 20.0
444	12 46 17	34 44.8	526	23 10 17	10 27.0
445	12 46 37	40 52.1	527	23 10 37	6 3.0
446	12 47 44	33 25.8	528	23 11 44	12 54.3
447	12 55 44	24 36.9	529	23 11 44	-3 0.0
449	13 9 12	36 33.0	530	23 16 12	0 1.0
450	13 12 30	35 8.0	531	23 21 30	9 23.6
451	13 22 4	36 51.0	532	23 24 4	11 5.0
453	13 23 48	33 16.0	533	23 25 48	8 30.0
454	13 24 30	26 50.7	534	23 26 30	3 14.0
455	13 28 18	31 32.0	535	23 29 18	25 35.0
456	13 30 48	37 26.9	536	23 29 48	2 9.0
459	13 32 54	34 17.2	537	23 31 54	4 42.1
461	13 45 4	34 24.0	538	23 33 4	1 52.7
462	13 49 18	40 27.0	540	23 44 18	0 43.4
463	13 53 40	18 36.7	541	23 53 40	7 14.6
464	13 53 45	38 48.9	542	23 54 45	-2 22.0
465	13 59 15	37 2.4	543	23 59 15	3 4.5
467	14 10 22	34 46.9	544	0 2 22	-1 46.6
468	14 13 46	41 13.2	545	0 7 46	25 39.0
469	14 16 6	34 35.0	546	0 12 6	-1 49.7
470	14 20 25	37 20.9	547	0 17 25	0 33.3
471	14 20 47	33 4.6	548	0 18 47	6 9.1
472	14 26 56	36 9.8	549	0 19 56	-1 53.4
473	14 30 52	57 4.0	550	0 24 52	1 10.9
474	14 33 6	48 52.8	551	0 26 6	30 17.0
475	14 37 3	37 1.1	552	0 28 3	8 12.0
477	14 39 3	53 42.9	553	0 33 3	2 54.6
478	14 40 6	35 39.0	554	0 36 6	3 41.0
479	14 52 36	18 14.0	555	0 43 36	-1 59.7
480	15 4 42	42 50.0	556	0 46 42	4 3.6
482	15 26 47	55 42.8	557	0 46 47	-2 58.0
484	15 29 38	54 51.5	558	0 49 38	-2 29.0
485	15 30 18	51 56.0	559	0 56 18	6 40.0
486	15 35 22	54 43.1	560	1 3 22	0 24.7

MARKARIAN GALAXIES

MRK	RA	DEC	MRK	RA	DEC
562	1 9 13	0 56.0	630	10 20 13	18 12.0
563	1 9 18	-1 55.0	631	8 43 18	36 37.0
564	1 11 12	7 31.2	632	10 41 12	16 9.2
565	1 13 28	4 2.0	633	10 54 28	37 50.0
566	1 16 30	4 4.0	634	10 55 30	20 45.0
567	1 16 43	4 19.0	635	11 20 43	30 12.9
568	1 19 49	8 47.6	636	11 33 49	16 15.0
569	1 20 2	1 38.0	637	11 36 2	21 15.5
571	1 33 26	0 25.0	638	11 39 26	38 16.2
572	1 41 5	11 55.0	639	11 40 5	24 10.7
573	1 41 23	2 6.0	640	11 44 23	21 33.0
575	1 45 53	12 22.0	641	11 49 53	35 10.4
576	1 46 34	5 23.0	642	11 49 34	23 54.3
577	1 46 50	12 16.0	644	11 54 50	23 38.9
579	1 49 21	7 2.0	645	12 1 21	23 59.0
580	1 50 21	6 42.9	646	12 3 21	35 27.5
581	1 50 57	6 25.7	647	12 3 57	21 30.3
582	1 55 36	2 50.0	648	12 4 36	24 28.6
583	1 56 54	9 41.4	649	12 34 54	26 28.6
584	1 57 51	2 25.8	650	12 35 51	27 24.2
585	2 0 54	2 19.0	651	12 36 54	28 35.4
586	2 5 14	2 28.5	652	12 36 14	28 16.8
587	2 8 0	5 38.0	654	12 40 0	34 22.0
588	2 8 36	3 32.8	656	12 43 36	27 20.0
589	2 11 9	3 52.0	657	12 46 9	27 27.0
590	2 12 0	-1 0.0	659	13 20 0	21 41.2
591	2 14 37	1 28.0	662	13 51 37	23 40.5
592	2 17 7	0 29.0	665	13 59 7	34 4.0
593	2 23 54	11 56.0	666	14 1 54	33 48.5
595	2 38 56	6 58.5	667	14 2 56	21 52.3
597	2 43 18	15 38.5	668	14 4 18	28 41.6
598	2 43 52	7 11.6	671	14 13 52	34 45.3
599	2 45 6	2 57.0	673	14 15 6	27 5.3
600	2 48 30	4 15.0	674	14 16 30	22 2.9
601	2 54 1	-2 58.0	682	14 26 1	27 28.4
602	2 57 12	2 34.0	684	14 28 12	20 30.5
603	3 6 26	-3 8.5	685	14 28 26	27 27.5
604	3 10 18	-5 27.3	686	14 35 18	36 47.2
605	3 13 7	-3 39.0	688	15 14 7	19 16.6
606	3 17 12	3 58.0	689	15 34 12	30 50.8
607	3 22 18	-3 13.0	691	15 44 18	18 2.0
609	3 22 58	-6 19.0	693	15 51 58	23 16.7
610	3 23 3	-6 18.0	694	15 59 3	16 35.3
611	3 23 41	0 23.0	695	16 0 41	16 5.9
612	3 28 10	-3 18.6	696	16 2 10	28 13.7
613	3 33 58	-4 51.9	699	16 22 58	41 12.0
614	4 8 29	-7 24.6	700	17 1 29	31 31.4
615	4 22 0	0 52.0	704	9 15 0	16 31.0
616	4 28 12	0 33.0	705	6 23 12	12 57.0
617	4 31 36	-8 40.7	708	9 39 36	4 54.0
618	4 34 0	-10 28.0	710	9 52 0	9 30.0
619	6 27 16	57 15.9	712	9 54 16	15 53.0
620	6 45 36	60 54.0	716	10 7 36	23 20.0
622	8 4 21	39 9.0	721	10 20 21	11 13.0
623	8 13 12	26 8.0	728	10 58 12	11 19.0
624	8 21 12	25 51.0	731	11 10 12	9 20.0
625	8 35 31	19 53.5	734	11 19 31	12 1.0
626	8 42 24	37 7.0	739	11 33 24	21 52.0
628	8 47 54	29 25.0	745	11 37 54	17 13.9
629	10 14 36	15 45.0	747	11 39 36	16 15.0

MARKARIAN GALAXIES

MRK	RA	DEC	MRK	RA	DEC
766	12 15 54	30 5.0	1127	22 59 54	26 47.0
771	12 29 30	20 25.0	1128	23 0 30	38 26.0
783	13 0 6	16 39.0	1129	23 4 6	9 45.0
786	13 14 24	12 49.0	1130	23 8 24	0 27.0
789	13 29 54	11 22.0	1131	23 25 54	-2 28.0
799	13 59 9	59 34.1	1132	23 35 9	31 21.0
813	14 24 54	20 3.0	1133	23 41 54	27 27.0
816	14 31 41	52 59.5	1134	23 44 41	29 10.7
817	14 34 54	59 1.0	1135	23 48 54	28 43.0
841	15 1 36	10 38.0	1136	23 48 36	35 23.0
845	15 6 13	51 38.2	1137	23 58 13	26 3.0
849	15 17 51	28 45.4	1138	0 11 51	7 59.0
854	15 24 37	43 34.3	1139	0 13 37	21 8.0
865	15 56 55	58 18.2	1140	0 13 55	24 32.0
871	16 6 18	12 28.0	1141	0 15 18	22 14.0
876	16 13 36	65 50.6	1142	0 18 36	21 41.0
877	16 17 56	17 31.6	1143	0 39 56	2 59.0
883	16 27 47	26 33.1	1144	0 39 47	2 58.0
885	16 29 43	67 29.1	1146	0 44 43	14 26.0
915	22 34 7	-12 48.3	1147	0 46 7	10 4.0
917	22 38 48	31 54.5	1149	0 53 48	-14 33.0
930	23 29 30	28 40.3	1150	0 59 30	34 51.0
938	0 8 34	-12 23.1	1151	1 6 34	-13 5.0
945	0 23 22	-3 41.8	1152	1 11 22	-15 7.0
955	0 35 2	0 0.3	1153	1 20 2	0 57.0
984	1 16 45	12 11.1	1154	1 22 45	-1 49.0
1014	1 57 16	0 9.2	1155	1 23 16	33 9.0
1018	2 3 43	0 31.5	1156	1 29 43	32 55.0
1027	2 11 28	4 56.5	1157	1 30 28	35 25.0
1058	2 46 47	34 46.9	1158	1 32 47	34 48.0
1066	2 56 40	36 37.3	1160	1 35 40	34 56.0
1073	3 11 43	41 51.0	1161	1 35 43	-9 25.0
1096	15 24 0	67 20.0	1162	1 36 0	29 23.0
1097	15 24 6	71 5.0	1163	1 37 6	31 0.0
1098	15 27 42	30 39.0	1164	1 38 42	32 38.0
1099	15 47 54	69 38.0	1165	1 40 54	27 58.0
1100	15 50 18	41 53.0	1166	1 46 18	12 50.0
1101	15 54 54	42 1.0	1167	1 53 54	31 27.0
1102	15 55 36	41 41.0	1168	1 55 36	3 13.0
1103	15 55 48	41 40.0	1169	1 55 48	2 11.0
1104	16 4 0	41 29.0	1170	1 55 0	37 20.0
1105	16 12 48	12 53.0	1171	1 58 48	31 38.0
1106	16 14 36	18 38.0	1172	2 3 36	-8 28.0
1108	16 48 48	28 56.0	1173	2 5 48	20 8.0
1109	16 51 36	63 12.0	1174	2 5 36	1 39.0
1110	16 51 48	69 1.0	1175	2 10 48	31 38.0
1112	16 55 18	28 16.0	1176	2 24 18	41 47.0
1114	16 58 42	32 45.0	1177	2 24 42	-13 20.0
1115	17 1 6	33 8.0	1178	2 24 6	-13 21.0
1116	17 36 0	85 47.0	1179	2 30 0	27 45.0
1117	17 38 36	39 16.0	1180	2 33 36	33 6.0
1118	17 49 43	24 29.6	1181	2 36 43	3 50.0
1119	17 50 54	37 45.0	1182	2 37 54	16 36.0
1120	17 55 0	40 15.0	1183	2 39 0	28 21.0
1121	18 9 30	31 51.0	1184	2 43 30	-5 51.0
1122	18 25 6	42 39.0	1185	2 44 6	15 35.0
1123	22 5 6	44 3.0	1186	2 45 6	15 43.0
1124	22 28 12	-14 27.0	1187	2 45 12	13 44.0
1125	22 47 6	19 9.0	1188	3 1 6	-1 5.0
1126	22 58 12	-13 11.0	1189	3 2 12	-2 32.0

MARKARIAN GALAXIES

MRK	RA			DEC	
1190	3	4	36	-2	18.0
1191	3	40	12	-6	32.0
1192	3	53	18	-9	42.0
1193	4	4	42	-10	18.0
1194	5	9	6	5	8.0
1195	6	40	0	78	5.0
1196	6	59	36	39	18.0
1197	7	2	54	28	23.0
1198	7	8	0	26	0.0
1199	7	20	30	33	31.0
1200	7	21	54	27	26.0
1201	7	22	36	30	4.0
1202	7	42	24	28	34.0
1203	7	44	48	28	27.0
1204	7	46	36	29	4.0
1205	7	53	0	16	42.0
1206	7	54	24	18	48.0
1208	8	1	12	8	50.0
1209	8	1	18	10	9.0
1210	8	1	30	5	15.0
1211	8	3	0	7	44.0
1212	8	4	0	27	16.0
1218	8	35	18	25	5.0
1220	8	51	48	17	53.0
1221	9	0	24	18	27.0
1222	9	0	36	20	52.0
1224	9	1	42	14	42.0
1228	9	12	13	19	54.3
1229	9	13	0	21	9.0
1230	9	14	6	25	38.0
1236	9	47	18	0	51.0
1237	9	47	24	44	34.0
1239	9	47	42	-1	22.0
1243	9	57	18	13	17.0
1246	10	7	30	77	58.0
1247	10	7	55	16	55.9
1259	10	35	54	-6	54.0
1298	11	26	42	-4	8.0
1302	11	36	24	3	51.0
1308	11	51	42	0	25.0
1310	11	58	42	-3	23.0
1320	12	16	42	-1	33.0
1344	13	6	48	-5	0.0
1347	13	20	24	8	25.0
1352	13	30	0	13	5.0
1354	13	30	36	9	46.0
1355	13	30	48	9	47.0
1365	13	52	6	15	17.0
1379	14	15	1	-7	11.2
1382	14	25	36	-1	27.0
1383	14	26	30	1	31.0
1387	14	41	30	16	41.0
1391	14	58	30	17	9.0

APPENDIX F

Graphs

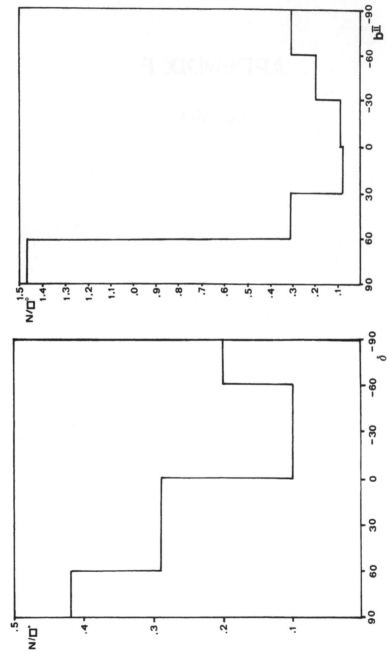

FIGURE 1 Number of Galaxies N with published Radial Velocity per square degree of sky as a function of Delination.

FIGURE 2 Number of Galaxies N with published Radial Velocity per square degree of sky as a function of Galactic Latitude.

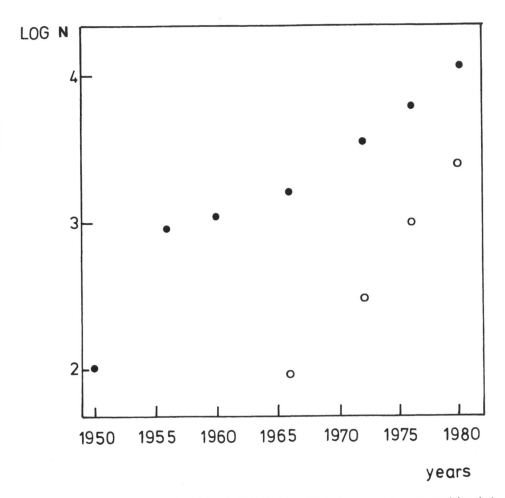

FIGURE 3   Integral number of published Radial Velocities of Galaxies measured versus year; (●) optical measurements, (○) radio measurements.